TRANSACTIONS

OF THE

AMERICAN PHILOSOPHICAL SOCIETY

HELD AT PHILADELPHIA
FOR PROMOTING USEFUL KNOWLEDGE

NEW SERIES—VOLUME XXIX
1936–1938

PHILADELPHIA:

THE AMERICAN PHILOSOPHICAL SOCIETY

104 SOUTH FIFTH STREET

1938

LANCASTER PRESS, INC., LANCASTER, PA.

CONTENTS OF VOLUME XXIX

TRANSACTIONS

OF THE

AMERICAN PHILOSOPHICAL SOCIETY

HELD AT PHILADELPHIA

FOR PROMOTING USEFUL KNOWLEDGE

NEW SERIES—VOLUME XXIX

DECEMBER, 1936

Hydrography of Monterey Bay, California. Thermal Conditions, 1929–1933
TAGE SKOGSBERG

PHILADELPHIA:

THE AMERICAN PHILOSOPHICAL SOCIETY

104 SOUTH FIFTH STREET

1936

LANCASTER PRESS, INC., LANCASTER, PA.

HYDROGRAPHY OF MONTEREY BAY, CALIFORNIA

By Tage Skogsberg

Hopkins Marine Station, Pacific Grove, California

TABLE OF CONTENTS

HYDROGRAPHY OF MONTEREY BAY, CALIFORNIA

By Tage Skogsberg

I. INTRODUCTION

History and Purpose of Survey. Acknowledgments

In the summer of 1928, Henry B. Bigelow, in conjunction with the California Department of Natural Resources, Division of Fish and Game, undertook a survey of the waters and plankton of Monterey Bay, California. In the course of this investigation, thirty-one well-spaced stations, scattered throughout the entire Bay, were occupied from June 30 until July 24. The water was examined for temperature, chlorinity, silicate, phosphate, nitrate, and oxygen. Quite a large number of plankton samples were secured and analyzed, mainly from a quantitative, but in part also from a qualitative point of view. In 1930 the results of this survey were published under the title of "Reconnaissance of the Waters and Plankton of Monterey Bay, July, 1928," by H. B. Bigelow and M. Leslie, in the Bulletin of the Museum of Comparative Zoology, Harvard College (Vol. 70:5).

Up to that time, this was the only investigation carried out for the specific purpose of reaching an understanding of the hydrobiological conditions in the Monterey Bay region. It should be noticed, however, that hydrographic work, more or less incidentally concerned with these waters, was done prior to Dr. Bigelow's survey. The Scripps Institution of Oceanography, of the University of California, had secured, ever since 1919, daily temperatures and samples of the surface waters just outside the Hopkins Marine Station, an institution maintained by Stanford University on the southern shore of Monterey Bay. The water samples were analyzed for chlorinity. The chlorinity data, as well as the temperature readings, were treated by the Scripps Institution in connection with similar material obtained from various localities along the west coast of North America and from other parts of the Pacific Ocean. This work of the Scripps Institution represented the most important of the preliminary investigations. During its extended oceanographic cruises, the U. S. S. "Albatross" visited repeatedly either Monterey Bay or the off-shore waters of this part of the coast. The work in this region consisted largely of sounding and dredging operations, but a few surface and bottom temperatures were also gathered. Noteworthy in this connection is also the United States Coast and Geodetic Survey which, in addition to its cartographic program, has investigated for many years the general features of the prevailing surface circulation along the west coast of North America. Other investigations pertaining to the general circulation along the west coast of this continent will be discussed later in this report. It may finally be mentioned that the hydrography of San Francisco Bay, thus of a locality not very far to the north of Monterey Bay, was examined by F. B. Sumner and co-workers (1914) and by R. C. Miller and co-workers (1928).

As will be seen from this cursory account, published information concerning the hydrography and planktonology of Monterey Bay and vicinity was very scant indeed until 1930. As a matter of fact, in nearly all respects the hydrography and biology of the open waters of this section of the California coast presented a perfectly virgin field at the time when Dr. Bigelow began his brief survey.   Moreover, between 1930 and the present time, no specific information on the hydrography of Monterey Bay has been published.

This situation is of course very deplorable, especially since the Monterey Bay region is one of the most important in the fishery industries of the United States, as will be realized from the following facts.   In the field of fish canning, which dominates the fishery industries of California, this state ranks next to the highest in the Union, the Territory of Alaska being the only section of the country to excel California in this respect.   Among the various fishing ports in California, Monterey ranks the second, Los Angeles occupying the highest position.   It should also be noted that Los Angeles is one of the largest fishing ports in the world.   A significant fact to be considered in this connection is that the Monterey fisheries are comparatively local, while the Los Angeles fishing industries depend to a large degree on the yield of distant waters, particularly on those off Lower California.   Furthermore, it should be emphasized that the California fisheries exploit mainly the pelagic fish species, such as the tunas and other mackerel species and the sardine.   Next to the tunas, the sardine dominates the fishing industries of this state.   Monterey is the largest landing port for the sardine, excelling in this respect even Los Angeles.   Finally, since of all the commercially important fishes, the pelagic species probably are the ones most dependent, in regard to their behavior, on the changes in the physico-chemical conditions of their environment, it becomes obvious that our very deficient knowledge of the hydrography of the Monterey Bay region is, to say the least, very unsatisfactory.   Indeed, the necessity for remedying this situation must be judged imperative.

For some time previous to the survey undertaken by Dr. Bigelow, plans for a hydrobiological investigation of Monterey Bay and neighboring waters had been under advisement as a part of the research program of the Hopkins Marine Station.   However, due to the inherent difficulties, this plan had not materialized.   The generous, cooperative spirit which Dr. Bigelow encountered among the officers of the California Department of Natural Resources induced the writer of this report to suggest to these men the continuation of the cooperation established by Dr. Bigelow between their organization and Stanford University, of which Hopkins Marine Station is a part.   In September, 1928, a formal agreement was reached, in accordance with which the two organizations, the California Department of Natural Resources and Stanford University, would contribute equally towards the establishment and maintenance of a hydrobiological survey of Monterey Bay and vicinity.   For further facts concerning this agreement, see an article by the present writer in California Fish and Game, 1930, volume 16:1.

The program of this survey may be divided into two principal parts, one dealing with the hydrography of Monterey Bay, the other with the biology of the open waters of this region.

In regard to the hydrography, there are two main aims; first, to reach a scientific knowledge and understanding of the water movements in this region and, secondly, to establish the seasonal and annual changes which characterize these waters.   The first point has proven itself very difficult because the circulation in and around the Bay is very complex and in addition mirrors, on a small scale, large-featured peculiarities of the Pacific Ocean.   A consequence of the latter characteristic is that the underlying causes of these local movements are, at least partly, to be sought in far removed regions of the open sea.   The seasonal and annual changes, on the other hand, have been studied quite successfully. Strange as it may seem, while in related fields, such as meteorology, continued and extended series of observations have long been deemed absolutely essential, this method of approach has hardly as yet become established in the field of oceanography, particularly in regard to the subsurface conditions. In our work, this principle has played the major part. Through observations and determinations, repeated as frequently as our resources permitted, we have attempted to obtain as continuous a picture as possible of the kaleidoscopic changes which occur in the hydrographic conditions throughout the years. The value of this type of work is easily discerned when we consider the fact that the seasonal as well as the general distribution, migration, and relative abundance of the organisms in the sea evidently are determined largely by the physico-chemical characteristics of the water. Of the various factors which regulate the distribution and migration of the organisms in the sea, the temperature is probably the most consequential; not only the mean temperatures of the various seasons, but also the extremes. The extreme temperatures during the spawning seasons evidently determine, at least to a very large extent, the success or failure of the spawning. Finally, a detailed knowledge of the hydrographic situation throughout the year is of course also of the greatest value to marine biological institutions, like the Hopkins Marine Station, where experimental work on marine organisms is carried out.

The rôle played in the upwelling of the deeper water towards the surface, by the deep submarine valley which bisects Monterey Bay, presents an interesting problem with which the survey has been concerned. Although this difficult question can by no means be considered as having been settled, we have at least made some contribution towards its solution.

In regard to the biological aspect of the program, which is concerned both with planktonological and fisheries ecological questions, it must suffice to mention in this connection that this part has been assigned so far nearly exclusively to students working towards advanced degrees at Stanford University. In the field of planktonology, unfortunately, the fauna and flora are so inadequately known that a fairly long period of largely descriptive work must precede the broader ecological and geographical problems. The same holds true in respect to the investigations of the deep sea organisms. Several biological reports have been concluded at this time and the results have partly been published.

On account of the extensive nature of the hydrographic material gathered in the course of the first five years of the survey, that is the period covered by the present report, it is unfortunately impossible to deal with the entire material in a single paper. The present report presents only the thermal conditions which obtained in Monterey Bay during this period. Another, and more serious, consequence of the extensiveness of the available data is that it is impossible to present the original readings and determinations in printed form. Even though all of the data were analyzed, only a very limited portion of them can be published, either in the form of graphs or condensed and selected in the form of tables. For the benefit of those specially interested in the hydrography of this region, it should be noted, however, that all of the original data are available, both in the field books and in tabulated form, at the Hopkins Marine Station, Pacific Grove.

This report is, by and large, descriptive. As was stated above, the Monterey Bay region is nearly virgin, from the point of view of hydrography. The crying demand at the present time is for factual information. Furthermore, the sound development of our knowledge in this as well as in other fields requires that description, based on abundant material, precede theoretical considerations. In chapter VIII an attempt is made to interpret the established thermal characteristics in terms of circulatory phenomena. This interpretation is fraught with difficulties due to the very imperfect state of the available information in respect to the water movements in this part of the sea.

The establishment and maintenance of the "Hydrobiological Survey of Monterey Bay" are due to the sympathetic aid and generous cooperation of a great number of persons. To all these, I wish to tender my deep appreciation. First and foremost among these, I wish to mention the executive officers of the California Department of Natural Resources, Division of Fish and Game, and the executive officers of Stanford University. Instrumental in establishing the cooperation between these two organizations, out of which the survey originated, were particularly President Ray Lyman Wilbur of Stanford University, President I. Zellerbach of the California Department of Natural Resources, and Mr. N. B. Scofield, in charge of the Bureau of Commercial Fisheries of this Department. In this connection should also be mentioned Dr. W. K. Fisher, who, in his capacity of director of the Hopkins Marine Station, has always held an attitude of helpful interest towards the survey.

Three of the patrol boats of the California Department of Natural Resources have been employed by the survey. During the first two years, a 33-foot launch, the "Steelhead" was, nearly exclusively, used for the routine work. The crew of this boat were Mr. N. Matthews and Mr. C. Rogers. On December 6 and 7, 1930, the 87-foot "Bluefin" (Captain W. Engelke) was put at our disposal for a 25-hour tidal investigation. Ever since the autumn of 1930 a 57-foot launch, the "Albacore," has been assigned to serve the survey; its permanent crew during these years were Captain Lars Wesseth and Engineer E. Greenleaf. To the crews of all of these boats, I am deeply indebted for their courteous attitude and fine cooperation. Particularly to Captain Wesseth, I wish to extend my sincerest thanks. On his unfailing care and skill we depended for the accurate location of our hydrographic stations, a task made especially difficult by frequent fogs and other vicissitudes, but always carried out with patience and cheerfulness.

Among those who have helped by managing the Nansen water bottles and other duties on board boat, I wish particularly to name Mr. E. C. Scofield, who helped during the first two years, and Dr. R. L. Bolin, who assisted me ever since Mr. Scofield's duties called him elsewhere. Both helped cheerfully, good weather or bad, Dr. Bolin even when the "Steelhead" ran onto a submerged rock and sprang a leak in a heavy fog, an accident from which we barely escaped without sinking. Dr. Bolin has also helped in handling some of the data. Among others who, on occasion, have stepped in to help us with the field work, I wish especially to mention Mr. J. B. Phillips, of the California Department of Natural Resources, Mr. D. Horsburgh, Mr. W. A. Dill, and Dr. E. W. Galliher. I am also under obligation to Dr. E. G. Moberg and Dr. G. F. McEwen of the Scripps Institution of Oceanography, La Jolla, California, for their kind criticism of the completed manuscript of this report. Many other persons have rendered their help to the survey, and to these I also offer my sincere appreciation, even though their names are not mentioned here.

PACIFIC GROVE, March, 1935

II. LOCATION. PHYSIOGRAPHY IN RELATION TO HYDROGRAPHY

Monterey Bay is located in central California, between latitudes 36° 36' N. and 36° 59' N., and longitudes 121° 47' W. and 122° 00' W., Greenwich meridian. It represents a comparatively shallow indentation in the coast, nearly semicircular in outline, about 23 miles [1] wide between the northern and southern headlands, and measuring only about 11–12 miles from its mouth to its innermost point, which is located about mid-way between the northern and southern extremities.

In regard to the bottom configuration (Fig. 1), the most outstanding feature is a deep, submarine valley or, as it is frequently termed, channel which runs approximately N. 70° E., dividing the Bay into two nearly equal halves. At the mouth of the Bay, this valley is about five miles wide; outside of this point it opens quite suddenly and soon begins to approximate in outline the general contour of the coast. Numerous irregularities characterize both the northern and the southern banks of this valley. One of these has the nature of a large side valley on the north side, trending in a N. 30° E. direction. The main channel terminates at the innermost point of the Bay, where the 100 fathom isobath comes within not more than about 1.5 miles from the shore line. The northern side valley, on the other hand, is not traceable to the coast; its 100 fathom isobath is removed from the beach by approximately six miles. At the mouth of the valley the 100 fathom isobath is removed from the northern headland by nearly nine miles, while its distance from the southern headland is only about four miles; and a very short distance to the south of the latter headland, only about one mile. Since the 100 fathom isobath is generally accepted to mark the continental edge, it will be seen from these data that the continental shelf is unusually narrow in this region. The bathymetry of the valley is evidently very complex. Extensive soundings have recently been made throughout the entire Bay by the United States Coast and Geodetic Survey, but this new material has not been embodied in the appended map; the features given in the map are not materially altered by the new observations. In regard to the configuration beyond the 500 fathom line, it must suffice to state that the channel gradually opens into the great oceanic depths, no threshold being present as in the case of the true fiords, and that it can be traced down to a depth of about 2000 fathoms. At the mouth of the Bay, its depth is about 700 fathoms; and at that place its southern bank is decidedly steeper than the northern.

The 100 fathom isobath on the north and south sides of the valley forms the outer boundary of two nearly equal flats. Of these the northern one appears to be very uniform in slope and configuration, with the exception that it is dented by the side valley noted above. The southern flat, on the other hand is somewhat more irregular. It slopes quite abruptly along the southern, rocky headland and more gently along the inner, sandy shore. In addition, it is characterized by some minor reefs as well as by an indistinct channel close to the southern shore, but this channel has no connection with the main valley and, moreover, appears to be in the process of disappearing due to silting, according to the latest soundings (see Galliher, 1932, chart).

The open shape of the Bay and the fact that it slopes directly into the great abysses of the ocean, of course, have very important consequences in regard to the hydrographic characteristics of the Bay. It should particularly be noted that the continental shelf is

---

[1] Nautical miles, as always in this report.

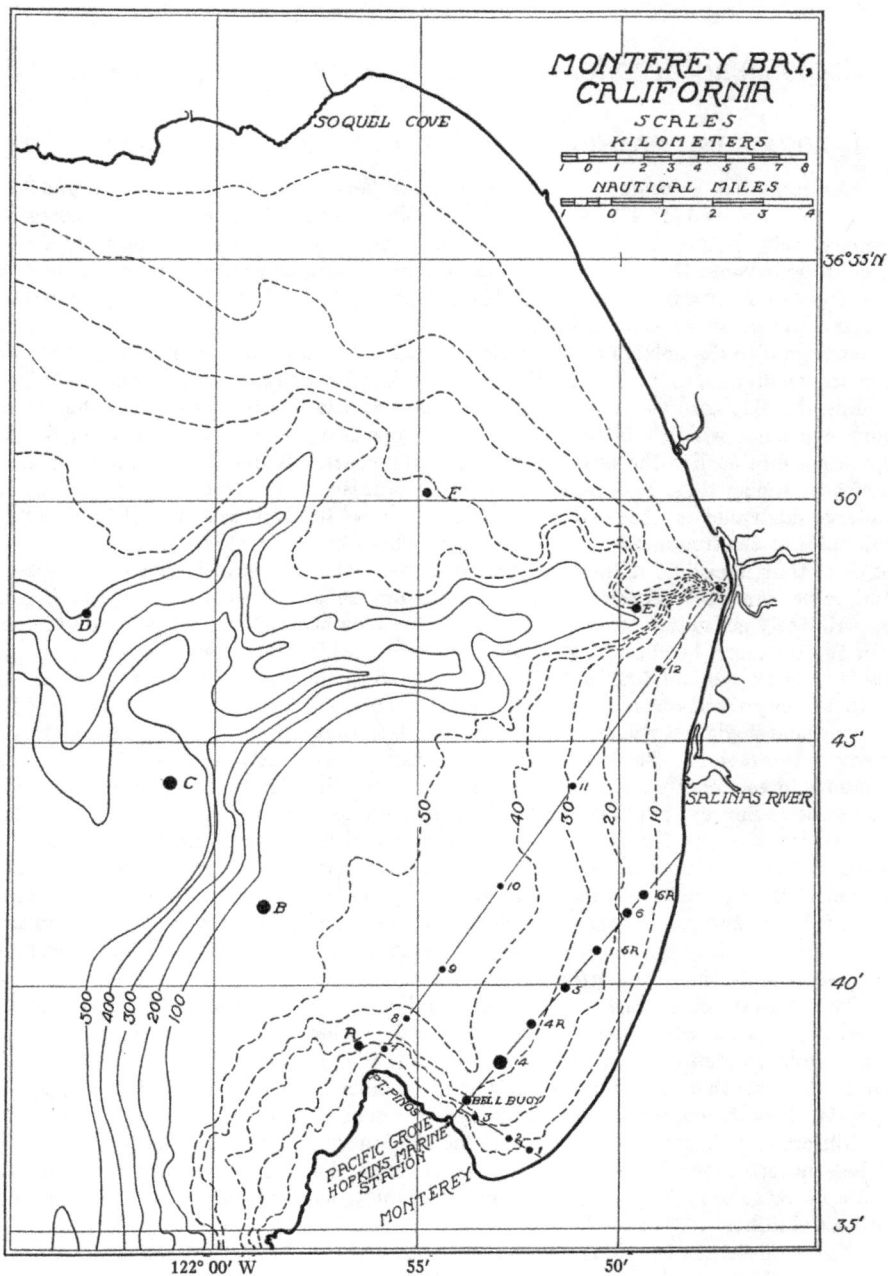

FIG. 1. Map of Monterey Bay. Depths in fathoms. Isobaths omitted beyond 500 fathoms. Greatest depth within area of map, about 800 fathoms. Large circles indicate stations treated most extensively in this report. Short line just off Monterey is station line no. 1. The middle line, running from Hopkins Marine Station, is the "key line," station line no. 2. The line from Point Pinos is station line no. 3.

very narrow in the southern part of the Bay where most of the hydrographic work was done and that there is a channel-like depression leading towards the place where during the first two years of the survey the key station was located.   Tides and coastal currents can sweep relatively unimpeded into the confines of the Bay, and the shape of the southern part of the Bay is conducive to the entrance of water of fairly deep origin.   Thus, instead of distorting beyond recognition the hydrographic features of the open sea, the Bay reflects them in a fairly close and direct manner.   A study of the hydrography of Monterey Bay therefore becomes rather a study of a section of the open sea than a study of a region in which the hydrographic peculiarities can be traced to local topographic and meteorological conditions. In a manner, this is very disconcerting and unsatisfactory from the point of view of the investigator whose resources are very limited.   The underlying causes of the observed changes, rhythms, and deviations from what may be termed the norms originate in far distant regions and are thus outside the personal reach of the investigator.   The fundamental explanation of the basic phenomena thus must, at least in several cases, be left in abeyance.   On the other hand, the wide communication between the Bay and the open sea also implies the gratifying consequence that the results obtained from a study of the hydrographic features within the Bay are not of necessity exclusively local in nature. Indeed, they may well prove to form the key to the understanding of the conditions prevailing in the coastal region of a fairly extended portion of California.   Furthermore, the narrowness of the continental shelf implies the great advantage that the upwelling of deep water towards the surface, a process characteristic of the entire California coast, and the effect of a submerged valley on these water movements can be studied within easy reach of a laboratory.

In regard to the distribution of the various types of bottom sediments and their relation to the water circulation, see the last section of chapter VIII.

III. METHODS AND MATERIAL

When Dr. Henry B. Bigelow undertook his hydrobiological reconnaissance of the waters of Monterey Bay, he opened up a region which, as noted above, had previously hardly been studied from the point of view of hydrography.   Since he had only slightly more than three weeks at his disposal, he did not have time to let his survey be preceded by a careful test of the field technique.   As a consequence, he naturally applied the universally accepted principle of placing a net of properly spaced hydrographic stations over the region under investigation.   Of these stations, only a few could be visited more than once.   The data thus gathered in the course of three weeks were remarkably consistent.   In other words, they appeared fully to justify the procedure.   As a consequence, Dr. Bigelow worked out hydrographic charts and profiles of the Bay in which all or most of the data were lumped together and treated as if they had been collected simultaneously, even though he was fully aware of the risk inherent in this procedure.

In spite of this apparently successful application, it was considered to be a matter of the utmost importance to submit this field technique to an exhaustive test before it was employed on a large scale in the course of our survey, before it was accepted as one of our standard methods of investigation.   For this purpose a key line of stations (Fig. 1, stations

"Bell Buoy"—6A) was laid out in the southern end of the Bay. The stations were to be examined, weather permitting, semi-weekly for the purpose of establishing the rate at which the hydrographic changes took place in this part of the Bay, and the results thus obtained were to be applied, although with great caution, to the balance of the Bay.

The outcome of the test expeditions was very startling indeed. It became evident beyond a peradventure that the field technique used by Dr. Bigelow should under no circumstances be applied to this region. The results demonstrated that the hydrographic situation of the southern end of the Bay was ever changing and that the changes were caused by forces outside the Monterey Bay vicinity. Examples of the high rate of these water movements are given in section B of chapter VIII. If a net of well-spaced stations were to be laid over the entire Bay, a fairly large number of boats would be necessary, since all the stations would have to be visited within a few hours to obtain comparable results. This, unfortunately, was beyond our possibilities.

Under these circumstances, the method employed in testing Dr. Bigelow's field technique was selected as the standard method of collecting our material during the first two years (1929–30) of the survey.

The small size and limited seaworthiness of the Steelhead, the boat placed at our disposal by the California Department of Natural Resources, forced us to restrict our field work to the southernmost portion of Monterey Bay. There three lines of stations were established (Fig. 1). One of these, line no. 2, was selected as the standard or "key line"; in other words, it was chosen to be examined regularly, twice a week, weather and other circumstances permitting. This line, seven miles long, extended from the neighborhood of the Hopkins Marine Station in a northeasterly direction to the inner shore of the Bay. It comprised seven stations. Between each two of six of these, the "Bell Buoy" station, stations 4, 4A, 5, 5A, and 6, the distance was one mile, while stations 6 and 6A were separated by only one-half mile. At stations 4, 5, and 6, the only ones to be visited during the first seven months of 1929, observations were made at each five meter level; at the remaining stations, only at each ten meter level. The maximum depth, at which observations were taken, was 50 meters, at stations 4 and 4A. Since the observational series was more complete at one of these two stations, station 4, this was made the first "key station" of the survey. In other words, this station we endeavored to visit even when the weather conditions were very trying. The total number of observations made during an excursion along this line was 45.

Another line of stations, line no. 1, was placed in the shelter of the southern headland of the Bay. This was to be examined only on days when weather conditions did not permit us to carry out the standard program. On this line, there were only three stations, viz., stations 1, 2, and 3. At each of these stations, observations were made at each five meter level. The deepest observational level was at 35 m.

Finally, a line of stations, line no. 3, was planned between the inner end of the large submarine valley of the Bay and Point Pinos. Seven stations were placed on this line (stations 7–13). Unfortunately, however, this task proved to be beyond the capacity of our small and slow craft. After we had attempted on two occasions, but without success, to cover this line in one day, this undertaking was discontinued.

During 1929, 421 stations were visited and the following data gathered:

Temperature	Chlorinity	Phosphate (PO_4)	Silicate (SiO_2)
3428	3301	3301	3301

Disregarding in this connection the chemical data, since these will be treated separately in a later report, the temperature observations were distributed in the following manner among these stations:

Stations	1	2	3	BB	4	4A	5	5A	6	6A	7	8	9	10	11	12	13
Number of observations..	24	41	52	173	1160	153	953	133	600	73	12	13	7	13	11	4	6

In 1930, only the key line was investigated. Station line no. 1 did not need to be visited, partly due to the fact that the seaworthy patrol boat "Albacore," of the California Department of Natural Resources, was used during several months. The routine collection of data in this year consisted of 4075 temperature readings and an equal number of analyses for chlorinity, phosphate, and silicate. The temperature readings were distributed as follows among 624 stations:

Stations	BB	4	4A	5	5A	6	6A
Number of observations....................	384	1045	552	910	440	516	228

On December 6 and 7, the "Bluefin," an 87-foot patrol boat of the California Department of Natural Resources, was anchored at station 4. In the course of a tidal period, 25 series of temperature and water samples were taken. The number of temperature readings was 275 and an equal number of water samples was analyzed for chlorinity, phosphate, and silicate.

On account of the fact that our results had demonstrated the desirability of investigating the hydrographic conditions in somewhat deeper strata than those occurring along the key line, a new key station was established in 1931. This station, station B (Fig. 1), was located about four miles from Point Pinos, in a northwesterly direction, only slightly inside the edge of the continental shelf. It should be noted that it was placed just outside the channel-like depression which occurs in the southernmost part of the Bay; ending in the vicinity of station 4, our previous key station. This situation was selected partly in order to make our new results comparable with those previously obtained. This station was also visited semi-weekly, weather permitting, and observations were made at each five meter level, as previously had been the practice at station 4.

The old station lines were discontinued and only station B was visited. This implied, in a manner, a reduction of our field work, a reduction made necessary by the other duties of the patrol boat "Albacore," which was used throughout the year.

Station B was visited 99 times in the course of the year and the following data were collected:

Temperature	Chlorinity	Phosphate	Silicate
2079	1332	1225	1685

It may finally be noted in this connection that current experiments, carried out with special floats (see section A of chapter VIII), were undertaken in 1931.

While in the first three years of the survey, the hydrographic work was limited to the waters in the vicinity of the Hopkins Marine Station, in 1932 the program was considerably extended on account of improved boating facilities. The entire Bay was submitted to investigation, and particular emphasis was placed on establishing the part played by the deep central submarine valley in the general hydrographic changes characteristic of the Bay.

Seven stations were established (Fig. 1). Two of these, stations 4 and B, had previously been occupied. One, station A, extending down to only 25 meters, was located just outside Point Pinos. This place was chosen because it often is characterized by very intensive water movements, water usually either entering or leaving the Bay around this Point. Two stations, extending down to 100 meters, were placed on the edge of the continental shelf; one of these, station D, was on the northern bank of the central valley, opposite station B; the other, station E, at the innermost end of this valley. One, station F, extending to a depth of 60 meters, was established near the middle of the northern flat, somewhat inside the inner end of the side valley to the central valley. The last one, station C, extending down to not less than 900 meters, was placed over the deepest part of the central valley, about midway between stations B and D. As will be seen from this account,

Fig. 2. View towards the stern of the "Albacore," to illustrate the arrangement adopted during the collecting operations. Photograph by J. B. Phillips.

three of these stations, *viz.*, stations B, D and E, in a manner triangulated the central channel, while two, stations 4 and F, were established in order to examine the conditions on the northern and southern flats.

These stations were visited once every week, weather and other circumstances permitting. Patrol duties of our boat, the "Albacore," necessitated the discontinuation of the semi-weekly expeditions. This change in our field work was of course unfortunate. However, even though it caused us to lose many of the details of the hydrographic changes, this program, nevertheless, proved itself sufficient for the recording of the main features of the larger pulses and thus, at least to a considerable degree, satisfactory. At stations A and B, temperature observations were made at every five meter level; at stations D, E, F, and 4, at every ten meters. At station C, water samples as well as temperature readings were taken; from 500–900 meters, at every 100 meter level; from 150–450 meters, at every 50 meters; and between the surface and 100 meters, at every ten meters. Our limited laboratory resources did not allow us to do chemical determinations for more than one station.

During this year, a total of 312 stations were occupied. In all, 3796 temperature readings were made, distributed among the various stations in the following manner:

Station	A	B	C	D	E	F	4
Number of observations....................	294	1052	985	462	440	287	276

Station C was visited 44 times. From water samples collected on these occasions, the following number of determinations were made: chlorinity, 916; phosphate, 891; silicate, 899.

The program initiated during 1932 was carried on during 1933 without any changes. The total number of stations that were visited was 331. At these, 4054 temperature readings were made, distributed as follows:

Station	A	B	C	D	E	F	4
Number of observations....................	295	1029	1127	495	506	314	288

On the 49 occasions when station C was occupied, water samples were collected from which the following number of determinations were made: chlorinity, 1147; phosphate, 894; silicate, 940.

As will be seen from this account, a total of 1788 stations were occupied during the first five years of the survey. The total number of recorded observations and determinations was 50589, distributed as follows:

Temperature	Chlorinity	Phosphate	Silicate
17707	11046	10661	11175

This may at first appear to be quite a large amount of material. However, even so, it must be considered fairly close to a minimum in view of the extraordinary instability and fluctuations characteristic of this region.

It may be of interest to note in this connection a few facts regarding our working arrangements on board the "Albacore," a boat not specially built for duties of this kind. The steel cable was run directly from the power winch over the meter wheel which was suspended from a steel frame in the form of a double, inverted V over the rear end of the boat (Fig. 2).

The man handling the metal water bottles was seated on a chair which was straddling the rail and furnished with a double side box for the brass messengers, the wrench for tightening the bottles onto the cable, and one citrate of magnesia bottle.   The metal water bottles were placed on a slanting rack within easy reach of the operator.   This arrangement proved very satisfactory, both from the point of view of speed and of comfort, during the long hours of operation, often carried on under fairly trying weather conditions.   It should be noticed that the bottle operator and the man keeping the records did not crowd each other, even though the space available was small.

The subsurface temperatures were taken by means of standard deep sea reversing thermometers, attached to the usual type of Nansen water bottles, and graduated to 0.1°. It was not considered advisable to attempt to obtain closer readings because the fluctuations in the hydrographic conditions were found to be so great that on most occasions we could not duplicate closer records in case we repeated the series.   The surface temperatures were taken with an ordinary thermometer, also graduated to 0.1°, and fastened within a large brass tube with an opening on one side and with the bottom closed, thus allowing the mercury bulb to remain under water while the readings were made (in other words, with the commonly used cup thermometer).   This was read in the same manner as the deep sea thermometers.

For the purpose of studying the tidal currents, special floats were constructed of such a type that they could be employed not only at the surface but at any depths in the Bay outside the deep central channel, i.e., within the upper one hundred meters.   These floats (Fig. 3), a modification of the type used by the Challenger Expedition (Krümmel, 1911, p. 422, Fig. 108), were constructed in the following manner.

Two plates of strong sheet iron, each about 81 cm. high and 91 cm. wide, were split vertically to the center and then fitted together at right angles.   They were held in this position by means of a strong sheet iron plate, 20 cm. square and furnished on its lower side with four pairs of small but strong iron wings set at right angles to each other, each pair forming a slot into which one of the four wings of the large sheet iron cross was fitted and fastened by means of a bolt and wing nut.   In order to obtain greater firmness of the float, the corners of the sheet iron plates may be fastened together by wire hooks in the manner shown in the figure of the Challenger floats referred to above.   To the middle of the top surface of the supporting plate was fastened a strong metal eye.   To this eye were attached glass balls protected and held by strong cord netting.   The number of these balls is determined by their size (those employed by us were 15 cm. in diameter and six of them were needed) and by the weight of the sheet iron cross and supporting plate.   Their buoyancy should not be quite sufficient to float these metal parts.

This main part of the float was suspended by a strong cord of adjustable length from what may be termed a position indicator of the following construction.   A strong iron rod, about 152 cm. long, was furnished with an eye at its lower end and with a metal pennant, about 35 cm. long, at the other.   The pennant was soldered to the rod.   At about its lower one third, this rod ran through the center of a can, made of strong sheet iron, to which it was soldered in an air-tight manner.   The can, which was about 75 cm. high, 19 cm. deep and 9 cm. wide, was stream-lined and fastened to the rod with its longer diameter parallel to the pennant.   The can should have enough buoyancy to float the iron rod and the excess weight of the float, i.e., the weight not compensated for by the buoyancy of the glass balls.   This excess weight keeps the position indicator in an upright position.

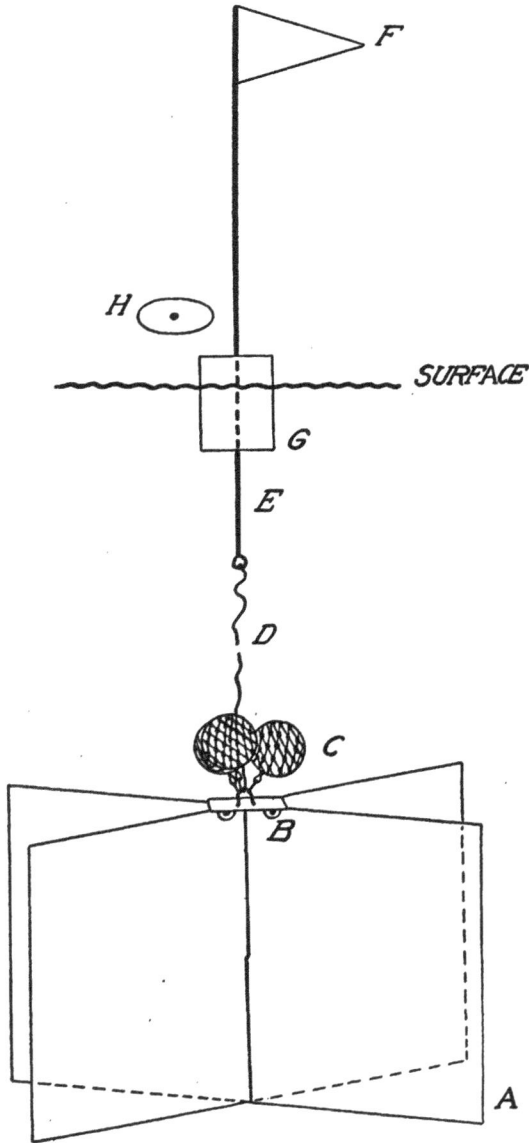

FIG. 3. Float used in study of tidal currents. A, sheet iron resistance; B, sheet iron plate to hold wings of resistance at right angles; C, glass balls to buoy the resistance; D, cord; E, iron rod of position indicator; F, metal pennant; G, float of position indicator in side view; H, the same in top view.

As will be seen from this description, the main principle in which this type of float differs from the one employed by the Challenger Expedition is that the float proper has a very large surface of resistance while the position indicator has a minimum volume, its resistance in addition being reduced by stream-lining in the direction of the wind.   During preliminary tests, these floats proved themselves very little affected by wind drift and all our later experiences strongly supported these observations.   The floats should be made in such a manner that they can easily be dismantled when not used, since otherwise they will occupy an unduly large deck space.   It should be noted that the use of glass balls for the buoyancy of the float proper proved very satisfactory.   Metal cans, which were first used, are unsatisfactory because they can not stand the water pressure even at a rather small depth unless they are made very heavy.   All the metal parts should be galvanized and red leaded.   The red color of the pennant, of course, greatly aids in locating the floats at a distance.

In regard to the treatment of the data, it is, of course, noticeable that no attempt was made to apply hydrodynamical analysis.   This was due to the fact that the results of such an analysis were considered altogether too uncertain on account of the extreme complexity and rapidity of the water movements in the Bay.   We have found that the Bay is characterized by the presence of a number of eddy-like and dynamically relatively independent bodies of water of variable size and direction of movement.   This, of course, renders the selection of the positions of the hydrographic stations to be used for hydrodynamic calculations very difficult, to say the least.   Furthermore, the water movements frequently are of such an intensity that the data collected at different stations separated by several hours' interval hardly become properly comparable from the point of view of such a method of analysis.   Finally, we have come to the conclusion that the primary forces behind these movements are to be sought outside the Bay within the confines of which all our work was done during the first five years of the survey.   Monterey Bay is apparently comparable to a big bowl into which the oceanic waters are being poured.

The degree of correctness of the temperature observations may be set at $\pm 0.1°$.   All observations were recorded in Centigrade.

IV. MONTHLY THERMAL CONDITIONS IN THE UPPER ONE HUNDRED METERS

Notwithstanding the fact that the hydrographic conditions in Monterey Bay are subject to considerable annual changes in the upper one hundred meters, the thermal features of the individual months are, on the whole, quite characteristic in this stratum. To be more specific, while the monthly average temperatures for this layer may exhibit relatively large differences from one year to another, the vertical thermal distribution is comparatively stable for the individual months, each month having a fairly pronounced pattern.   In addition, there is a distinct and characteristic seasonal progression.   The unfolding of the seasons is particularly expressed in the thermal gradients.   The amplitudes of thermal variation at and near the surface also exhibit a certain seasonal trend, although in a more irregular and obscure manner.   For the sake of emphasis, it may be noted even at this early stage in our discussion that the seasonal progression of the thermal gradients, even though it strikingly simulates the one occurring in the temperate regions, nevertheless is brought forth in a manner nearly diametrically opposite to that typical of temperate waters.   Since this seasonal progression, by and large, is restricted to the upper one hundred meters, this stratum is treated separately from the lower layers, even though the transition towards the depths is quite gradual.

The following descriptions are based upon data obtained at stations 4(1929–30), B(1931–3) and C(1932–3).   At station 4, observations extended down to 50 meters only; and at station B and C, to 100 meters or beyond.   Station B was used for computing the averages and amplitudes of variation during the period of 1931–3, while the data from station C were utilized as the basis for the graphs of the thermal gradients of 1932–3 because it was deemed desirable to include in the graphs for these two years also the data from between 100 and 250 meters.   Details in regard to the locations of these stations are given in chapter II.   Stations 4 and B are quite closely comparable; and station C, although somewhat removed from these, is not too far away to make a comparison permissible. The discussed thermal gradients are shown in Figs. 4–6, while the amplitudes of thermal variation are presented in these and in Fig. 20.   In regard to the information pertaining to the amplitude of thermal variation, it should be noted that more weight should be placed on the data from the years 1929–31 because from these years nearly twice as much material was available.   The expeditions during these years were semi-weekly instead of weekly as in the later years.

The remaining stations were, of course, also analyzed in a manner similar to that applied to the standard stations.   However, the differences between these and the standard stations were not large enough materially to effect the averages and extremes, and thus their exclusion, for the sake of economy, appears fully justifiable.   The various monthly descriptions were prepared according to a uniform pattern in order to facilitate comparison.

An accurate knowledge of what may be termed the typical or "normal" hydrographic conditions of a region is, of course, of the greatest value to the comparative oceanographer, for instance in his construction of isabnormals.   For this reason it was deemed advisable to present in an appendix of tables (chapter XI) the monthly gradients for each of the five years under investigation.   It is in part on these tables that Figs. 4 to 6 are based.   In these tables are also presented the total monthly averages for these five years, values which may be considered as "normal" for this region.   It is of course fully realized that these calculations are only approximate, based as they are both on a very limited number of observational days and on a small number of years.   The reason for presenting the data in spite of this limitation is twofold: first, there are no subsurface records covering an extended period from this region published up to the time of the present writing; and, secondly, in the course of the five years the amplitude of thermal variation was found to be extremely narrow, even at the surface, indicating that the present results are quite close to the actual conditions.   This statement is borne out by the close similarity between the total averages based on five and on three years; in most cases the differences amount to only a small fraction of one degree.

January: The most outstanding features of this month were the low thermal gradient and the relatively narrow amplitude of thermal variation.

The gentle slope of the thermal gradient was most pronounced in 1931, in which year the average temperature for this month remained unchanged from the surface to a depth of 45 meters and the decrease from the surface to 75 meters was only 0.2°.   The thermal gradients for this month were as follows from 1929 to 1933:

	Surf.—50 m.	Diff.	Surf.—100 m.	Diff.
1929.........................	12.3°−11.2°	1.1°		
1930.........................	12.7°−12.5°	0.2°		
1931.........................	13.2°−13.1°	0.1°	13.2°−12.1°	1.1°
1932.........................	11.7°−11.3°	0.4°	11.7°−10.1°	1.6°
1933.........................	11.4°−10.9°	0.5°	11.4°−10.2°	1.2°
Average.....................	12.3°−11.8°	0.5°	12.1°−10.8°	1.3°

Thus the temperature decreased, on the average, only 0.5° in the upper 50 meters, and 1.3° in the upper 100 meters. It should also be noticed that, even though the thermal gradient was always low, nevertheless, its degree of inclination was somewhat variable from year to year. This variation was not correlated with the annual changes in the temperature, as may be seen from the fact that the gradient was very low both in 1931 and in 1932, although the former year was comparatively warm and the latter cold.

While the thermal gradient was always quite low, the monthly average temperatures varied decidedly from year to year, as will be seen from the following values:

	1929	1930	1931	1932	1933	Average
0– 50 m.........	11.8° .	12.7°	13.2°	11.6°	11.2°	12.1°
50–100 m........			12.8°	10.7°	10.5°	11.3°

The total amplitudes of thermal variation from 1929 to 1933 were 2.4° (13.6°−11.2°), at the surface; 3.4° (13.6°−10.2°), at 50 m.; 3.8° (13.2°−9.4°), at 100 m.; and for the averages for 0–50 m. and 50–100 m., 2.8° (13.6°−10.8°) and 3.6° (13.4°−9.8°) respectively. For each of the years of this period, the following values of thermal variation were established:

	Surf.	50 m.	100 m.
1929..............	0.5° (12.6°−12.1°)	0.8° (11.6°−10.8°)	
1930..............	0.6° (13.1°−12.5°)	1.2° (13.3°−12.1°)	
1931..............	0.7° (13.6°−12.9°)	0.8° (13.6°−12.8°)	1.7° (13.2°−11.5°)
1932..............	0.6° (12.0°−11.4°)	1.4° (12.0°−10.6°)	0.8° (10.6°− 9.8°)
1933..............	0.6° (11.8°−11.2°)	1.2° (11.4°−10.2°)	1.7° (11.1°− 9.4°)
Average...........	0.6°	1.1°	1.4°

For the average temperatures at 0–50 m., these five years exhibited the following amplitudes, 1.0°, 0.9°, 0.7°, 1.1° and 0.8° (average 0.9°); the amplitudes for the averages of 50–100 m. were 0.9°, 1.0°, and 1.4° (average 1.1°).

As will be seen from these data, the range of thermal variation was very small, particularly at the surface. It gradually increased towards the depth, even as far down as to 100 m. where it was more than twice that of the surface. This increase took place, on the whole, quite regularly, although some minor irregularities did occur. The underlying causes of this striking and aberrant peculiarity will be discussed later in this report (chapter VIII).

The local winter cooling of the surface waters was usually too slight to be noticed with our methods of observation. The effects of this phenomenon were expressed in the averages only in one of the five years, *viz.*, in 1930, when the average for the five meter level was 0.1° above that for the surface. The most extreme cooling of the surface, when compared with the five meter level, was 0.2°, which was recorded in 1930. The greatest depth to which this process was traced was 15 meters. However, due to the uniform density of the superficial strata during this month, convectional cooling undoubtedly did penetrate somewhat farther down, at least on occasions, although mixing obliterated the phenomenon beyond easy recognition. The observations on which these statements are based were usually made between 8 a.m. and 9 a.m. Later in the day insolation nearly always completely canceled this process.

February: A very marked resemblance still lingered in the hydrographic picture of this month, when compared with that of the preceding one. Nevertheless, a distinct progressive change could usually also be observed. In two of the years under consideration (1929 and 1932), there were no changes in regard to the thermal gradient; it may be noted, however, that in 1932, the positions of the January and February gradients were somewhat different. Moreover, as will be seen from the description given above, in 1931 the month of January was characterized by an unusually low gradient down to a depth of 75 meters. In February of that year, the same low gradient occurred, even though it did not extend below 35 meters; indeed, on one day the thermal difference between the surface and 100 meters was only 0.1° (14.1° − 14.0°). Just as in January, the variations in the slopes of the gradients were not correlated with the changes in the temperatures of the superficial strata.

The progressive changes were expressed in a twofold manner, *viz.*, in a very slight increase in the slope of the thermal gradient and in a similarly slight increase in the range of the thermal variation at the surface. A much more pronounced difference was present in the form of a very decided increase in the amplitude of thermal variation in the deeper strata. However, even though this difference between January and February in all probability is always present, it does not fall within the category of the progressive seasonal changes.

During the years under consideration, the thermal gradients for February were as follows:

	Surf.—50 m.	Diff.	Surf.—100 m.	Diff.
1929........................	12.1° − 11.0°	1.1°		
1930........................	13.0° − 11.9°	1.1°		
1931........................	13.9° − 13.5°	0.4°	13.9° − 12.4°	1.5°
1932........................	11.1° − 10.7°	0.4°	11.1° − 9.9°	1.2°
1933........................	10.9° − 10.2°	0.7°	10.9° − 9.2°	1.7°
Average....................	12.2° − 11.5°	0.7°	12.0° − 10.5°	1.5°

The decrease in the temperature from the surface to 50 meters and to 100 meters was thus, on the average, 0.7° and 1.5° respectively. This increase over the January values, although extremely slight, is significant since it indicates a general trend. The variations in the inclination of the gradients from year to year was somewhat less than in January, but this difference undoubtedly is of no consequence whatsoever.

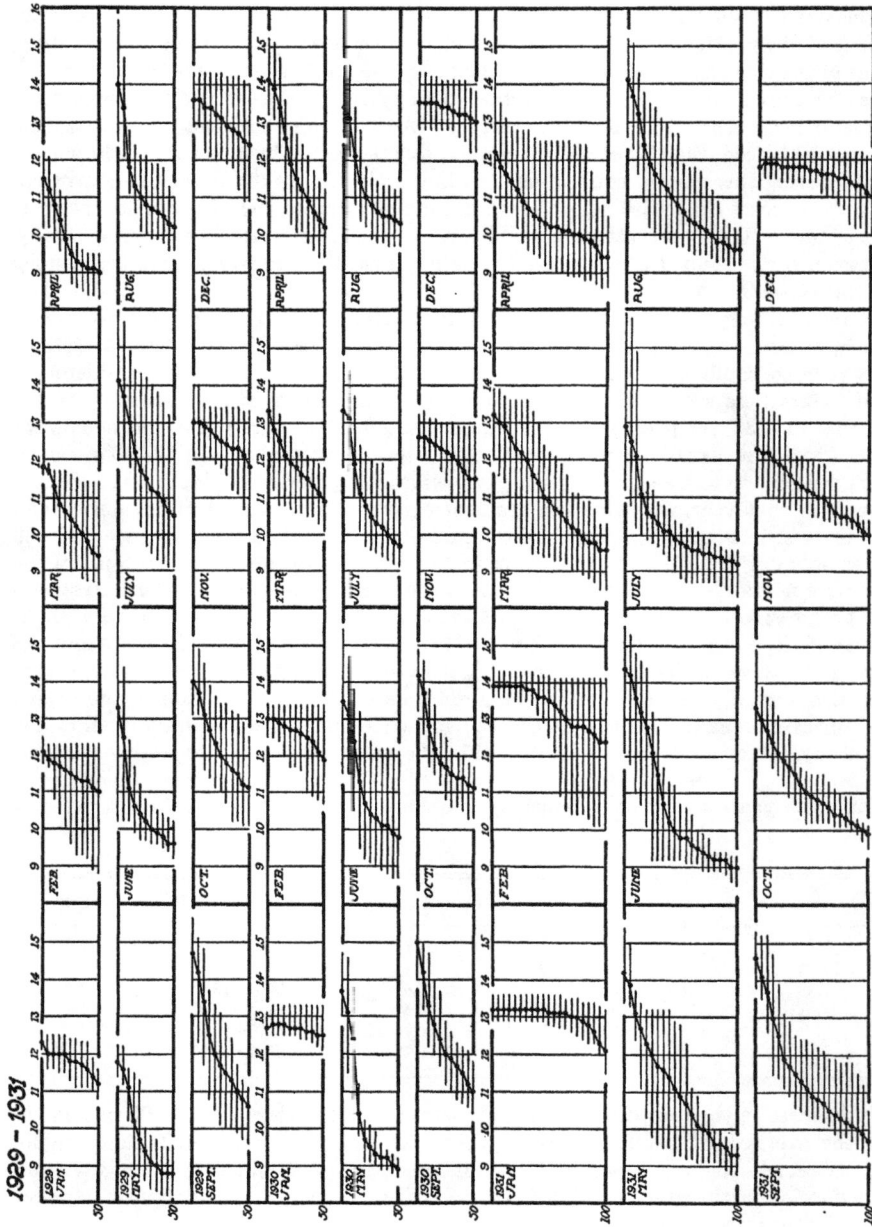

FIG. 4. Monthly thermal gradients and amplitudes of monthly thermal variation, in 1929–31. Dots indicate monthly averages; horizontal lines, amplitudes of monthly variations. Observations made at 5 meter intervals. In 1929 and 1930 the upper 50 meters were investigated; in 1931, the upper 100 meters.

The annual variations in the average monthly temperatures at the various levels are illustrated by the following values:

	1929	1930	1931	1932	1933	Average
0– 50 m.........	11.5°	12.6°	13.8°	11.0°	10.5°	11.9°
50–100 m.........			12.8°	10.2°	9.7°	10.9°

The averages for this month thus were on the whole lower than those for January. Again, even though the observed differences are very slight, they are significant as indicators of a general trend.

The total amplitudes of thermal variation from 1929 to 1933 were 3.9° (14.4°−10.5°), at the surface; 5.4° (14.2°−8.8°), at 50 m.; 5.5° (14.0°−8.5°), at 100 m.; and for the averages for 0–50 m. and 50–100 m., 4.8° (14.3°−9.5°) and 5.3° (14.0°−8.7°) respectively. Some of the corresponding values for the individual years were as follows:

	Surf.	50 m.	100 m.
1929.............	0.7° (12.5°−11.8°)	3.5° (12.3°− 8.8°)	
1930.............	0.9° (13.4°−12.5°)	2.7° (13.4°−10.7°)	
1931.............	0.8° (14.4°−13.6°)	1.7° (14.2°−12.5°)	4.0° (14.0°−10.0°)
1932.............	0.7° (11.5°−10.8°)	0.9° (11.2°−10.3°)	1.1° (10.6°− 9.5°)
1933.............	0.8° (11.3°−10.5°)	2.0° (10.8°− 8.8°)	1.5° (10.0°− 8.5°)
Average..........	0.8°	2.2°	2.2°

For the average temperatures of 0–50 m., these five years showed the following amplitudes, 2.3°, 1.6°, 0.8°, 0.5°, and 1.5° (average 1.3°); the amplitudes for the averages of 50–100 m. were 3.5°, 1.3°, and 1.8° (average 2.2°).

These data demonstrate that there was a decided increase in the amplitude of thermal variation in February over the preceding month. This phenomenon was expressed nearly consistently during the five year period, the only exception from the rule being the upper 50 meters in 1932. Even in that year, however, the surface followed the general trend which is of interest since only at and near the surface there was in this respect a true annual cycle. The increase in the amplitude of variation towards the depth was particularly pronounced in 1929 (to a depth of at least 50 m.) and in 1931 (to a depth of at least 100 m.).

Local winter cooling was observed even more seldom than in January. On a few occasions the surface was found to be 0.1° colder than the five or even the ten meter level. However, there can be but little doubt that cooling took place on a fairly extensive scale. The effects of the process easily escape detection unless specially subjected to investigation. Just as in January, the observations were made between 8 a.m. and 9 a.m. During the later hours of the day the results of the cooling were obliterated.

March: The general seasonal progression, which was but slightly foreshadowed in February, became clearly expressed during March. Through the lowering of the temperatures at the deeper levels, the thermal gradient assumed a decidedly more pronounced slope. This lowering was very consistent, being characteristic not only of the month as a whole but also of the individual days. Indeed, a low gradient was found on no occasion in

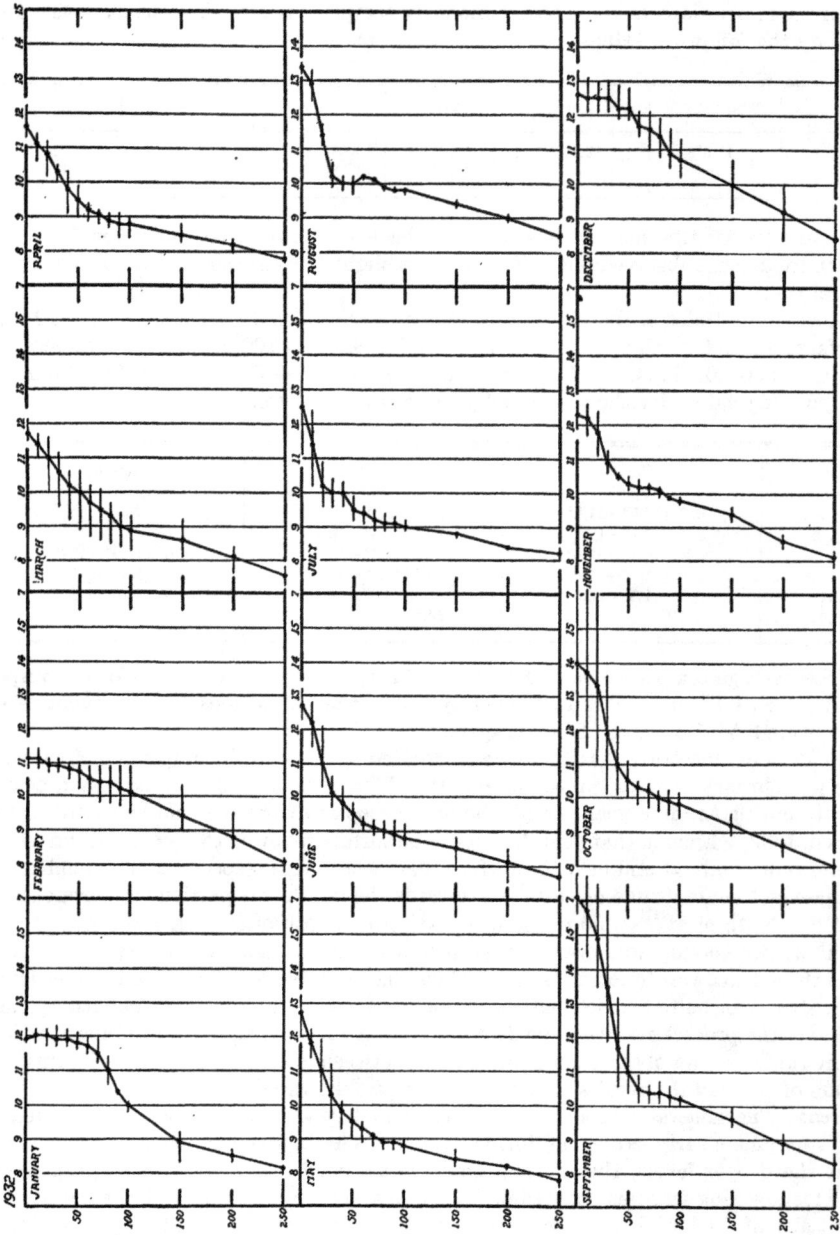

FIG. 5. Monthly thermal gradients and amplitudes of monthly thermal variation, in 1932. Dots indicate monthly averages; horizontal lines, amplitudes of monthly variations. In upper 100 meters, observations were made at 10 meter intervals.

this month during the investigated five year period. Much less uniformity could be distinguished in regard to the amplitude of thermal variation.

An examination of the monthly thermal gradients showed the following averages from 1929 to 1933:

	Surf.—50 m.	Diff.	Surf.—100 m.	Diff.
1929	11.8°− 9.4°	2.4°		
1930	13.3°−10.9°	2.4°		
1931	13.2°−10.9°	2.3°	13.2°−9.6°	3.6°
1932	11.3°− 9.9°	1.4°	11.3°−9.1°	2.2°
1933	11.1°− 9.6°	1.5°	11.1°−9.0°	2.1°
Average	12.1°−10.1°	2.0°	11.9°−9.2°	2.7°

From this table it will be seen that, on the average, the temperature decreased from the surface to 50 meters and from the surface to 100 meters not less than 2.0° and 2.6°, respectively. This was a very pronounced decrease indeed when compared with the previous two months. The variations in the inclination of the gradients from year to year were of the same magnitude (1.0° from the surface to 50 m.) as in January. There was no correlation between this variation and the annual changes in the temperature; *e.g.*, even though 1929 was cold and 1930 was warm their gradients were the same (2.4°).

The average monthly temperature varied from year to year in the following manner:

	1929	1930	1931	1932	1933	Average
0– 50 m.	10.5°	11.9°	12.1°	10.6°	10.3°	11.1°
50–100 m.			10.1°	9.4°	9.3°	9.6°

The averages for this month thus were consistently and distinctly lower than those for February, and the difference was more pronounced in the layer between 50 m. and 100 m. than in the upper 50 meters. In other words, the gradual lowering of the temperatures, which was characteristic of the seasonal progression in this region during the earlier part of the year, and which was barely foreshadowed in February, was well marked in March.

The total amplitudes of thermal variation during the investigated five year period were 3.4° (14.1°−10.7°), at the surface; 3.6° (12.2°−8.6°), at 50 m.; 2.1° (10.3°−8.2°), at 100 m.; and for the averages for 0–50 m. and 50–100 m., 3.8° (13.3°−9.5°) and 2.6° (11.0°−8.4°), respectively. Some of the corresponding data for the individual years were as follows:

	Surf.	50 m.	100 m.
1929	0.7° (12.3°−11.6°)	2.8° (11.4°− 8.6°)	
1930	1.6° (14.1°−12.5°)	1.7° (11.8°−10.1°)	
1931	1.9° (13.9°−12.0°)	2.8° (12.2°− 9.4°)	1.8° (10.3°−8.5°)
1932	0.9° (11.7°−10.8°)	2.4° (11.0°− 8.6°)	1.6° (9.8°−8.2°)
1933	0.7° (11.4°−10.7°)	1.4° (10.3°− 8.9°)	1.7° (10.2°−8.5°)
Average	1.2°	2.2°	1.7°

For the average temperatures of 0–50 m., these five years yielded the following amplitudes, 1.8°, 1.5°, 3.0°, 1.7°, and 1.2° (average 1.8°); the amplitudes for the averages of 50–100 m. were 2.1°, 1.7°, and 1.5° (average 1.8°).

These data exhibit a decided lack of uniformity.   For instance, in 1929 the amplitude of thermal variation maintained about the same characteristics as in the preceding month; *i.e.*, the variation was slight at the surface and showed a strong increase towards the depth. The same was true in 1932, although in a less striking manner.   On the other hand, in 1930 there occurred a decided increase in the range of variation at and near the surface and a corresponding decrease in the deeper strata, resulting in a fairly wide and more or less uniform variability throughout the upper 50 meters.   The slight variability recorded for 1933 in the upper 50 meters was probably partly real, the entire year being cold and apparently stable, partly apparent only, due to a comparative scarcity of observations.   The surface waters exhibited, on the average, a distinct seasonal progression in the range of thermal variation, March being more variable than February, with a range of 1.2° as compared with 0.8°.   The only year in which this rule did not hold was 1933, in which the March range was 0.7° as compared with 0.8° for February.   At 50 m. the average range was the same for these two months, but the individual years showed little or no regularity, three of the five years being less variable in March than in February, the remaining two being the reverse.   The same irregularity occurred in the variability at the 100-meter level. Generally speaking, if there were a seasonal tendency in the deeper strata, it was in the direction of a decrease in the thermal variation towards the summer months.

No indications of local winter cooling were observed.

April: The gradual increase in the thermal gradient continued.   As before, this increase was nearly always due to the progressive cooling of the deeper strata, not to a rise in the temperature of the surface waters.   In regard to the amplitude of thermal variation, April also exhibited features, when compared with March, which the latter month displayed in relation to February.   In other words, the seasonal progression was consistent in the entire hydrographic situation.

From 1929 to 1933, the thermal gradients for the month of April were as follows:

	Surf.—50 m.	Diff.	Surf.—100 m.	Diff.
1929	11.5°— 9.0°	2.5°		
1930	14.1°—10.2°	3.9°		
1931	12.2°—10.2°	2.0°	12.2°—9.4°	2.8°
1932	11.2°— 9.4°	1.8°	11.2°—8.7°	2.5°
1933	11.2°— 8.9°	2.3°	11.2°—8.4°	2.8°
Average	12.0°— 9.5°	2.5°	11.5°—8.8°	2.7°

The increase in the thermal gradient from the surface to 50 meters (2.5° as compared with 2.0° in March) was quite marked, while the increase in the gradient from the surface to 100 meters was but slight (2.7° against 2.6° in March).   In other words, the cooling of the subsurface layers which proceeded from the depth towards the surface was, peculiarly enough, very slight in the stratum between 50 and 100 meters, while it still was progressing quite strongly in the uppermost 50 meters.   The only year in which the gradient did not exceed that of the previous month was 1931, a year very irregular in several other respects.

While the variation in the steepness of the gradients was extremely slight from the surface to 100 meters, it was more pronounced (2.1°) in the upper 50 meters than in any of the previous months. The greatest deviation from the normal occurred in the warm year of 1930; but no correlation can be discerned between the variability and the actual temperature.

The variations in the monthly temperatures are displayed by the following averages:

	1929	1930	1931	1932	1933	Average
0– 50 m.........	9.9°	11.9°	11.0°	10.3°	9.9°	10.6°
50–100 m........			9.9°	8.9°	8.6°	9.1°

With the exception of the year 1930, the month of April thus was consistently cooler than March; and the difference between these two months was on the average the same for the uppermost 50 meters and for the stratum between 50 and 100 meters. In 1930, the average temperature in the upper 50 meters was identical for these two months.

The total amplitudes of thermal variation from 1929 to 1933 were 5.0° (15.2° − 10.2°), at the surface; 4.2° (12.5° − 8.3°), at 50 m.; 2.3° (10.5° − 8.2°), at 100 m.; and for the averages for 0–50 m. and 50–100 m., 3.4° (12.9° − 9.5°) and 3.4° (11.8° − 8.4°) respectively. The corresponding values for each of these five years were as follows:

	Surf.	50 m.	100 m.
1929..............	2.0° (12.2° − 10.2°)	1.2° (9.5° − 8.3°)	
1930..............	2.0° (15.2° − 13.2°)	1.8° (11.2° − 9.4°)	
1931..............	3.5° (14.6° − 11.1°)	3.4° (12.5° − 9.1°)	1.9° (10.5° − 8.6°)
1932..............	0.4° (11.4° − 11.0°)	0.8° (9.8° − 9.0°)	0.4° (8.8° − 8.4°)
1933..............	0.8° (11.7° − 10.9°)	0.4° (9.2° − 8.8°)	0.3° (8.5° − 8.2°)
Average..........	1.7°	1.5°	0.9°

In regard to the average temperatures of 0–50 m., these years exhibited the following amplitudes: 0.6°, 1.7°, 2.7°, 0.6°, and 0.6° (average 1.2°); and the amplitudes for the averages of 50–100 m. were 3.0°, 0.5°, and 0.4° (average 1.3°).

There was hardly any similarity in the thermal variation between this and the preceding month. The peculiarity of a strong increase in the variability towards the depth displayed by the earliest months of the year was not present any more. When an increase did occur, it was always slight (1932). Sometimes (1930, 1931), the variability was about the same throughout the upper 50 meters; and sometimes (1929, 1933), there was a slight decrease. The surface continued to show an increase in variability (1.7° as compared with 1.2° in March), the only year deviating from this rule being 1932 which exhibited an abnormal stability, with a range of only 0.4° for the entire month. For the five year period the variation at the surface was also decidedly greater in April than in March, 5.0° against 3.4°. In regard to the deeper waters it should be noted that in general the variability was decreasing, a seasonal phenomenon indicated by the observations in March. A particularly slight variability was to be recorded at the 100 meter level but also at the 50 meter level the decrease in variability was fairly striking (1.5° as compared with 2.2° in March). However, it should be emphasized that the individual years behaved very differently in this respect; at the 50 meter level only three of the five years displayed a decrease and at the 100 meter level only two of the three years, a very unsatisfactory basis for a generalization.

At this time insolation undoubtedly caused a considerable local heating of the surface

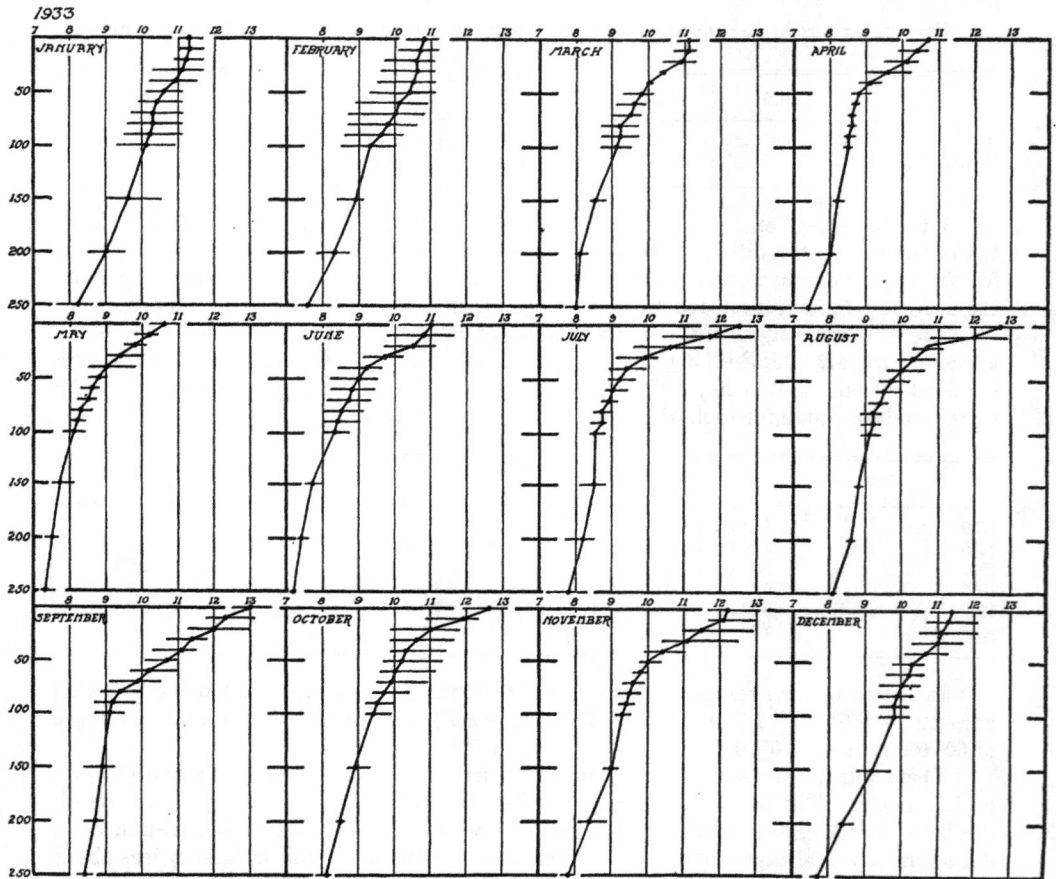

Fig. 6. Monthly thermal gradients and amplitudes of monthly thermal variation, in 1933. Dots indicate monthly averages; horizontal lines, amplitudes of monthly variations. In upper 100 meters, observations were made at 10 meter intervals.

waters but the effects of this process were hidden by the circulatory phenomena which caused a lowering of the surface temperatures.

May: This month was characterized particularly by a decided increase in the thermal gradient. However, at this time the phenomenon was only to a slight extent due to the cooling of the deeper waters; its cause was largely to be found in the progressive warming of the superficial strata.   While in this respect the month of May fell in line with the typical seasonal progression, in regard to the amplitude of thermal variation it was peculiarly aberrant, being each year abnormally constant for this season.   To a certain extent, this month may be regarded as a turning point in the unfolding of the hydrographic rhythm.

The monthly thermal gradients varied in the following manner during the five years under investigation:

	Surf.—50 m.	Diff.	Surf.—100 m.	Diff.
1929	11.8°− 8.8°	3.0°		
1930	13.7°− 8.9°	4.8°		
1931	14.2°−10.9°	3.3°	14.2°−9.3°	4.9°
1932	12.8°− 9.4°	3.4°	12.8°−8.8°	4.0°
1933	10.6°− 8.7°	1.9°	10.6°−8.0°	2.6°
Average	12.6°− 9.3°	3.3°	12.5°−8.7°	3.8°

Thus the increase in the thermal gradient from the surface to 50 meters (3.3° as compared with 2.5° in April) was even greater than in April; and it occurred in four out of five years, the unusually cool and stable year of 1933 forming the exception.   Although the increase was due largely to a rise in the surface temperature, the 50 meter level continued to register a slight cooling (average temperature, 9.3° as compared with 9.5° in April). Sometimes the cooling was very pronounced, as in 1930 when a drop from 10.2° to 8.9° was recorded.   In three of the five years the deeper waters of this stratum continued to cool during this month; in one year (1932) the temperature at the 50 meter level remained constant; and only in 1931, an aberrant year, a rise in temperature was recorded at this level.   In the stratum between 50 and 100 meters, in which the process of cooling showed a distinct retardation in April, the cooling was still going on in May, but in this month the process either reached or was fairly close to the turning point.   It is of interest to note that, in spite of the last fact, the average lowering of the temperature from 50 to 100 meters was even more pronounced in May than in April.

The average monthly temperatures varied from 1929 to 1933 in the following way:

	1929	1930	1931	1932	1933	Average
0– 50 m.	9.8°	10.4°	12.3°	10.8°	9.7°	10.6°
50–100 m.			9.8°	9.0°	8.2°	9.0°

The May averages for the five year period thus were nearly identical with those for the month of April.   In regard to the individual years, in three of the five years May had lower averages than April.   In 1932, the entire stratum from the surface to 100 meters was warmer in May, while in 1931 only the upper 50 meters exhibited this feature.

The total amplitudes of thermal variation from 1929 to 1933 were as follows: at the surface, 5.2° (15.4°−10.2°); at 50 m., 4.6° (12.8°−8.2°); at 100 m., 1.8° (9.6°−7.8°); and for the averages for 0–50 m. and 50–100 m., 4.1° (13.3°−9.2°) and 2.3° (10.3−8.0°), respectively.   The following are the corresponding values for each of these five years:

	Surf.	50 m.	100 m.
1929.............	0.8° (12.2°−11.4°)	1.3° (9.5°−8.2°)	
1930.............	1.7° (14.7°−13.0°)	0.5° (9.2°−8.7°)	
1931.............	2.0° (15.4°−13.4°)	3.5° (12.8°−9.3°)	0.8° (9.6°−8.8°)
1932.............	1.1° (13.4°−12.3°)	0.8° (9.8°−9.0°)	0.4° (9.0°−8.6°)
1933.............	0.6° (10.8°−10.2°)	0.6° (9.1°−8.5°)	0.6° (8.4°−7.8°)
Average..........	1.2°	1.3°	0.6°

The average temperatures of 0–50 m. displayed the following amplitudes in each of these five years: 1.2°, 1.1°, 2.5°, 0.6°, and 0.5° (average 1.2°); for 1931, 1932, and 1933 the amplitudes for the averages of 50–100 m. were 1.3°, 0.4°, and 0.7° (average 0.8°).

From these data it will be clear that the variability showed no regularity within this month.   Sometimes, as in 1933, there was a slight and uniform variability throughout the upper 100 meters.   At other times, as in 1932, there was a slight but regular decrease in the variability towards the 100-meter level.   The middle layer (around 50 m.) sometimes was more variable than both the surface and the 100-meter level (1931); or, finally, as in 1929, there was a fairly marked, though irregular, increase from the surface downwards.   Remarkably enough, the surface as well as the 50- and the 100-meter levels displayed on the average less variability than in April.   The month of May in this respect showed the most striking deviation from the seasonal rhythm in the variability of the surface layer.   Only in May 1932 did the surface register greater variability than in April.   On the other hand, if the entire five year period instead of the individual year is considered, then May was more variable than April in regard to the upper 50 meters.   Indeed, the constancy within each year and the strong variability from year to year were among the most outstanding features of the surface waters in the month of May.   This may be connected with the fact that May represents a transitional period in the hydrography.   The variability of the 50–100-meter water decreased when compared with that of the previous month; and in this respect it was well in harmony with the possible pattern indicated by the month of April (see above, month of April).

The results of progressive local heating of the surface was gradually becoming evident (average of surface temperature 2.6° as compared with 2.0° in April).   However, out of the five years under consideration only three registered advance in surface temperature (1929, 1931, 1932); in the remaining years strong circulatory movements hid this phenomenon.

June: The increase in the thermal gradient continued but was fairly slight.   It was due nearly exclusively to the rise in the temperatures of the superficial strata.   At the 50-meter level, the cooling had come to an end except in years of very unusual hydrographic conditions; but at the 100-meter level, a slight lowering of the temperatures frequently took place, the cooling in all probability being the rule rather than the exception.   Thus, for all practical purposes, this month represented the definite end of the subsurface cooling in the upper 50 meters, a process which dominated the hydrography of this stratum during the earlier part of the year.   In regard to the variability, as well as in respect to the thermal gradient, the month of June may be said to have fitted the seasonal pattern quite well.

The following variations were displayed by the monthly thermal gradients in the course of the five year period:

	Surf.—50 m.	Diff.	Surf.—100 m.	Diff.
1929	13.3°−9.6°	3.7°		
1930	13.5°−9.8°	3.7°		
1931	14.4°−9.8°	4.6°	14.4°−9.0°	5.4°
1932	12.9°−9.5°	3.4°	12.9°−8.6°	4.3°
1933	11.0°−8.8°	2.2°	11.0°−8.1°	2.9°
Average	13.0°−9.5°	3.5°	12.8°−8.6°	4.2°

When compared with May, the increase in the thermal gradient from the surface to the 50-meter level thus was small (3.5° as compared with 3.3° in May); and it was present in only three out of the five years (1929, 1931, and 1933); in 1932 the monthly gradients for May and June were identical; and in 1930 the gradient was unusually large in May and showed a moderate compensatory decrease in June. The rise in the surface temperature was moderate to slight and it was recorded for four out of five years. At the 50-meter level, there was a moderate to slight rise in the temperature in all of the five years except in 1931 in which year this temperature was abnormally high in May. The 100-meter level, on the other hand, exhibited a slight cooling in two out of the three years on record. In other words, while no cooling could normally be observed above the 50-meter level, there was still traceable a very distinct aftermath of this interesting and fundamental seasonal trend in the somewhat deeper strata.

The following monthly averages were recorded between 1929 and 1933:

	1929	1930	1931	1932	1933	Average
0– 50 m.	10.6°	11.1°	12.0°	10.9°	9.9°	10.9°
50–100 m.			9.3°	8.9°	8.4°	8.9°

The thermal averages for the upper 50 meters in June were thus slightly higher in four out of the five years than those in the previous month, while the averages for 50–100 meters were slightly lower in two out of the three years. In other words, the gradual lowering of the temperatures in the stratum from the surface to 50 meters which characterized the early months of the year (January, February, March, and April, 12.1°, 11.9°, 11.1°, and 10.6°, respectively) and which came to a standstill in May (average 10.6°), gave place to a warming process in June. The process was quite irregular and was not obvious unless monthly averages rather than daily readings were used.

During the investigated period the following total amplitudes of thermal variation were recorded: at the surface, 5.8° (16.0°−10.2°); at 50 m., 3.8° (12.1°−8.3°); and at 100 m., 1.5° (9.3°−7.8°); and for the averages for 0–50 m. and 50–100 m., 3.5° (12.9°−9.4°) and 1.7° (9.7°−8.0°), respectively. Some of the corresponding values for the individual years were as follows:

	Surf.	50 m.	100 m.
1929..............	5.8° (16.0°−10.2°)	1.0° (10.2°−9.2°)	
1930..............	3.9° (15.4°−11.5°)	3.4° (12.1°−8.7°)	
1931..............	3.4° (15.5°−12.1°)	1.5° (10.7°−9.2°)	0.8° (9.3°−8.5°)
1932..............	0.5° (13.2°−12.7°)	0.9° (10.0°−9.1°)	1.0° (9.1°−8.1°)
1933..............	1.4° (11.6°−10.2°)	1.3° (9.6°−8.3°)	0.9° (8.7°−7.8°)
Average..........	3.0°	1.6°	0.9°

The average temperatures of 0–50 m. displayed during this period the following amplitudes: 1.2°, 2.0°, 2.6°, 0.6°, and 1.1° (average 1.5°); and in the last three years of this period the amplitudes for the averages of 50–100 m. were 0.8°, 0.5°, and 0.9° (average 0.7°).

In regard to the surface waters, June displayed the greatest total monthly variability (5.8°); furthermore, together with July it exhibited the greatest average monthly variability (3.0°). At the same time, the individual years differed very strikingly from each other in this respect. Thus, for instance, in 1929 the record for this month was 5.8°, a value equal to the total amplitude for this month during the entire five year period. On the other hand, June of 1932 showed only an amplitude of 0.5°, thus a value equal to the lowest obtained for any month during the entire five year period, and decidedly lower than the corresponding record for May. On the average, the amplitude of thermal variation decreased with depth, but in this respect there was no regularity among the various years. The variability decreased strongly (1929, 1931) or slightly (1933) with the depth, or it showed a slight increase (1932).

Local heating of the surface waters was progressing slowly; four out of the five years showed a rise in temperature and the June average for the five years was 13.0° as compared with 12.6° for May.

July: By and large, the hydrographic situation of July showed a very striking similarity to that of the previous month. This statement is true not only in regard to the thermal gradient but also concerning the annual changes in the average temperatures and the type of thermal variation. As stated above, the month of June represented the end of the subsurface cooling in the upper 50 meters except under very unusual conditions. This statement is borne out by the fact that during the entire five year period water of a temperature of 8° extended up to the 50-meter level in July on only one occasion, *viz.*, on July 2, 1931, a year in several respects of very unusual features. In the stratum between 50 and 100 meters, water of this temperature was becoming increasingly rare; indeed, this month may be considered to have represented the termination of the subsurface cooling of this stratum, at least under what may be termed "normal" years.

From 1929 to 1933, the following monthly thermal gradients were recorded:

	Surf.—50 m.	Diff.	Surf.—100 m.	Diff.
1929........................	14.1°−10.5°	3.6°		
1930........................	13.3°− 9.7°	3.6°		
1931........................	12.9°− 9.8°	3.1°	12.9°−9.2°	3.7°
1932........................	13.2°− 9.4°	3.8°	13.2°−9.0°	4.2°
1933........................	12.2°− 9.2°	3.0°	12.2°−8.7°	3.5°
Average.....................	13.1°− 9.7°	3.4°	12.8°−9.0°	3.8°

From these data it will be seen that the thermal gradient of the upper 50 meters had decreased in three out of the five years and that even for the entire five year period it registered a very slight decrease (3.4° as compared with 3.5° for the month of June). The gradient of the upper 100 meters showed a decrease in two out of three years; and for the entire investigated period a moderate decline was observed (3.8° against 4.2°). A rise in the surface temperature was found in three of the five years (1929, 1932, and 1933); and a slight decrease occurred in 1930 and quite a pronounced cooling in 1931. At the 50-meter level there was a moderate warming in 1929 and 1933; and in 1931 the average temperatures at this level were identical for July and June, while in 1932 July showed a very slight cooling. The irregularity of the variations is clearly shown by these examples. At 100 meters, there was a distinct warming in all of the three years investigated; the averages at this level for July and June were 9.0° and 8.6°, respectively. This was the most pronounced difference between these two months which in other respects were so similar. This warming, of course, strongly supports the opinion expressed above that July represented the last month of subsurface cooling in the stratum from 50 to 100 meters, even though the 8° water, which seemed to form such an excellent indicator of this cooling process, was still far from being rare in this stratum.

The average monthly temperatures varied from year to year in the following manner:

	1929	1930	1931	1932	1933	Average
0– 50 m.........	11.9°	11.0°	10.9°	10.8°	10.3°	11.0°
50–100 m.........			9.4°	9.1°	8.8°	9.1°

The total averages for the five year period thus were very slightly higher for July than for June, the differences, however, although significant of a general trend, being nearly negligible (11.0° and 9.1° as compared with 10.9° and 8.9°). In the upper 50 meters, not less than three out of the five years were colder in July than in June. The most interesting feature is that the 50–100-meter stratum was consistently warmer in July than in June. Another feature perhaps worthy of notice in this connection is the relatively slight variation from year to year, the maximum range in the upper 50 meters being 11.9° − 10.3°.

From 1929 to 1933 the total amplitudes of thermal variation were as follows: at the surface, 4.8° (16.0° − 11.2°); at 50 m., 3.9° (12.7° − 8.8°); and at 100 m., 1.5° (9.6° − 8.1°); and for the averages for 0–50 m. and 50–100 m., the records yielded 4.1° (14.0° − 9.9°) and 1.4° (9.8° − 8.4°), respectively. The individual years showed the following values:

	Surf.	50 m.	100 m.
1929.............	4.0° (16.0° − 12.0°)	3.6° (12.7° − 9.1°)	
1930.............	2.8° (14.6° − 11.8°)	1.9° (11.0° − 9.1°)	
1931.............	4.7° (15.9° − 11.2°)	1.5° (10.3° − 8.8°)	1.3° (9.6° − 8.3°)
1932.............	3.0° (14.2° − 11.2°)	0.4° (9.6° − 9.2°)	0.1° (9.0° − 8.9°)
1933.............	1.3° (12.8° − 11.5°)	0.4° (9.4° − 9.0°)	1.0° (9.1° − 8.1°)
Average..........	3.2°	1.6°	0.8°

The average temperatures of 0–50 m. exhibited in these five years the following amplitudes: 3.5°, 1.3°, 1.8°, 1.6°, and 0.7° (average 1.8°); and in the last three years, 1931 to 1933, the amplitudes for the averages of 50–100 m. were 1.2°, 0.2°, and 0.8° (average 0.7°).

While July displayed the greatest average monthly variability of the surface temperatures (3.2°), the total variation in this layer during the five year period was distinctly less than that recorded for June (4.8° against 5.8°); however, even so, it was comparatively high. The same was true in regard to the differences in the amplitudes from one year to another. In general the amplitude of thermal variation decreased quite substantially with depth. However, the irregularities in the subsurface thermal variation from year to year were very striking; e.g., in 1929, the amplitude at 50 m. was not less than 3.6° while in 1932 and 1933 it was only 0.4° at the same level; furthermore, at 100 m., 1932 displayed an extraordinarily narrow amplitude (0.1°) while the remaining of the three investigated years were what may be termed normal in this respect.

Local heating of the surface waters undoubtedly went on, although at a comparatively slow rate due to prevailing overcast sky conditions. However, due to circulatory phenomena, this process was frequently hidden; just as often as not, the surface temperatures decreased rather than increased from June to July.

August: In some respects the thermal conditions in August simulated those of July; e.g., in regard to the thermal gradient in the upper 50 meters and in the gradual decrease in the thermal variation towards the depth. The most striking difference which was established in the course of the investigated five year period was the distinctly warmer water in August when compared with July, in which regard this month fitted very well into the pattern of the seasonal temperature progression. The best illustration of this warming process was to be found in the fact that not only did 8° water not extend up to the 50-meter level but this type of water was found above the 100-meter level on only one day in the five-year period, *viz.*, on August 7, 1933, when it penetrated to the 80-meter level.

The monthly thermal gradients varied as follows during the analyzed five year period:

	Surf.—50 m.	Diff.	Surf.—100 m.	Diff.
1929..........................	14.0°−10.2°	3.8°		
1930..........................	13.4°−10.3°	3.1°		
1931..........................	14.1°−10.6°	3.5°	14.1°−9.6°	4.5°
1932..........................	13.6°−10.4°	3.2°	13.6°−9.7°	3.9°
1933..........................	13.0°− 9.6°	3.4°	13.0°−9.0°	4.0°
Average......................	13.6°−10.2°	3.4°	13.6°−9.4°	4.2°

In the upper 50 meters the thermal gradient thus was very constant from one year to another, just as in the month of July; in addition, the total average for the five year period was identical in July and August (3.4°); and no distinct trend could be distinguished, the variations apparently being random in nature. In regard to the monthly gradient from the surface to 100 meters, the one for August was somewhat greater than that for July (4.2° as compared with 3.8° in July); this was true for two out of the three years as well as for the entire period. This difference was not the expression of any general trend; for instance, the gradient for August was also characteristic for June. A rise in the surface temperature over July was recorded for four out of the five years as well as for the entire five-year period (13.6° against 13.1°); and this held true also for the 50-meter level, in which case the total averages were 10.2° and 9.7°, respectively.   The same distinctive rise in the temperature at 100 m. that was recorded for July also occurred in August (9.4° compared with 9.0° in July). In other words, the subsurface warming which was initiated in the previous months was continued in August.   The disappearance of the 8° water which was noted above was particularly striking.

The annual variations in the monthly averages were as follows during the five years under consideration:

	1929	1930	1931	1932	1933	Average
0– 50 m........	11.3°	11.3°	12.0°	11.5°	10.9°	11.4°
50–100 m........			9.9°	9.9°	9.2°	9.7°

While the differences in the total averages for the five year period were nearly negligible in the case of a comparison between the months of June and July, these averages for August were found to be distinctly higher than those for July (11.4° and 9.7°, as compared with 11.0° and 9.1° in July).   In regard to the upper 50 meters, four out of the five years were warmer in August than in July; 1929 formed the only exception.   In the stratum between 50 and 100 meters, August was always found to be the warmer of these two months.   The slight variability of the thermal conditions of this month from one year to another was also noteworthy, just as in the case of July, the range in the upper 50 meters being only from 12.0° to 10.9°, thus even less than in July.

The total amplitudes of thermal variation were as follows from 1929 to 1933: at the surface, 4.2° (16.0° − 11.8°); at 50 m., 2.6° (11.9° − 9.3°); and 1.3° (10.2° − 8.9°) at the 100 meter level; the corresponding values for 0–50 m. and 50–100 m. were 3.6° (13.6° − 10.0°) and 1.9° (10.9° − 9.0°), respectively.   The records for the individual years displayed the following results:

	Surf.	50 m.	100 m.
1929.............	3.2° (15.7° − 12.5°)	1.4° (11.0° − 9.6°)	
1930.............	1.9° (14.5° − 12.6°)	1.5° (11.2° − 9.7°)	
1931.............	2.6° (15.2° − 12.6°)	2.2° (11.9° − 9.7°)	1.0° (10.2° − 9.2°)
1932.............	4.2° (16.0° − 11.8°)	2.0° (11.5° − 9.5°)	1.1° (10.2° − 9.1°)
1933.............	1.1° (13.6° − 12.5°)	0.7° (10.0° − 9.3°)	0.2° (9.1° − 8.9°)
Average..........	2.6°	1.6°	0.8°

The average temperatures of 0–50 m. displayed in these five years the following amplitudes: 1.3°, 1.3°, 2.4°, 3.6°, and 0.3° (average 1.8°); and in 1931, 1932, and 1933, the amplitudes for the averages of 50–100 m. were 1.5°, 1.3°, and 0.5° (average 1.1°).

The amplitude of thermal variation in the surface waters which, by and large, had been increasing ever since the beginning of the year now began to show a distinct decline.   The season was beginning, in a manner, to pass its apex.   The decline was expressed not only in the total monthly amplitude but also in the average amplitude (2.6° for August; 3.2° for July).   The total amplitude began to show a decrease in July.   It should also be noted that a decrease was recorded for four out of the five years, 1932 forming the only exception. The variations decreased from the surface towards the depth, and this rule did not exhibit a single exception during the period under examination.   In other respects the variation in the deeper strata was decidedly irregular.   At 50 meters it was less than in July only in two out of the five years; and in the remaining three years (1931, 1932, and 1933) the variation was seasonally subnormal in July.   At the 100-meter level a decrease occurred in two out of three years.   An abnormally small variation was recorded at 100 meters in 1933 (0.2°).

To what an extent the warming of the water in this month was due to local heating can not be decided as yet.   However, in all probability the recorded rise in temperature was to a very large extent the result of peculiarities in the conditions of circulation.

September: This was in some respects the most easily distinguished month of the year from the point of view of its thermal characteristics.   Thus for instance, the temperature gradient had a steeper slope than in any other month; and in regard to the average temperatures, September represented the very peak of the seasonal progression, these temperatures being higher than in any of the remaining months.   It should be noted that in spite of the fact that warm water prevailed, nevertheless, on very rare occasions 8° water did ascend as high as to the 80-meter level, as happened on September 25, 1933, *i.e.*, during an unusually cold year.   Even though water of this kind was otherwise never reported during September, it seems to have prevailed during the last two weeks of this month of 1933, in the stratum just above the 100-meter level.   In regard to the thermal variation, September showed a fairly close resemblance to the preceding month.

Our records yield the following monthly thermal gradients from 1929 to 1933:

	Surf.—50 m.	Diff.	Surf.—100 m.	Diff.
1929	14.7°—10.6°	4.1°		
1930	15.0°—11.0°	4.0°		
1931	14.6°—10.9°	3.7°	14.6°— 9.7°	4.9°
1932	15.9°—10.9°	5.0°	15.9°—10.2°	5.7°
1933	13.5°—10.0°	3.5°	13.5°— 9.1°	4.4°
Average	14.7°—10.7°	4.0°	14.7°— 9.7°	5.0°

As will be seen from these data, even though the gradients did exhibit a moderate amount of variation, they always had a more or less pronounced slope.   In some years (*e.g.*, 1932) the gradients even had an extraordinarily steep inclination, considering the thermal characteristics of this general region.   An increase in the slope of the gradients was characteristic for the upper 50 and 100 meters not only for the entire five year period but it was also displayed by every one of the component years.   The same remarkable uniformity in the change of the conditions from August to September was recorded for the progressive warming of the water.   Throughout the entire upper 100 meters and in every one of the five years investigated, the temperature of September rose over that of August.   The average difference between these two months at the surface, at 50 m. and at 100 m. was more than usually well marked: 14.7°, 10.7°, and 9.7°, as compared with 13.6°, 10.2°, and 9.4°, respectively.

From 1929 to 1933 the September averages displayed the following variations:

	1929	1930	1931	1932	1933	Average
0– 50 m.	12.1°	12.4°	12.4°	13.8°	11.4°	12.4°
50–100 m.			10.2°	10.5°	9.6°	10.1°

Just as in the case of the previous two months, the differences between the total averages, for the five year period, of September and August were quite marked, particularly in regard to the upper 50 meters (12.4° and 10.1° on the one hand and 11.4° and 9.7° on the other). Moreover, not only were the total averages for September higher, but every one of the five years investigated yielded higher values for this month both in the upper and the lower 50-meter stratum—a quite unique consistency of change. It may also be noted that September showed a distinctly higher variability from one year to the next than that found in the previous two months, 1932 and 1933 representing the extremes, the former being remarkably warm, the latter unusually cool.

During the five year period, the following total amplitudes of thermal variation were recorded: at the surface, 3.7° (16.7°−13.0°); at 50 m., 2.7° (12.3°−9.6°); and at 100 m., 1.9° (10.5°−8.6°); the corresponding data for 0–50 m. and 50–100 m. were 4.3° (15.2°−10.9°) and 2.6° (11.7°−9.1°), respectively. An examination of the individual years yielded the following values:

	Surf.	50 m.	100 m.
1929.............	2.3° (15.6°−13.3°)	1.5° (11.1°− 9.6°)	
1930.............	2.1° (16.1°−14.0°)	1.4° (11.8°−10.4°)	
1931.............	1.5° (15.3°−13.8°)	2.0° (12.3°−10.3°)	1.2° (10.5°− 9.3°)
1932.............	2.1° (16.7°−14.6°)	0.7° (11.2°−10.5°)	0.1° (10.3°−10.2°)
1933.............	1.3° (14.3°−13.0°)	0.8° (10.4°− 9.6°)	0.9° (9.5°− 8.6°)
Average..........	1.9°	1.3°	0.7°

The average temperatures of 0–50 m. exhibited the following amplitudes in each of these years: 2.0°, 1.5°, 1.8°, 2.3°, and 0.6° (average 1.6°); and in the last three years of this period, the amplitudes for the averages of 50–100 m. were 1.8°, 0.6°, and 0.8° (average 1.1°).

The decrease in the average monthly variation of the surface temperature, which began in August, was again very conspicuous (2.6° for August as compared with 1.9° for September). The total amplitude for the five year period, which began to decline in July, also exhibited a pronounced decrease (4.2° and 3.7°, respectively). In regard to the individual years, a decided irregularity prevailed; only three out of the five years (1929, 1931, and 1932) showed a decrease while the remaining two exhibited a slight increase. There was, on the average, a distinct decrease in the variation towards the depth. However, in this respect, too, there was a random variation from year to year. In 1932 the 100 meter level was remarkably stable, the amplitude being not more than 0.1°.

The rise in temperature was evidently due rather to circulatory causes than to local insolation.

October: This month was characterized by the fact that a regression had set in from the summer condition, as climaxed by September, to the situation prevailing during the winter months. The most outstanding changes were: first, that the water was becoming cooler and, secondly, that the thermal gradient was assuming a gentler slope. On the other hand, the thermal variation at the surface was of an aberrant type; it was too great for the seasonal pattern. This peculiarity was probably due to the fact that October was in a measure a period of transition; a swaying back and forth of the seasonal climax might have caused an unusual degree of thermal instability. (It should be mentioned, as will be further discussed later in chapter VIII, that the thermal progression through the seasons to a very large extent was caused by circulatory phenomena rather than by local heating and cooling.)   It may finally be noted that water of 8°, which characterized the latter weeks of September, 1933, was recorded on only one day, *viz.*, on October 23, 1933, when it was located not more than about three meters above the 100 meter level.

The monthly thermal gradients were found to be as follows from 1929 to 1933:

	Surf.—50 m.	Diff.	Surf.—100 m.	Diff.
1929.........................	14.0°−11.1°	2.9°		
1930.........................	14.2°−11.1°	3.1°		
1931.........................	13.3°−10.9°	2.4°	13.3°−9.9°	3.4°
1932.........................	14.0°−10.5°	3.5°	14.0°−9.7°	4.3°
1933.........................	12.2°−10.1°	2.1°	12.2°−9.5°	2.7°
Average......................	13.5°−10.7°	2.8°	13.2°−9.7°	3.5°

There was thus both in the upper fifty and in the upper one hundred meters a consistent decrease in the thermal gradients from September to October, and this decrease was expressed both in the total averages and in each and every one of the investigated years.   The amount of variation in the gradients of both strata was moderate (3.5°−2.1°, in the upper 50 m.; 4.3°−2.7°, in the upper 100 m.).   These variations evidently were random in nature and not correlated with other temperature conditions.   The recorded decrease in the gradients was brought forth in all cases nearly exclusively by a lowering of the superficial temperatures rather than by a change in the temperatures of the deeper strata.   The total average of the five-year period at the surface fell from 14.7° in September to 13.5° in October, and every one of the examined years displayed a thermal decline.   The temperatures at the deeper levels moved less, and their changes apparently were random.   The total averages for 50 and 100 m. were found to be the same for September and October (10.7° and 9.7°, respectively).   At 50 m., three out of the five years yielded a slight thermal rise; and at 100 m. this was true for two out of three years.

The annual variations in the October averages, from 1929 to 1933, were as follows:

	1929	1930	1931	1932	1933	Average
0– 50 m.........	12.3°	12.1°	12.0°	12.2°	10.8°	11.9°
50–100 m........			10.3°	10.0°	9.8°	10.0°

From these averages it will be seen that the differences between September and October were not very great; in addition, the observed deviations were fairly erratic.   In the total averages for the five year period, the October water was very slightly cooler both in the upper fifty meters and in the stratum between 50 and 100 meters (11.9° and 10.0° for October, as compared with 12.4° and 10.1° for September).   There was little regularity as far as the individual year was concerned.   Four out of five years were cooler in the upper 50 meters, while in the lower stratum (50–100 m.) two out of three years exhibited a thermal rise.   By and large, the variability was very slight in October; thus, from 1929 to 1932, inclusive, the thermal averages for the upper 50 meters varied only from 12.0° to 12.3°, a remarkable stability, indeed.

For this month the following total amplitudes of thermal variation were recorded in the course of the analyzed five year period: at the surface, 5.2° (16.4° − 11.2°); at 50 m., 3.0° (12.5° − 9.5°); and at 100 m., 1.9° (10.8° − 8.9°). The corresponding data for 0–50 m. and 50–100 m. were 4.0° (14.0° − 10.0°) and 1.7° (11.0° − 9.3°), respectively. Each one of these years yielded the following values:

	Surf.	50 m.	100 m.
1929.............	1.9° (15.0° − 13.1°)	2.4° (12.5° − 10.1°)	
1930.............	1.7° (14.9° − 13.2°)	1.4° (11.7° − 10.3°)	
1931.............	2.3° (14.9° − 12.6°)	1.3° (11.5° − 10.2°)	0.6° (10.1° − 9.5°)
1932.............	4.5° (16.4° − 11.9°)	1.3° (11.2° − 9.9°)	1.1° (10.3° − 9.2°)
1933.............	1.8° (13.0° − 11.2°)	1.8° (11.3° − 9.5°)	1.9° (10.8° − 8.9°)
Average..........	2.4°	1.6°	1.2°

The average temperatures of 0–50 m. exhibited the following amplitudes in each of these years: 2.1°, 1.2°, 1.1°, 3.1°, and 1.6° (average 1.8°); in 1931, 1932, and 1933, the amplitudes for the averages of 50–100 m. were 0.7°, 1.1°, and 1.7° (average 1.2°).

While during the investigated five-year period there was a steady decrease in the thermal variation of the surface water from June to September (total amplitudes of these months, 5.8°, 4.8°, 4.2°, and 3.7°, respectively), October showed an abnormally high variability (total 5.2°). This increase was expressed not only by the total amplitudes but also by the total average (2.4°, as compared with 1.9° for September) and by three out of the five years, 1929 and 1930 registering a slight decrease. There was, on the average, a moderate decrease in the thermal variations towards the depth; but a pronounced difference prevailed among the various years in this respect. For instance, in 1933 there was about the same variation throughout the upper 100 meters while in 1932 the decrease towards the depth was strong. A greater variability was characteristic not only of the surface waters but also of the subsurface strata (at 50 and 100 m., 1.6° and 1.2° in October as compared with 1.3° and 0.7° in September). At 50 m., three out of the five years and at 100 m. two out of the three years showed an increase over September.

The lowering of the temperature evidently was due to circulatory rather than to local cooling phenomena.

November: The regression from the summer to the winter situation, which was begun and carried well under way during October, was continued at approximately the same rate in November. Thus when the hydrographic conditions of September and November are compared, the changes are very striking, indeed. However, even so the winter conditions did not obtain full development in the latter month. As was noted in the case of October, the most outstanding of these seasonal changes were: first, that the water became cooler and, secondly, that the thermal gradient became gentler. Contrary to the month of October, November was also characterized by a decided decrease in the range of thermal fluctuations of the surface waters. The water of 8°, which was noted in the discussion of the last couple of months, was completely absent from the upper one hundred meters, according to our records. The lowest temperature observed at 100 m. was 9.3° C.

The following variations in the monthly thermal gradients for November were recorded from 1929 to 1933:

	Surf.—50 m.	Diff.	Surf.—100 m.	Diff.
1929	13.0°−11.8°	1.2°		
1930	12.6°−11.5°	1.1°		
1931	12.3°−11.1°	1.2°	12.3°−10.0°	2.3°
1932	12.5°−10.5°	2.0°	12.5°− 9.8°	2.7°
1933	12.0°−10.3°	1.7°	12.0°− 9.4°	2.6°
Average	12.5°−11.0°	1.5°	12.3°− 9.7°	2.6°

Thus in regard to the thermal gradient the relation between this month and November was about the same as the one between October and September. In other words, there was a consistent decrease in the thermal gradient from October to November both in the upper fifty and in the upper one hundred meters, and this decrease was displayed in each and every one of the five years under investigation. The variation in the gradients of these strata was small, decidedly smaller than in October (2.0°−1.1° in the upper stratum, 2.7°−2.3° in the lower). As usual, these variations appear to have been random and not correlated with other temperature conditions. The decrease in the gradient was always due exclusively to the lowering of the surface temperature. The average surface temperature sank from 13.5° in October to 12.5° in November, when the entire five year period is considered; and for every one of these five years a decrease was recorded. On the other hand, a slight rise was found to have occurred at the 50-meter level (average for the five year period, 10.7° in October to 11.0° in November); in one of these years (1932), this temperature remained unchanged while in all the remaining years a rise was observed. At the 100-meter level the changes were negligible, the total averages for the three year period remaining unchanged (9.7°).

The November averages displayed, from 1929 to 1933, the following variations:

	1929	1930	1931	1932	1933	Average
0– 50 m	12.5°	12.1°	11.7°	11.3°	11.0°	11.7°
50–100 m			10.5°	10.2°	9.8°	10.2°

Just as in the case of the thermal gradients, November occupied in regard to the thermal averages a position to October which was about the same as the one which the latter month occupied to September. That is to say, the total averages of November and October for the five year period were but slightly different (11.7° and 11.9° for the upper 50 meters; 10.2° and 10.0° for 50–100 meters). Also, there was hardly any regularity in regard to the individual year. Concerning the upper 50 meters, two years (1931, 1932) were colder in November than in October; two (1929, 1933) were warmer; and one (1930) registered no change. The water from 50 to 100 meters was different in so far as no year was colder in November than in October; two years (1931, 1932) were warmer, and one (1933) was characterized by these months having identical averages. The variability of November was moderate.

From 1929 to 1933 the following total amplitudes of thermal variation were recorded: 3.1° (14.0°−10.9°), for the surface; 3.6° (13.3°−9.7°), for the 50-meter level; and 1.1° (10.4°−9.3°), for the 100-meter level; for 0–50 m. and 50–100 m. the corresponding values were, 3.0° (13.4° − 10.4°) and 1.5° (11.0° − 9.5°), respectively.   The individual years showed the following values:

	Surf.	50 m.	100 m.
1929.............	1.9° (14.0° − 12.1°)	3.3° (13.3° − 10.0°)	
1930.............	1.2° (13.2° − 12.0°)	2.4° (12.9° − 10.5°)	
1931.............	2.2° (13.5° − 11.3°)	1.5° (12.0° − 10.5°)	0.6° (10.4° − 9.8°)
1932.............	1.5° (13.2° − 11.7°)	0.7° (10.9° − 10.2°)	0.8° (10.2° − 9.4°)
1933.............	2.1° (13.0° − 10.9°)	0.9° (10.6° − 9.7°)	0.4° (9.7° − 9.3°)
Average..........	1.8°	1.8°	0.6°

The average temperatures of 0–50 m. displayed in each of these five years the following amplitudes; 1.6°, 1.7°, 1.6°, 0.9°, and 1.5° (average 1.5°); in 1931, 1932, and 1933, the amplitudes for the averages of 50–100 m. were 0.7°, 0.5°, and 0.8° (average 0.7°).

The steady decrease in the amplitude of thermal variation of the surface water present from July to September, but interrupted by the abnormally high variability in October, was continued in November.   The total amplitude for the five year period was 3.1° in November as compared with 3.7° in September; and the total averages showed the same phenomenon, 1.8° in November as compared with 1.9° in September.   Three of the investigated years (1929, 1930, and 1932) registered a decrease in the surface temperatures, when compared with September, and the remaining two showed an increase.   It is interesting to note that in November the total average amplitude of variation was the same at the surface and at 50 m. (1.8°) and that the total amplitude at 50 m. was even greater than at the surface (3.6°, as compared with 3.1°).   This is distinctly a winter feature.   However, in this respect the individual years behaved decidedly differently, the deviations evidently being random.   On the other hand, at the 100-meter level there was a very narrow amplitude (total, 1.1°; total average 0.6°), thus less than in October.   In no year was the amplitude at 100 m. greater in November than in October.

Local winter cooling began, even though it happened seldom that the temperature was lower at the surface than below.   In the course of five years it occurred only once (November, 10, 1930) that the surface temperature registered 0.1° less than the reading obtained at 5 m.

December: Remarkable changes characterized December when compared with November.   These changes were only partially in line with what may be termed the normal annual rhythm of the temperate regions; partially they were more or less unique for this general region of the California coast.   The thermal gradients manifested a striking decrease, in some years even obtaining a slope about equal to the gentle ones of January. This, however, does by no means imply that the surface water had become colder, as was the case in the previous months.   On the contrary, it was the deeper waters which had turned warmer on the average.   Also, the superficial waters tended to become less and the deeper strata more variable.   It may finally be noted that, according to our records, the water of 8° was never as high up as at the 100-meter level.   The lowest temperature at 100 m. was 9.2°.

The December monthly thermal gradients displayed the following variations from 1929 to 1933:

	Surf.—50 m.	Diff.	Surf.—100 m.	Diff.
1929.	13.6°−12.4°	1.2°		
1930.	13.5°−13.0°	0.5°		
1931.	11.8°−11.7°	0.1°	11.8°−11.0°	0.8°
1932.	12.4°−11.8°	0.6°	12.4°−10.6°	1.8°
1933.	11.3°−10.6°	0.7°	11.3°− 9.7°	1.6°
Average.	12.5°−11.9°	0.6°	11.8°−10.4°	1.4°

The strong decrease in the thermal gradients which was so strikingly characteristic of the previous several months was also one of the outstanding features of December. There was in this regard a nearly consistent decrease from November to December both in the upper 50 and in the upper 100 meters. The only exception to this rule was formed by the upper 50 meters in 1929, in which case the November and December gradients were identical. In one year (1931), the gradient of the upper 50 meters was extraordinarily slight (0.1°), equalling the gentlest gradient for this layer observed in January. In the same year the December gradient for the upper 100 meters was the lowest on record for all months during the entire five year period (0.8°), the next lowest gradient occurring in January 1931 (1.1°). Indeed, both for the upper 50 and the upper 100 meters, the total average gradients were nearly identical for December and January (0.6° and 1.4°, as compared with 0.5° and 1.3° for January). In other words, as far as the thermal gradients were concerned, December had reached the full development of the winter conditions. The variations in the gradients from year to year were somewhat larger than in November and of about the same magnitude as in January; and they were apparantly random in nature. The decrease in the gradients was due to a very large extent to a rise in the subsurface temperatures and not to a lowering of the surface temperatures. Indeed, the total average surface temperatures for December and November were identical (12.5°) in the period under investigation. The surface temperatures in December registered a decrease in three out of five years; and in 1929 and 1930 they exhibited a pronounced rise, the average December temperatures of these years being 13.6° and 13.5°, as compared with 13.0° and 12.6° in November. At 50 meters, the total average of December rose to 11.9° from 11.0° in November; and for 100 meters the corresponding values were 10.4° and 9.7°. The advance in the temperatures at 50 and 100 meters occurred in every one of the five years.

The thermal averages for the upper 50 meters and for the stratum from 50 to 100 meters manifested the following variations from 1929 to 1933:

	1929	1930	1931	1932	1933	Average
0– 50 m.	13.1°	13.3°	11.8°	12.2°	10.9°	12.3°
50–100 m.			11.4°	11.2°	10.1°	10.9°

The remarkable fact that the deeper waters were warmer in December than in November was thus also expressed in quite a striking manner by these averages. In regard to the upper 50 meters, not less than four out of the five years displayed a rise in December; and in the remaining year (1933), the averages of these two months were nearly identical (10.9° and 11.0°, respectively). The total averages for this stratum were 12.3° in December and 11.7° in November.   In the case of the layer between 50 and 100 meters, all the three investigated years exhibited a rise; and the December total average was well above the one for November (10.9° and 10.2°, respectively).   The variability in December was moderate but greater than in November.

The total amplitudes of thermal variation manifested in the course of the investigated five year period the following values: 3.6° (14.3° − 10.7°), at the surface; 4.0° (14.0° − 10.0°), at 50 m.; and 2.9° (12.1° − 9.2°), at 100 m.; and for 0–50 m. and 50–100 m., the corresponding data were 3.8° (14.2° − 10.4°) and 2.8° (12.2° − 9.4°), respectively.   The individual years of the investigated period displayed the following amplitudes:

	Surf.	50 m.	100 m.
1929.............	1.4° (14.3° − 12.9°)	3.1° (14.0° − 10.9°)	
1930.............	1.5° (14.3° − 12.8°)	1.6° (13.8° − 12.2°)	
1931.............	1.1° (12.3° − 11.2°)	0.8° (12.2° − 11.4°)	2.1° (12.1° − 10.0°)
1932.............	1.8° (13.6° − 11.8°)	0.8° (12.2° − 11.4°)	1.6° (11.1° − 9.5°)
1933.............	1.0° (11.7° − 10.7°)	1.2° (11.2° − 10.0°)	0.9° (10.1° − 9.2°)
Average..........	1.4°	1.5°	1.5°

The December average temperatures of 0–50 m. showed the following amplitudes in each of these five years: 2.1°, 1.3°, 0.8°, 1.4°, and 0.9° (average 1.3°); in the last three years of this period, the amplitudes for the averages of 50–100 m. were 1.7°, 0.9°, and 1.2° (average 1.3°).

The decrease in the amplitude of thermal variation of the surface waters which was discussed in the treatment of the previous month was expressed by the total averages (1.4°, as compared with 1.8° for November) and by three out of the five years (1929, 1931, 1933). The remaining years exhibited a slight increase in the December variability (from 1.2° in November to 1.5° in December, 1930; from 1.5° to 1.8°, in 1932).   On the other hand, the total amplitude showed an increase in December (3.6°) when compared with November (3.1°).   Just as in November, the total amplitude at 50 m. was even greater than at the surface (4.0° as compared with 3.6°).   The total average amplitudes in December were about the same at the surface, 50 and 100 meters (1.4°, 1.5°, and 1.5°, respectively).   These features are characteristic of the winter.   The individual years behaved decidedly differently.   In 1929 there was a strong increase towards the depth (1.4° at the surface, 3.1° at 50 m.); in 1931 there was a moderate and irregular increase (1.1° at the surface; 0.8° at 50 m.; and 2.1° at 100 m.); and in other years the variations were fairly uniform although somewhat irregular in the upper 100 meters.   At the 100 meter level, the increase over the November conditions was consistent in all of the three years under investigation.

Local winter cooling occurred undoubtedly to a very considerable extent, although the process was not often very evident.   On a number of occasions the temperature at the surface was 0.1° lower than that at a depth of 5 m., and on one morning the surface temperature was not less than 0.4° below the 5-meter reading.   Just as in the other winter months, the readings were made in the early morning; later in the day these thermal abnormalities disappeared.   The intensive and peculiar circulatory phenomena during this month brought about the apparent paradox of a rise in temperature from one day to the next, accompanied by a steady and progressive cooling process.

V. ANNUAL THERMAL CONDITIONS IN THE SUPERFICIAL STRATA

At least to a certain extent, the annual hydrographic rhythm of this region may be deduced from a study of the account of the monthly thermal conditions as presented in chapter IV.   However, this phenomenon deserves a special and detailed treatment, not only due to its great local significance but also because of its unusual and extraordinarily characteristic type which makes it interesting to the hydrographer in his investigations of certain quite broad problems.   The following treatment of this subject begins with a discussion of the conditions prevailing during each of the five years 1929 to 1933 inclusive; thereafter comes a presentation of what may be termed the normal annual rhythm, based on the data from these five years.   Owing to the difficulties inherent in the latter phase of the subject, this "normal" rhythm must, of course, be accepted with due criticism.   However, in spite of its admittedly tentative nature, I hope that this norm will prove to be useful in the evaluation of future hydrographic data obtained in Monterey Bay as well as in this general region.

The data on which the discussion is based were taken at stations 4 (1929 and 1930) and B (1931–1933).   They are graphically presented in Figures 7–12.   In these graphs the time scale is horizontal, each base line representing an entire year with January 1 at the left extreme and December 31 at the right.   The depths are given by the ordinates.   The temperatures are represented by symbols, each degree from 8° to 15° having a special pattern, from black (8°) to blank (15°).   Temperatures of less than 8° and higher than 15° were not differentiated except through lines of distributional limitation.   This restriction in the number of symbols was chosen in order to simplify the graphs as much as possible and was allowed due to the very rare occurrence of these extreme temperatures.   At the two key stations the temperatures were read at each 5 meter level.   When the limits of the full degrees were located between these 5 meter levels, their positions in the graphs were estimated to the nearest meter by interpolation carried out on the assumption that the thermal changes were progressing at a uniform rate.   Of course, this assumption is arbitrary and far from correct in all cases.   However, the errors introduced in this manner are undoubtedly nearly always too small to influence the graphs to an appreciable degree.

A. *Thermal Conditions in the Upper 50 Meters, in 1929* (Fig. 7): During January and in the first week of February, the water over the southern flats of Monterey Bay, for the purpose of description, may be said to have been "typically" at 12°.   However, on two occasions this water was forced out of the Bay by the influx of moderately cold water.   One of these cold waves reached its maximum at about the end of the first week of January and came to a conclusion on January 17.   At the maximum of this influx, water of 11° reached within about two meters of the surface; and 10° water covered the bottom up to a depth of 45 m.   At the end of the influx, the 12° water occupied not less than the upper 45 m.   The second influx of cool water was of about the same magnitude and thermal character; it lasted between January 17 and February 2.   On February 5, water of 11°, of the type which had been forced out completely at the end of the last cool wave, reappeared along the bottom; it immediately began to increase rapidly in volume until, by February 11, it had entirely replaced the 12° water, covering the southern flats of the Bay to the exclusion of other kinds of water.   This may be said to represent the end of the typical winter situation and the foreshadowing of the striking spring phenomena.   From that time on, until the first part of

Fig. 7. Thermal conditions in the upper 50 meters at station 4, in 1929. For the interpretation of the graph, see the second paragraph of chapter V.

June, the 12° water was absent from this part of the Bay except for brief, intermittent visits when it came in along the surface, attaining at the most a thickness of about five meters, thus always remaining of subordinate importance.

The water of 11°, which entered the Bay on February 5, continued its complete dominance uninterruptedly until February 25, when a cold wave of 8°–10° suddenly entered the Bay, soon rising to within less than ten meters of the surface.   Cold though it was, this wave was of the same short duration as the two previously described cool waves of the month of January.   It came to a complete and sudden end on March 9, when water of 11° once more entirely filled this part of Monterey Bay.   However, on this occasion this type of water did not linger very long.   Almost immediately the cold water began to return.   Water of 9°–10° first made its appearance along the bottom about March 10; it increased steadily in volume until on April 11, *i.e.* during a period of one month, when water of 10° had risen to the very surface.   In other words, the short cold wave at the end of February and the beginning of March was nothing but the precursor of a huge influx of this type of water into this region.   This extensive influx of cold water, with temperatures ranging from 8° to 10°, lasted until July 3 when it was forced out completely by a mass of warm water; on that day no water even of 10° was recorded at our key station.   In other words, this large wave of cold water filled the southern part of the Bay from about March 12 to July 3, *i.e.*, for a period of not less than sixteen weeks.   This statement should not be understood to imply that the water remained more or less stationary in the region during the entire time.   On the contrary, there was a continuous renewal of the water throughout the period, although the incoming water maintained the same characteristics and in all probability came from the same source as the one which it replaced.   The constant renewal was clearly reflected and demonstrable in the steady changes occurring in the relative proportions of the water of different degrees of temperature, from 8° to 10°.   Thus, for instance, the 10° water which, as we have seen above, reached the surface in the beginning of April (around the 11th) did so again on June 3, while between these occasions it was depressed repeatedly to a variable extent, even to a level below a depth of 20 m.   Furthermore, on two occasions in June, the 9° water was completely forced out of this part of the Bay, only to return immediately in large volume.   And finally, not less than four very cold pulses were recorded within this extensive, cold influx; these pulses were characterized by the fact that they contained 8° water in greater or less proportion.   The largest of them, which reached its climax in the first part of May when 8° water reached within only 14 meters of the surface, lasted for not less than three weeks.   The remaining three were of decidedly shorter duration, two weeks or less, and of varying magnitudes; one of them occurred in the latter part of March and the first few days of April; the others in the middle of April and in the latter part of May, respectively; during the one in the middle of April the 8° water rose only to the 40 meter level.

The breaking of the dominion of the cold water (on July 3) had been heralded by the influx along the surface of unusually warm water.   This influx began as early as at the end of the first week of June, but its volume increased at a very slow rate; indeed, at the end of this month the 12° water had not penetrated, except occasionally, below the ten meter level. However, although this forerunner was of small bulk, it was very pronounced in thermal features.   For instance, on June 24, when the 10° water came within nine meters of the surface, the surface temperature was not less than 16.0°, *i.e.*, very close to the maximum record during the investigated five year period.   The warm wave that brought the large influx of cold water to a sudden end was of a short duration; it began to retreat a couple of days after it reached its climax (July 11).   To demonstrate how intensive the change of water was, it must suffice to note that while the average temperature for the entire body of water from the surface to 50 meters was 11.1° both on July 1 and July 29, on July 11 it was not less than 14.0°; and on the last mentioned day 12.7° water was recorded at the 50 meter level, a very unusual condition at this time of the year in the Monterey Bay region.

The incoming water had distinctly the appearance of a cold wave.   This wave, however, was of moderate duration and intensity; it lasted from July 14 to August 19 and it contained no 8° water, and the 10° water did not extend beyond the 12 meter level.   This cold wave, too, was composed of minor pulses.   Two of these, each lasting 10–14 days, were characterized by the elevation of the 9° water to or even slightly above the 35 meter level. While during this time the cold water dominated the lower strata, the superficial layers were distinguished by variable and irregular thermal conditions; at the very surface the temperature extremes were 12.4° and 15.5°, and the changes were not correlated with the deeper pulses, at least such a correlation could not be established.   In all probability the lower and the higher strata were independent of each other in their respective movements, a condition quite characteristic of this region.   Indeed, this independence was evident in a very striking manner from the end of the first week in June until the latter part of November, and in all probability it occurred all through the year (see section A of chapter VIII dealing with current experiments).

During the balance of the year, the lower strata in the southern part of the Bay may be said to have been distinguished by intermittent cold waves which by and large became less and less cold and at the same time progressively of shorter duration.   The first of these cold waves began on August 20, that is immediately following the cessation of its predecessor, and it ended on September 12; it was thus somewhat briefer than the latter.   Water of 10° reached at the maximum of the wave within 14 meters of the surface, and for a very short time, 9° water entered into the composition of the wave.   Noteworthy in this connection is the fact that this was the last time that water as cold as 9° was recorded in 1929; the last day on record was September 2.   The second wave extended over the last 18 days of September; thus it began immediately after the termination of the preceding one; and at its maximum, water of 10° reached as high as to the 30 meter level (on September 26).   The remaining cold waves, five in number, were even less significant; they lasted only about five to ten days each.   In the case of all but one of them, the 10° water did not become elevated above the 35 meter level.   During the fifth, a brief one lasting only about five days, this water reached as high as to the 25 meter level (on October 31).   Two of these waves occurred in the first half of October and one at the very end of this month; one was recorded towards the middle of November and the last towards the end of December.

During the last six months of the year, the 11° water expressed on a bigger scale the same rhythmic behavior as did the 9° and 10° water.   Apparently it was largely dominated by the deeper pulses.   Of course, the same similarity of rhythm reappeared in the higher strata, although progressively less distinctly towards the surface.   After all, the deeper layers form the substratum on which the superficial waters move!   In general, however, the surface waters exhibited a very pronounced independence of the deeper ones in respect to their thermal changes, an independence similar to the one noted above.

The warm and variable conditions that characterized the superficial strata in the latter part of July and the first three weeks of August continued during the last weeks of the latter month. These warm waters were followed by the main warm pulse of the year, a pulse which dominated the upper 25 meters during the entire month of September. During this influx, 12° water penetrated as far down as to the 37 meter level and 14°–15° prevailed at and near the surface. At the end of the month, the warm flow appeared to wane, but in the beginning of October it again became intensified; and it reached its maximum about October 7 when 12° water extended below the 50 meter level. Shortly following this maximum, a very marked and rapid change set in, and the warm flow may be said to have ceased by the middle of October. During the last part of October, in November, and in the first week of December, the more superficial strata were fairly monotonously dominated by 12°–13° water. In spite of this apparent monotony, intensive hydrographic changes continued to characterize the region. For instance, from November 22 to November 29, the temperature of the entire upper 50 meters changed from 12° to 13°, indicative of a total exchange of water. It should finally be noted that conditions during the latter part of December were very changeable. The most extraordinary and startling change, considering the season of the year, took place in the middle of the month; at that time a body of 14° water moved in and filled the southern part of the Bay down to the bottom, only to be forced out presently by a decidedly cool wave, the lowest temperature of which was 10°.

B. Thermal Conditions in the Upper 50 Meters, in 1930: The cool wave that characterized the last two weeks of 1929 was waning before the end of that year. It came to a complete termination shortly after the beginning of the new year. By January 6, water of 13° filled the shallow southern portion of Monterey Bay down to the bottom, *i.e.*, somewhat below the 50 meter depth. However, this warm water prevailed for but a few days. It was succeeded by a peculiarly uniform type of water. From January 12 to the end of the month, the Bay was filled with 12° water, the lowest bottom temperature being 12.2°, the highest surface temperature 12.9°. Such an extended dominance of monotonous thermal conditions is evidently very rare in this general region. During February the situation changed in an auspicious manner. This month was characterized by the first cool wave of the year. This wave, which occupied the first half of the month, came in two pulses of which the later was decidedly the stronger. It was but moderately cool; water of 10° extended, at the maximum, up to the 35 meter level; 11° water reached the 15 meter level. During the last week of the month, conditions again underwent a radical change, this time in the opposite direction; 13° water entered the Bay and for a few days (February 20–24) dominated the situation entirely. This warm water maintained itself in the Bay until the end of March, but it was slowly and steadily encroached upon from below by the influx of colder water, the precursor of the principal cold wave of the year.

In the last week of March, this cold influx obtained quite an intensity; 10° water reached to not less than within seven meters of the surface. However, the pulse was of short duration; by the beginning of April, this type of water (10°) was completely gone from the southern part of the Bay. From that time on, the principal cold wave grew quite regularly and steadily until June 9 when it reached its maximum development, with 10° water within 7 meters and 9° water within 13 meters of the surface. Shortly following this climax, the wave was brought to an abrupt, even though temporary, cessation; on June 17, the bottom water (50 m.) registered as high as 11°. Just as during the previous year, the principal

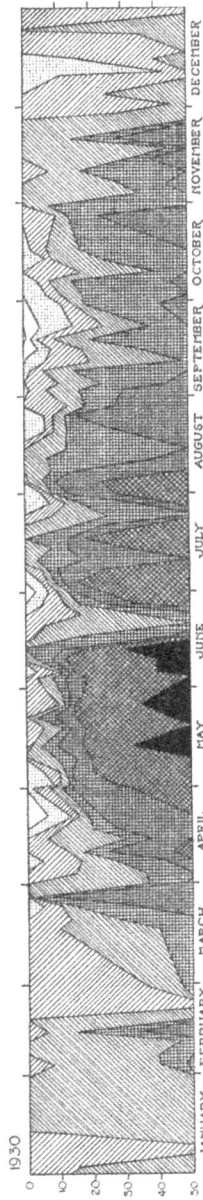

FIG. 8. Thermal conditions in the upper 50 meters at station 4, in 1930. For the interpretation of the graph, see the second paragraph of chapter V.

cold influx was not characterized by a slowing down of the hydrographic changes.   Within it, we could distinguish a sequence of individual pulses, of which those containing 8° water were the most conspicuous.   Of the 8° pulses, three could be discerned, *viz.*, two in the last three weeks of May and one during the first half of June; each of them lasted about ten to twelve days; and on no occasion did this type of water become elevated above the 30 meter level.   As will be seen from previous statements, this phase of the cold water period extended over three months.   Thus it was of a considerable duration, even though decidedly shorter than the corresponding phase during the previous year.   Furthermore, the temperatures were less extreme in 1930 than in 1929.

Just as in 1929, the interruption of the supremacy of the principal cold wave (about June 17) had been preceded, near the surface, by a flow of decidedly warm water.   This warm water could, indeed, be traced as far back as to March 24, when 14° was recorded for the upper 5 meters.   The entire month of April was distinguished by an unusually strong influx of warm water, during which water of 14° penetrated to a depth of 13 meters and the surface stratum (6 m.) registered 15°, the latter a very rare temperature for this season of the year.   Characterized by a constant change in volume, this warm, superficial flow continued to pass through the southern part of the Bay until the middle of June when it was forced out completely for a couple of days just before its sudden swelling and consequent conquest of the principal cold wave; however, during this time, 14° water was of rare occurrence and 15° was never recorded.   The swelling of the warm water, about June 17, although simulating the corresponding wave in 1929 in its sudden appearance, strength, and comparatively brief duration, did not equal this wave in regard to warmth.   The last statement is best demonstrated by the fact that the highest average temperature for the entire body of water (surf.-50 m.) was 12.5°, on June 20, 1930, while the corresponding record for 1929 was not less than 14°.   However, the intensity of the change is shown by a comparison with the corresponding averages on either side of this maximum; on June 9 and 12, 1930, this average was 9.8° and on July 3, 10.6°.   During the maximum flow of the warm water, 12° was recorded at the 50 meter level, and in the course of the waning of the phenomenon, 14° and even 15° water appeared at the surface.   In this, as well as in the preceding year, the development of the hydrographic changes near the surface seemed to take place quite independently of the occurrences at the deeper levels.

The similarity with 1929 was further emphasized by the fact that, after the lapse of a week, the warm water gave place to cold water and that this appeared in the form of a definite wave.   During this wave, which lasted exactly one month, until July 22, 10° water rose to within 10 meters and 9° water to within 17 meters of the surface; just as during the succeeding waves, 8° water did not enter the Bay.   Two very distinct pulses of about equal size and duration could be distinguished within this wave.

From the termination of this wave until the end of November, the deeper strata, just as in 1929, were characterized by a series of cold pulses.   Eight of these were recorded.   The first one, which lasted from July 25, *i.e.*, but a few days after the end of its precursor, till August 14, was by far the most conspicuous; at its maximum, water of 10° was found but 10 meters below the surface and 9° water was recorded as high as at the 24 meter level.   It should be noted that this was the last time that 9° water occurred in the Bay in 1930.   The next wave, extending from August 15 till September 7, was also of considerable size; 10° water rose to the 15 meter level.   The remaining six waves were somewhat irregularly spaced between September 22 and November 25.   During the third one, in the latter part of October, water of 10° rose not less than to the 23 meter level, but in most of the others only to about the 40 meter level or less.   The 11° water reflected on a greater scale and in an irregular manner the movements of the colder water, indicating that it, at least as a rule, was part of the deeper flow.

In regard to the hydrographic developments in the superficial strata from the time of the large, warm influx that terminated the principal cold wave, the following general statements must suffice.   During July and August, constant, although quite moderate, changes were recorded; warm pulses were coming and going.   Three of these were very distinct, *viz.*, one at the beginning and one at the end of July and one in the latter part of August.   Even during these pulses, 13° water but slightly penetrated below the 10 meter level and 14° was rare.   The situation was decidedly changed in September.   This month was dominated by a single, large influx of warm water.   During this flow, 12° water on one occasion was found as deep as near the 50 meter level, 14° water was recorded on every day, and at the climax of the wave, 15° and even 16° was found to prevail at and near the surface.

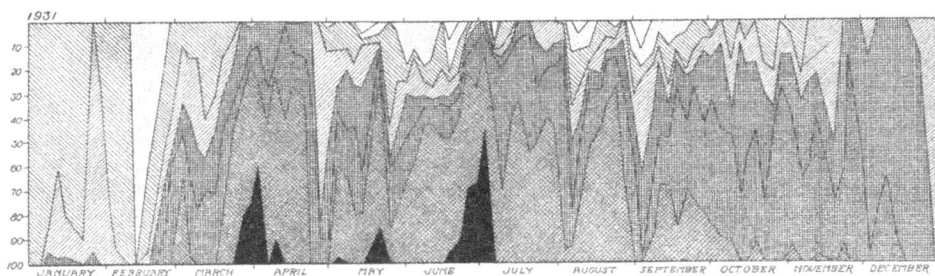

FIG. 9.   Thermal conditions in the upper 100 meters at station B, in 1931.   For the interpretation of the graph, see the second paragraph of chapter V.

This warm influx, which may be said to have begun in the last few days of August, gradually waned in October and was brought to a complete termination at the end of the latter month. Of course, as usual this major wave was composed of a series of pulses.

From the beginning of November, there was a persistent, although irregular, increase in the volume of the 12° water which by the end of the month completely filled the southern end of the Bay.

The month of December was decidedly warm.   It may be said to have been dominated by 13° water, but this was sometimes encroached upon by 12° water, which type completely filled this part of the Bay on December 24, and by 14° water which formed the same peculiar pulse as it did at this time of the previous year.   The latter occurrence was a peculiar coincidence of no particular significance other than that at this time warm pulses may be expected.

C. *Thermal Conditions in the Upper 100 Meters, in 1931* (Fig. 9): The year was ushered in by remarkably uniform conditions; from the surface to a depth of 100 meters, the temperature ranged only from 13.6° to 13.2°.   This 13° water dominated the situation until February 10.   However, its dominance was not unchallenged; a mildly cold wave, composed of two distinct pulses, entered the Bay, one pulse around January 12, when 12° water rose to a level of 62 meters, the other around January 26, when this type of water reached the surface for a day, only to be completely forced out shortly afterwards.   A slight amount, below 95 m., of 11° water also entered into the composition of this wave.   In general the hydrography of January thus was quite monotonous; the surface temperatures varied only by 0.7°, from 13.6° to 12.9°.   However, even though the changes were very slight when expressed in degrees of temperature, nevertheless, they were both sudden and thorough.   In the middle of February, an unusual occurrence took place.   Water of 14° suddenly appeared, penetrated for a brief interval as far as to the 100 meter level, having at the same time an extraordinarily low gradient (14.1°, at the surface and 14.0°, at 100 m.), and was then completely forced out again.   This brief irregularity in the annual rhythm foreshadowed a long series of more or less deviating occurrences which set this year apart from the other years under consideration in this report.   Even though this region in general was characterized by an extraordinary lack of stability, pulse following pulse in the superficial strata, nevertheless, the year 1931 must be characterized as unusually unstable.   The major hydrographic features of the region which ordinarily could be discerned without much difficulty were so distorted and broken up that they nearly defied recognition.   The situation at first glance appeared totally topsy-turvy, without a trace of regularity.   It should be noted that, this apparent confusion notwithstanding, the normal rhythm was present.

The most striking deviation from the normal rhythm was to be found in the circumstance that the cold water period did not exhibit a principal wave, to employ the phraseology applied above.   Just as in the previous two years, the beginning of the cold water period occurred in February.   The water which forced out the warm pulse in the middle of that month had the nature of a large, cold wave, a wave which grew strongly, although irregularly, until the middle of April, after which it declined rapidly until the last day of the month when the 10° water had been forced down to the 90 meter level.   In other words while in 1929 the cold spring wave came to a sudden, although temporary, termination as late as in July, and in 1930 in June, this break took place much earlier in 1931.   In the development of the cold wave, the 14° water left even the surface at the end of February and the 13° and 12° water during the last few days of March; 11° water dominated the superficial strata throughout the first three weeks of April.   The cold wave came as a series of pulses, two of which were quite distinct from each other.   The first of these two was in a manner but a forerunner of the second; it was decidedly the smaller, lasting only slightly more than two weeks, and its coldest water, of 9°, rose to about the 67 meter level.   In the course of the second major pulse, 8° water entered the Bay, rising at its maximum to the 60 meter level (April 2).   The water of 9° ascended on March 30, when it reached its climax, to within 18 meters of the surface, and 10° was recorded within two meters of the surface in the middle of April.   Thus, in spite of the smallness when compared with the corresponding phenomenon of the previous years, the cold wave obtained quite an impressive development even in this aberrant year.

The warm wave which terminated the cold one just described did not only come unexpectedly and with suddenness, it also developed into large proportions; indeed, it became the largest warm influx of the year.   It continued slightly more than two months, until the end of June.   Like all large-sized water movements in the region, it developed in a series of pulses, and just as in the preceding cold wave, two major pulses could be distinguished. These major pulses were separated from each other on May 21, when 11° water again rose to the 10–15 meter layer. The first pulse was distinguished by the fact that on one occasion (April 27), 12° water penetrated as deep as 83 meters; 13° was once recorded as low as at 17 meters; and 14° and sometimes even 15° water occupied the surface layer; the last temperature was rare in May. The second pulse was the larger, even though the 12° water never reached below the 56 meter depth; 13° was found as far down as at the 43 meter level, and particularly the 14° water was bulkier.

During the development of this warm influx in the more superficial strata, another cold wave was gradually, though erratically, advancing in the deeper layers. This wave lasted not less than three months, until the end of July when 9° water was forced down to the 95 meter level. Its climax was obtained in July, that is, in the very middle of the summer. If we were to distinguish a principal cold wave of this year, a thing difficult to do, this wave must be so assigned. It might be said to have been composed of two main pulses, although this statement may be considered as fairly arbitrary since some of the other pulses also were quite distinct. The first of these main pulses occupied nearly the entire month of May. It contained a small proportion of 8° water which rose, when it was highest, to the 85 meter level; and 9° water attained at the same time a level within 23 meters of the surface, only to be forced down immediately. The second pulse, which lasted more than two months, June and July, was not only of longer duration, but it was also of a heavier and more persistent nature. However, even so, 8° water appeared only during a period of about three weeks; it reached its maximum on July 2, shortly after which it was completely forced out of the Bay, never to return during the year. Even at this maximum, the 8° water reached only the 45 meter level; that is, it was very much less developed than during the previous two years. For all practical purposes, water of 9° dominated the stratum from 50 to 100 meters during this cold water influx; when best developed, the 9° water extended to a level within 15 meters of the surface; and 10° was recorded twice at a depth of only 4 meters.

In July, the very surface water did not, as a rule, become cold, even though the cold wave reached its climax in that month; water as cold as 11° was only rarely present at the surface. In the first half of the month, there was a brief, warm wave with 15° water occupying the upper 9 meters and with a very sudden decrease in temperature between 10 and 15 meters, 11° water being present at the latter depth. During the last 10 days of July, 12° water was entering the Bay along the surface, thus ushering in the warm influx which was going to cause a break in the cold wave which had dominated the last three months. This warm flow occupied nearly the entire month of August; it reached its maximum in the first ten days of the month when 15° water occurred in the upper five meters and 12° water penetrated to a depth of nearly 50 meters. A nearly identical, although somewhat larger, warm wave covered the month of September. This corresponded to the influx which in most years was the largest warm water pulse of the year. In 1931, however, it did not occupy this position; it did not attain the size of the one which occurred in May and June. This was another of the most outstanding differences between this and other years.

From the end of July to the end of November, the deeper strata were characterized by a nearly continuous series of cold pulses. Of these the first two were, as far as the 9° water was concerned, of considerable magnitude; the last four, on the other hand, were nearly abortive. The first of them occupied nearly the whole month of August; water of 9° rose to the 27 meter level, 10° to the 14 meter level at the climax of the wave (August 24).   The second one, although of somewhat longer duration, five weeks, was less severe, the 9° water being confined below a depth of about 70 meters and 10° water below about 30 meters. During October and November, the stratum from 25 to 100 meters was largely dominated by 10° and 11° water moving back and forth in a series of distinct pulses.   The upper 25 meters were similarly characterized by 12° and 13° water.   The month of December was distinguished by a fairly monotonous hydrographic situation.   Water of 11° was preeminent until the last week of the month when it was suddenly replaced by 12° water.   A slight and mildly cool wave occurred in the beginning of the month.

As will be seen from this description, even in this aberrant year two of the most outstanding hydrographic features were present.   Firstly, there was a large, cold wave covering most of the months except the winter months and showing a gradual decrease in the autumn. Secondly, the month of September was distinguished by a heavy flow of warm water.

D. *Thermal Conditions in the Upper 100 Meters, in 1932* (Fig. 10): The hydrographic development in this year had a clear-cut pattern, decidedly in contrast to the irregular and

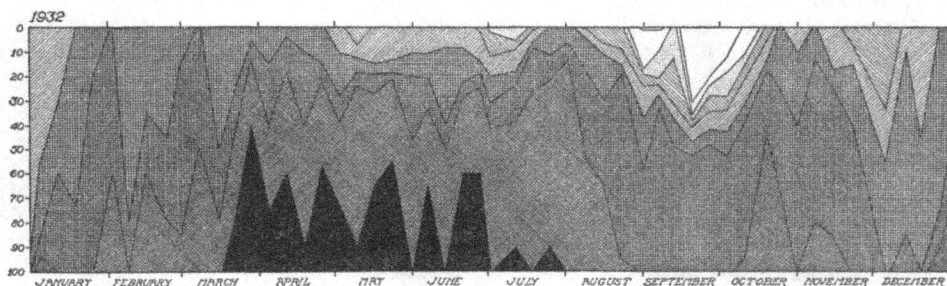

Fig. 10.   Thermal conditions in the upper 100 meters at station B, in 1932.   For the interpretation of the graph, see the second paragraph of chapter V.

confusing features characteristic of 1931.   The difference was probably partly apparent only, and due to the fact that in 1932 the Survey abandoned the semi-weekly schedule of sampling for weekly expeditions.   Undoubtedly it was partly real; that is, the water changes in the course of the year very strongly gave the impression of being less erratic, more directive.

The wave of 12° water which filled the southern part of the Bay during the last few days of 1931 began to retreat immediately after the beginning of the new year, and by January 18 the last traces of this water had disappeared from the region, not to return until one of the last days of April.   Throughout this entire period, the surface temperature was nearly consistently 11°, an exceptionally monotonous condition.   In the deeper strata, cold pulses began on the very first day of the year.   The first of these, lasting till February 8, was moderately cold and may be designated as the forerunner of the principal cold wave. It reached its climax in the beginning of February when 10° water was elevated to the surface for a very short interval and 9° water reached the 60 meter level.   After its termination, the cold water returned directly.   If water of 9° be considered as the indicator of the next cold wave, then this began in the first half of February and lasted not less than until the end of August without any interruption, as far as our records show; that is, it continued nearly seven months.   For about five months, this cold water dominated the moderately deep parts of the southern regions of the Bay.   On the other hand, if 10° water be used as the indicator, then this cold wave lasted uninterruptedly from the first day of the year until the beginning of December.   This fact, of course, emphasizes better than anything else the extraordinarily persistent cold conditions prevailing in the region during this year.

The principal 9° wave rose irregularly from February 8 until the end of March, when it reached within 15 meters of the surface.   It occupied about the same level at the end of July.   During this interval of four months it was characterized by rhythmic elevations and subsidences but it never went below 50 meters; in other words, it was remarkably steady. During August, this wave declined steadily until the end of the month, when the 9° water disappeared from the southern end of the Bay.   Water of 8° was present in the Bay most of the time from March 15 until the end of July.   As usual, it exhibited distinct pulses, eight of which were recorded.   At the end of March, this water attained its highest level, viz., 40 meters; it thus reached an elevation somewhat higher than in 1931 but decidedly lower than in 1929–30.   During this period, i.e., from February 8 until the end of August, the water of 10° showed the same general configuration as that of 9°.   The most prominent exception from this rule occurred in the beginning of March when the 10° water increased disproportionally, even reaching the surface for a short interval.   At the time of the two maxima of the 9° water, the 10° water reached within 5 meters of the surface, and when the 9° water had its minimum, the 10° water also was at its lowest, being depressed to the 40 meter level.   From the end of July, when both the 9° and the 10° waters reached their maxima, the latter type of water suffered a moderate decline, being depressed about 50 meters, to the 55 meter level.   From this minimum, the 10° water again rose in an irregular manner until the first week in November when it attained its last climax of the year, at an approximate level of 14 meters.   In the course of the next four weeks, its level was steadily forced down and on December 5 it was temporarily absent from the southern part of the Bay.   In the middle of December, a short pulse of 10° water returned to a level of 85 meters, and during the last part of the month, this type of water again rose suddenly to a level of about 50 meters, the latter pulse also containing a considerable proportion of 9° water.   In other words, this year, in contradistinction to the preceding three years, remained comparatively cold until the very end.

In the superficial strata, as was noted above, 12° water was absent from about the middle of January until the end of April.   At the latter time, a moderately warm wave entered the Bay, persisting until the later part of July.   It was largely confined to the upper 10–15 meters and reached its maximum depth of 20 meters on July 1.   It consisted largely of 12° water, but just before its termination, in the middle of July, some 14° water also appeared.   After the cessation of this warm wave, there was a short period, of about ten days, when 11° water occurred at the surface.   Then followed a very spectacular episode. One of the first days of August, water of 12° appeared at the surface, ushering in by far the largest warm wave of the entire five year period under investigation.   This wave extended over somewhat less than three months, until the last week of October, and was characterized by very unusual thermal conditions.   It is true that the 12° water did not penetrate to a very great depth, less than to the 50 meter level; but, on the other hand, temperatures as high as 16.7° were recorded and, at the climax of the flow, 16° water occupied not less than the upper 30 meters.   As usual, September was distinguished by the highest temperatures; and the wave was not simple but composed of pulses, two of which were quite distinct.   It .was a period of clear, transparent, and blue water of rare beauty for this region.   At the completion of this wave, 11° water once more ascended to the surface but again this was of but short duration; a moderately warm wave forced it out.   This warm influx, which reached its climax in the beginning of December and terminated a few days before the end of the year, was represented largely by 12° water which at the maximum of the flow reached a depth of 55 meters, thus deeper than during the principal warm wave; at this phase 13° water had entered the Bay and attained a thickness of not less than 30 meters.   This wave was made up of at least three pulses: one minor and somewhat detached, in the last week of October; one, the largest, in the beginning of December; and the last, of medium size, in the middle of this month.

Of all the five years under investigation, this year, 1932, was perhaps closest to what might be termed the normal regional rhythm.

E. *Thermal Conditions in the Upper 100 Meters, in 1933* (Fig. 11): The cool wave which entered the southern portion of the Bay a few days before the close of 1932 continued to

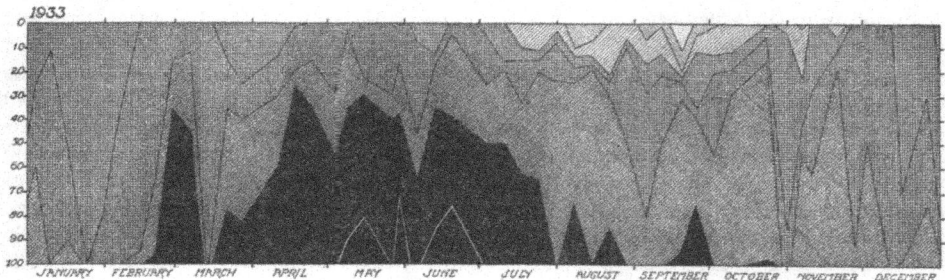

Fig. 11.   Thermal conditions in the upper 100 meters at station B, in 1933.   For the interpretation of the graph, see the second paragraph of chapter V.

increase during the first nine days of January, at the end of which period the 10° water reached to a level within but 12 meters of the surface.   Following this climax, the wave began to recede in the course of the next fourteen days until it was completely gone from the region, the 10° water returning, however, immediately.   Besides 10° water, the first cold wave of the year also contained 9° water which rose as high as to the 60 meter level, a condition which was most unusual for this time of the year.   Meanwhile the temperature of the superficial waters was 11°.   During the decline of the cold wave, this type of water (11°) increased until January 24 when water of 11.2°–11.1° occupied the upper one hundred meters.   During the last week of January and the first two weeks of February, the 11° water was gradually forced out of Monterey Bay by water of 10°; on February 14, this new water occupied nearly the entire upper one hundred meters.   Following the arrival of the 10° water fairly closely, the first cold wave of the year made its appearance.   This wave was of comparatively short duration, about five weeks, from February 6 to March 12, but at the same time very severe for the region.   On March 6, water of 9° was within 12 meters of the surface, and on February 27, water of 8° was recorded at a depth of only 35 meters.

During this cold influx, the surface waters registered 10°.   This brief, cold period was but the herald of the main thermal feature of the year.   If, as was done for 1932, water of 9° be considered as the indicator of the next cold wave, then this feature lasted without interruption from March 13 until October 29, that is not less than seven and one-half months. On the other hand, if 10° water be thus assigned, then the principal cold wave of the year extended from January 24 until the very last day of December; indeed, it may nearly equally well be said to have lasted throughout the entire year.   The cold conditions in 1932 were thus accentuated in 1933.

The 9° water which entered the Bay on March 13 increased irregularly in volume until the first week of May when it occurred but 3 meters below the surface; it attained about the same level in the middle of June, and on October 23, thus very shortly before the termination of this phase, it reached the 15 meter level.   Throughout this period, pulses occurred, and there were two distinctly low points: one in the beginning of June when this water had its upper limit at the 45 meter level, and the other in the first part of September when it was lowered to the 80 meter level, after which it again rose irregularly to a last high level mark. Quite a large proportion of this cold influx consisted of water of less than 9°.   The principal pulse of 8° water entered the Bay in the middle of March, i.e., shortly following the arrival of the 9° wave.   This pulse continued for a period of not less than four and one-half months, or until the end of July.   It displayed a fairly steep rise and decline, and its maximum elevation was attained in the middle of April when it was found at a depth of 26 meters.   It exhibited two low levels during the prime of its development, one at 55 meters, in the beginning of May, the other at 65 meters, on one of the first days of June.   In the middle of this four and one-half month period, i.e., in May and June, even 7° water was recorded as part of this cold wave; indeed, on one occasion, this type of water was found as high as at the 68 meter level.   It may be noted in this connection that this was the only time that the records of the Survey show water of such a low temperature to occur in the Bay above the 100 meter level.   After the termination of the principal pulse of 8° water, four minor pulses of this type of water were observed to enter the Bay, viz., two in August, one in the latter part of September, and one towards the end of October.   On none of these occasions was the 8° water reported above the 75 meter depth.

When the 9° water left the southern parts of the Bay at the end of October, it was absent only for a very short time.   In the very beginning of November, it returned as a strong wave that reached within 20 meters of the surface on November 20, and continued until December 10.   It consisted of at least two pulses, being depressed to a depth of 92 meters on November 27, after which it again reached the 50 meter level.   Following the termination of this wave, there was during the last part of December a slight pulse of this type of water, reaching, when at its highest, the 75 meter level.   In other words, as stated above, the 9° water persisted for all practical purposes until the very end of the year.

In regard to the superficial strata, the following remarks must suffice.   From the middle of March until the middle of April, water of 11° occupied the surface; when at its maximum, this water had a thickness of 25 meters.   From the middle of April until the beginning of July, that is for two and one-half months, the surface recorded 10°, except for a few days in the beginning of June.   In other words, during the first six months of the year, the surface temperatures ranged within 10° and 11°.   The main warm influx of the year began on July 1. Although it never did attain large dimensions, nevertheless, it lasted nearly four months, or until the latter part of October.   It consisted of distinct pulses, the most important of which occurred in the middle of September when 14° on one occasion reached to a depth of 10 meters.   The warm water maximum thus followed the normal schedule even in this remarkably cold year.   The last few days of October and the larger part of November were occupied by a mildly warm wave; 11° water at one time reached a depth of 85 meters. During the last 14 days of December, 11° water dominated the upper strata, the first two weeks having been characterized by a strong development of 10° water.

Like the preceding year, this year thus had a rhythm quite close to the normal, even though in many respects the two years were very different from each other.

F. Normal Annual Rhythm in the Upper 100 Meters.   (Figs. 12 and 13): Even a comparatively cursory inspection of the graphs illustrating the thermal changes that took place in Monterey Bay in the course of the five years treated above will suffice to demonstrate two fundamental facts: first, there is in this region a distinct, annual rhythm in the thermal conditions; and secondly, to a different extent in different years, this rhythm is warped and

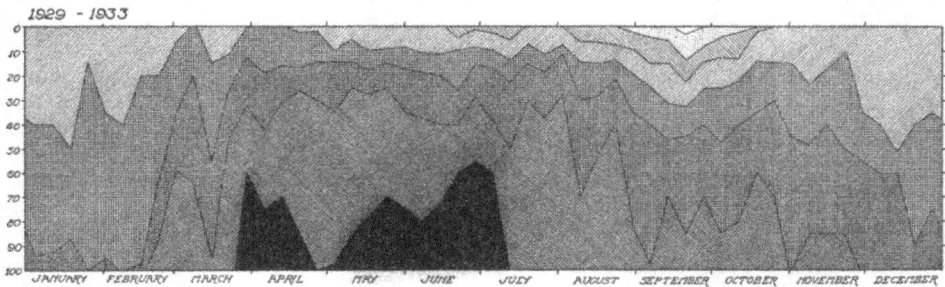

Fig. 12.   Normal Annual Rhythm of the temperatures in the upper 100 meters, in Monterey Bay, obtained by the combination of the thermal rhythms of 1929–33.   For the interpretation of the graph, see the second paragraph of chapter V and the third paragraph of section *F* of this chapter.

obscured by secondary, interfering phenomena.   Of course, it would be of the greatest interest and value, from the point of view of a general understanding of this region, to obtain a clear-cut picture of this rhythm.

In the following attempt at establishing the ideal rhythm, the procedure was based on the assumption that within each year the effects of the interfering forces were distributed at random and that the distribution as a consequence varied from year to year.   By combining all the five years available, there would thus be at least a tendency for the unessential details to eliminate each other, even though the limited data were not sufficient to smooth them out completely.   In order to obtain smoothness by this method, material from a much greater number of years would, of course, be required.   The results obtained in this manner seemed fully to justify the method.   The composite picture exhibited quite smooth features and at the same time maintained what at first glance had appeared to be the fundamental seasonal phenomena.

Fig. 12 represents graphically the average thermal distribution in the south part of Monterey Bay, from 1929 to 1933.   The averages were calculated after the data had been classified by the week, and all the years were given equal weight.   The time axis in the graph thus was divided into 52 units, representing the weeks, and the thermal averages were marked on the ordinates above the middle of each of these units.   The method of interpolation described in the second paragraph of this chapter was also applied to this graph. Of course, it should be kept in mind that, while the calculations for the upper fifty meters were based on five years, the values for the lower fifty meters were derived from the data of three years only, a number altogether too small for the desired purpose.   For this and other reasons, the seasonal pattern of the 8° water obtained in this manner must be considered as quite unreliable and subject to amendments.   It is, unfortunately, not possible to present tables of the primary data used for the computations.   The averages are to be found in chapter XI.

As will be seen from this graph, there are three outstanding features which to a greater or less extent dominate the annual hydrographic rhythm in the Monterey Bay region:

1. The most extraordinary and striking of these features is the wave of cold water which controls the subsurface strata during the greater part of the year.   If this wave be

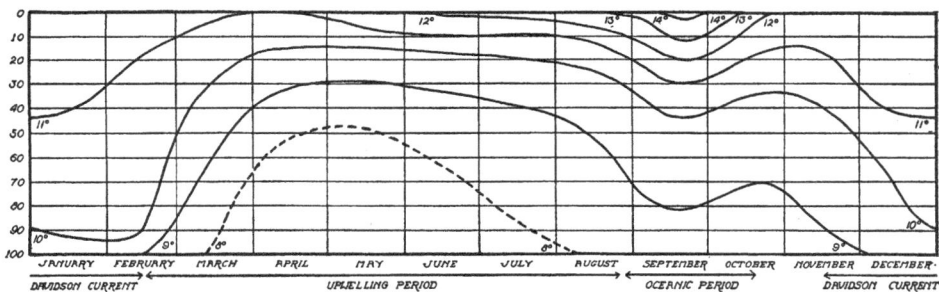

FIG. 13.   Normal Annual Rhythm of the temperatures in the upper 100 meters, in Monterey Bay, obtained by optical smoothing of the curves in Fig. 12.   The rhythm of the 8° water is represented by a broken line to indicate its uncertainty.   For the interpretation of the graph, see the explanation to Fig. 12.

assumed to be limited to water of 9° or less, then it usually enters the Bay at the 100 meter level in the middle of February.   Its rise is comparatively rapid; the maximum elevation, at about the 30 meter level, is reached in May.   On the other hand, its decline is quite slow; it subsides below the 100 meter level as late as in the beginning of December.   It should be noted that during a period of about eight weeks, around September, the declining curve is quite depressed by the warm wave which is the prominent feature of the autumn.   The first part of the cold wave is nearly cut off from the principal part, forming, as it were, a heralding pulse of about four weeks.   This feature of the graph is in all probability accidental, its maintenance being due simply to the very limited nature of the material available. The same explanation probably applies to the broken and irregular autumn part of the 9° curve.   The wave of 8° water which is largely confined below the 50 meter level is very irregular in the graph and is unquestionably distorted beyond recognition by the scarcity of data.   Its bimodal type is probably not an expression of the actual conditions.   It should be noted that out of the last three of the investigated years, only 1933 had this type of water well developed; 1931 was even quite deficient in this respect.   In all probability the typical wave of 8° water is unimodal and in general of about the same configuration as the 9°

wave although much shorter and proportionally higher.   Its normal duration probably is from about the middle of March until the end of July.   Its maximum elevation is very uncertain, but in all likelihood it does not rise normally very much above the 50 meter level, except when wind conditions force this water towards the coast.   Of all the features of the normal rhythm, this is by far the most uncertain; the conditions found in 1933 possibly represent the closest approximation to normalcy, although on a somewhat exaggerated scale.

The 10° water apparently is present normally within the southern part of the Bay throughout the entire year.   When it is at its minimum, *i.e.*, when it is not elevated, it seems to be confined below the 90 meter isobath.   However, this condition is maintained only during a very short period, *viz.*, throughout January and the first half of February.   During the period of elevation, the 10° water has about the same rhythm as the 9° water, indicating that, at least as a rule, these two types of water are closely associated in their movements. In the middle of February, the 10° water rises with remarkable abruptness from below the 90 meter level, attaining its maximum elevation, about the 15 meter level, at the end of April, after which its position is kept with but slight change until the end of July.   In the short period of six weeks, this water is elevated on the average not less than 80 meters. From midsummer until the end of the year, the level of this water declines, its curve exhibiting a distinct depression around September in connection with the influx of warm water noted above.   Under normal conditions, the 10° water does not reach the surface.

In more than one sense, the 11° water is of a type transitional between the cold water of the depths and the typical surface waters.   Its most important season extends from the first part of December until the middle of February (see below, under point 3).   From the middle of February, it is forced up towards the surface and to a large extent even out of the Bay.   During a couple of weeks in April, it typically occupies the surface.   Soon, however, local insolation raises the temperature of the surface to 12°, a condition which prevails in May and the first part of June.   From the middle of June until the end of July, the surface temperatures normally vary from 12° to 13°.

2. The feature in the annual thermal rhythm which is next in importance to the cold and deep wave discussed above, is a warm influx along the surface which takes place in the autumn.   This phenomenon, which has its maximum in September, occurs every year but its size is very variable.   During the span of about two months, the temperatures from the surface to a depth of about 30 to 40 meters are above the normal for these strata, causing the depression in the isotherms noted above.   Water of 12° extends to a depth of about 30 meters and at the surface the temperature is normally not less than 15°.   It should furthermore be noted that this water is of a deeper blue and of a greater transparency than is the water of other seasons.   As will be seen from the descriptions given above, warm pulses do occur in an irregular manner throughout the entire year.   This influx occupies a special position on account of its persistence from year to year and due to the fact that in most years it is of a fairly large size.

3. The last of the three main features of the rhythm is decidedly less conspicuous to persons not familiar with this general region.   It is the condition which prevails during December, January, and the first half of February.   It has two outstanding traits.   First, the thermal gradient is decidedly more gentle than in other seasons.   On the average, the temperature decreases less than two degrees from the surface to a depth of 90 meters.

Second, the temperatures are quite high, considering the fact that we are concerned with the middle of the winter season.   Between 30–40 meters and 90 meters, the temperature is normally 11°, and above this stratum it is 12°.   In order to emphasize the fact that we are dealing with a very distinct hydrographic season it may be timely to note even in this connection that this water is characterized by a peculiar plankton.

As will be observed from this description, the seasonal rhythm of the sea in this region is both distinct and remarkable, utterly different from the one characteristic under what may be termed the typical conditions prevailing in the north Atlantic Ocean.

VI. COMPARISONS OF THE ANNUAL THERMAL CONDITIONS IN THE SUPERFICIAL STRATA

In the following descriptions and discussions, the subject matter will be dealt with from several different angles, and to each of these a special section will be devoted.   This mode of presentation, although of necessity to a certain extent repetitious, was chosen largely on account of two reasons: first, from an economic necessity it was impossible to publish the entire original data, and second, future investigators undoubtedly will be concerned with the thermal conditions in this region from different points of view and these should, of course, if possible be met by those who have the data at hand.   The treatment is of a very general nature; to a very large extent the reader will have to depend on his own analyses of the graphs in which the data are embodied.   As in the case of the previous discussions, the material was limited to that gathered at the two key stations, stations 4 and B.   This procedure, which was decided upon for reasons of economy, does but slightly infringe upon the applicability of the results to the entire region under investigation; the thermal differences among the various stations were quite small, as is shown elsewhere in this report.

Some of the following comparisons may at first appear premature in view of the decided limitations of the material available, the investigation covering only five years.   However, the fact that the thermal variations occurring throughout these five years were very slight, about 6.5° at the surface, at least indicates that even a considerably extended series of observations would not have altered materially the general features presented in this report.

First the temperature conditions at the surface, 50 meters, and 100 meters will be presented, both by means of the actual observations and by the monthly averages.   Next will be given the relative volumes of the different kinds of water, as characterized by their degrees of temperature; and an attempt will be made at determining the deviations of the various monthly averages from the calculated normal thermal conditions.   After that follows a discussion of selected amplitudes of thermal variations.   Finally, a brief discussion of the temperature conditions at the surface during the years 1919–34 will be added, based on data collected at the Hopkins Marine Station for the Scripps Institution of Oceanography, La Jolla, California.

A. Thermal Conditions at the Surface, 50 Meters, and 100 Meters: In Figures 14 and 15 the thermal variations at these three levels are presented in so far as observations are available.   During the first three years of the survey, 1929 to 1931, recordings were made on the average semi-weekly, while in 1932 and 1933 they were made, as a rule, only once a week. Unquestionably, to a certain degree it is deplorable that our resources did not allow more closely-spaced data to be gathered.   However, the observations presented will show, at least in a general manner, the extent of the changeability characteristic of these waters, and, what is more important, they apparently are quite sufficient for comparisons among the various years. In regard to the first of these two aspects, it may be noted that a comparison between the first three years on the one hand and the last two on the other will, at least partly, demonstrate the increase in the apparent fluctuations consequent upon an increase in the number of observations. In studying the graphs, the reader should always keep in

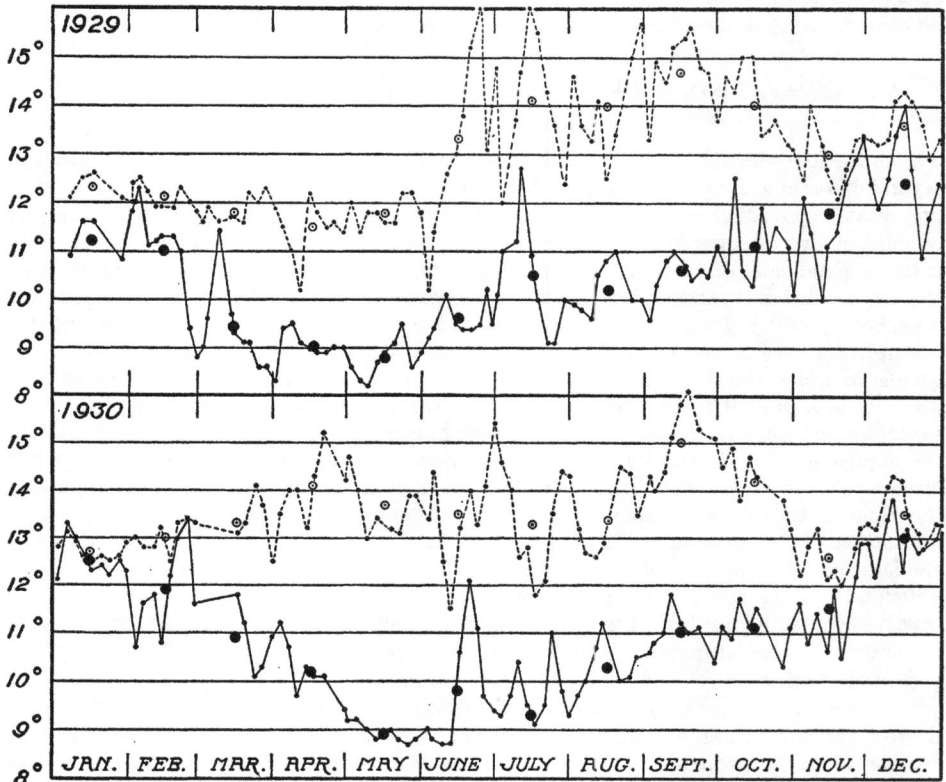

FIG. 14.   Temperatures at the surface and 50 meters at station 4, in 1929 and 1930. For further explanations, see the second paragraph of section A of chapter VI.

mind that he is concerned with a body of water that was in a constant state of flux during the period of investigation. The circulation was so intense, indeed, that even had the temperature readings been taken within the short interval of one hour, duplications of results would, as a rule, not have been forthcoming. Only on comparatively rare occasions did water enter the Bay which was part of a large and thermally uniform body, causing the temperatures to remain approximately constant for as long a period as 24 hours. This happened most frequently when truly oceanic waters approached the coast.

In the graphs, the time axes are horizontal and the thermal ones are vertical.   The months are indicated by short division lines.   The curves of the surface temperatures are represented by broken lines, those of the temperatures at the 50 and 100 meter levels by solid lines, of which the ones of the latter level are somewhat heavier.   The individual readings are shown by solid circles.   The monthly thermal averages at the surface are shown by open circles and those of the 50 and 100 meter levels by solid circles and squares, respectively.   No attempt at smoothing was made, since such a procedure would not have brought forth anything beyond the trends exhibited by the monthly averages.

A study of the curves will show that the thermal development of the surface waters was quite different from one year to another.   Thus, for instance, while, in 1933, the consecutive temperature readings followed the seasonal trend fairly closely, in other years the very opposite was true.   The most extreme example of the latter condition was probably presented by 1931, in which year the curve oscillated up and down in, to all appearances, a completely random manner.   This irregularity of the thermal evolution throughout the year is evidently the rule rather than the exception in the Monterey Bay region.   It is interesting to observe that the seasonal irregularities sometimes exhibited a persistent trend over a considerable period, indicating that they were caused by large-sized water movements. A striking example of such an occurrence was the gradual warming of the surface during June, 1929.   In that month, not less than six consecutive readings showed the same trend, the temperature rising from 10.2° to not less than 16.0°, a rise approaching the entire range of thermal variation observed in the course of the investigated five year period.   On other occasions, the thermal behavior was quite the opposite.   For instance, from the last week of June to one of the first days of July, 1929, the temperature rose from 13.1° to 14.8°, only to drop immediately to 12.0°.   Another feature of interest is that while in most instances the change from one reading to the next lay within one degree, in some cases the change was very sudden, indeed.   Thus, in the beginning of July, 1931, the temperature rose within a couple of days from 12.1° to 15.9°, that is, about two-thirds of the entire amplitude of thermal variation observed at the surface during the five year period.   All these peculiarities give us a powerful impression of the extreme instability of these waters, of their constant renewal, and thus of the fact that this part of the coast does not have what might be termed endemic waters.

Great irregularities were also exhibited by the 50 meter thermal curves.   However, in general the seasonal trends were more closely followed by the individual readings at this level than at the surface.   The most striking feature of the observed irregularities was that they occurred almost independently of those at the surface.   Thus, for instance, while the surface temperatures in March, 1929, remained fairly unchanged, the 50 meter thermal curve showed a pronounced and unseasonal rise and fall.   Following this event, when the surface temperature fell, that at 50 meters rose.   On rare occasions, the thermal changes at this level were both strong and sudden, approximating in magnitude those at the surface. Perhaps the most extreme case of this kind was the rise from 9.3° to 12.8° that took place within a couple of days in the last part of May, 1931.   Another interesting example occurred in July, 1929, when the 50 meter temperature sank from 12.7° to 9.1° within less than a week, a drop nearly equalling the total range established for this month in the course of the five year period.

The 100 meter curves were also characterized by their irregularities.   At the same time, the individual readings exhibited seasonal consistency, perhaps even better than those

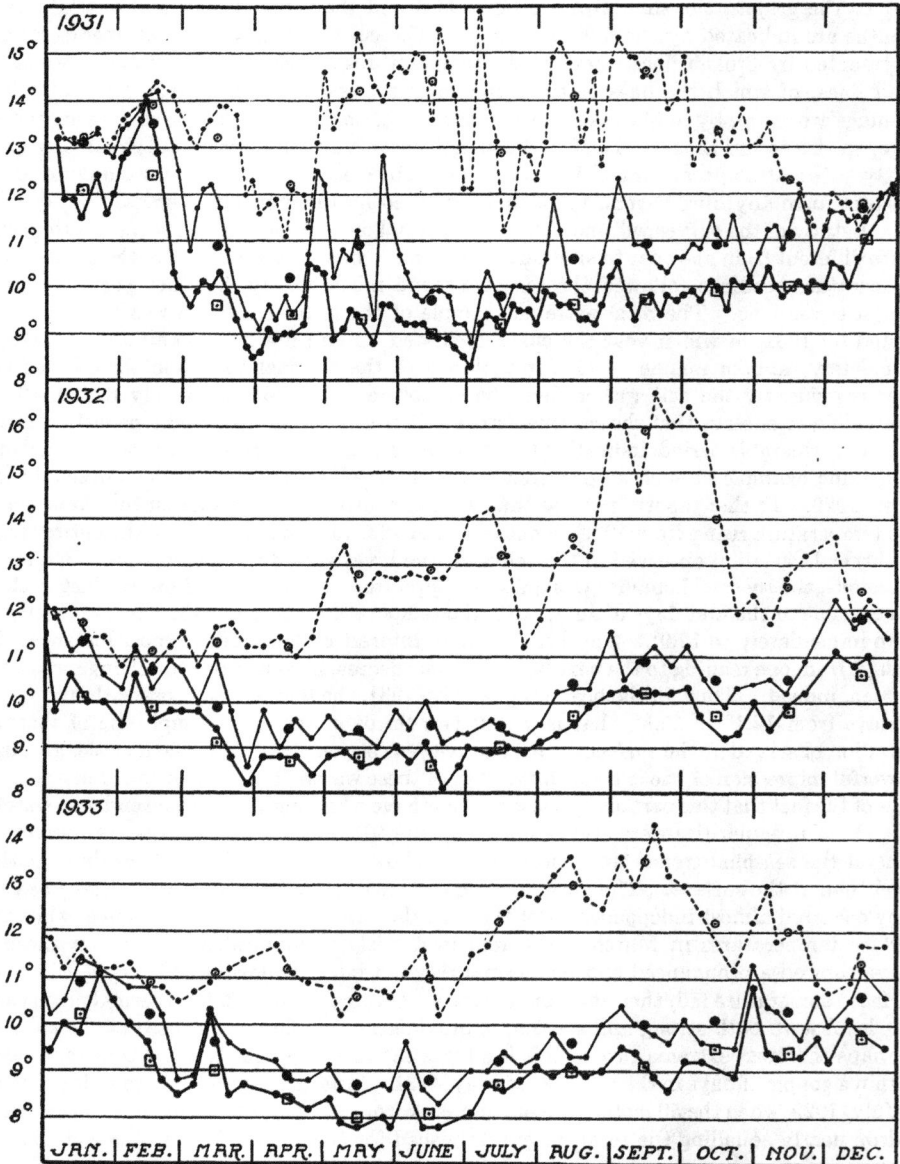

FIG. 15.   Temperatures at the surface, 50 and 100 meters at station B, in 1931–33.   For further explanations, see the second paragraph of section A of chapter VI.

of the 50 meter level.   Generally speaking, the curves of these two levels showed a fairly pronounced parallelism, although the thermal movements at 100 meters, on the whole, were not quite so extreme as those at 50 meters.   An excellent example of the smaller amplitudes of the changes at 100 meters was found during the period from the beginning of March to the end of June, 1931.   Sometimes the changes at 100 meters were just as large, or even slightly larger, than those at 50 meters; for examples, see the first three months of 1933. Even though the thermal trends at these levels usually were more or less parallel, occasionally they were opposite.   Thus, near the middle of March, 1932, the temperature rose at 50 meters, while it fell at 100 meters.

The fact that the temperatures, as a rule, moved in a fairly parallel manner at 50 and 100 meters strongly indicates that the waters of these levels usually moved in the same directions.   Conversely, the different trends of the surface temperatures, when compared with those of 50 and 100 meters, suggests the independence of the most superficial strata relative to the underlying ones.   The relative independence of the superficial strata in relation to the somewhat deeper ones, as well as the great instability of the entire body of water down to a depth of 100 meters, are perhaps the most interesting of the facts that can be obtained from a study of these graphs.   For more detailed features, the reader is referred to the graphs.

The monthly averages are represented by Fig. 16.   In this, the time axis is again horizontal, and the thermal one is vertical.   In the upper part of the graph, the monthly values of the five years are given consecutively.   The years are separated by complete lines, the months by short lines at the top and bottom.   The surface temperatures are indicated by small, solid circles, connected by dots; the averages for the entire body of water between the surface and 50 meters, by triangles, connected by dots and dashes; the 50 meter temperatures, by squares, with dashes; and the 100 meter ones, by open circles with continuous lines. The same symbols are employed in the right graph of the bottom row, representing the total averages for the five year period.   In the case of the remaining four bottom graphs, the symbols for the various temperatures are the same as in the top row; thus the left one of the graphs represents the surface temperatures of the five years superimposed upon each other; the next one, the averages for the entire body of water between the surface and 50 meters, treated in like manner, *etc.*   On the other hand, the lines connecting the temperature values are used differently.   They indicate the various years; dotted line indicates the year 1929; dots and dashes, 1930; dashes, 1931; fine, continuous line, 1932; and heavy, continuous line, 1933.

For the sake of clarity of presentation, it seems most advisable to begin the discussion with the graph in the right bottom corner, representing the total monthly averages of the five year period, that is, the averaged monthly averages.   In all probability, this graph shows a fairly close picture of the normal annual rhythm in this region.   In the following table, which gives these "normal" values, the months are indicated by Roman numerals.

	I	II	III	IV	V	VI	VII	VIII	IX	X	XI	XII
Surf..........	12.3°	12.2°	12.1°	12.0°	12.6°	13.0°	13.1°	13.6°	14.7°	13.5°	12.5°	12.5°
0–50 m.......	12.1°	11.9°	11.1°	10.6°	10.6°	10.9°	11.0°	11.4°	12.4°	11.9°	11.7°	12.3°
50 m.........	11.8°	11.5°	10.1°	9.5°	9.3°	9.5°	9.7°	10.2°	10.7°	10.7°	11.0°	11.9°
100 m........	10.8°	10.5°	9.2°	8.8°	8.7°	8.6°	9.0°	9.4°	9.7°	9.7°	9.7°	10.4°

The "normal" yearly averages were as follows: 12.8°, at the surface; 11.5°, for the entire body of water between the surface and 50 meters; 10.5°, at 50 meters; and 9.5°, at 100 meters.

As will be seen from these values, the normal monthly surface temperatures exhibited during these five years a very slight decrease, from 12.3° to 12.0°, between January and April. From April until September, they rose steadily to the maximum of 14.7°, after which they declined to 12.5° in November and December. The corresponding temperatures for the body of water between the surface and 50 meters sank from 12.1° to 10.6° from January to April and May, rose steadily between the latter month and September, when they attained 12.4°, declined from that time until November, to 11.7°, and finally rose to close to the maximum of the year in December. The 50 meter curve was simpler. From January to May, it exhibited a decline from 11.8° to 9.3°, after which it rose to its maximum of the year, 11.9°, in December. The 100 meter curve was of the same general shape, but it had its maximum in January and its minimum in June; the January and December values were 10.8° and 10.4°, respectively, and the minimum was 8.6°. These are, indeed, extraordinary curves, clearly demonstrating the unusual nature of this region.

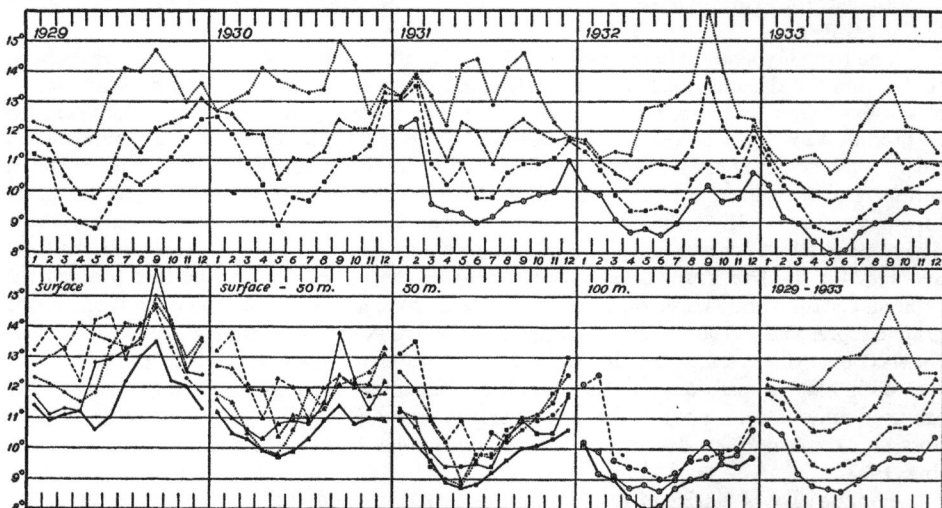

FIG. 16. Monthly average temperatures at the surface, 50 meters, and 100 meters, at stations 4 and B, in 1929–33. For further explanations, see p. 61.

The thermal features of the individual years, which are illustrated by the remaining graphs of Fig. 16, are shown by the following tables:

1929

	I	II	III	IV	V	VI	VII	VIII	IX	X	XI	XII
Surf..........	12.3°	12.1°	11.8°	11.5°	11.8°	13.3°	14.1°	14.0°	14.7°	14.0°	13.0°	13.6°
0–50 m.......	11.8°	11.5°	10.5°	9.9°	9.8°	10.6°	11.9°	11.3°	12.1°	12.3°	12.5°	13.1°
50 m.........	11.2°	11.0°	9.4°	9.0°	8.8°	9.6°	10.5°	10.2°	10.6°	11.1°	11.8°	12.4°

The following yearly averages were obtained: at the surface, 13.0°; for the body of water between the surface and 50 meters, 11.4°; and at 50 meters, 10.5°.

1930

	I	II	III	IV	V	VI	VII	VIII	IX	X	XI	XII
Surf..........	12.7°	13.0°	13.3°	14.1°	13.7°	13.5°	13.3°	13.4°	15.0°	14.2°	12.6°	13.5°
0–50 m	12.7°	12.6°	11.9°	11.9°	10.4°	11.1°	11.0°	11.3°	12.4°	12.1°	12.1°	13.3°
50 m..........	12.5°	11.9°	10.9°	10.2°	8.9°	9.8°	9.7°	10.3°	11.0°	11.1°	11.5°	13.0°

For the entire year, the following values were obtained: at the surface, 13.5°; for the body of water between the surface and 50 meters, 11.9°; and at 50 meters, 10.9°.

1931

	I	II	III	IV	V	VI	VII	VIII	IX	X	XI	XII
Surf..........	13.2°	13.9°	13.2°	12.2°	14.2°	14.4°	12.9°	14.1°	14.6°	13.3°	12.3°	11.8°
0–50 m........	13.2°	13.8°	12.1°	11.0°	12.3°	12.0°	10.9°	12.0°	12.4°	12.0°	11.7°	11.8°
50 m..........	13.1°	13.5°	10.9°	10.2°	10.9°	9.8°	9.8°	10.6°	10.9°	10.9°	11.1°	11.7°
100 m........	12.1°	12.4°	9.6°	9.4°	9.3°	9.0°	9.2°	9.6°	9.7°	9.9°	10.0°	11.0°

The following yearly averages were calculated: 13.3°, at the surface; 12.1°, for the surface to 50 meters; 11.1°, at 50 meters; and 10.1°, at 100 meters.

1932

	I	II	III	IV	V	VI	VII	VIII	IX	X	XI	XII
Surf..........	11.7°	11.1°	11.3°	11.2°	12.8°	12.9°	13.2°	13.6°	15.9°	14.0°	12.5°	12.4°
0–50 m.......	11.6°	11.0°	10.6°	10.3°	10.8°	10.9°	10.8°	11.5°	13.8°	12.2°	11.3°	12.2°
50 m..........	11.3°	10.7°	9.9°	9.4°	9.4°	9.5°	9.4°	10.4°	10.9°	10.5°	10.5°	11.8°
100 m........	10.1°	9.9°	9.1°	8.7°	8.8°	8.6°	9.0°	9.7°	10.2°	9.7°	9.8°	10.6°

The annual averages were: surface, 12.7°; surface to 50 meters, 11.4°; 50 meters, 10.3°; and 100 meters, 9.5°.

1933

	I	II	III	IV	V	VI	VII	VIII	IX	X	XI	XII
Surf..........	11.4°	10.9°	11.1°	11.2°	10.6°	11.0°	12.2°	13.0°	13.5°	12.2°	12.0°	11.3°
0–50 m.......	11.2°	10.5°	10.3°	9.9°	9.7°	9.9°	10.3°	10.9°	11.4°	10.8°	11.0°	10.9°
50 m..........	10.9°	10.2°	9.6°	8.9°	8.7°	8.8°	9.2°	9.6°	10.0°	10.1°	10.3°	10.6°
100 m........	10.2°	9.2°	9.0°	8.4°	8.0°	8.1°	8.7°	9.0°	9.1°	9.5°	9.4°	9.7°

The corresponding values for the year as a whole were: surface, 11.7°; surface to 50 meters, 10.6°; 50 meters, 9.7°; and 100 meters, 9.0°.

From a study of the graphs in the top row, it is evident that the surface curves of three out of the five years, *viz.*, 1929, 1932, and 1933, were quite similar in shape to what has been described above as the "normal" surface temperature curve; the differences were rather in the general levels of the curves and in details than in fundamentals of shape.   On the other hand, the surface curves of the two remaining years deviated in important features.   In 1930, this curve was distinctly bimodal.   The temperature rose from January to April, that is, it trended in a direction opposite to the "normal" curve; then it declined until July, rising sharply from August to September, after which it exhibited the normal, rapid decline to November.   In 1931, the curve was trimodal; in other words, it showed quite an excessive lack of regularity of rhythm.   The temperature rose from January to February, then sank strongly to April; the remaining part of the curve showed two high points, one in June, the other in September; following the last month, the usual, rapid decline occurred.

Whether the decline during the first four months of the year, present in the "normal" surface curve, is characteristic of this region, is difficult to decide at this time.   This trend was expressed in four out of the five years, and this fact appears to point strongly in the direction of the affirmative.   On the other hand, in case we study the surface temperatures from 1919 to 1934 as presented in the last section of the chapter, we find that the early months of the year are extremely variable, and that an affirmative answer to this question would be quite unjustifiable.   The thermal rise from November to December which distinguished 1929 and 1930 from the remaining years must be attributed to an exaggeration of a hydrographic phenomenon characteristic of that part of the year.   The "normal" value for December should therefore in all probability not be identical with that for November but somewhat lower.

The differences occurring in the surface curves from one year to another are strongly brought forth in the left-hand graph of the bottom row, in which these curves are superimposed.   This graph demonstrates that the most pronounced deviations occurred in the first half of the year, that the maximum temperatures always were present in September, and that the similarity among the curves from September to November was very striking.   In general, the surface curves exhibited a more pronounced individuality from year to year than the curves of the somewhat deeper strata.   It is recommended to compare these results with those presented in the last section of the chapter.

In regard to the curves representing the monthly temperatures of the entire body of water between the surface and 50 meters, it should be noted that the one of 1933 showed, in respect to shape, a very remarkable similarity to that of the "normal" situation.   To a somewhat less extent, the same was true for the curve of 1932, the most pronounced difference being found in the relative heights of the maxima in September, the one of 1932 being by far the higher.   In 1930, the curve was comparatively deep in May, while its peak in September was less developed.   These differences were still stronger in 1929, in which year the peak of September was totally absent.   The differences in respect to the September maxima are very interesting and will be discussed in chapter VIII.   The curve for 1931 was by far the most aberrant; it exhibited the same peculiar trimodal shape as did the surface curve of that year; the modes had the same location, except the middle one which occurred in May instead of in June.   As will be seen from these brief remarks, as well as from a comparison between the left two graphs in the bottom row of Fig. 16, the curves of the body of water between the surface and 50 meters, although quite different from one year to another, were somewhat less different mutually than were the surface curves.   Of course, the deviation in the thermal levels were fairly pronounced.

The curve for the monthly temperatures at 50 meters in 1933 also showed a very strongly marked similarity in shape to the "normal" curve of this level. The one for 1932 differed particularly in the flatness of the bottom part, the temperatures remaining nearly constant from April to July, and by the presence of a peak in September, evidently connected with the particularly high temperature at this place in the curve discussed in the last paragraph. The curves for 1929 and 1930 were angular at the bottom, the one for the former year also having the atypical rise in July, just as the curve of the water between the surface and 50 meters. The thermal pattern of 1931 exhibited a much greater regularity and approach to normalcy than in the more superficial strata; the mode in September was suppressed. The superimposition of the five 50 meter curves in the middle graph of the lower row of Fig. 16 reveals a remarkably uniform behavior of the temperatures at this level from year to year. This is, indeed, peculiar, considering the fact that, after all, we are still dealing with comparatively superficial waters subject to the vicissitudes of meteorological phenomena.

The only comment that needs to be made in regard to the curves of the 100 meter level is that they showed a strikingly uniform seasonal pattern, that the temperatures in January and February of 1931 were abnormally high, and that in September of 1932 there was an atypical elevation connected with the exceptionally heavy influx of warm water in the higher strata. The similarity between the "normal" curves of the 50 and 100 meter levels is, indeed, conspicuous.

The most noteworthy general features revealed by these data are, firstly, that the seasonal trends at 50 and 100 meters, as well as the trends at somewhat higher levels, were nearly opposite to those characteristic of what may be termed the normal, temperate seas; secondly, that the rhythms at the 50 and 100 meter levels were subject to but relatively minor changes from year to year; and, thirdly, that the surface waters on the one hand and those of the lower levels on the other presented nearly diametrically different thermal trends, strongly suggesting that the movements of these waters were more or less independent of each other. The data presented in the first section of the chapter revealed the hydrographic instability of the region. The monthly averages, on the other hand, demonstrate that, in spite of this instability, the seasonal pattern was quite firmly established.

In respect to the yearly averages presented in this section, suffice it to state in this connection that they clearly show the changes in the general thermal level which took place from one year to another. This interesting and important aspect of the hydrographic situation of the region will be dealt with in the last section of the chapter.

B. Relative Amounts of Water of Different Degrees of Temperature in the Upper 50 Meters: An excellent method of comparing the thermal conditions of different years and localities is to calculate the relative volumes occupied by waters of different degrees. This was done for the years 1929–33, and the results were presented graphically in Figs. 17 and 18. In these, as well as in the following tables, the values were rendered in percentages in order to make them more readily comparable with those from other stations and regions. However, since the maximum depth investigated for this purpose throughout the entire five year period happened to be 50 meters, these percentages can very readily be translated into meters, 10% corresponding to 5 meters. The original data were taken at stations 4 and B.

As was done in the previous section, beginning will be made by presenting what may be considered the "normal" distribution of the temperatures, obtained by averaging the available values of the five years.   This "normal" situation is graphically represented by Fig. 17. On account of the fact that certain temperatures were omitted from this figure, the data will also be given in the form of a table.   In this, the months are indicated by Roman numerals. The value 0.0 refers to present but in a quantity of less than 0.1%.

Temp.	I	II	III	IV	V	VI	VII	VIII	IX	X	XI	XII
16°						0.0	0.0	0.2	5.3	1.0		
15°				0.3	0.3	0.8	1.1	2.0	3.5	1.5		
14°	20.8	6.6	0.5	1.4	1.6	4.7	3.7	3.5	10.7	3.6	0.3	6.1
13°	20.8	20.5	8.3	3.7	8.4	6.0	6.2	10.0	14.2	11.9	10.3	24.7
12°	32.8	16.9	13.5	8.6	11.2	11.7	11.1	15.0	17.4	19.2	32.7	24.1
11°	37.4	27.1	29.5	17.3	14.1	17.1	17.1	18.7	28.1	39.1	29.9	32.8
10°	9.0	23.5	27.3	29.3	21.7	22.9	29.0	33.6	17.3	17.6	23.8	12.3
9°		3.7	17.7	30.3	28.9	31.5	31.6	17.0	3.6	6.1	3.1	
8°		1.8	3.2	7.5	13.8	5.2	0.2					

In Fig. 17 the time axis is horizontal, and each block represents one month, as indicated below by numbers.   The temperatures are symbolized in the following manner: 8°, black; 9° oblique crosshatching; 10°, straight crosshatching; 11° and 12° are omitted for the sake of emphasis of the remaining ones; 13°, oblique, parallel lines; 14°, vertical, parallel lines; 15°, dots; and 16°, blank.   For the sake of simplicity in the visualization of the actual conditions, the temperatures are placed in the relative positions which they occupy in nature.   Each horizontal division corresponds to 10% or 5 meters.

The treatment of the data in this table and graph should not be confused with the one underlying Fig. 12.   In Fig. 17, as already noted, the percentage of each thermal degree during the five years is represented, while in Fig. 12 we are concerned with the average temperatures.   Thus, for instance, in December, water of 13° and 14° was present in the upper 30 meters in fair amounts in the course of the five years, but there was not enough of it recorded to raise the five year average above 12°.   As a consequence, in Fig. 17 the upper part of the December column is occupied by 14° and 13°, while in Fig. 12 the upper 30 meters show a temperature of 12°.   The two graphs, therefore, are complementary rather than repetitious, even though they in many respects exhibit similar features, for instance, the characteristic rhythm of the colder waters.

Fig. 17 reveals that the lower temperatures, from 8° to 10°, occupied, from 1929 to 1933, annual volumes which formed symmetrical curves, indeed, nearly typical probability curves.   However, the modes of the 9° and 10° curves were somewhat displaced toward the early part of the year and that of the 8° curve toward the later part of the year.   Water of 8° sometimes began to appear above the 50 meter level as early as in February, reached its maximum development in May, and sometimes persisted into July.   Its maximum average percentage was 13.8, while its index of total frequency, as expressed by the sum of all its monthly percentages, was 31.7.   Water of 9° sometimes also came into the Bay, above the 50 meter level, as early as in February, but its frequency in that month was very low, being only slightly higher than that for the 8° water.   It obtained its greatest volume in July, and at least in some years it continued until November, even though its average frequency from September to the end of November was very low.   Its maximum frequency, 31.6%, occurred in July, but there was only a slight difference among the percentages of the four months of April, May, June, and July.   Its index of total occurrence, expressed in the same manner as in the case of the 8° water, was 174.0, thus decidedly higher than that for the 8° water.   The combined indices of the 8° and 9° waters thus totalled 205.7, and the maximum of these combined waters occurred in May.   Water of 10° was ordinarily found throughout the year.   Nevertheless, it was represented abundantly only from February to August, inclusive, September, October, and November also showing fairly high values.   The lowest

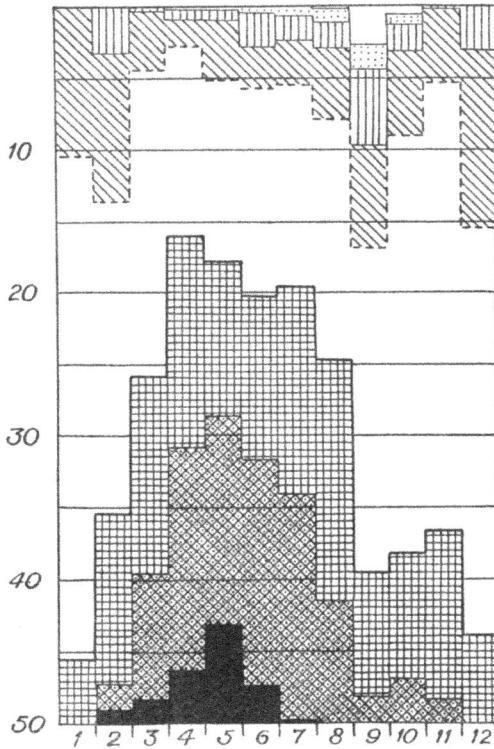

Fig. 17.   Relative amounts of water of different degrees of temperature in the upper 50 meters, at stations 4 and B.   "Normal" condition, obtained by averaging the data from 1929–33.   For further explanations, see the third paragraph of section *B* of chapter VI.

month was January, with 9.0%; and the highest was August, with 33.6%; the index of total occurrence was 267.9.   The combined indices of 8° to 10° water therefore totalled 473.6. The peak of these combined waters occurred in April, not in May as in the case of the previous two types.

The reason for the introduction of the indices of total occurrence was that they yielded an excellent standard for the comparison of the volumes of the different types of water that entered the Bay in the course of the various years.

At this place it may be appropriate to note the indices for the 11° and 12° waters, even though these were excluded from the graphs.   The index for the 11° water was 308.7, the highest of all.   This degree may thus, in a sense, be considered as the most representative for the upper 50 meters of the Bay.   The corresponding value for 12° was also high, *viz.*, 214.2, in other words, intermediate between those of the 9° and 11° waters.

Water of 13° was also found in every month of the year.   According to the available data, it was present in the largest volumes in December, January, and February, with its maximum of 24.7% in December.   During the remaining months, it usually occurred in less than 10%.   Whether this condition is representative for this region is questionable.   In all probability it was due to the limitation of our data.   When, in the course of the five years, the 13° water entered the Bay, it frequently did so in conjunction with still warmer water. At the same time, it was the most abundant of the warmer surface waters, having an index of total occurrence of not less than 145.0.   In respect to the remaining types of water, the

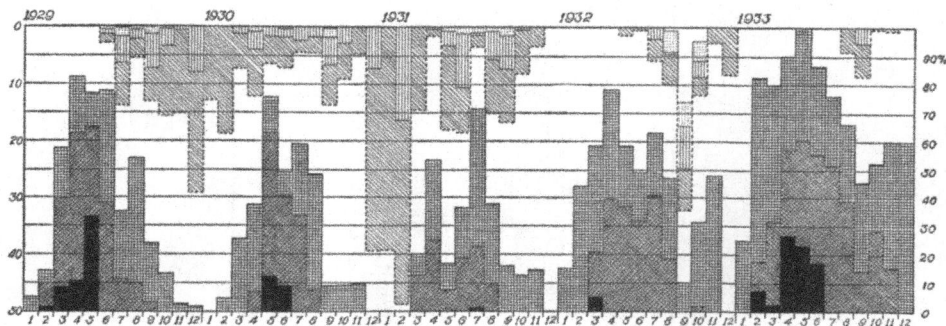

Fig. 18.   Relative amounts of water of different degrees of temperature in the upper 50 meters, at stations 4 and B, from 1929 to 1933.   For further explanation, see the third paragraph of Section *B* of chapter VI.

treatment can be more summary.   The 14° water had an index of only 42.7, was recorded in all months except January, had its maximum of 10.7% in September, and in the remaining months occurred in most instances in less than 5%.   In other words, its volume was always small and its occurrence erratic.   Water of 15° was recorded only in seven months, from from April to October, inclusive; its maximum, 3.5%, was found in September; and its index was only 9.5.   Still more seldom water of 16° was found.   It occurred in five months, June-October, usually only as traces, with its maximum of 5.3% in September; its index was 6.5.

Although this picture undoubtedly is only an approximation of the "normal" conditions, we shall now compare it with each of the investigated years, in order to obtain an idea of the degree of variation from year to year which characterized these waters in 1929–33. The five years are represented serially in Fig. 18, and this graph was constructed on the same principles as the one depicting the "normal" situation.   The following tables give the relative amounts of the different kinds of water in each of the five years.

1929

Temp.	I	II	III	IV	V	VI	VII	VIII	IX	X	XI	XII
16°						0.2	0.2					
15°						0.9	3.3	1.1	2.7	0.9		
14°						1.7	9.1	3.3	11.8	5.8	1.3	16.0
13°						2.8	15.1	6.6	11.6	24.6	29.0	42.5
12°	34.0	35.0	1.6	1.3	2.9	6.2	14.0	13.9	14.9	22.4	46.3	31.6
11°	61.0	51.0	41.1	16.5	20.7	10.7	23.3	21.4	35.4	33.2	21.3	8.4
10°	5.0	4.2	17.8	19.3	11.8	39.1	23.6	43.8	20.4	13.1	2.2	1.6
9°		8.4	31.2	52.0	31.6	38.2	11.3	10.0	3.3			
8°		1.6	8.4	10.9	33.1							

1930

Temp.	I	II	III	IV	V	VI	VII	VIII	IX	X	XI	XII
16°									0.3			
15°				1.7		0.7			3.3			
14°			2.4	6.3	3.3	3.1	4.7	3.4	9.5	5.7		14.6
13°	25.6	37.6	12.0	16.3	9.5	10.4	4.2	5.7	14.3	12.6	9.5	64.4
12°	74.4	47.0	34.8	21.7	6.5	20.0	11.8	13.4	26.2	20.6	59.3	21.0
11°		11.2	25.6	16.8	5.5	16.7	20.4	29.0	37.5	52.0	21.5	
10°		4.3	25.2	30.6	12.2	8.9	25.6	41.0	9.0	9.1	9.7	
9°				6.6	50.5	31.4	33.4	7.4				
8°					12.5	8.9						

1931

Temp.	I	II	III	IV	V	VI	VII	VIII	IX	X	XI	XII
16°												
15°					1.4	2.2	2.0	1.3	3.0			
14°		32.9		0.7	4.9	18.9	0.9	10.0	11.8	0.9		
13°	78.6	64.9	29.3	2.4	29.7	16.0	4.0	18.9	18.5	15.3	6.8	
12°	21.4	2.3	31.3	20.0	24.3	12.2	9.2	17.8	19.5	25.4	29.5	32.5
11°			19.3	23.6	22.9	14.7	13.0	14.4	31.3	45.8	49.3	67.5
10°			7.1	28.2	9.1	17.6	48.7	27.6	16.0	12.7	14.5	
9°			12.9	25.1	7.7	18.5	21.3	10.0				
8°							1.1					

1932

Temp.	I	II	III	IV	V	VI	VII	VIII	IX	X	XI	XII
16°								1.0	26.0	4.8		
15°								7.5	8.5	6.8		
14°							4.0	15.0	15.0	5.6		
13°					2.8	1.0	7.5	10.5	15.0	6.4	5.3	16.5
12°	34.0				22.4	20.0	10.5	12.0	12.5	15.2	16.0	35.5
11°	51.0	56.0	41.5	22.0	21.6	30.0	15.0	21.0	13.0	30.0	30.7	48.0
10°	15.0	44.0	38.0	38.5	35.6	17.5	22.5	28.5	10.0	29.2	48.0	
9°			15.5	39.5	17.6	31.5	40.5	18.5		2.0		
8°			5.0									

1933

Temp.	I	II	III	IV	V	VI	VII	VIII	IX	X	XI	XII
16°												
15°												
14°									5.5			
13°								8.5	11.5	0.5	1.0	
12°							10.0	18.0	14.0	12.5	12.5	
11°	75.2	17.5	20.0	10.0		13.3	14.0	7.5	23.5	34.6	26.5	40.0
10°	24.8	65.0	48.5	32.7	39.6	31.3	24.4	27.0	31.0	24.0	44.5	60.0
9°		10.0	29.0	30.6	37.2	38.0	51.6	39.0	14.5	28.4	15.5	
8°		7.5	2.5	26.7	23.2	17.3						

The next table presents the thermal indices for these years, calculated according to the method given above.

Temp.	1929	1930	1931	1932	1933
16°		0.3		31.8	
15°	8.9	5.7	9.9	22.8	
14°	49.0	53.0	81.0	25.6	5.5
13°	132.2	222.1	284.4	65.0	21.5
12°	224.1	356.7	245.4	178.1	67.0
11°	344.0	236.2	301.8	379.8	282.1
10°	201.9	175.6	181.5	326.8	452.8
9°	186.0	129.3	95.5	165.1	293.8
8°	54.0	21.4	1.1	5.0	77.2
8°–9°	240.0	150.7	96.6	170.1	371.0
8°–10°	441.9	326.3	278.1	496.9	823.8

In regard to most of the facts brought forth by these data, it will be necessary for the reader to obtain them by a direct study of the tables and Fig. 18. Only selected points will be treated in the following discussion.

The values show that even when the years are taken individually, the colder waters, from 8° to 10°, form at least in some instances fairly regular curves with their peaks located slightly towards the earlier part of the year. Most symmetrical were the curves of 1929, 1930, and 1932; 1931 was quite aberrant, as it was also in several other regards; and in 1933 the cold water was atypically bulky during the last months of the year, giving to the 10° curve a pronounced skew, while the 9° curve still exhibited a distinct symmetry.

When comparing the indices, for the sake of establishing the relative prominence of the colder waters from one year to another, we gain the following results. The year 1929 was colder than normal in regard to the deeper strata; both the 8° and 9° indices were above normal, and the total index of these waters was 240.0 against 205.7, the normal. For this reason the influx of deep cold water during that year should in all probability be considered as above normal, even though the total index of the waters of 8° to 10° was somewhat less than normal, 441.9 as compared with 473.6. It should be noted that the 10° water on some occasions evidently had a superficial rather than a deep origin. In 1930, all these waters were less bulky than normal, the total index for 8° to 10° being only 326.3 against 473.6.

1931 was by far the warmest of the five years in regard to the deeper waters; its index of the 8° to 10° waters was as low as 278.1.   While in 1932 the 8° water was present only in a very small quantity, the 9° water was close to normalcy, and this was also true in respect to the combined waters of 8° to 10°.   The year 1933, finally, was by far the coldest of all. This statement applies to all the three kinds of water, the total index of which was not less than 823.8.

In 1929, the 8° water occurred above the 50 meter level from February until May, inclusive, reaching its peak in May, with the highest value recorded during the five year period, after which it disappeared with conspicuous suddenness.   The 9° water extended from February to September, inclusive, with its maximum in April.   The 10° water was found throughout the entire year, reaching its maximum development in the late summer. In 1930, the 8° water was limited to May and June, the 9° water to April-August, with a pronounced maximum in May, and the 10° water was absent from the months of January and December and was heavily represented either in March-April or in August.   In 1931, the small amount of 8° water did not enter the Bay until July, the 9° water was limited to March-August, having no pronounced peak, and the 10° water occurred from March until November, inclusive, with two strong maxima, one in April and the other in July.   In 1932, the small amount of 8° water was recorded in March, the 9° water was present in March-August and in October, with no prominent peaks, and the 10° water was absent only from December, being fairly evenly distributed throughout this period, except in September when it attained its lowest point.   In 1933, the 8° water occurred during February-June, with its maximum in April; the 9° water was absent only from January and December and was heavily represented throughout the summer, with a prominent peak in July, and the 10° water reached an extraordinary development throughout the whole year.

From this description, three fundamental features of the annual hydrographic rhythm become evident: first, the colder water of the shallower regions of the Bay exhibited a very marked pattern from year to year, their rise and fall occurring like a huge, wave-like movement; second, the total bulk of this rhythmic phenomenon varied from one year to another, the lower strata in some years being distinctly colder than in others, and third, even though the rhythm always was repeated annually, nevertheless its pattern was characterized by conspicuous variations.

While 1929 was comparatively cold in regard to the deeper strata, its warmer surface waters were quite close to normalcy, if the year as a whole be taken into consideration.   It should be noted, however, that the first five months were relatively cold, no water of temperatures higher than 12° being present.   From this it follows, of course, that the latter part of the year was thermally above normal, the month of December being characterized by a remarkable volume of the 13° and 14° waters.   The warm condition in the superficial strata persisted throughout 1930 which year was above normal in the 13° water (222.1, as compared with the normal of 145.0), close to normalcy in the 14° water (53.0 against 42.7, the normal), and very slightly below normalcy in the 15° to 16° water.   The month of December was again remarkable on account of the very heavy influx of the 13° and 14° waters; indeed, in that year this phenomenon was even more conspicuously developed than in 1929, the 13° water reaching a percentage of not less than 64.4, the highest value for warm water in this month during the five year period.   This warm wave was maintained in January of 1931, and attained a striking maximum in February when the combined percentages of the 13° and

14° waters totalled not less than 97.8%. The year 1931 was markedly above normal in respect to the 13° and 14° waters (284.4 and 81.0, as compared to the normal of 145.0 and 42.7) and normal in the 15° water. It should be noted, however, that the warm condition in the superficial layers was gradually declining from October and that in December 12° was recorded at the surface. This temperature was also recorded in January, 1932, but during the next three months, 11° prevailed. The last eight months of 1932 were characterized by a warm wave of quite a regular increase and decline. This phase obtained its maximum development in September when thermal conditions were recorded unique for the five year period. The 16° water attained in that month the record value of 26.0% and the 16° to 14° not less than 49.5%. At the peak development of this influx, a remarkable effect of the hydrographic conditions on the local fisheries was observed, an occurrence which will be further noted in chapter IX of this report. During 1933, the surface waters, generally

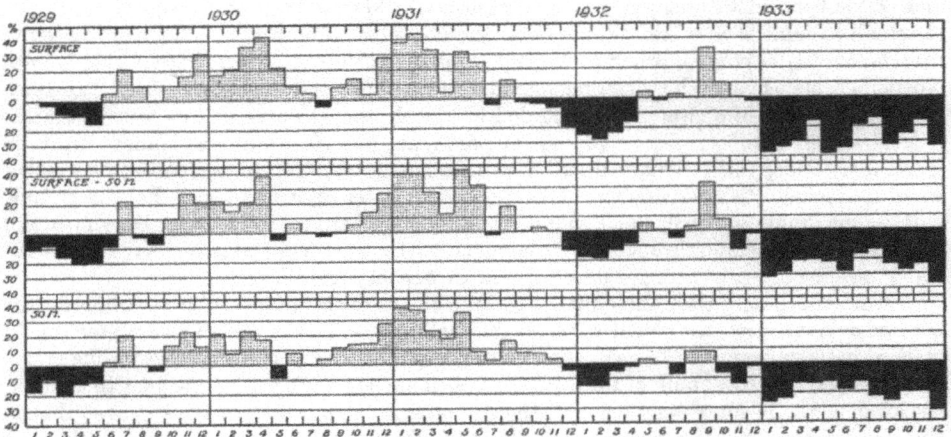

FIG. 19.   Deviations from the "normal" thermal conditions at stations 4 and B, from 1929 to 1933.
For further explanation, see the next paragraph.

speaking, were very cold; only from August to November, inclusive, did the temperature rise above the 12° level and only in September, the warmest month, did 14° water enter the Bay and then in quite small quantities.

Since the phenomenon of deviation from the thermal normalcy has an important bearing on many problems, this aspect has been treated in still another manner. The results of these calculations are presented in the following tables and represented graphically in Fig. 19. In this graph, the five years are arranged serially, the years are separated by entire lines, and the months indicated by short lines above and below. In the upper row of graphs the monthly temperatures are treated; in the middle row, the monthly averages for the body of water between the surface and 50 meters; and in the lower row, the monthly temperatures at 50 meters. The base lines of these three rows of graphs represent the monthly five year averages fitted together linearly. In other words, the values of these lines change from month to month; for instance, from January to March 1929, the base line for the surface temperatures changes from 12.3° to 11.8°, since, as will be seen from the first section of this chapter, the averages at the surface for the first three months of that year were 12.3°, 12.1°, and 11.8°, respectively.   The deviations were calculated in percentages of the total five year thermal range for each month as given in the next section of this chapter. For example, in February, 1929, the deviation at the surface was − 0.1° (five year average, 12.2°; average for 1929, 12.1°) and the total five year range at that level was 3.9°; the percentage deviation thus was − 2.6%.   This method of presentation was chosen particularly for the sake of facilitating comparisons between the conditions at the key stations and other stations in this general region as well as in other waters.   The method also lends emphasis to the phenomenon, which is desirable in view of the fact that the range of variation in terms of degrees is extraordinarily narrow.   Minus deviations were represented in the graph in black and plus deviations in stippling.

In the following tables, the deviations are given both in degrees, bottom row of each year, and in percentages, top row.   When no sign precedes the value, the deviation is positive; the negative ones alone are marked.   Roman numerals indicate the months.

Surface:	I	II	III	IV	V	VI	VII	VIII	IX	X	XI	XII
1929.............	0.0	−2.6	−8.8	−10.0	−15.4	5.2	20.8	9.5	0.0	9.6	16.1	30.6
	0.0	−0.1	−0.3	−0.5	−0.8	0.3	1.0	0.4	0.0	0.5	0.5	1.1
1930.............	16.7	20.5	35.3	42.0	21.2	8.6	4.2	−4.8	8.1	13.5	3.2	27.8
	0.4	0.8	1.2	2.1	1.1	0.5	0.2	−0.2	0.3	0.7	0.1	1.0
1931.............	37.5	43.6	32.4	4.0	30.8	24.1	−4.2	11.9	−2.7	−3.8	−6.5	−19.4
	0.9	1.7	1.1	0.2	1.6	1.4	−0.2	0.5	−0.1	−0.2	−0.2	−0.7
1932.............	−25.0	−28.2	−23.6	−16.0	3.8	−1.7	2.1	0.0	32.4	9.6	0.0	−2.8
	−0.6	−1.1	−0.8	−0.8	0.2	−0.1	0.1	0.0	1.2	0.5	0.0	−0.1
1933.............	−37.5	−33.4	−29.4	−16.0	−38.5	−34.5	−18.8	−14.3	−32.4	−25.0	−16.1	−33.3
	−0.9	−1.3	−1.0	−0.8	−2.0	−2.0	−0.9	−0.6	−1.2	−1.3	−0.5	−1.2

Surface—50 m.	I	II	III	IV	V	VI	VII	VIII	IX	X	XI	XII
1929.............	−10.7	−8.3	−15.8	−20.6	−19.5	−8.6	22.0	−2.6	−7.0	10.0	26.7	21.1
	−0.3	−0.4	−0.6	−0.7	−0.8	−0.3	0.9	−0.1	−0.3	0.4	0.8	0.8
1930.............	21.4	14.6	21.1	38.2	−4.9	5.7	0.0	−2.6	0.0	5.0	13.3	26.3
	0.6	0.7	0.8	1.3	−0.2	0.2	0.0	−0.1	0.0	0.2	0.4	1.0
1931.............	39.3	39.6	26.3	11.8	41.5	31.4	−2.4	16.7	0.0	2.5	0.0	−13.2
	1.1	1.9	1.0	0.4	1.7	1.1	−0.1	0.6	0.0	0.1	0.0	−0.5
1932.............	−17.9	−18.8	−13.2	−8.8	4.9	0.0	−4.9	2.6	32.6	7.5	−13.3	−2.6
	−0.5	−0.9	−0.5	−0.3	0.2	0.0	−0.2	0.1	1.4	0.3	−0.4	−0.1
1933.............	−32.1	−29.2	−21.1	−20.6	−22.0	−28.6	−17.1	−13.9	−23.3	−27.5	−23.3	−36.8
	−0.9	−1.4	−0.8	−0.7	−0.9	−1.0	−0.7	−0.5	−1.0	−1.1	−0.7	−1.4

50 m.	I	II	III	IV	V	VI	VII	VIII	IX	X	XI	XII
1929..............	−17.6	−9.3	−19.4	−11.9	−10.9	2.6	20.5	0.0	−3.7	13.3	22.2	12.5
	−0.6	−0.5	−0.7	−0.5	−0.5	0.1	0.8	0.0	−0.1	0.4	0.8	0.5
1930..............	20.6	7.4	22.2	16.7	−8.7	7.9	0.0	3.8	11.1	13.3	13.9	27.5
	0.7	0.4	0.8	0.7	−0.4	0.3	0.0	0.1	0.3	0.4	0.5	1.1
1931..............	38.2	37.0	22.2	16.7	34.8	7.9	2.6	15.4	7.4	6.7	2.8	−5.0
	1.3	2.0	0.8	0.7	1.6	0.3	0.1	0.4	0.2	0.2	0.1	−0.2
1932..............	−14.7	−14.8	−5.6	−2.4	2.2	0.0	−7.7	7.7	7.4	−6.7	−13.9	−2.5
	−0.5	−0.8	−0.2	−0.1	0.1	0.0	−0.3	0.2	0.2	−0.2	−0.5	−0.1
1933..............	−26.5	−24.1	−13.9	−14.3	−13.4	−18.4	−12.8	−23.1	−25.9	−20.0	−19.4	−32.5
	−0.9	−1.3	−0.5	−0.6	−0.6	−0.7	−0.5	−0.6	−0.7	−0.6	−0.7	−1.3

These data again demonstrate that during the first five months of 1929, the temperatures at the surface by and large were subnormal.   From June, 1929, until August, 1931, inclusive, they were generally speaking above normal.   From September, 1931, until the end of 1933, they were again subnormal, except during the summer and autumn of 1932, in which year the month of September was remarkable on account of its extraordinarily high temperatures.   The entire year 1933 in particular was very cold; its average monthly minus deviation was not less than 27.4% of the total five year range.

The averages for the body of water between the surface and 50 meters, as well as those for the 50 meter level, showed the same conditions with minor variations.

Although in a certain measure the available information indicates that in the Monterey region there are large-sized pulses of warm and cold water, each of which may extend over a considerable number of months, nevertheless, it is as yet not sufficient to allow any conclusions in regard to the nature and amplitudes of the pulses.   Observations covering at least a twenty-five year period will be needed for this purpose.   See also the last section of this chapter.

The yearly averages given in the first section of this chapter also illustrate the differences among the various years.   Thus, for instance, the yearly surface temperatures for the five investigated years were in order as follows: 13.0°, 13.5°, 13.3°, 12.7°, and 11.7°.

C. Monthly Amplitudes of Thermal Variation: As has been indicated repeatedly in the preceding part of this report, the amplitudes of thermal variation in the Monterey Bay region are quite extraordinary.

The results of the analysis of the material from this point of view are given in Figs. 20 and 21.   In the former figure, the five years are presented serially, the surface conditions in the upper row, the averages for the body of water between the surface and 50 meters in the middle row, and the situation at the 50 meter level in the lower row.   The years and months are indicated by short lines between these rows, the years by complete lines and the months by short lines and numbers.   The total and average ranges are repeated each year for the sake of ease of comparison, the former in broken lines, the latter in heavy continuous lines. The ranges of the various months are represented by fine continuous lines and by stippling.

Most of the data obtained in this analysis were given in chapter IV on the Monthly Thermal Conditions, *viz.*, the monthly amplitudes for the surface, 50 and 100 meters, the average monthly ranges for these levels, and the total ranges for the five year period.   For the sake of emphasis, the total ranges will be repeated in this connection.   Another table will deal with the ranges for the averages of the body of water between the surface and 50 meters, since this was not completely presented before.   The first table gives the total ranges.   As usual, the months are indicated by Roman numerals.

1929–1933

	I	II	III	IV	V	VI	VII	VIII	IX	X	XI	XII
Surface........	13.6°	14.4°	14.1°	15.2°	15.4°	16.0°	16.0°	16.0°	16.7°	16.4°	14.0°	14.3°
	11.2°	10.5°	10.7°	10.2°	10.2°	10.2°	11.2°	11.8°	13.0°	11.2°	10.9°	10.7°
	2.4°	3.9°	3.4°	5.0°	5.2°	5.8°	4.8°	4.2°	3.7°	5.2°	3.1°	3.6°
0–50 m........	13.6°	14.3°	13.3°	12.9°	13.3°	12.9°	14.0°	13.6°	15.2°	14.0°	13.4°	14.2°
	10.8°	9.5°	9.5°	9.5°	9.2°	9.4°	9.9°	10.0°	10.9°	10.0°	10.4°	10.4°
	2.8°	4.8°	3.8°	3.4°	4.1°	3.5°	4.1°	3.6°	4.3°	4.0°	3.0°	3.8°
50 m........	13.6°	14.2°	12.2°	12.5°	12.8°	12.1°	12.7°	11.9°	12.3°	12.5°	13.3°	14.0°
	10.2°	8.8°	8.6°	8.3°	8.2°	8.3°	8.8°	9.3°	9.6°	9.5°	9.7°	10.0°
	3.4°	5.4°	3.6°	4.2°	4.6°	3.8°	3.9°	2.6°	2.7°	3.0°	3.6°	4.0°

The next table presents the yearly extremes and ranges, by the month, for the averages of the body of water between the surface and 50 meters.

	I	II	III	IV	V	VI	VII	VIII	IX	X	XI	XII
1929....	12.3°	12.3°	11.5°	10.1°	10.4°	11.2°	14.0°	12.0°	12.9°	13.5°	13.4°	14.2°
	11.3°	10.0°	9.7°	9.5°	9.2°	10.0°	10.5°	10.7°	10.9°	11.4°	11.8°	12.1°
	1.0°	2.3°	1.8°	0.6°	1.2°	1.2°	3.5°	1.3°	2.0°	2.1°	1.6°	2.1°
1930....	13.3°	13.4°	12.4°	12.5°	11.1°	12.1°	11.8°	11.8°	13.2°	12.7°	13.0°	14.1°
	12.4°	11.8°	10.9°	10.8°	10.0°	10.1°	10.5°	10.5°	11.7°	11.5°	11.3°	12.8°
	0.9°	1.6°	1.5°	1.7°	1.1°	2.0°	1.3°	1.3°	1.5°	1.2°	1.7°	1.3°
1931....	13.6°	14.3°	13.3°	12.9°	13.3°	12.9°	11.7°	13.4°	13.6°	12.5°	12.5°	12.2°
	12.9°	13.5°	10.3°	10.2°	10.8°	10.3°	9.9°	11.0°	11.8°	11.4°	10.9°	11.4°
	0.7°	0.8°	3.0°	2.7°	2.5°	2.6°	1.8°	2.4°	1.8°	1.1°	1.6°	0.8°
1932....	12.0°	11.2°	11.4°	10.5°	11.1°	11.3°	11.6°	13.6°	15.2°	14.1°	11.8°	13.2°
	10.9°	10.7°	9.7°	9.9°	10.5°	10.7°	10.0°	10.0°	12.9°	11.0°	10.9°	11.8°
	1.1°	0.5°	1.7°	0.6°	0.6°	0.6°	1.6°	3.6°	2.3°	3.1°	0.9°	1.4°
1933....	11.6°	11.0°	10.7°	10.3°	9.9°	10.5°	10.7°	11.0°	11.7°	11.6°	11.9°	11.3°
	10.8°	9.5°	9.5°	9.7°	9.4°	9.4°	10.0°	10.7°	11.1°	10.0°	10.4°	10.4°
	0.8°	1.5°	1.2°	0.6°	0.5°	1.1°	0.7°	0.3°	0.6°	1.6°	1.5°	0.9°

When we consider these data as a whole, the most striking feature is the exceptional narrowness of the amplitudes in view of the latitude, 36° 40′ N., in which these waters are located.   For instance, at the surface where variations usually are the largest, the greatest amplitude observed throughout the five year period was only 6.5°, 16.7° in September, 1932, and 10.2° recorded on a number of occasions.   This characteristic of the Monterey Bay region has as an important biological consequence the fact that generally speaking the thermal conditions play a very subordinated role in these waters in regard to the distribution of the bottom organisms from season to season and from one year to another. Furthermore, while in other regions it sometimes happens that extreme thermal conditions bring about a mass destruction of littoral organisms, such destruction probably never occurs in this general locality.

In regard to the amplitudes at the surface, it should first be noted that the maximum total monthly value on record is 5.8°; this was obtained in June. There appeared to be a distinct seasonal rhythm in the ranges, a rhythm expressed especially well by the five year averages (heavy continuous line) but also quite evident in the total ranges. It was expressed in a remarkably low range for January (average 0.6°), a general increase from this month until June–July, when the maximum was reached, and a fairly evenly graded decrease towards December, in which month the range was still distinctly higher than in January (average 1.4°). As far as the conformity of the individual years to this rhythm is con-

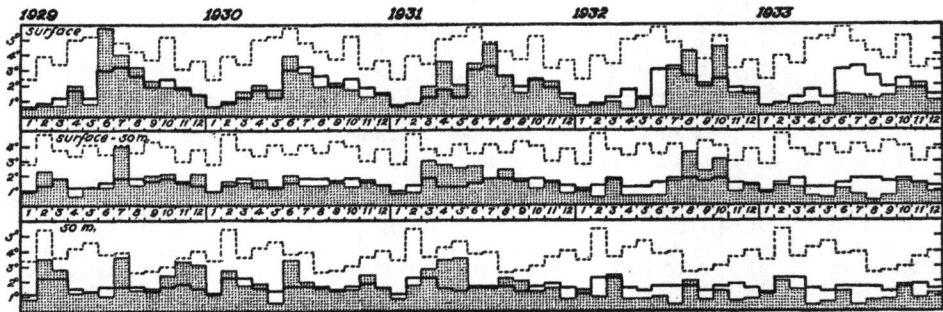

FIG. 20.   Amplitudes of thermal variation, at stations 4 and B, from 1929 to 1933.   For further explanation, see the beginning of section C of chapter VI.

cerned, it should be noted that the years 1929 to 1932 inclusive, exhibited the rhythm very clearly, while in 1933 it was completely absent. It should also be noted that 1930 showed the closest approximation to normalcy; that 1929 was quite typical, except for June which was unusually variable covering the entire maximum monthly range for the five year period (5.8°); and that of the remaining three years, 1933 was extraordinarily constant, its maximum range being only 2.1° and its average monthly range 1.1°.

The maximum range for the averages of the body of water between the surface and 50 meters during the five year period was 6.0°, thus only 0.5° less than the corresponding value for the surface, a very small difference indeed, indicating that the surface waters are only to a subordinated extent influenced by local meteorological factors. The maximum total monthly range was 4.3°, recorded in September. It would be an exaggeration to state that a seasonal rhythm is present; only an extremely faint indication of the surface rhythm may be traced in the averages. For all practical purposes, the monthly amplitudes were on the average constant throughout the year, even though the individual years exhibited a fair variability with a preponderance of the higher ranges in the summer months. Again, the year 1933 was decidedly subnormal in variability. That the total five year ranges should not show any periodicity is readily understood when we consider that they should be intermediate between those of the surface and 50 meters and these were, as we shall see presently, complimentary rather than similar.

For the 50 meter level, the greatest amplitude during the five years was 6.0°, from 14.2° in February, 1931, to 8.2° in May, 1929. The highest recorded total monthly range was greater than the corresponding one for the 0–50 meter water, *viz.*, 5.4°, obtained in February. In the total five year ranges, there appeared to be at least an indication of a seasonal rhythm, with the greatest amplitudes in the beginning and end of the year and the minimum during the summer. The average conditions, however, did not conform to this pattern; and in all probability the rhythm was apparent only and due to the limitations of our data. The averages showed about the same conditions as those characterizing the waters between the surface and 50 meters. At this level, the year 1933 was also abnormally constant.

From these facts it will be evident that when the variability of the surface waters is compared with that of the 50 meter level, the latter variability is found to have been the more pronounced, on the average, during the winter and spring, while in the summer and autumn the reverse was true. This characteristic is graphically presented in Fig. 21 which

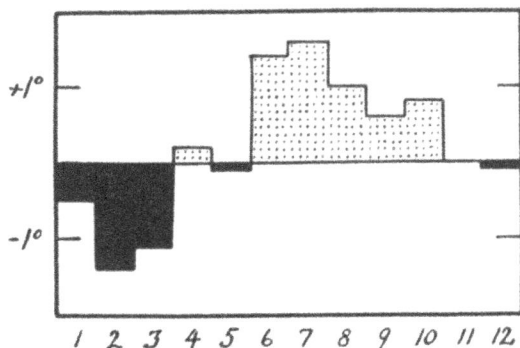

FIG. 21. Differences in the amplitudes of the monthly thermal variations between the surface and 50 meters. Average condition for the period of 1929–33. Minus deviations, shown in black, indicate that the amplitudes at 50 meters were greater than those at the surface. Plus deviations, shown in stippling, indicate that the surface temperatures were more variable.

shows the average differences in the variation between the surface and 50 meters for the five year period. From this figure it will be seen that, on the average, the February deviation at 50 meters was 1.4° greater than that at the surface, while in July the surface variation was the greater by 1.6°. When taken by themselves, these values appear to be quite small, but when they are considered in the light of the extremely small total variation for the five year period they lose their insignificance. The fact that the variability sometimes was greater at 50 meters than at the surface, of course, appears to indicate that the forces that caused the variation on these occasions were subsurface in origin rather than superficial. However, that this was not always the case will be shown in section *D* of chapter VIII.

Considering the significance of this phenomenon in judging the hydrographic rhythm in this general region, it seems advisable to enable the reader to reach a more detailed information in regard to the behavior of the individual years in this respect. This information is embodied in the following table, in which the thermal variation at the surface and that at 50 meters are compared. Minus signs signify that the 50 meter level was the more variable; the differences are expressed in degrees.

	I	II	III	IV	V	VI	VII	VIII	IX	X	XI	XII
1929....	−0.3°	−2.8°	−2.1°	0.8°	−0.5°	4.8°	0.4°	1.8°	0.8°	−0.5°	−1.4°	−1.7°
1930....	−0.6°	−1.8°	−0.1°	0.2°	1.2°	0.5°	0.9°	0.4°	0.7°	0.3°	−1.2°	−0.1°
1931....	−0.1°	−0.9°	−0.9°	0.1°	−1.5°	1.9°	3.2°	0.4°	−0.5°	1.0°	0.7°	0.3°
1932....	−0.8°	−0.2°	−1.5°	−0.4°	0.3°	−0.4°	2.6°	2.2°	1.4°	3.2°	0.8°	1.0°
1933....	−0.6°	−1.2°	−0.7°	0.4°	0.0°	0.1°	0.9°	0.4°	0.5°	0.0°	1.2°	−0.2°
Av......	−0.5°	−1.4°	−1.1°	0.2°	−0.1°	1.4°	1.6°	1.0°	0.6°	0.8°	0.0°	−0.1°

As will be seen from this table, there were very decided differences among the various years. At the same time, there were also striking consistencies. Thus, for instance, in January, February, and March, the 50 meter level was always the more variable while the reverse was always true in July and August. The situation expressed by the averages is presumably fairly characteristic of these waters, and the many exceptions in all probability simply furnish another example of the many forces which all the time were interfering with the unfolding of what I have termed in this report the "normal" rhythm.

D. *Surface Temperatures, by the Shore, from 1919 to 1934:* The data presented in this section were courteously put at the disposal of the Hydrobiological Survey by the Scripps Institution of Oceanography, La Jolla, California. They were gathered at the very shore line just outside the Hopkins Marine Station, where readings were made every morning at about eight o'clock. As a consequence of the fact that these temperatures were taken at a place where the water was subjected to the vicissitudes of land heating and cooling, as well as to the thermal effects of run-off water during the winter and early spring months, they undoubtedly are, at least in part, more extreme than those gathered at some distance from shore. Another factor to be considered in this connection is that small eddies sometimes are formed close to shore. Through these, water entering the Bay is prevented from touching the shore, and as a consequence, the shore temperatures on these occasions sometimes become less extreme than those taken farther out. Apart from these exceptions, the Scripps data should be more representative, since they are based on daily readings, while those of the survey refer to weekly, or at the most semi-weekly, measurements. Even though the Scripps data are not directly comparable with those of the survey, nevertheless, they are of the greatest interest in this connection, since they throw light on certain aspects of the thermal rhythm revealed by the latter values.

In Fig. 22, the continuous line represents the monthly total averages for the 1919–33 period, while the dotted and broken lines represent the extreme monthly averages during this interval. The total monthly average curve is quite similar to the corresponding one for 1929–33 (Fig. 16). It differs from this mainly in its lower maximum, as well as in having lower values at the beginning and at the end of the year. In these respects it is presumably somewhat more representative than the 1929–33 curve, since the maximum temperature for September, 1932, was apparently unusually high and the December temperatures in 1929 and 1930 and the January temperatures in 1930 and 1931 probably were also higher than normal.   It should also be noted that, while, due to the latter fact, there is a drop in the 1929–33 curve from January to April, this peculiarity is not present in the 1919–33 curve.   On the other hand, some of the extreme monthly averages in Fig. 22 are in all probability either too high or too low, on account of factors noted above.

In regard to Fig. 23, which represents the monthly averages from 1919 to 1934 inclusive, it should first of all be observed that it confirms the fact, revealed by our data, that the annual rhythm at the surface is extremely variable.   Not only is the annual amplitude of

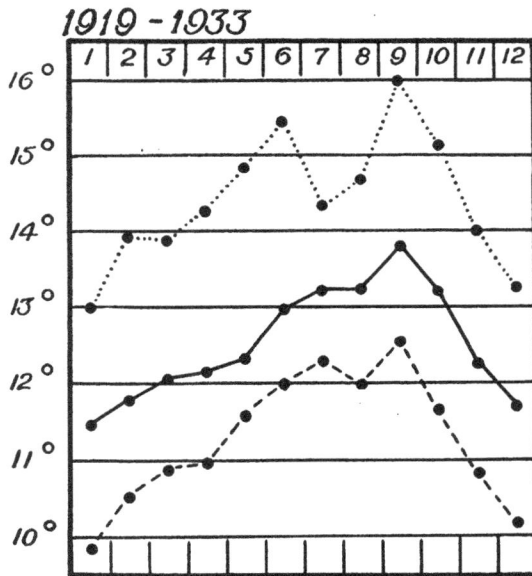

FIG. 22.   Monthly surface temperatures by the shore at Hopkins Marine Station.   Averages for 1919–33. For further explanation, see the second paragraph of section *D* of chapter VI.

thermal variation subject to great changes (note, for instance, that the total ranges for 1928 and 1932 were about 1.5° and 5.5, respectively), but the annual minimum and maximum are located in different months, from one year to the next.   The minimum may be found in any month from December (*e.g.*, in 1919) to April (1929); and the maximum may shift from April (1926) to November (1927).   Perhaps better than any fact shown by our data, these shifts illustrate the relative independence of the water and air temperatures (see the introductory part of chapter VIII).   It should be emphasized, however, that in not less than seven out of the sixteen years, the maximum was to be found in September, the month in which it was always found by us.   Comparison between the part of the curve covering 1929–33 and Fig. 16 should be made; it will bring forth some quite interesting differences in details, differences explainable in terms of the factors that were noted above.

The heavy smooth curve in Fig. 23 demonstrates the variations in the annual averages, which were as follows:

1919: 12.70°	1923: 12.13°	1927: 12.33°	1931: 13.81°
1920: 12.53°	1924: 11.60°	1928: 12.39°	1932: 12.78°
1921: 12.22°	1925: 12.25°	1929: 12.88°	1933: 12.17°
1922: 11.81°	1926: 12.50°	1930: 13.48°	1934: 13.30°

From these values it will be seen that 1931, which was the warmest year investigated by us, was the warmest year during the entire sixteen year period. On the other hand, 1933, which was by far our coldest year, was rather intermediary than exceptionally cold. Furthermore, although there evidently was during this period in the Monterey Bay region an alternation of relatively cool and warm years, nevertheless, the changes did not occur in a simple, long-range cycle. Such a simple cycle was, of course, not to be expected. If the long-range changes be of rhythmical, repetitive nature, then this must be of a complexity

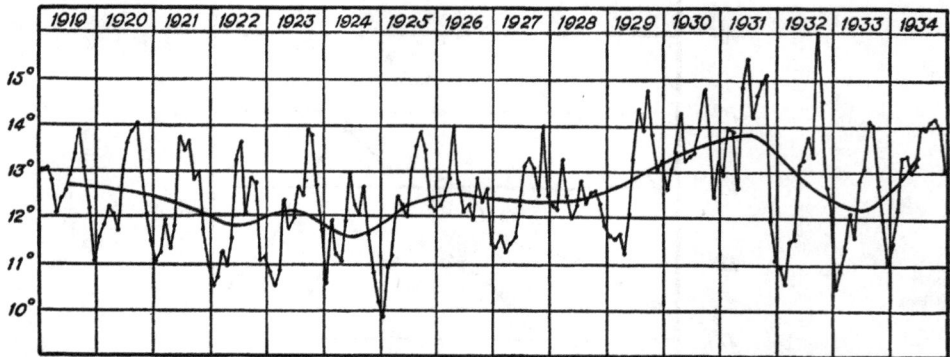

FIG. 23.   Monthly surface temperatures by the shore at Hopkins Marine Station, from 1919 to 1934. For further explanation, see section *D* of chapter VI.

which can not as yet be established on account of the limitation of our material. No indication of a repetition could be found in the 1919–34 interval. The small amplitude of the fluctuation in the annual averages is noteworthy; the lowest annual average (1924) was only 2.21° below the highest (1931).

Fig. 24 shows the deviation of each month, during the 1919–34 interval, from the normal as represented by the total monthly averages during the 1919–33 period (not the 1919–34 period). It was constructed essentially on the same principles as Fig. 19, although the deviations were given in degrees and not in percentages of the total amplitudes of monthly variation. The most striking feature of the thermal development illustrated by this graph is that the years were by no means always either entirely above or entirely below normal; on the contrary, brief warm and cold periods often alternated in an irregular manner. Thus 1926 was warm during the first five months and decidedly cold during the balance of the year, except in November which was somewhat above normal. The most conspicuous exception from this irregularity occurred between June, 1929, and October, 1931, when not less than 29 consecutive months were above normal. It should be noted in this connection

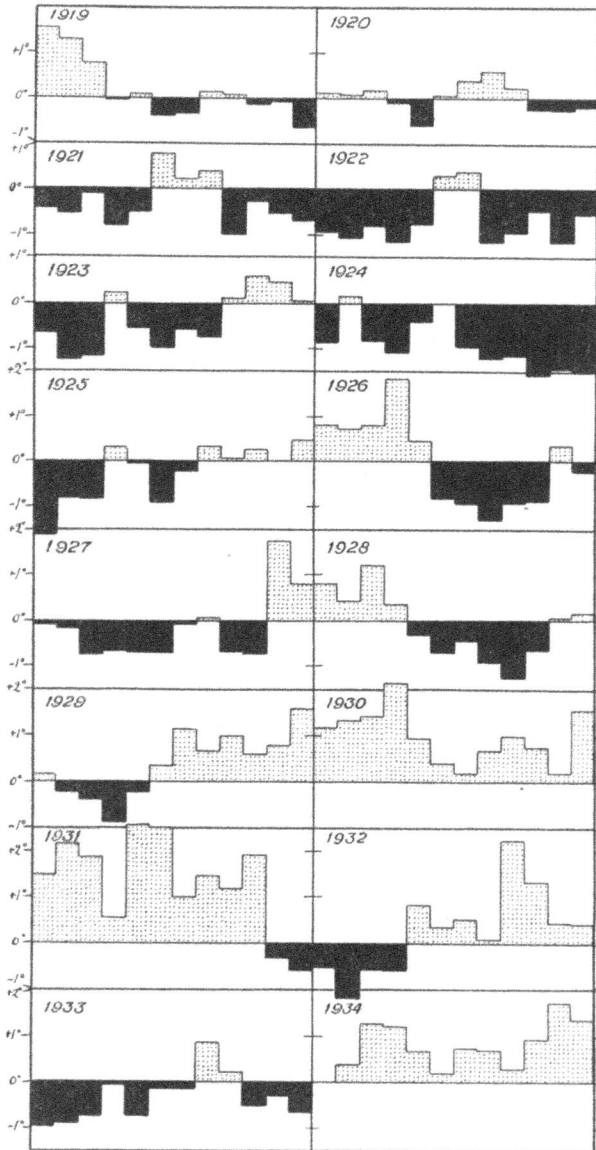

FIG. 24.   Monthly thermal deviations, during the 1919–34 interval, from the "normal" as represented by the total averages during the 1919–33 period (not the 1919–34 period).   For further explanation, see the last paragraph of section *D* of chapter VI.

that, according to the data of the survey, this period was not quite so long, nor quite so uniform (Fig. 19).   Another, although not quite unbroken, series occurred from the latter part of 1920 to the middle of 1925, during which period cold conditions prevailed, with the climax in 1924 which was by far the coldest year on record.   The year 1933, which was so uniformly cold, according to our data, was not only less extreme, according to the shore temperatures, but August and September were actually above normal.   In this case, the survey observations are probably more representative.   These irregularities in the thermal development strongly suggest that the factors underlying the changes in the Monterey Bay region are very complex, the correctness of which assumption will be brought forth in the discussion of the water movements in this region presented in chapter VIII.

VII. THERMAL CONDITIONS BELOW ONE HUNDRED METERS

Observations pertaining to the levels below 100 meters were unfortunately not taken during the first three years of the survey, for reasons given in chapter III on Methods and Material.   Thus thermal data for the deeper waters are available only from 1932 and 1933. This, of course, decidedly limits their value for a generalized characterization of the Monterey region.   However, in spite of its limitation, the present material is likely to be quite representative on account of the circumstance that the variability of the lower strata unquestionably is much less than that of the upper 100 meters, and this, as we have seen from the preceding chapter, is remarkably narrow.

First the monthly conditions between 100 and 250 meters will be treated separately on account of the fact that between these levels seasonal changes were still fairly clear-cut. Next the situation between 250 and 900 meters will be presented.   And, finally, the annual rhythm from the surface to 900 meters, as it unfolded itself at station C, will be considered.

A. Monthly Thermal Conditions Between 100 and 250 Meters: In this stratum there were, as noted above, fairly distinct differences among the various months.   However, important restrictions must be applied to this statement: first, in spite of the differences obtained from month to month, the general shapes of the curves showed in most instances very marked mutual similarities; and, secondly, the differences between the same months of the two years frequently were more pronounced than the seasonal changes.   The latter circumstance, of course, demonstrates that more material will be necessary before we can give reliable characterizations of the monthly conditions of these waters.   For the sake of brevity, the essentials of the various months will be given in condensed, tabular form, after which follow some remarks and comparative notes.   Concerning the amplitudes of variation at each level, it should be noted that observations were never made at less than weekly intervals and that on some occasions not even this minimum could be accomplished on account of weather difficulties and other adverse happenings.   As a consequence, at least in some instances, the thermal constancy suggested by the following values is apparent only.   The gradients pertain to the monthly averages.   The data are represented graphically in Figs. 5 and 6.

The tables show that the monthly ranges of variation changed towards the depth in a very irregular manner.   If we consider all the months of the two years together, we find that the 150 meter level exhibited a greater variability than the 250 meter level in 13 out of the 24 months; in other words, there was no distinct preponderance of a greater variability at the higher levels.   In addition, there was no indication of a seasonal regularity in this

January

Range	100 m.	150 m.	200 m.	250 m.
1932.......	0.3°(10.1°–9.8°)	0.9°(9.2°–8.3°)	0.4°(8.7°–8.3°)	0.2°(8.2°–8.0°)
1933.......	1.6°(10.9°–9.3°)	1.5°(10.5°–9.0°)	1.0°(9.5°–8.5°)	1.0°(8.6°–7.6°)
Average.....				
1932.......	10.0°	8.9°	8.5°	8.1°
1933.......	10.1°	9.6°	9.0°	8.2°
Gradient....	100–150 m.	150–200 m.	200–250 m.	100–250 m.
1932.......	1.1°(10.0°–8.9°)	0.4°(8.9°–8.5°)	0.4°(8.5°–8.1°)	1.9°(10.0°–8.1°)
1933.......	0.5°(10.1°–9.6°)	0.6°(9.6°–9.0°)	0.8°(9.0°–8.2°)	1.9°(10.1°–8.2°)

February

Range	100 m.	150 m.	200 m.	250 m.
1932.......	1.4°(10.9°–9.5°)	1.3°(10.3°–9.0°)	1.0°(9.5°–8.5°)	0.4°(8.2°–7.8°)
1933.......	1.5°(10.0°–8.5°)	0.7°(9.1°–8.4°)	0.9°(8.7°–7.8°)	0.9°(8.2°–7.3°)
Average.....				
1932.......	10.1°	9.4°	8.8°	8.1°
1933.......	9.3°	8.9°	8.3°	7.6°
Gradient....	100–150 m.	150–200 m.	200–250 m.	100–250 m.
1932.......	0.7°(10.1°–9.4°)	0.6°(9.4°–8.8°)	0.7°(8.8°–8.1°)	2.0°(10.1°–8.1°)
1933.......	0.4°(9.3°–8.9°)	0.6°(8.9°–8.3°)	0.7°(8.3°–7.6°)	1.7°(9.3°–7.6°)

March

Range	100 m.	150 m.	200 m.	250 m.
1932.......	1.0°(9.3°–8.3°)	1.1°(9.2°–8.1°)	0.8°(8.4°–7.6°)	0.6°(7.8°–7.2°)
1933.......	0.8°(9.5°–8.7°)	0.4°(8.8°–8.4°)	0.3°(8.3°–8.0°)	0.5°(8.2°–7.7°)
Average.....				
1932.......	8.9°	8.6°	8.1°	7.5°
1933.......	9.1°	8.5°	8.1°	8.0°
Gradient....	100–150 m.	150–200 m.	200–250 m.	100–250 m.
1932.......	0.3°(8.9°–8.6°)	0.5°(8.6°–8.1°)	0.6°(8.1°–7.5°)	1.4°(8.9°–7.5°)
1933.......	0.6°(9.1°–8.5°)	0.4°(8.5°–8.1°)	0.1°(8.1°–8.0°)	1.1°(9.1°–8.0°)

April

Range	100 m.	150 m.	200 m.	250 m.
1932.......	0.6°(9.0°–8.4°)	0.5°(8.8°–8.3°)	0.4°(8.4°–8.0°)	0.3°(8.0°–7.7°)
1933.......	0.2°(8.6°–8.4°)	0.3°(8.4°–8.1°)	0.6°(8.2°–7.6°)	0.7°(7.7°–7.0°)
Average.....				
1932.......	8.8°	8.5°	8.2°	7.8°
1933.......	8.5°	8.2°	8.0°	7.4°
Gradient....	100–150 m.	150–200 m.	200–250 m.	100–250 m.
1932.......	0.3°(8.8°–8.5°)	0.3°(8.5°–8.2°)	0.4°(8.2°–7.8°)	1.0°(8.8°–7.8°)
1933.......	0.3°(8.5°–8.2°)	0.2°(8.2°–8.0°)	0.6°(8.0°–7.4°)	1.1°(8.5°–7.4°)

May

Range	100 m.	150 m.	200 m.	250 m.
1932........	0.4°(9.0°–8.6°)	0.5°(8.7°–8.2°)	0.1°(8.2°–8.1°)	0.4°(8.0°–7.6°)
1933........	0.5°(8.4°–7.9°)	0.6°(8.1°–7.5°)	0.5°(7.8°–7.3°)	0.4°(7.6°–7.2°)
Average.....				
1932........	8.8°	8.4°	8.2°	7.8°
1933........	8.1°	7.7°	7.5°	7.3°
Gradient....	100–150 m.	150–200 m.	200–250 m.	100–250 m.
1932........	0.4°(8.8°–8.4°)	0.2°(8.4°–8.2°)	0.4°(8.2°–7.8°)	1.0°(8.8°–7.8°)
1933........	0.4°(8.1°–7.7°)	0.2°(7.7°–7.5°)	0.2°(7.5°–7.3°)	0.8°(8.1°–7.3°)

June

Range	100 m.	150 m.	200 m.	250 m.
1932........	0.6°(9.2°–8.6°)	1.0°(8.8°–7.8°)	1.0°(8.5°–7.5°)	0.8°(8.1°–7.3°)
1933........	0.7°(8.7°–8.0°)	0.5°(8.0°–7.5°)	0.3°(7.6°–7.3°)	0.1°(7.2°–7.1°)
Average.....				
1932........	8.8°	8.5°	8.1°	7.7°
1933........	8.3°	7.7°	7.4°	7.2°
Gradient....	100–150 m.	150–200 m.	200–250 m.	100–250 m.
1932........	0.3°(8.8°–8.5°)	0.4°(8.5°–8.1°)	0.4°(8.1°–7.7°)	1.1°(8.8°–7.7°)
1933........	0.6°(8.3°–7.7°)	0.3°(7.7°–7.4°)	0.2°(7.4°–7.2°)	1.1°(8.3°–7.2°)

July

Range	100 m.	150 m.	200 m.	250 m.
1932........	0.3°(9.2°–8.9°)	0.1°(8.8°–8.7°)	0.0°(8.4°–8.4°)	0.4°(8.4°–8.0°)
1933........	0.4°(8.8°–8.4°)	0.7°(8.8°–8.1°)	0.8°(8.5°–7.7°)	1.1°(8.3°–7.2°)
Average.....				
1932........	9.0°	8.8°	8.4°	8.2°
1933........	8.5°	8.5°	8.2°	7.8°
Gradient....	100–150 m.	150–200 m.	200–250 m.	100–250 m.
1932........	0.2°(9.0°–8.8°)	0.4°(8.8°–8.4°)	0.2°(8.4°–8.2°)	0.8°(9.0°–8.2°)
1933........	0.0°(8.5°–8.5°)	0.3°(8.5°–8.2°)	0.4°(8.2°–7.8°)	0.7°(8.5°–7.8°)

August

Range	100 m.	150 m.	200 m.	250 m.
1932........	0.1°(9.8°–9.7°)	0.2°(9.5°–9.3°)	0.2°(9.1°–8.9°)	0.6°(8.8°–8.2°)
1933........	0.5°(9.4°–8.9°)	0.2°(8.9°–8.7°)	0.2°(8.7°–8.5°)	0.4°(8.2°–7.8°)
Average.....				
1932........	9.8°	9.4°	9.0°	8.5°
1933........	9.1°	8.8°	8.6°	8.1°
Gradient....	100–150 m.	150–200 m.	200–250 m.	100–250 m.
1932........	0.4°(9.8°–9.4°)	0.4°(9.4°–9.0°)	0.5°(9.0°–8.5°)	1.3°(9.8°–8.5°)
1933........	0.3°(9.1°–8.8°)	0.2°(8.8°–8.6°)	0.5°(8.6°–8.1°)	1.0°(9.1°–8.1°)

September

Range	100 m.	150 m.	200 m.	250 m.
1932........	0.3°(10.3°–10.0°)	0.6°(10.0°–9.4°)	0.6°(9.2°–8.6°)	0.8°(8.6°–7.8°)
1933........	0.8°(9.5°– 8.7°)	0.8°(9.2°–8.4°)	0.5°(8.9°–8.4°)	0.4°(8.6°–8.2°)
Average.....				
1932........	10.2°	9.6°	8.9°	8.3°
1933........	9.1°	8.9°	8.7°	8.4°
Gradient....	100–150 m.	150–200 m.	200–250 m.	100–250 m.
1932........	0.6°(10.2°– 9.6°)	0.7°(9.6°–8.9°)	0.6°(8.9°–8.3°)	1.9°(10.2°–8.3°)
1933........	0.2°(9.1°– 8.9°)	0.2°(8.9°–8.7°)	0.3°(8.7°–8.4°)	0.7°(9.1°–8.4°)

October

Range	100 m.	150 m.	200 m.	250 m.
1932........	0.8°(10.2°–9.4°)	0.6°(9.5°–8.9°)	0.5°(8.8°–8.3°)	0.4°(8.2°–7.8°)
1933........	0.7°(9.9°–9.2°)	0.6°(9.3°–8.7°)	0.2°(8.6°–8.4°)	0.4°(8.2°–7.8°)
Average.....				
1932........	9.8°	9.2°	8.6°	8.0°
1933........	9.4°	8.9°	8.5°	8.1°
Gradient....	100–150 m.	150–200 m.	200–250 m.	100–250 m.
1932........	0.6°(9.8°–9.2°)	0.6°(9.2°–8.6°)	0.6°(8.6°–8.0°)	1.8°(9.8°–8.0°)
1933........	0.5°(9.4°–8.9°)	0.4°(8.9°–8.5°)	0.4°(8.5°–8.1°)	1.3°(9.4°–8.1°)

November

Range	100 m.	150 m.	200 m.	250 m.
1932........	0.2°(9.9°–9.7°)	0.4°(9.6°–9.2°)	0.4°(8.8°–8.4°)	0.5°(8.3°–7.8°)
1933........	0.4°(9.5°–9.1°)	0.4°(9.2°–8.8°)	0.8°(8.9°–8.1°)	0.4°(8.0°–7.6°)
Average.....				
1932........	9.8°	9.4°	8.6°	8.1°
1933........	9.3°	9.0°	8.4°	7.8°
Gradient....	100–150 m.	150–200 m.	200–250 m.	100–250 m.
1932........	0.4°(9.8°–9.4°)	0.8°(9.4°–8.6°)	0.5°(8.6°–8.1°)	1.7°(9.8°–8.1°)
1933........	0.3°(9.3°–9.0°)	0.6°(9.0°–8.4°)	0.6°(8.4°–7.8°)	1.5°(9.3°–7.8°)

December

Range	100 m.	150 m.	200 m.	250 m.
1932........	1.1°(11.3°–10.2°)	1.5°(10.7°–9.2°)	1.6°(10.0°–8.4°)	1.9°(9.4°–7.5°)
1933........	0.8°(10.2°– 9.4°)	0.9°(9.7°–8.8°)	0.4°(8.7°–8.3°)	0.5°(8.0°–7.5°)
Average.....				
1932........	10.7°	10.0°	9.2°	8.4°
1933........	9.8°	9.2°	8.4°	7.7°
Gradient....	100–150 m.	150–200 m.	200–250 m.	100–250 m.
1932........	0.7°(10.7°–10.0°)	0.8°(10.0°–9.2°)	0.8°(9.2°–8.4°)	2.3°(10.7°–8.4°)
1933........	0.6°(9.8°– 9.2°)	0.8°(9.2°–8.4°)	0.7°(8.4°–7.7°)	2.1°(9.8°–7.7°)

phenomenon.   In all instances the ranges were very narrow; at 150 meters, the average monthly ranges were 0.7° (1932) and 0.6° (1933); at 200 meters, 0.6° (1932) and 0.5° (1933); and at 250 meters, the same value, 0.6°, was obtained for both years.   The maximum monthly range was 1.5° at 150 meters, 1.6° at 200 meters, and 1.9° at 250 meters.   Again it should be noted that these values in all probability are somewhat too small, although not much so.

The fact that the variability did not decrease from the 150 meter level to the 250 meter level and that the maximum range was even higher at the latter level are intimately connected with the circulatory perculiarities in this region (see chapter VIII).   The absence of seasonal regularity in the variations was to be expected since such a regularity did not occur even at the 50 meter level.

In regard to the gradients, it may be stated that by and large the smallest ones were found from April to July, inclusive, from which period they increased toward the beginning and the end of the year.   This rule is especially evident when we study the gradients between 100 and 250 meters; during the mentioned four months, the gradients did in no case exceed 1.1°, while at the beginning and end of the year they were about 2.0°.   As a rule, the temperature changed in a very uniform manner towards the depth, a fact which is clearly demonstrated by the yearly average gradients for 100–150 meters, 150–200 meters, and 200–250 meters; in 1932, these were in all cases the same, viz., 0.5°; and in 1933, they were 0.4°, 0.4°, and 0.5°, respectively.

From this summary account, and from the facts presented in chapter IV, it will be evident that although seasonal changes in the gradients occurred in the course of the investigated period, both in the upper 100 meters and in the waters between 100 and 250 meters, nevertheless, the changes did not take place in a similar manner in these two strata. Indeed, they were nearly diametrically opposite.   In the upper 100 meters, the gradients were smallest in December and January, while in the lower stratum this condition was attained in the early summer.   Between 100 and 250 meters, the January and December gradients were the steepest.   This phenomenon stands in close relation to the fundamental circulation of this region.

Perhaps the most distinct seasonal rhythm was displayed by the monthly averages. This condition will be discussed in the last section of this chapter.   The yearly averages for the 150 meter level were 9.1° in 1932, and 8.7° in 1933; for 200 meters, the corresponding values were 8.6° and 8.3°; and for 250 meters, 8.0° and 7.8°.   (It may be interesting to note in this connection that the yearly average for the 100 meter level at station C was 9.6° in 1932 and 9.1° in 1933, while at stations 4 and B the corresponding value for the entire five year period was 9.5°, a very close agreement, indeed.)   The extreme temperatures recorded in the two years were: 10.7°–7.5°, at 150 meters; 10.0°–7.2° at 200 meters; and 9.4°–7.0° at 250 meters.

B. *Thermal Conditions Between 250 and 900 Meters:* The thermal situation below the 250 meter level was markedly constant.   This constancy is especially conspicuous when the annual averages and ranges of variation at the several investigated levels are considered. This statement may be verified from the following table in which these values for 1932 and 1933 are presented.

Depth	1932		1933	
	Range	Av.	Range	Av.
300 m............	1.4°(8.3°–6.9°)	7.5°	1.0°(7.8°–6.8°)	7.4°
350 m............	1.0°(7.5°–6.5°)	7.1°	1.1°(7.6°–6.5°)	7.0°
400 m............	1.1°(7.3°–6.2°)	6.7°	1.2°(7.4°–6.2°)	6.7°
450 m............	0.8°(6.7°–5.9°)	6.3°	1.4°(7.0°–5.6°)	6.3°
500 m............	0.9°(6.4°–5.5°)	6.0°	1.2°(6.5°–5.3°)	5.9°
600 m............	0.7°(5.7°–5.0°)	5.3°	0.7°(5.5°–4.8°)	5.2°
700 m............	0.6°(5.1°–4.5°)	4.8°	0.8°(5.2°–4.4°)	4.7°
800 m............	0.5°(4.7°–4.2°)	4.4°	0.6°(4.6°–4.0°)	4.4°
900 m............	0.7°(4.5°–3.8°)	4.1°	0.5°(4.3°–3.8°)	4.1°

As will be seen from these data, the annual averages were identical in these two years at four out of the nine investigated levels, and in no instance did the differences in these values exceed 0.1°. The average difference in thermal ranges between these years amounted only to 0.2° and the maximum, at the 450 meter level, was 0.6°. The differences between the highest temperatures of the two years at each level averaged 0.2°, with a maximum of 0.5° (at 300 m.); the corresponding values for the lowest temperatures were 0.1° and 0.3°.

The annual thermal gradients are shown by the following table, in which, for the sake of comparison, the data from 100–300 meters are also included.

Meters	100–200	200–300	300–400	400–500	500–600	600–700	700–800	800–900
1932......	1.2°	0.9°	0.8°	0.7°	0.7°	0.5°	0.4°	0.3°
1933......	0.9°	0.8°	0.7°	0.8°	0.7°	0.5°	0.3°	0.3°

Thus, in 1932, there was an almost even annual gradient between 200 and 900 meters, decreasing on the average by 0.1° for each 100 meters towards the depth. Moreover, the gradient between 100 and 200 meters was only 0.3° greater than the one for 200–300 meters. It was thus only in the upper 100 meters that the annual gradient exhibited a pronounced difference from that of the deeper strata. In 1933, the situation was slightly different. The annual gradient between the surface and 400 meters was somewhat less, and the thermal progression towards the depth was not quite so uniform, although the difference was insignificant.

The conditions discussed above are represented graphically in Fig. 25. In this the 1932 values are indicated by small, solid circles, the averages being connected by a continuous line. The 1933 symbols are open circles with small dots, and the averages are connected by a broken line. At each level, the averages and the extremes are entered.

Notwithstanding the great similarity of these two years, when considered from the point of view discussed above, there were distinct differences between them in regard to the monthly averages. Since these differences appear to be significant in conjunction with the general circulation in this region, they will be dealt with here in some detail. The data

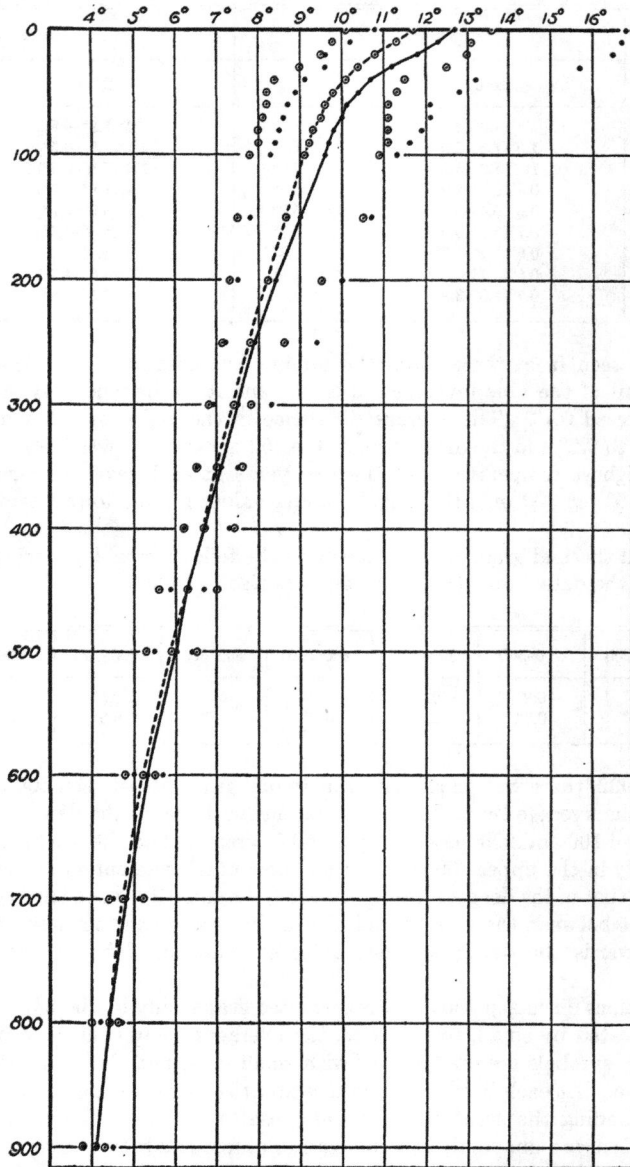

FIG. 25.   Annual thermal gradients and ranges of thermal variation from the surface to 900 meters at station C, in 1932 and 1933.   For further explanation, see section *B* of chapter VII.

are given in the following table, and represented graphically in Fig. 26.   This figure consists of twelve sub-figures, one for each month, arranged in order from top to bottom.   The depths, from 100 to 900 meters, are arranged horizontally and indicated by short lines at

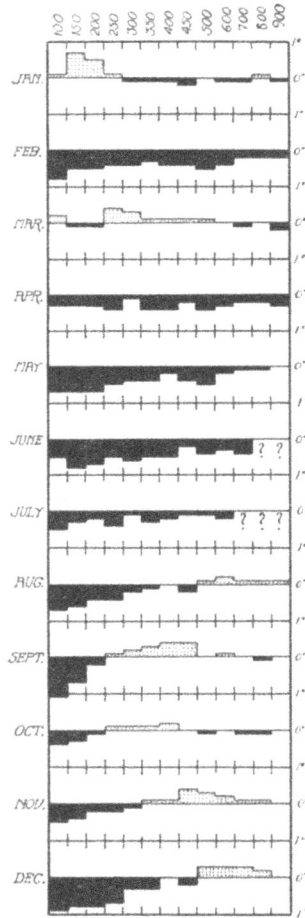

FIG. 26.   Comparison between the monthly thermal averages of 1932 and 1933, at station C.   For further explanation, see section *B* of chapter VII.

the top and bottom of each sub-graph.   Minus deviations show that 1933 was the colder, and they are represented in black while the plus deviations are in stippling.   The space above and below each base line equals one degree.   Unfortunately, there was no information for 800 and 900 meters in June, 1932, and for 700–900 meters in July of that year.   The length of the cable available in those months did not permit us to reach those levels.   In the table, the months are shown by Roman numerals; and only minus deviations are furnished with signs.

Through this simple type of comparative treatment of the monthly averages, certain water movements which occurred in the course of 1932 and 1933 are brought into focus, movements which otherwise might easily have escaped our attention on account of the fact that their long periods and low thermal amplitudes tend to make them very elusive.

In January, 1933, the water at 150 and 200 meters was warmer than in 1932, while at the remaining investigated levels, these years were quite similar, 1933 being only slightly the cooler.   In February, on the other hand, 1933 was distinctly colder than the preceding year, and the difference was not restricted to the superficial strata but extended at least down to a depth of 900 meters, the differences gradually decreasing towards the depth. The month of March exhibited differences of a somewhat uncertain significance.   Apparently, there was during that month a water movement tending towards the development of an unusually cold year along the west coast of North America, but, for some unknown causes, this movement became abortive, the break in the rhythm taking place at the end of that month.   It appeared as if the forces underlying the water movements that caused these thermal changes had been prematurely spent.   The coldness of March, 1932, was shown by the fact that even though this month in 1933 was quite cold, nevertheless it was,

	100 m.	150 m.	200 m.	250 m.	300 m.	350 m.	400 m.	450 m.	500 m.	600 m.	700 m.	800 m.	900 m.
I													
1932.......	10.0°	8.9°	8.5°	8.1°	7.6°	7.2°	6.8°	6.5°	6.0°	5.4°	4.8°	4.3°	4.1°
1933.......	10.1°	9.6°	9.0°	8.2°	7.5°	7.1°	6.7°	6.3°	6.0°	5.3°	4.7°	4.4°	4.0°
Diff........	0.1°	0.7°	0.5°	0.1°	−0.1°	−0.1°	−0.1°	−0.2°	0.0 °	−0.1°	−0.1°	0.1°	−0.1°
II													
1932.......	10.1°	9.4°	8.8°	8.0°	7.6°	7.1°	6.9°	6.6°	6.2°	5.5°	4.9°	4.6°	4.2°
1933.......	9.3°	8.9°	8.3°	7.6°	7.2°	6.8°	6.5°	6.2°	5.7°	5.1°	4.7°	4.4°	4.0°
Diff........	−0.8°	−0.5°	−0.5°	−0.4°	−0.4°	−0.3°	−0.4°	−0.4°	−0.5°	−0.4°	−0.2°	−0.2°	−0.2°
III													
1932.......	8.9°	8.6°	8.1°	7.5°	7.2°	6.9°	6.5°	6.2°	5.9°	5.3°	4.9°	4.5°	4.3°
1933.......	9.1°	8.5°	8.0°	7.9°	7.5°	7.0°	6.6°	6.3°	6.0°	5.3°	4.8°	4.5°	4.1°
Diff........	0.2°	−0.1°	−0.1°	0.4°	0.3°	0.1°	0.1°	0.1°	0.1°	0.0°	−0.1°.	0.0°	−0.2°
IV													
1932.......	8.8°	8.5°	8.2°	7.8°	7.4°	7.1°	6.9°	6.3°	6.0°	5.4°	4.8°	4.5°	4.3°
1933.......	8.5°	8.2°	7.9°	7.4°	7.3°	6.7°	6.5°	6.1°	5.6°	5.1°	4.6°	4.3°	4.0°
Diff........	−0.3°	−0.3°	−0.3°	−0.4°	−0.1°	−0.4°	−0.4°	−0.2°	−0.4°	−0.3°	−0.2°	−0.2°	−0.3°
V													
1932.......	8.8°	8.4°	8.2°	7.8°	7.5°	7.2°	6.9°	6.4°	6.1°	5.3°	4.8°	4.5°	4.1°
1933.......	8.1°	7.7°	7.5°	7.3°	7.1°	6.8°	6.7°	6.0°	5.6°	5.1°	4.7°	4.4°	4.1°
Diff........	−0.7°	−0.7°	−0.7°	−0.5°	−0.4°	−0.4°	−0.2°	−0.4°	−0.5°	−0.2°	−0.1°	−0.1°	0.0°
VI													
1932.......	8.8°	8.5°	8.1°	7.7°	7.4°	7.1°	6.8°	6.3°	6.0°	5.3°	5.0°		
1933.......	8.3°	7.7°	7.4°	7.2°	6.8°	6.6°	6.3°	6.1°	5.6°	5.0°	4.6°	4.4°	4.1°
Diff........	−0.5°	−0.8°	−0.7°	−0.5°	−0.6°	−0.5°	−0.5°	−0.2°	−0.4°	−0.3°	−0.4°		

	100 m.	150 m.	200 m.	250 m.	300 m.	350 m.	400 m.	450 m.	500 m.	600 m.	700 m.	800 m.	900 m.
VII													
1932........	9.0°	8.8°	8.4°	8.2°	7.5°	7.3°	6.9°	6.6°	6.2°	5.5°			
1933........	8.5°	8.5°	8.2°	7.8°	7.4°	7.0°	6.7°	6.5°	6.1°	5.3°	4.6°	4.3°	4.1°
Diff........	−0.5°	−0.3°	−0.2°	−0.4°	−0.1°	−0.3°	−0.2°	−0.1°	−0.1°	−0.2°			
VIII													
1932........	9.8°	9.4°	9.0°	8.5°	7.7°	7.2°	6.8°	6.6°	6.0°	5.3°	4.9°	4.5°	4.1°
1933........	9.1°	8.8°	8.6°	8.1°	7.5°	7.1°	6.8°	6.4°	6.1°	5.5°	5.0°	4.6°	4.2°
Diff........	−0.7°	−0.6°	−0.4°	−0.4°	−0.2°	−0.1°	0.0°	−0.2°	0.1°	0.2°	0.1°	0.1°	0.1°
IX													
1932........	10.2°	9.6°	8.9°	8.3°	7.6°	7.1°	6.7°	6.3°	6.0°	5.3°	4.8°	4.4°	4.1°
1933........	9.1°	8.9°	8.7°	8.4°	7.8°	7.4°	7.1°	6.7°	6.0°	5.4°	4.8°	4.3°	4.1°
Diff........	−1.1°	−0.7°	−0.2°	0.1°	0.2°	0.3°	0.4°	0.4°	0.0°	0.1°	0.0°	−0.1°	0.0°
X													
1932........	9.8°	9.2°	8.6°	7.9°	7.4°	7.1°	6.5°	6.1°	5.8°	5.2°	4.8°	4.4°	4.1°
1933........	9.4°	8.9°	8.5°	8.0°	7.5°	7.2°	6.7°	6.1°	5.7°	5.2°	4.7°	4.3°	4.1°
Diff........	−0.4°	−0.3°	−0.1°	0.1°	0.1°	0.1°	0.2°	0.0°	−0.1°	0.0°	−0.1°	−0.1°	0.0°
XI													
1932........	9.8°	9.4°	8.6°	8.0°	7.5°	7.0°	6.5°	6.0°	5.6°	5.0°	4.7°	4.3°	4.0°
1933........	9.3°	9.0°	8.4°	7.8°	7.4°	7.1°	6.6°	6.4°	5.9°	5.2°	4.8°	4.4°	4.0°
Diff........	−0.5°	−0.4°	−0.2°	−0.2°	−0.1°	0.1°	0.1°	0.4°	0.3°	0.2°	0.1°	0.1°	0.0°
XII													
1932........	10.7°	10.0°	9.2°	8.4°	7.6°	7.2°	6.8°	6.5°	5.8°	5.1°	4.6°	4.3°	4.0°
1933........	9.8°	9.2°	8.4°	7.7°	7.3°	6.9°	6.8°	6.3°	6.1°	5.4°	4.9°	4.5°	4.0°
Diff........	−0.9°	−0.8°	−0.8°	−0.7°	−0.3°	−0.3°	0.0°	−0.2°	0.3°	0.3°	0.3°	0.2°	0.0°

particularly between 250 and 500 meters, warmer than in 1932. In 1933, on the other hand, the water movement which brought forth the decline of the temperatures in this region had a stronger influence behind it; the movement was sustained. The situation which characterized February, and which perhaps was foreshadowed in January, continued in April, May, June and July. In those four months, the temperatures in 1933 were distinctly and consistently lower than in 1932, probably as far down as to the 800 meter level. (As noted above, data were, unfortunately, not obtainable from the greatest depth for June and July.) Just as in February, the differences decreased, generally speaking, towards the depth, a situation partly due to the fact that the total amplitudes of thermal variation followed this pattern. In August, 1933, this movement began to subside; from 500 meters down to 900 meters, that month was slightly warmer than the corresponding period of the previous year. This development continued until October, when 1933 was colder only in the upper 200 meters. In November and December, 1933, the upper layers again began to become colder in relation to 1932. This cooling had the appearance of proceeding from the surface towards the depth, instead of *vice versa*, as in the early part of the year. However, the changes in the relation between the two years that occurred between August and December were due only in part to the decline in the water movement causing the cooling in this region in 1933; it was also connected with a corresponding increase in a similar deep movement in 1932, anticipatory to the condition in the beginning of 1933. Further attention to this phenomenon will be given both in the next section of this chapter and in chapter VIII.   In the latter place, the difficulty and uncertainty inherent in the whole situation will be discussed.

C. *Annual Thermal Rhythm from the Surface to 900 Meters:* Although the account will be much more abbreviated, in the following presentation, very largely the same principles will be applied as in the case of the treatment of the upper 100 meters.   In other words, first will be given the thermal development throughout the two years, as it is known from the actual observations (Fig. 27).   Next, this development will be presented from the point of view of the monthly averages (Fig. 28).   And, finally, the vertical fluctuations of the various types of water will be traced (Fig. 29).   For the sake of economy, the actual temperature records will not be produced; the monthly averages were given in the preceding section of this chapter.   Only a few outstanding features of the graphs will be noted; thus the reader must to a very large extent rely on his own study of the presented material.

Fig. 27 was drawn according to the same principles as Figs. 14 and 15.   Thus the readings of the surface temperatures were connected by a broken line, the remaining ones by continuous lines, of which those of the 50, 150, 250, 350, and 450 meter levels were made somewhat lighter than the others.   In order to facilitate the tracing of the fluctuations, the spaces between every other 100 meter stratum was stippled, between 100 and 200 and 300 and 400 meters, *etc.*

In regard to the upper 100 meters, it may be said that the data agreed very closely with those obtained at station B.   Since it is of interest to establish just how far this agreement extended it may be advisable to present the deviations in a tabular form.   As we shall see later, the degree of deviation does not have significance only from the point of view of the general thermal conditions in the various parts of the Bay but also for an understanding of the fundamental circulation.   For the sake of simplicity, the deviations are those calculated from the monthly averages, not those from the actual readings; this simplification was proved to be justified.   Most of the averages are to be found elsewhere in this report (section *A* of this chapter and chapter XI).   Minus signs show that station B was the colder.   Roman numerals indicate the months.

	I	II	III	IV	V	VI	VII	VIII	IX	X	XI	XII
Surf.												
1932..............	−0.2°	0.0°	−0.4°	−0.4°	0.1°	0.2°	0.7°	0.2°	−0.2°	0.0°	0.2°	−0.2°
1933..............	0.1°	0.1°	0.0°	0.5°	0.0°	0.0°	−0.3°	0.3°	0.5°	−0.4°	−0.2°	−0.1°
50 m.												
1932..............	−0.5°	0.0°	−0.1°	−0.1°	−0.1°	0.0°	−0.1°	0.4°	−0.9°	0.0°	0.2°	−0.4°
1933..............	0.3°	−0.2°	−0.2°	0.1°	−0.1°	−0.2°	0.0°	−0.1°	−0.7°	0.1°	0.3°	0.3°
100 m.												
1932..............	0.1°	−0.2°	0.2°	−0.1°	0.0°	−0.2°	0.0°	−0.1°	0.0°	−0.1°	0.0°	−0.1°
1933..............	0.1°	−0.1°	−0.1°	−0.1°	−0.1°	−0.2°	0.2°	−0.1°	0.0°	0.1°	0.1°	−0.1°

In 1932 the surface was colder at station B than at station C in five months, warmer in five, and of the same temperatures in two; and in 1933, the corresponding values were: four, five, and three.   The greatest difference was 0.7°, the usual one 0.2° or less, and the minus and plus variations were distributed in what appeared to be a random manner.   Thus

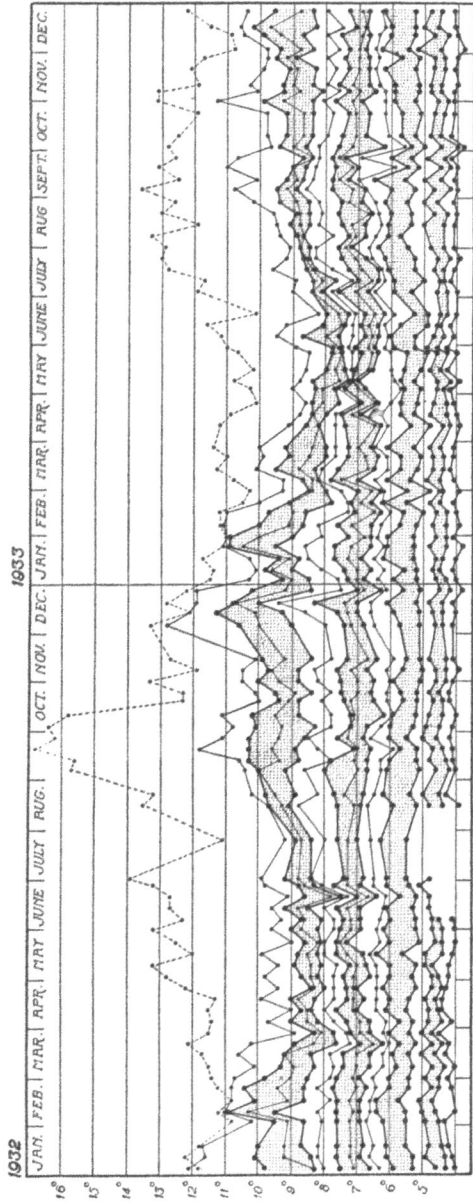

Fig. 27. Thermal fluctuations between the surface and 900 meters, at station C, in 1932 and 1933. For further explanation, see the second paragraph of section C of chapter VII.

during the spring months, which were characterized by a steadily recurring circulatory phenomenon, station B seems to have been somewhat cooler in 1932, but this was not the case in 1933 when this phenomenon was unusually strong.   At the 50 and 100 meter levels, station B apparently was, on the whole, slightly cooler than station C, but the differences in

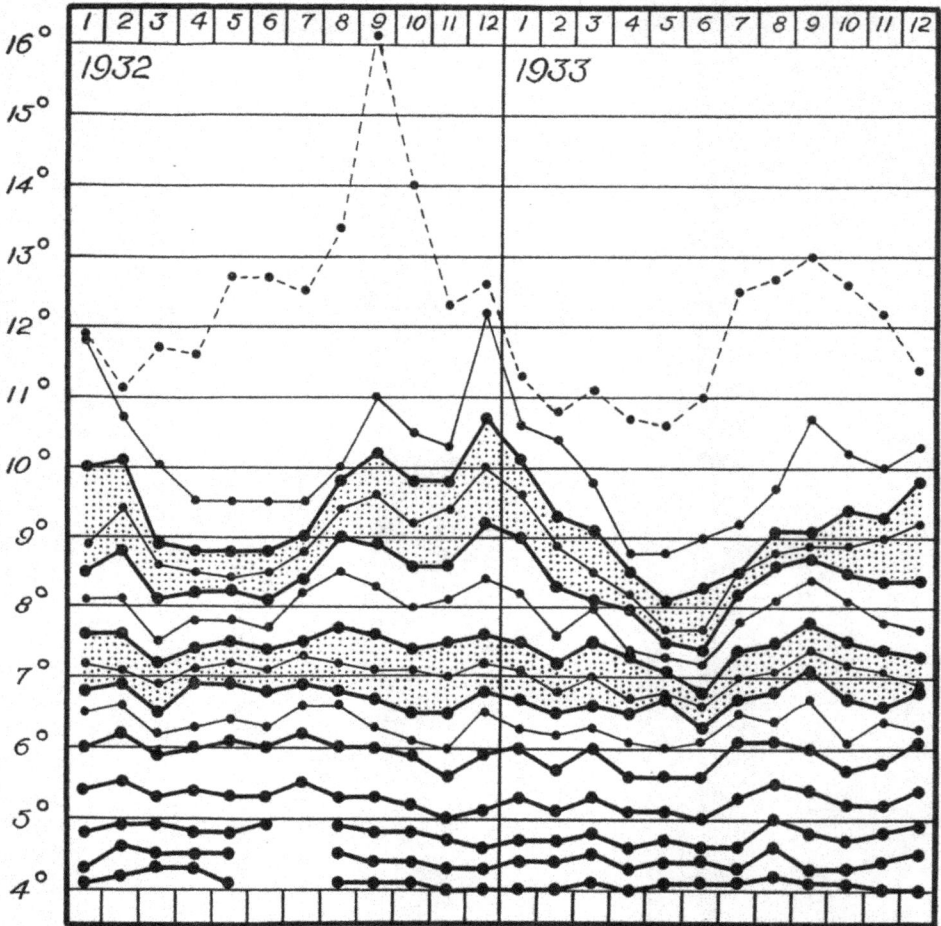

· Fig 28.   Monthly thermal averages between the surface and 900 meters, at station C, in 1932 and 1933.   For further explanation, see section C of chapter VII.

most cases did not exceed 0.2°.   It may also be noted that by and large the minus deviations were more common during the first six months of the year.   This condition will be further discussed in the next chapter, as will also the fact that in both 1932 and 1933 the month of September exhibited higher temperatures at station C than at station B.

Fig. 27 also shows that below the 100 meter level, the temperature observations varied, generally speaking, from one expeditionary day to the next.   Another characteristic of the deeper waters was that the thermal changes, in a large measure, moved quite parallel to each other; in other words, these waters appear to have moved, mutually, in the same directions.

As was shown in a previous chapter, the temperatures at the 50 and 100 meter levels of the key stations were, generally speaking, lowest in May and June and highest in the beginning and end of the year.   The same was true at station C, as will be seen from Fig. 28, a graph largely made on the same pattern as Fig. 27 but representing the monthly averages instead of the individual observations.   This striking peculiarity was also characteristic of the 150 and 200 meter levels, at station C, both in 1932 and 1933, but it was ever so much more pronounced in 1933.   At a depth of 250 meters, the temperatures exhibited, in 1932, a slight decline and rise from February to August, inclusive, after which an equally slight thermal depression took place, corresponding to a similar one in the strata above.   In 1933, on the other hand, this level exhibited nearly as well as the one at 200 meters the phenomenon of declining temperatures from January to June, inclusive, followed by a decided rise.   At a depth of 300 meters and more, 1932 apparently was nearly, or completely, lacking in seasonal rhythm.   This was not the case in 1933.   In that year, the 300 meter level had about the same rhythm as the 200 meter level, and even at the greater depths, that rhythm could be discovered although it was characterized by a steadily decreasing amplitude towards the depths, until at the lowest levels hardly any traces of it could be discerned. Thus, briefly stated, the seasonal rhythm below the 100 meter depth was the same in these two years.   It was of moderate proportions in 1932 and therefore did not penetrate, or at least could not be clearly traced, below the 250 meter depth; in 1933 it was extreme and could be detected correspondingly deeper.

Another view of the seasonal phenomena is presented in Fig. 29, in which the vertical fluctuations of the temperatures in 1932 and 1933 are given.   The graph is constructed on the principles described in the beginning of chapter V.   The months are indicated by short, vertical lines at the top and bottom of the figure, and the two years are separated in a similar manner by somewhat longer lines.   The 8° water, as in Fig. 7, *etc.*, is symbolized in black, and the 15° and 16° are both without symbols and thus only separated by an isotherm. It was not judged necessary or advisable to indicate temperatures of less than 8° by special symbols; the space just below the 8° water is, of course, 7° water, the one next below that, the 6° water, and the lowest two, the 5° and 4° waters.   Since two whole years were included and the depths much greater than in the previous graphs, necessity demanded that both the horizontal time scale and the vertical depth scale be considerably reduced.   The depth from 750 to 900 meters was omitted on account of the fact that 3° water rose above the 900 meter level only on rather few occasions.

By comparing this graph with Figs. 10 and 11, it will be seen that the vertical fluctuations of the temperatures at station C and B agreed in all essentials in those two years, although those at the latter station were, on the whole, relatively somewhat more extreme. This difference should be expected in case the changes were due to the oscillations of wavelike bodies of water against the continental shelf.   Station B, as noted above, is located just inside the edge of the continental shelf.   However, while it is interesting and instructive to establish with full certainty that the seasonal rhythm was fundamentally the same over

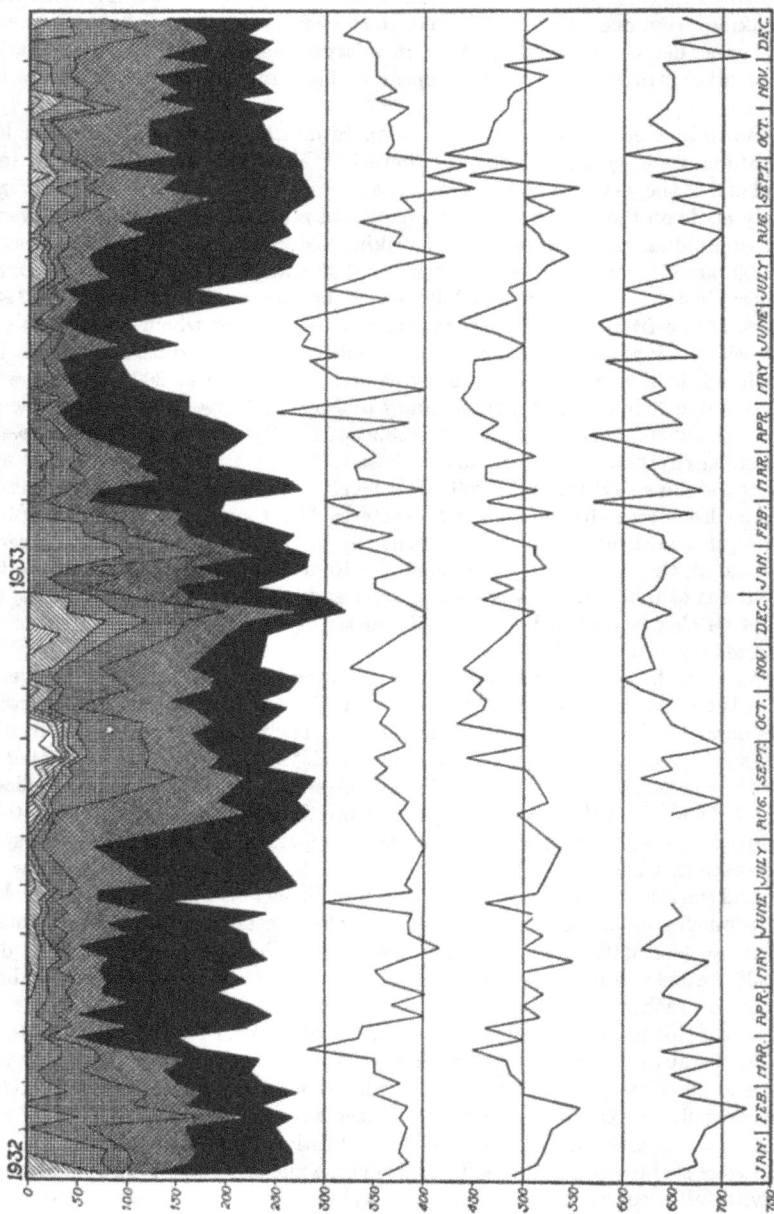

FIG. 29. Thermal conditions in the upper 750 meters at station C, in 1932 and 1933. For the interpretation of the graph, see section C of chapter VII.

the great depths as over the continental shelf, indeed even over the shallows next to the beach, the most significant features revealed by Fig. 29 pertain to the conditions below the 100 meter level during the different parts of the year.

The last aspect can probably best be visualized by first studying the seasonal progression of the 8° water (symbolized by black). In the beginning of 1932, this water occupied, roughly speaking, a stratum about 100 meters thick, around the 200 meter level. This thickness and general level was largely maintained until the middle of March when quite a sudden change in level, of about 75 meters, occurred. This second phase, which was of short duration, was followed by an increase in the volume of this water and by a gradual and slow sinking of its general level. This third phase came to an end in the latter part of July when the vertical extension of this water was about 175 meters and its general level about 160 meters. Following this phase, very radical changes took place; not only did the 8° water decrease its thickness to less than 100 meters, but it also subsided almost entirely below a depth of 200 meters. This state of affairs was attained by the middle of August. During the remaining part of 1932, the thickness of this water was roughly between 75 and 100 meters; in other words, it approximated the condition in the earliest part of the year. Its general level, on the other hand, showed a peculiar rise to within 200 meters of the surface; i.e., it suffered an elevation of about 50 meters. This situation was reached in the middle of October. After this secondary maximum, the general level fell to a depth of about 275 meters, that is, to the greatest depth in the year, indeed deeper than at any other time during the entire investigated two year period.

It should, of course, be noted that this description, as well as the one which follows, is held in the most general terms, and that as a consequence even some quite interesting details have been omitted. I am particularly referring to some remarkable changes of short duration in the thickness of this water layer. For instance, a very conspicuous fluctuation of this kind was recorded in the earlier part of February, 1932, when the thickness decreased suddenly from about 100 meters to approximately 50 meters, only to regain its previous dimensions shortly afterwards. This pronounced change evidently was closely associated with an equally sudden increase and decrease in the volume of the more superficial strata, a phenomenon which will be further discussed in the next chapter.

During the last weeks of 1932, the 8° water again rose, reaching a general level of about 200 meters, at the same time obtaining a thickness of approximately 75 meters. In other words, this water had at the beginning of 1933 the same general level and nearly the same bulk as at the beginning of the previous year.

In the first two weeks of January, 1933, the general level of the 8° water subsided to a depth of approximately 250 meters. This, however, was only in the nature of a temporary displacement. It was followed by a prolonged period of extraordinary elevation extending until the last week of May when the general level of this water was found within not more than about 60 meters of the surface. In other words, in the course of this period, the general level of this cold water rose, on the average, approximately 1.5 meters each day. The rate of elevation was, as may be expected, not uniform; however, the deviations from uniformity of motion were, on the whole, decidedly of a subordinate and rather insignificant nature.

In comparison, the rise of the general level of this water in the first part of 1932 was relatively sudden, approximating not less than three meters each day. On the other hand, the relatively high rate of motion was compensated for by the less extreme extent of the change due to its comparatively short duration.

In the last week of May, 1933, the general level of this water suffered a short depression to slightly below a depth of 100 meters. Its maximum elevation was nearly regained in the middle of June. This temporary break in the rhythm was caused by a sudden accumulation of water of superficial origin against the coast. From the middle of June until the middle of August, the general level of the 8° water subsided again to a depth of about 200 meters, the change of depth thus progressing at approximately twice the rate as during the rise in the spring months, *i.e.*, at about three meters per day. Finally, from the middle of August until the end of the year, this general level remained, broadly speaking, unchanged, that is around 200 meters, a level apparently characteristic for this water during the first and the last parts of the year.

In respect to the thickness of the 8° water in the course of 1933, the following few remarks must suffice. The changes which occurred were somewhat less sustained than in 1932; there were repeated fluctuations from a thickness of about 75 meters to one of around 200 meters. The latter value was found both in the beginning of March and around August 1st. During October, November, and December, the thickness averaged approximately 100 meters; that is, the condition was regained which evidently prevails ordinarily in the first and last months of the year.

The vertical fluctuations of the 7° water can perhaps best be described by emphasizing the trends of the lower isotherm of this water. From January until the middle of July, 1932, this isotherm was found, broadly speaking, around a depth of 375 meters. It should be noted, however, that in March it was lifted slightly above the 300 meter level, but this movement, which occurred simultaneously with the principal elevation of the 8° water, was not sustained, a fact already noted in conjunction with the discussion of Fig. 26. Another elevation took place in June although shorter and less extreme. From the end of July until the end of the year, there was a moderate rise and fall of the 7° water, a pulse of only slightly smaller magnitude than the corresponding one of the 8° water. In the middle of December, the lower isotherm of the 7° water was found at the 400 meter level. From that time until the middle of June, 1933, this isotherm rose in an irregular, but fairly consistent, manner to somewhat above the 300 meter level. The upper isotherm of this water, which even in the early summer months of 1932 did not penetrate, except on a few occasions, beyond the 200 meter depth, was for a short time in May, 1933, elevated to slightly above the 100 meter level. Between the middle of June and the middle of September, 1933, this water subsided, its lower limit reaching, in September, even as far down as 450 meters below the surface. Following this depression, another and somewhat less extreme elevation occurred lasting throughout the remaining months of the year. From this very general description will be seen that the 7° water rose and fell, in 1933, in a manner very similar to that of the 8° water, although the vertical extent of its motion was by no means so extensive. In 1932, on the other hand, the vertical shifts of the 8° water were mirrored by the 7° water generally speaking only during the latter part of the year. In the first seven months, with a couple of exceptions, the general level of the 7° water remained quite unchanged.

Concerning the thickness of the 7° water, it may simply be noted that, on the whole, it oscillated between 125 and 150 meters, only occasionally decreasing to 100 meters or somewhat less. The principal exception to this rule occurred in the late spring and the early summer of 1933 when this water reached the remarkable dimension of 200 meters, indeed, on one occasion of not less than 275 meters.

In these two years the 6° water occupied a position relative to the 7° water that resembled very much the one which the latter water occupied relative to the 8° water. The details of the changes may perhaps best be learned from a study of Fig. 29. It must suffice to note that possibly the most obvious deviation from the stated rule was recorded in the latter part of 1933 when the lower isotherm of the 6° water fell while the corresponding limit of the 7° water rose.

The lower isotherm of the 5° water was nearly always above the 700 meter level throughout the entire investigation period. Pronounced changes in level still were recorded. On the whole, this isotherm underwent changes similar to those of the lower isotherm of the 6° water. During the middle part (about 12 months) of the investigated period, its general level was somewhat higher than during the first six months, and during the last five months of 1933 there occurred a rise and subsidence.

In regard to the water of 4°, it may be sufficient to state that its lower isotherm usually was just below the 900 meter depth and that only on comparatively few occasions did it rise above this level.

VIII. WATER MOVEMENTS UNDERLYING THE THERMAL RHYTHM

As will be seen from the previous chapters, the most outstanding thermal characteristics of the waters of Monterey Bay were as follows during the investigated period, from the beginning of 1929 to the end of 1933.

1. There was a distinct annual rhythm of unusual and striking features (Figs. 12, 13, 29). This rhythm was expressed particularly by the following three phases:

   a. Low temperatures dominated the subsurface strata during the larger part of the year. Above the 100 meter isobath, the cold water period extended from the middle of February to the end of November.

   b. During the latter part of August, through September and the larger part of October, waters of quite high temperatures occupied the superficial strata.

   c. In December, January, and the first half of February, waters of very low gradients and of relatively high temperatures were found down to a depth of from 50 to 100 meters.

As a result of this fundamental and peculiar sequence of events, the annual temperature curves of the 50 and 100 meter levels possessed the unusual characteristics of having their minima in May and June, respectively, and their maxima in the beginning and at the end of the year.

The surface waters, on the other hand, which only to a lesser extent were influenced by this rhythm, had annual curves which expressed a distinct resemblance to the conditions prevailing in what may be termed the typical temperate regions of the northern hemisphere.

2. The thermal patterns at the 50 and 100 meter levels were reproduced much more closely from year to year than was the thermal situation in the surface waters.

3. The average temperatures for the body of water between the surface and 50 meters followed the conditions at 50 meters more closely than they did the superficial ones. Furthermore, the trends of the 50 and 100 meter values exhibited a very pronounced agreement.

4. In spite of the fact that the general thermal patterns of the 50 and 100 meter levels were reproduced from year to year, nevertheless, the vertical shifts of the temperatures of the cold waters above the 100 meter level varied decidedly from one year to another.

5. The seasonal rhythm, both at the surface and in the subsurface waters, to a depth of 100 meters, was more or less obscured throughout all the five years by secondary, interfering phenomena, distributed apparently at random at all times. These phenomena were not only decidedly erratic in their occurrence, but they were sometimes quite extensive in respect to the degree of change from the normal progress of the rhythm as well as of considerable duration. The latter feature evidently expressed water movements of comparatively large proportions.

6. From the point of view of changes as expressed in degrees of temperature, this region was found to be remarkably stable. On the other hand, if the frequency of thermal changes is considered, the opposite was true during these five years. The temperatures were found to be ever changing. Indeed, on most occasions it probably would have been impossible to duplicate a set of readings in the upper 50 meters within as short a time as one hour.

7. In some months of each year, the thermal fluctuations were more extensive at 50 meters, or even at 100 meters, than they were at the surface (Fig. 21).

8. The thermal gradients from the surface to 100 meters showed a seasonal progression closely simulating the one characteristic of the typical temperate zone of the northern hemisphere. However, it was brought forth in a manner nearly diametrically opposite to the one present in the temperate zone. The increase in the gradients from January to the late summer was caused only partially by a rise in the superficial temperatures. To a large extent, in the case of several months exclusively, it was due to a lowering of the temperatures at the deeper levels. Conversely, the lowering of the gradients towards the end of the year was caused not only by a decrease in the surface temperatures but also by a decided increase in the temperatures of the deeper strata. This general process, of course, was a corollary of the rhythm described above.

9. The seasonal progression of the gradients between 100 and 250 meters was very different from that characteristic of the upper 100 meters; indeed, it was nearly diametrically the opposite. In the former stratum, the lowest gradients were found in the early summer, while the January and February gradients were the steepest.

10. Seasonal thermal changes of an amplitude decreasing with depth were observed as far down as measurements were made.

11. Irregular, non-seasonal, fluctuations in the vertical distribution of the temperatures occurred constantly throughout the entire investigated body of water.

12. The thermal conditions even of the deepest strata differed from one year to another (Fig. 26).

The problem now arises: What factors caused, more or less directly, these thermal peculiarities of the Monterey Bay waters?

When we attempt to answer this question, we must, of course, first of all bear in mind that we are not faced with a single problem but rather with a complex of more or less independent problems, the features of which blend into an apparent entity. Some of these problems can be understood fairly well, at least partially, at the present time, while others will remain obscure to varying degrees pending further investigations.

Even a comparatively cursory review of the thermal characteristics embodied in the twelve points given above will suffice to show that they pertain largely to the subsurface waters and that they are of such a nature that they can not possibly be referred to changes in the insolation and to thermal interaction between the atmosphere and the water.

The only possible exceptions to this general rule are certain aspects of the surface waters, particularly the fact that the normal annual curve of the superficial temperatures exhibited a strong resemblance to the corresponding one of what may be termed the typical temperate regions of the northern hemisphere (1).* An additional feature to consider in this connection is that the superficial rhythm was more variable from year to year than the rhythm of the subsurface waters, a condition which indicates that the factors underlying the thermal changes at the surface were less sustained than those causing the changes in the deeper strata (2).*

It goes without saying that insolation profoundly influences the surface waters of this region, even though its efficacy is strongly reduced by the circumstance that during the months when the sun is highest, there are very frequent "high fogs," dense and thick clouds blanketing the coastal waters. Important factors in the regulation of the water temperatures unquestionably are also evaporation and radiation. In regard to the direct interaction between water and air it should be noted that cooling at and near the surface was established during the winter months, as will be seen from chapter IV. For instance, on one occasion, in December, the temperature at the surface was not less than 0.4° below that of the 5 meter level. Furthermore, due to the extremely low thermal gradients in the upper 100 meters during the winter months, the cooling process may have effects fairly far down. This will be evident from the fact that, even with our limited method of measuring this process, cooling from the surface was recorded as far down as 15 meters.

These facts notwithstanding, the thermal rhythm of the surface waters was evidently brought forth only to a limited extent by insolation and the direct thermal interaction between the water and the atmosphere. This statement is borne out by Fig. 30 in which the average monthly air and surface temperatures during the investigated five year period are given for the purpose of comparison. In this graph, the changes in the water temperatures are represented by a continuous line, and those in the air temperatures by a broken line. The months are indicated below by numbers. It should be noted that, although the temperatures of the air were taken at a station located only a few hundred yards from the beach, nevertheless they were undoubtedly somewhat extreme when compared with the conditions at stations 4 and B, where the water temperatures were recorded. In the following, the air temperatures are considered as an indicator of the amount of local insolation, even though this is justified only to a limited extent.

As will be seen from the data included in this graph, the differences between the water and air temperatures were, on the whole, very great, and during about six months each year the water exercised a cooling effect on the air. The correlation between the thermal rhythm of the local air and that of the water was fairly restricted and when it was quite pronounced, as in 1932 and 1933, it was apparent only, or perhaps better expressed, indirect. The justification for this statement may be obtained from the conditions which prevailed in 1931, in which year the correlation was very slight; note, for instance, the drop in the water temperature from February to April, as well as the one occurring in July. On the other hand, fairly striking and detailed changes in the air temperatures probably were connected with the water temperatures. For example, the temporary rise in the air temperature which took place in December, 1929, was presumably in part due to the concomitant rise in the water temperature. In regard to the two years, 1932 and 1933, in which the

* Refers to the point of this number in the list given above.

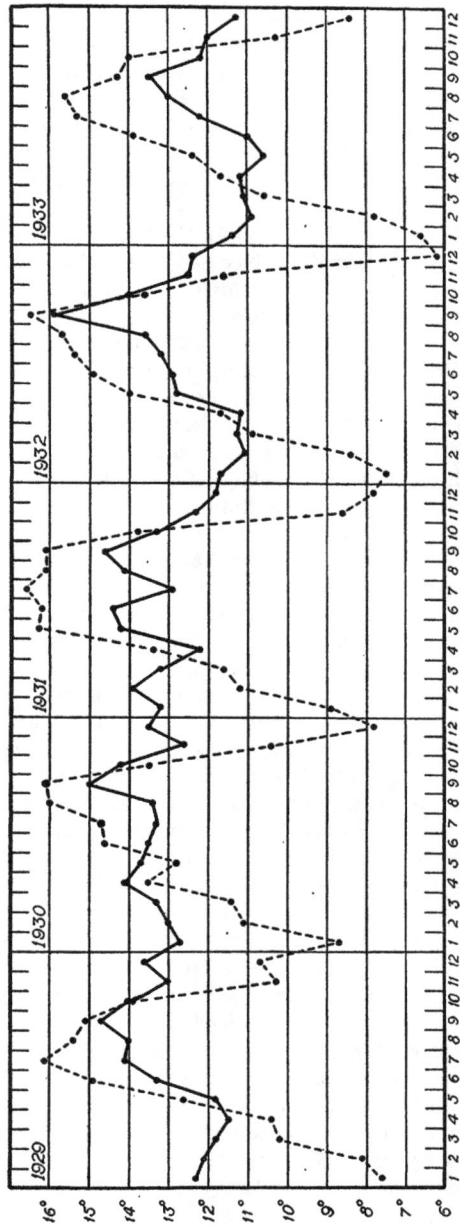

FIG. 30.   Monthly temperatures of the air (at Monterey) and of the surface waters (at stations 4 and B), from 1929 to 1933, inclusive. The air temperatures were based on the records of the U. S. Weather Bureau, station Hotel Del Monte, near Monterey, and reproduced by the courtesy of this organization.   For further explanation of the graph, see the introductory part of chapter VIII.

correlation between the air and water temperatures was the closest, it should particularly be noticed that the relation between the air and water maxima differed decidedly from one year to the next.   While in 1932, these maxima occurred in the same month and were very close, viz., 16.5° and 15.9°, in 1933, they were located in successive months and quite well spaced, viz., 15.6° and 13.5°.   In 1929, the air maximum was in July and that of the water in September.   These examples of differences in the time of the maxima illustrate the significant rule that while the air maxima shifted from one year to another between July and September, the water maxima were apparently always located in September (see also the last section of Chapter VI).   Finally, it may be mentioned in passing that by and large the general thermal level of the air temperatures was quite dependent on the general thermal level of the water, as will be seen from the following values representing the annual averages of the air and water temperatures.

	1929	1930	1931	1932	1933
Air...........	12.1°	12.6°	13.1°	12.2°	11.7°
Water..........	13.0°	13.5°	13.3°	12.7°	11.7°

If insolation and the local thermal interaction between the air and the water were only to a limited extent responsible for the thermal rhythm of the surface water, then what forces were behind this rhythm?   The answer to this question is the same as in the case of the remaining thermal features of this region: the thermal changes were brought forth by water movements which dominated not only the deeper strata but also the very surface. In the course of this investigation, it became overwhelmingly evident that the Bay is a bowl through which the water of the open sea is constantly passing, and moreover, passing on most occasions at a considerable speed.   The last statement is particularly well illustrated by the fact noted under the sixth of the twelve points, viz., that it was, as a rule, impossible to duplicate a set of temperature readings even within as short a time as one hour.

For the purpose of clarity in the presentation of these water movements, it may be advisable to deal first with the phenomenon noted under the fifth point of the above list, viz., the minor irregularities which were found in the annual rhythm of both the surface and deeper waters, down to the 100 meter level.   Afterwards, the remaining features will be discussed, as far as possible in the order which they occupy in the list.

In dealing with these irregularities, the tide will first be eliminated as a causative factor. Following the tidal section, the nature of the interfering water movements will be described and at least an attempt will be made to find their direct causes.

A. Effects of the Tides on the Thermal Rhythm: On examining the striking and curious rhythm which the hydrography of Monterey Bay displayed in the course of the five years, one naturally asks himself whether the established irregularities in the principal pulses, at least to some extent, may have been related to the tidal phenomena.   Even though on numerous occasions the changes were so pronounced as to suggest sudden and more or less complete replacements of the water within comparatively few hours, this question is fully justifiable, as will be seen from the following facts.   The waters off the central California coast are decidedly heterogeneous both in respect to their origin and their thermal features. The coastal water in the strict sense is usually more or less cold and either of deep or of northern origin.   Outside this belt, we meet the California Current, a branch of the Japan Current system.   Its waters, which also come from the north, are of varying temperatures but usually warmer than the coastal waters.   Finally, there are the oceanic waters of westerly origin and high temperatures.   The volumes of the coastal water and the California Current are subject to decided fluctuations.   As a consequence of the band-like arrangement of these waters and their constant changes in volume, the hydrographic picture may change more or less rapidly as one travels from the Bay in a westerly direction.   Even during the spring months, when the volume of the coastal waters in the strict sense is comparatively large, water about 4°–5° warmer than within the Bay may be located at a distance from the mouth of the Bay of only around twenty miles or even somewhat less.

For these and other reasons, it was judged advisable to present in this report a brief account of the tidal conditions in the Bay, although the tides were outside the scope of our program during the past years.

In this account, I shall first give some of the outstanding features of the tidal cycles; then follows a description of some drift experiments carried out in order to establish the extent of the tidal oscillations within the southern part of the Bay and the relation between the tidal phases and the tidal streams; and, finally, an example will be given to demonstrate the time relation between the hydrographic observations and the tidal phases.   It is, of course, fully realized that the drift experiments were of a preliminary nature.   However, in spite of their inconclusiveness, these experiments strongly suggest tidal stream conditions so complex and unusual that they not only fully justify their inclusion but also call for further experiments and careful theoretical considerations.   Indeed, so far I have not been able to find in the literature an account showing similar tidal stream conditions in any other part of the world, although in all probability they do exist.

The following condensed characterization of the tides in Monterey Bay is based on the tide tables of the Pacific coast issued by the U. S. Coast and Geodetic Survey.   In these tables, San Francisco (at the Golden Gate, where the tides are comparatively unaltered by local factors) is the tidal reference station of Monterey.

The Monterey tides agree in all essentials with those at the Golden Gate, but they are one hour and fifteen minutes earlier and 0.1 foot lower.   In other words, they are of the mixed type; i.e., there are on each tidal day two high and two low waters usually of more or less striking inequality.   The inequality is particularly expressed by the two low tides (Figs. 32, 34, and 35).   For instance, in Fig. 34, representing the tides of October 9, 1931, the two high tides are equal, while the low waters differ by 1.2 feet.   In Fig. 32 (September 29, 1931), both the high and the low waters are unequal; the former differ by 0.9–1.1 feet, and the latter by 1.7 feet.   In Fig. 35, the tidal curves for the first four months of 1929 are represented.   (For a discussion of this figure, see the last part of this section.)   On January 24, in this figure, the two high tides differ by 2.0 feet, the two low ones by not less than 4.5 feet.   Another important feature of the tides is that their mean range is 3.8 feet.   The mean diurnal range, i.e., the range between the higher high and the lower low tides, is 5.5 feet.   Finally, the maximum daily range is more than eight feet.   It should be noted that the datum to which the heights are referred is the mean lower low water, as is the case along the entire west coast of the United States.

From these facts it will be evident that the Monterey Bay tides are of an intermediate magnitude.   It may be noted in passing that they play a comparatively subordinate rôle in human activities for the following reasons: first, the foreshore is very narrow, due to the steep nature of the coast; and, secondly, on account of the wide mouth of the Bay, the tidal streams are not strong enough to interfere with shipping and with most types of fishing.

In the investigation of the problem of the tidal circulation in the Bay, the direct method of float experiments was employed rather than the hydrodynamical method. The reason for not applying the latter method is given in the next but last paragraph of chapter III. The construction of the floats is described in chapter III.

The floats were employed in two sets of experiments, one carried out on September 29, 1931 and the other on October 9, 1931. On account of the fact that the two sets yielded somewhat different results, each of them will be treated separately.

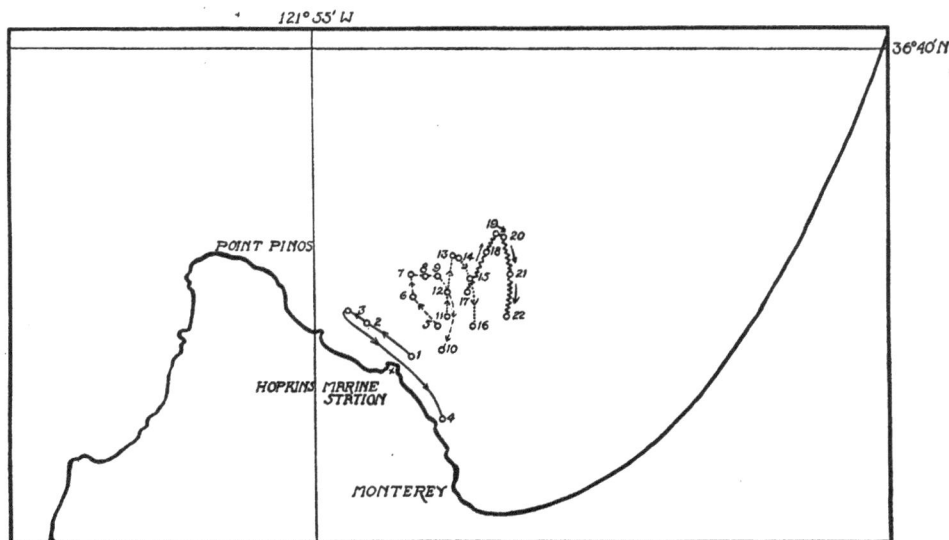

FIG. 31.   Drift of four surface floats in the southern part of Monterey Bay, on September 29, 1931.   For further explanation, see section A of chapter VIII.

On September 29, the air and the ocean were calm until noon. In the afternoon there was a light N.W. breeze and a fairly heavy westerly swell. Four surface floats were set out at and just following the maximum flooding of the tide. One of these was placed just outside the Hopkins Marine Station, the remaining ones serially in a N.E. direction, at about right angles to the southern coast line, the outermost at a distance from the coast of about one mile (Fig. 31). The float nearest to the shore moved out, toward the sea, at a speed of 0.30 mi./h. (Fig. 31, sta. 1–2); then it slowed down to 0.15 mi./h. (sta. 2–3) and came to a standstill (sta. 3) from 11.25 a.m. to 1.15 p.m., apparently due to the tide being slack and to the formation of a local eddy; finally, it began to move in toward Monterey, reaching an average speed of at least 0.27 mi./h. (sta. 3–4). The last speed is somewhat uncertain on account of the fact that the float was temporarily lost and when found it was caught at the edge of a kelp bed. The remaining floats all behaved mutually in a very similar manner. They first moved out toward the sea at a speed of 0.20–0.23 mi./h.; then they slowed down, coming to rest at the turn of the tide; after this they began to move in toward Monterey at a gradually increasing speed, the maximum speed (sta. 9–10, 15–16, 21–22) being 0.25–0.36 mi./h. At the end of the experiments, the speed was apparently decreasing again, evidently

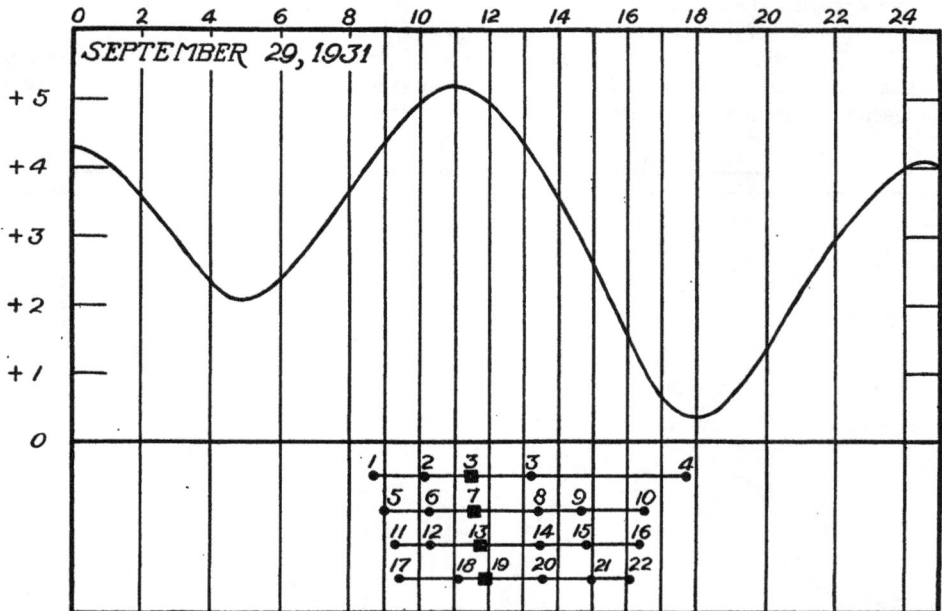

Fig. 32.   Relation between the tidal phases and the movements of the four surface floats employed on September 29, 1931, in the southern part of Monterey Bay. The horizontal time axis covers 25 hours, beginning at midnight, and the hours are shown above by figures. The height of the tide is given in feet on the left side of the graph; the height of the two low tides thus were + 2.1 feet and + 0.4 foot. The four horizontal lines in the lower section of the graph represent the time intervals during which the four floats were drifting. Their circles and squares represent the occasions when the positions of the floats were determined and the numbers correspond to the numbered stations in Fig. 31. The squares show the time when the movements of the floats were reversed. Example: Float no. 1 was set out at 8:40 a.m. and its position was next determined at 10:10 a.m.; it remained stationary from 11:25 a.m. to 1:15 p.m. at station 3; it was taken up at station 4 at 5:40 p.m.; and it reached its outermost position just following the turn of the tide which occurred at about 11 a.m.

in connection with the oncoming slack of the tide. It should be noted that all the floats were taken up only a rather short distance inside the position where they were set out.

The main results of this series thus were as follows:

First, there was no or very little lag in the tidal stream at the turn of the tide. This, of course, should be expected on account of the very wide mouth of Monterey Bay.

Secondly, the surface water moved out on the flood-tide and in on the ebb-tide (Fig. 32). This was contrary to expectations, and, as far as I know, no similar condition has been

previously recorded.   A possible explanation of this phenomenon may be that the bulk of
the tidal water came in and left the Bay along or fairly close to the bottom.   In other
words, the water moving in along the bottom during the flood-tide raised the water level
in the inner parts of the Bay, and thus caused the surface water to run off toward the sea.
During the ebb-tide, the situation was reversed.   In this case we must assume, however,
that the bottom water was banked against another body of water moving either in a
northerly or southerly direction.   In regard to the question as to why the bulk of the tidal
water should have moved in such a manner, no suggestion can be made at this time.

Thirdly, to all appearances the surface water exhibited in addition to the tidal oscillation
a slow movement toward Point Pinos from the inner part of the Bay.   This is indicated by

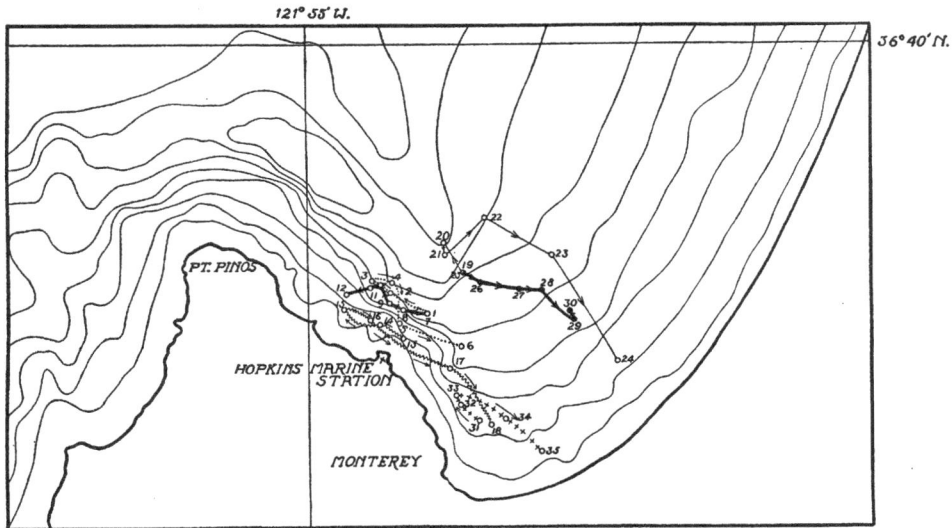

FIG. 33.   Drift of four surface floats, one 10 fathom and one 20 fathom float in the southern part of Monterey
Bay, on October 9, 1931.   For further explanation, see section A of chapter VIII.

the fact that the incoming flow was but slightly more extensive than the outgoing in spite
of the fact that the experiment covered a much larger portion of the ebbing of the tide than
of the flooding.   It may be noted in passing that this type of slow flow appears to occur
most of the time and that it is closely connected with the irregularities in the remarkable
rhythm under discussion in this report.

Fourthly, the entire tidal oscillation was of such a small magnitude that it could not be
expected to have had any appreciable effects on our records.

On October 9, 1931, the ocean was calm the entire day.   There was no wind until noon,
when a gentle west wind began which continued nearly till the end of the experiments.   Six
floats were set out (Figs. 33 and 34).

One of these floats was set out at the surface in the inner part of the Bay, about 0.5
mile from Monterey Harbor, at the time of high tide (Figs. 33, sta. 31; course indicated by

+ signs).   During the entire ebbing of the tide, it moved out at a slow speed (maximum, 0.15 mi./h.).   At the turn of the tide, it came to rest and then began to move in the opposite direction at a speed gradually increasing from 0.16 mi./h. (sta. 33–34) to 0.25 mi./h. (sta. 34–35).

Another surface float was set out at about 0.3 mile in a N.E. direction from the Hopkins Marine Station (course indicated by zig-zag line), approximately at the same time as the one treated in the last paragraph.   At first it moved out toward the sea at a speed of 0.31–0.29 mi./h. (sta. 13–15); it came to a standstill when the tide was about one-third out; then it began to move in towards Monterey at a gradually increasing speed (sta. 15–16, 0.07 mi./h.; sta. 16–17, 0.40 mi./h.); its speed was decreasing when the experiment ended (sta. 17–18, 0.34 mi./h.).   It should be noted that it did not change its direction at the turn of the tide from ebbing to flooding.

Two floats were set out simultaneously at about high tide somewhat outside the last one.   One of these, a surface float, behaved in a manner very similar to the one described in the last paragraph (course indicated by dotted line), although it did not attain quite so high a speed during the last part of the experiment (sta. 5–6; 0.29 mi./h.).   The other float (course indicated by heavy, unbroken line and open circles), was set at a depth of ten fathoms; it moved out toward the ocean during the entire period of the experiment regardless of the tidal phases.   It had its maximum speed during the beginning of the ebbing (sta. 7–8; 0.16 mi./h.) and its lowest speed at the lowest ebb (sta. 8–9, 0.13 mi./h.; sta. 9–10, 0.07 mi./h.; sta. 10–11, 0.05 mi./h.); when the tide began to flood, peculiarly enough, it did not continue to lose speed but actually gain in speed (sta. 11–12, 0.10 mi./h.).

The last two of the six floats were put out simultaneously at about high tide, at a distance of approximately one mile in a N.E. direction from the Hopkins Marine Station. One of these floats (course indicated by a fine, unbroken line) was set at the surface; it first moved out toward the ocean at a speed of 0.21 mi./h. (sta. 19–20); shortly afterwards it began to move towards Monterey at a slow speed (sta. 20–21, 0.09 mi./h.); then it moved in a N.E. direction at a decidedly higher speed (sta. 21–22, 0.27 mi./h.), after which it began to move toward the beach at a gradually increasing speed, regardless of the turn of the tide between the ebb and the flood (sta. 22–23, 0.39 mi./h.; sta. 23–24, 0.59 mi./h.).   The other float (course shown by heavy, unbroken line and solid circles), set at a depth of twenty fathoms, moved onshore during the entire ebbing of the tide at speeds varying between 0.12 and 0.21 mi./h.; at the turn of the tide at the ebb, it began to move out toward the sea at a very slow speed (sta. 29–30, 0.03 mi./h.).

The chief results of this series thus were as follows:

First, in the inner region of the Bay, just outside Monterey Harbor, the tidal streams coincided fairly closely with the tidal phases (Fig. 34), the water moving out at the surface during ebbing and in during flooding, without any distinct lag at the turn of the tide.   To all appearances, this region was outside the general, non-tidal circulation.

Secondly, outside this inner region, at a depth of ten fathoms, the water appeared to have a consistent, outward drift.

Thirdly, at a depth of twenty fathoms, in the outer region, the water moved according to the tidal phases (Fig. 34), but in directions opposite to those that should be expected; i.e., it moved in during the ebb-tide and out during the flood-tide.

Fourthly, in the outer region the surface water was decidedly out of rhythm with the tidal phases (Fig. 34), which is remarkable in view of the wide mouth of the Bay.   It moved in during the larger part of the ebb-tide and did not turn at the turn of the tide.

Fifthly, the different water strata moved decidedly independently of each other.

Sixthly, the tidal oscillations were by no means large enough to justify an assumption that the tide could have appreciably affected our records.

Considered together, these two sets of experiments demonstrated first of all that the tidal oscillations in Monterey Bay were of such a minor amplitude that they could not have

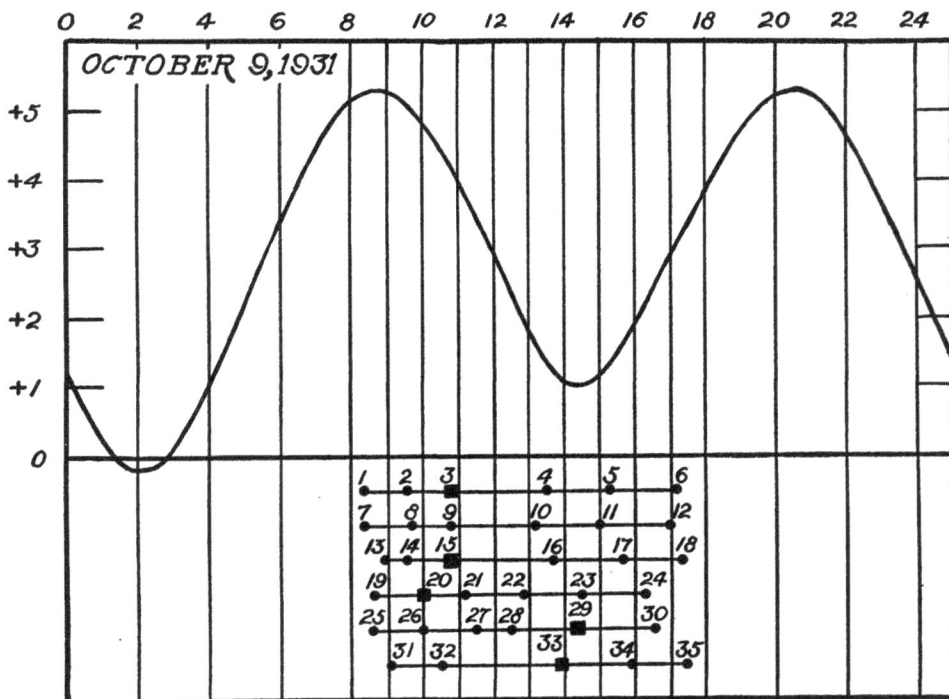

Fig. 34.   Relation between the tidal phases and the movements of the six floats employed on October 9, 1931, in the southern part of Monterey Bay.   For the explanation of the graph, see the explanation to Fig. 32.

affected our thermal records to an appreciable extent.   In other words, the irregularities characteristic of the annual thermal rhythm can not be referred to the tidal oscillations.   In addition, the experiments revealed that the tidal circulation in the southern part of the Bay is not only very complex but also variable.   The variations are probably due not only to the characteristic rhythmic changes inherent in the nature of the tides but also, and to a large extent, to the constantly recurring changes in what may be termed the more funda- mental circulation in the Bay.   This circulation, as is noted in the next section of this chapter, is partly in the form of large and apparently mostly open eddies, in which the water moves either in a clock-wise or in an anticlock-wise direction, depending on the direction of the water movements outside the Bay, and partly in the form of a general northerly or southerly drift.   Furthermore, the tidal streams, as should be expected from the open nature of the Bay, are usually not very strong, the maximum speed probably being approximately one mile per hour.   The maximum speed observed in the course of the experiments was, as will be seen from the above data, 0.59 miles per hour.   It should finally be noted that the peculiarity of relatively thin water strata moving in different directions from each other seems to be characteristic of this general region.   Indeed, in some places of this neighborhood, this phenomenon is so pronounced that it causes considerable difficulties to the fishermen when they set out their very deep sardine nets.

In order to establish the time relation between our routine hydrographic observations and the tidal phases, a graph was constructed representing the main features of the tidal conditions in Monterey Bay during the first four months of 1929.   The employed data are the predictions in the tide table for the Pacific coast issued by the U. S. Coast and Geodetic Survey.   In this graph (Fig. 35) the time scale is horizontal and the heights of the tide (in feet) are represented on the vertical scale.   The four tidal predictions, *i.e.*, the higher and lower high and the higher and lower low waters, were marked on the graph on each of the days of this four month period.   In this manner the general monthly trends came to be represented by four curves, two high and two low water curves.   Heavy, vertical lines were drawn to represent the days on which hydrographic work was undertaken.   Each of these lines joins two of the four tidal curves, indicating in this manner between which tidal extremes the observations were made.   The arrows on these vertical lines show the directions of the tidal movements, and the dots, surrounded by circles, indicate the height of the tide at the time of observation.   For instance, on January 7, the observations were made at a + 1.9 foot ebb-tide, between the higher high and the lower low tide; and on January 17, they were made at the time the tide began to turn from the higher low to the lower high tide.

The method of calculating the height of the tide at the time of taking the hydrographic data is based on the assumption that the tidal rise and fall correspond to a simple harmonic motion (see Harvey, 1928, p. 89).   Whether this is justified or not is, of course, not established since the actual tidal conditions in Monterey Bay have not as yet been subjected to a detailed analysis.   The application of this method was assumed to be permissible because Monterey Bay is very open and slopes into deep water at its mouth, thus but a few miles from the places where our key stations were located.   Some distorting elements probably do occur, but it should be noted that for several years the official predictions of high and low tides have been observed at Hopkins Marine Station to follow the actual conditions very closely.

As will be seen from this graph, the hydrographic observations were made at times distributed in relation to the tides in a perfectly random manner.   Thus an equal number of observational series (16 out of 32) were taken on the ebbing and the flooding tides, and 15 series were taken above and 17 series below the + 3.0 foot level.   Lastly, and most important, if the data presented in this graph are applied to Fig. 7, in which the thermal conditions in the southern part of Monterey Bay during 1929 are pictured, it will readily be established that the striking hydrographic changes which occurred in this region during the first four months of that year were in no manner connected with the tidal phases.

It should also be noted that a series of twenty-five hourly observations, thus covering a whole tidal cycle, undertaken on December 6 and 7, 1930, fully substantiated the conclusion drawn from the facts presented in this section, *viz.*, that the irregularities in the hydrographic rhythm were not markedly affected by tidal phenomena.

B. Water Movements Underlying the Irregularities in the Thermal Rhythm and the Causes of these Movements: In the previous section it was demonstrated that, as far as available evidence goes, the irregularities in the hydrographic rhythm of the superficial

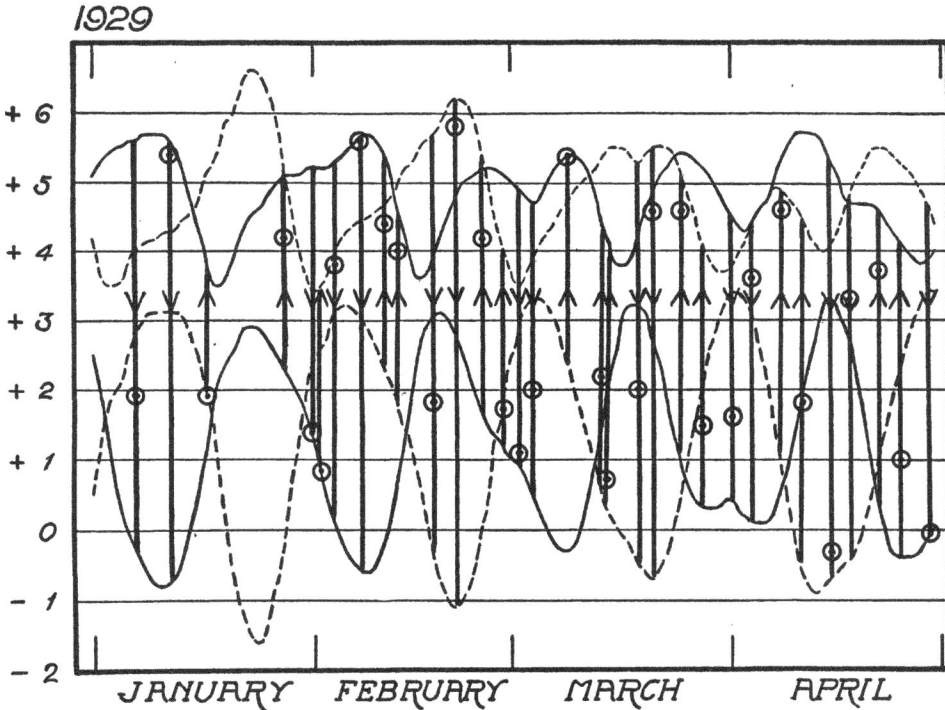

Fig. 35.   Relation between the tidal phases and the times at which the routine hydrographic observations were made during the first four months of 1929.   For the explanation of the graph, see the last part of section *A* of chapter VIII.

strata which characterized Monterey Bay in each of the five investigated years were not, at least appreciably, affected by the tidal oscillations.

The question now presents itself: If the irregularities were not caused by the tides, then of what nature were the water movements which underlay them?

The answer to this question was partly indicated, in passing, in the previous section. Through the float experiments it was established that, superimposed upon the tidal oscillations, there occurred, on September 29 and October 9, 1931, a slow and consistent movement of the superficial strata.   Furthermore, all available evidence suggests that such a slow water movement is characteristic of the Bay at all times, or nearly so, and that it is in the form of large and apparently mostly open eddies moving either clock-wise or anticlockwise, depending on the direction of the current just outside the Bay, and partly in the form of a general northerly or southerly current.

In order to illustrate the nature of this slow water movement, a few hydrographic profiles made in the course of 1930 in the southern portion of Monterey Bay (station line no. 2) will be presented and discussed. It should be noticed that during 1929 and 1930 a very large number of such profiles were made along this line. The results of these profiles, which can not be published for reasons of economy, agree fully with those obtained from the ones to be discussed.

As will be seen from chapter III, station line no. 2 was located across the S.E. corner of Monterey Bay. Its length was approximately seven miles. At its southern end (to the left in the profiles), the shore is rocky, and the bottom slopes steeply to the deepest part of the section, about 60 meters, from where it gradually rises in an irregular manner towards the sandy shores at the northern end. On this line, seven stations were established. The total number of observations (solid circles in the graphs) made along this line was 45, and these were fairly uniformly spaced between the surface and the 50 meter level.

Figs. 36–8 represent the thermal changes that took place along this section from November 26 to December 1, 1930. These profiles were chosen as the first examples not on

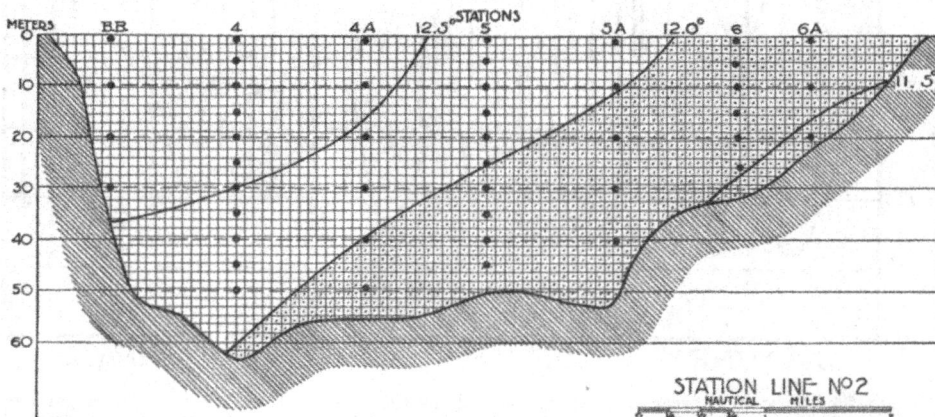

FIG. 36.   Hydrographic section, along station line number 2, in the southern part of Monterey Bay, on November 26, 1930.   For a discussion of the figure, see the beginning of section B of chapter VIII.

account of the fact that they demonstrated more pronounced changes than usually occurred. Their choice was simply caused by the circumstance that they illustrated the changes very clearly, because their isotherms had a very decided slant instead of being more or less horizontal as was usually the case in this region.

On November 26 (Fig. 36), a moderate portion of the southern end of the section was occupied by water of 12.9°–12.5°; and the rest of the section was filled with cooler water, the coldest being less than 11.5°. The 12.0° isotherm divided the section into two nearly equal halves. Two days later (Fig. 37), water of less than 12.5° occupied only a small portion of the section, *viz.*, along the bottom between stations 5 and 6A; and there was no water of temperatures less than 12.0°. The southern corner at this time contained water of more

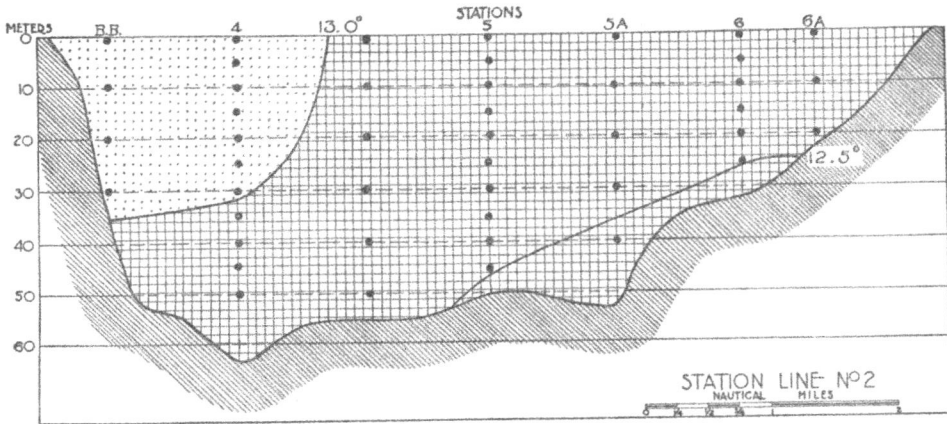

FIG. 37. Hydrographic section, along station line number 2, in the southern part of Monterey Bay, on November 28, 1930. For a discussion of the figure, see the beginning of section *B* of chapter VIII.

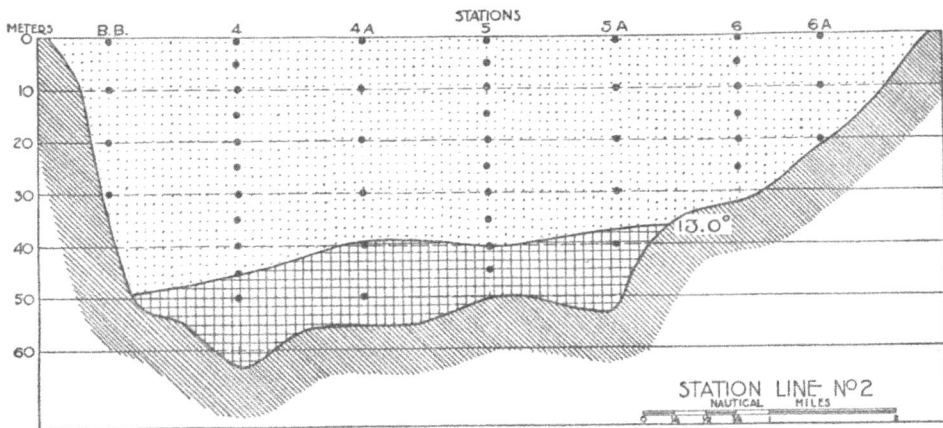

FIG. 38. Hydrographic section, along station line number 2, in the southern part of Monterey Bay, on December 1, 1930. For a discussion of the figure, see the beginning of section *B* of chapter VIII.

than 13° reaching a maximum depth of 35 meters. Finally, on December 1 (Fig. 38), the 13° water occupied by far the larger part of the section, and water of slightly less than 13.0° was restricted to near the bottom in the deepest portion of the section.

From this brief description it will be evident that during the short interval of less than a week, the hydrographic conditions of the southern end of the Bay changed fundamentally. Indeed, it is unquestionably fully justifiable to conclude that there was a complete exchange of water in the region between November 26 and 28, as well as between the latter day and December 1. In addition, it is evident from these sections that the general direction of the drift was from the south to the north, in an anticlock-wise manner. It should also be noted that the drift, although eddy-like, was of an open nature. This is evident from the fact that the conditions characteristic of November 26 did not return even in an approximate manner. In further support of this statement, it should be noted that, following December 1, the hydrographic picture changed entirely.

Figs. 39–44 illustrate the thermal changes in the southern part of Monterey Bay from July 14 to July 31, 1930. On July 14 (Fig. 39), the region below 30 m. was occupied by

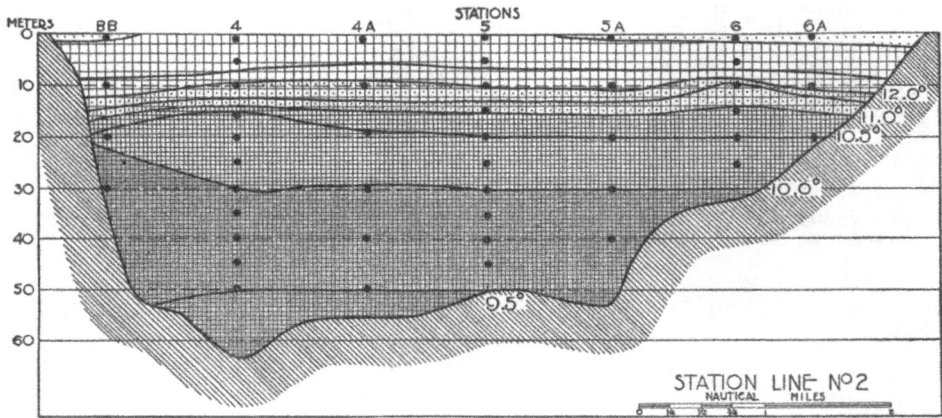

FIG. 39. Hydrographic section, along station line no. 2, in the southern part of Monterey Bay, on July 14, 1930. For a discussion of the figure, see the beginning of section B of chapter VIII.

water of 9°; above this, water of 10° formed a layer about 15 m. thick; above a depth of 15 m. the temperatures were largely between 11.0° and 12.9°, but a thin surface layer of 13° water was found both in the southern and northern parts of the investigated section. On July 17 (Fig. 40), colder water, of 8°, covered the bottom in the deepest part of the section, below 50 m.; the 9° water had swelled until it had reached on the average the 20 m. level; the layer of 10° water had shrunk to an average thickness of less than 10 m.; while 11° water extended to the surface in a large portion of the southern end of the section, 12° water had maintained itself at an average thickness of 10 m. at the surface of the northern half; and the thin, superficial 13° water had disappeared. On July 21 (Fig. 41), the 12° water which previously was restricted to the northern half of the section had swelled; it now occupied the entire surface and had an average thickness of slightly more than 15 m.; the 11° and 10° waters had also increased in bulk, forming a stratum about 20 m. deep, and the 9° water, on the other hand, had decreased correspondingly; it filled the deeper portion of the section, below about the 35 m. level, the 8° water having disappeared entirely. This general trend had progressed further on July 24 (Fig. 42). On that day, water of 14° had entered along the surface of the northern half of the section, reaching the maximum thickness of about 10 m.; 13° water extended over the entire section, with an average thickness of 10 m.; the

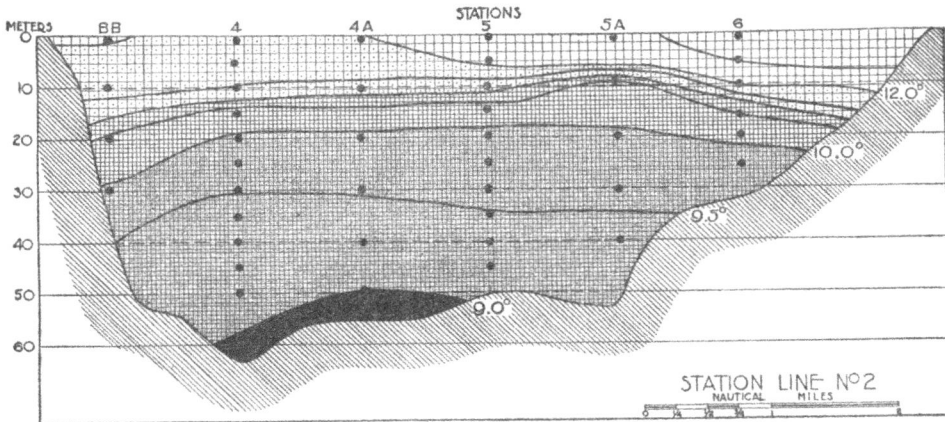

FIG. 40. Hydrographic section, along station line no. 2, in the southern part of Monterey Bay, on July 17, 1930. For a discussion of the figure, see the beginning of section *B* of chapter VIII.

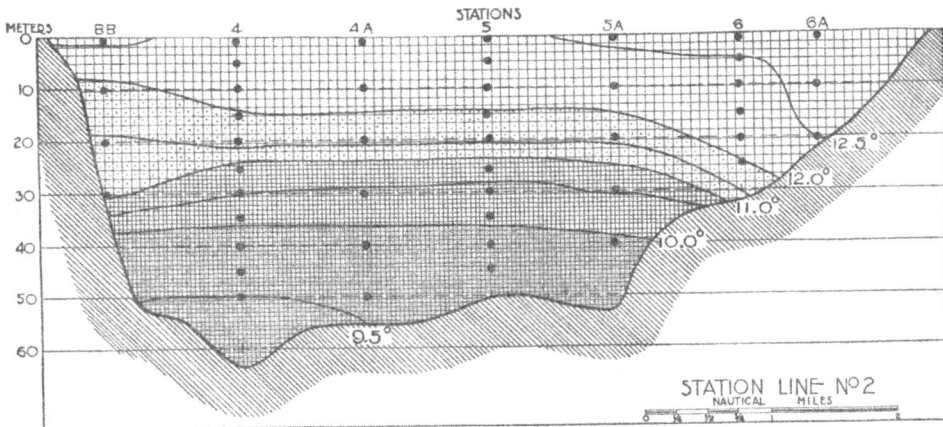

FIG. 41. Hydrographic section, along station line no. 2, in the southern part of Monterey Bay, on July 21, 1930. For a discussion of the figure, see the beginning of section *B* of chapter VIII.

12°–11° water occupied the bulk of the region, having an average thickness of nearly 40 m.; the 10° water filled only the very deepest part, below 50 m., and the 9° water had disappeared completely. On July 28 (Fig. 43), the hydrographic conditions were to a certain extent reverting toward an earlier state although a complete reversal was not realized. The 14°

water, which previously was in the northern half of the section, now was found in the southern half where it occupied about the same volume as before; the cold bottom water had increased, 9° water filling the deepest region, below about 45 m., and the 10° water extended

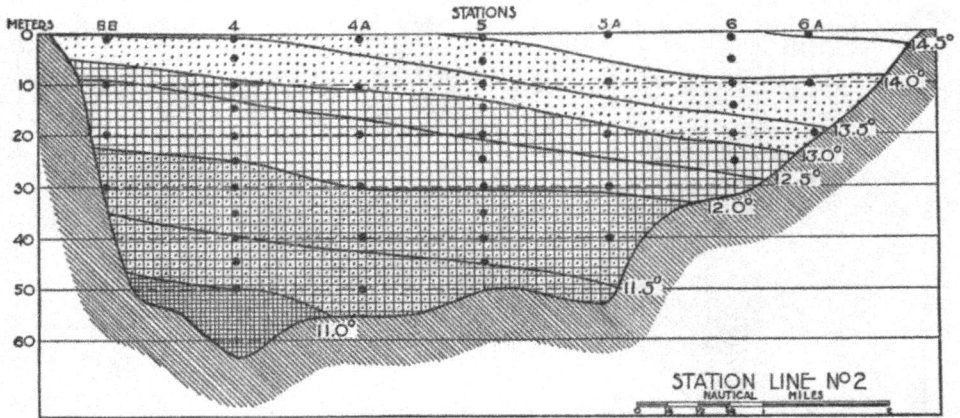

FIG. 42.   Hydrographic section, along station line no. 2, in the southern part of Monterey Bay, on July 24, 1930.   For a discussion of the figure, see the beginning of section B of chapter VIII.

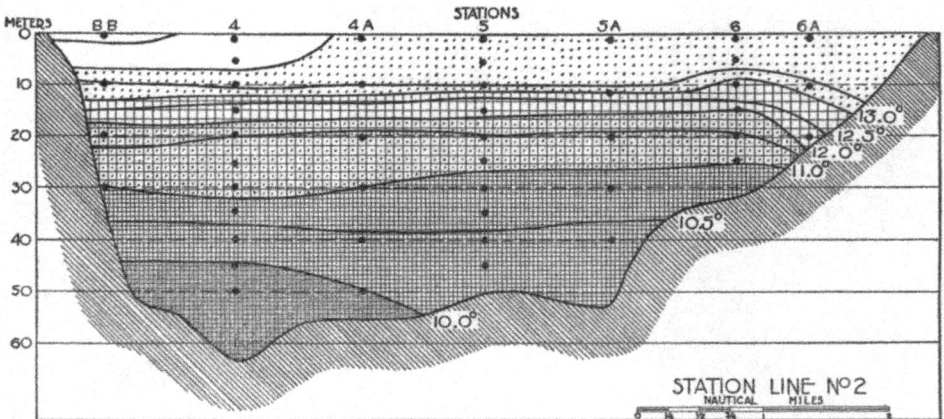

FIG. 43.   Hydrographic section, along station line no. 2, in the southern part of Monterey Bay, on July 28, 1930.   For a discussion of the figure, see the beginning of section B of chapter VIII.

all over the section below 30 m.   This increase in the cold water was continued in the profile of July 31 (Fig. 44).   On that day the 9° water had about the same upper limit as the 10° water had on July 28; the 10° water had an average thickness of about 15 m.; and since the surface was almost entirely covered with water of 14°, attaining an average depth of about 5 m., the strata of intermediate temperatures (11°–13°) were very thin and thus presented very steep gradients. Even 15° water occurred in the southern end of the section.

From a comparative study of these profiles, it will be seen that during the latter half of July, 1930, the water in the southern part of Monterey Bay was in a state of constant flux. We can again infer that the water was renewed nearly every day, either entirely or at least to a very large extent. It was indeed transient; no water that may be termed endemic was probably ever present during the investigated period. The water movements were accompanied by decided variations in the relative thickness of the different thermal strata. As a consequence, the thermal gradients sometimes were quite gentle and sometimes very steep, forming fairly pronounced thermoclines. The general direction of the water movements is very difficult to establish with certainty from the given data. It should be noted

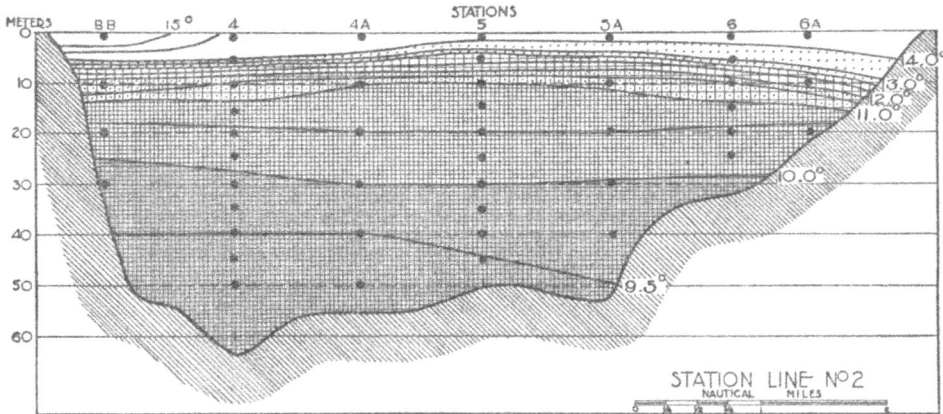

FIG. 44.   Hydrographic section, along station line no. 2, in the southern part of Monterey Bay, on July 31, 1930. For a discussion of the figure, see the beginning of section *B* of chapter VIII.

that, as far as available evidence goes, we are concerned with movements caused by pressure from the region outside the Bay, not with flows brought forth by hydrodynamic gradients within the Bay. Dynamic calculations would thus not produce the correct answers to the question. A comparison of the various profiles indicates, however, that the direction was variable: sometimes clock-wise, from the north to the south, thus opposite to the one prevailing from November 26 to December 1, 1930, and sometimes anticlock-wise. The flow probably was in the form of one or two open, eddy-like curves. There was a certain tendency from July 24 to July 31 to revert to the condition of July 14, but the reversal was not complete and probably did not indicate a return of water which had previously left the Bay.

It should be noted in this connection that the changes in the direction of this type of flow could be observed from the shore, without the employment of special apparatus, throughout the larger part of the year when the Monterey canneries emptied quantities of fish offal into the Bay water. Sometimes the refuse-laden water trended toward the north from Monterey, along the inner sandy shores; sometimes it moved westward, toward Point Pinos, and out into the open sea.   In addition, this flow was often marked by characteristic and well defined current rips.

The exchange of water was also demonstrated by the ever changing composition, qualitative as well as quantitative, of the local plankton.   These changes were usually altogether too sudden and complete to be accounted for in any other manner than by the replacement of the entire population.   The plankton also demonstrated in a conclusive manner, by the differences in its composition, the relative independence of the various eddy-like regions, as well as the variations in the relative sizes of these.   Thus, for instance, on one occasion, when "red water" was observed, the sharpness of the demarcation between two eddy-like bodies was expressed by the extremely sudden transition between the "red" and the "normal" water.   Moreover, the fact that the taxonomic composition of the plankton of the different places in the Bay sometimes remained mutually different but quite similar at each locality for a considerable time indicated that the eddy-like bodies, at least on these occasions, remained relatively unaltered for prolonged intervals.   For example, for quite a long period, dinoflagellates predominated, for some unknown reason, in the micro-plankton in the N.E. part of station line no. 2, while at the same time diatoms prevailed in the waters just off the Hopkins Marine Station.

Lastly, it should again be emphasized that the water movements were greatly compli-cated by the circumstance that the various strata moved in different directions, a fact established in the course of the drift experiments.   This peculiarity was in all probability rather the rule than the exception.

Comparatively little can be said with absolute certainty in regard to the factors which caused these water movements in Monterey Bay.   In order to obtain a definite and well founded knowledge in this field, intensive investigations, both hydrographical and meteor-ological, must be carried out within a large area of the California coastal and off-shore regions.   The following remarks must suffice at present.

As will be seen from a later section of this chapter (Fig. 45), it is a matter of knowledge that there is in central California a coast-wise current moving, by and large, in a southerly or in a northerly direction, the former direction predominating.   A change from either of these directions to the other will, of course, result in a corresponding change in the circulation in Monterey Bay.   The open, eddy-like flow, noted above, represents simply a more or less deep bend in a side-branch of the general coastal flow.   In the case of closed eddies, on the other hand, the movement is clock-wise with a general northerly current and anticlock-wise with a general southerly current.   This relation between the circulation in an open bay and the general coastal currents is of course evident from the elemental principles of hydro-dynamics.   From these summary statements, it will be evident that the presence of and the fluctuations in a general coastwise water movement are at the root of the changes in the current system of the Bay.

The changes in the direction of the coastwise current, in their turn, evidently are closely connected with the changes in the direction of the prevailing winds.   It is the experience among the practical seafaring men of this region that the turn of the wind from the northerly to the southerly quarters, and *vice versa*, will bring forth a corresponding change in the direction of the coastwise flow of the water; see also below, the section treating of the Davidson Current.   It may also be noted in this connection that in the course of the several years of the Hydrobiological Survey it became increasingly clear that the directions of sustained and strong winds have a pronounced effect on the temperatures of the water. Thus, for example, if, following a day of low temperatures, a southerly or south-westerly storm occurred for some days, then we could expect with nearly complete certainty that the temperatures of the water would be raised more or less markedly. However, the currents were, by no means, always connected with the local winds. This was in agreement with previous observations. In the United States Coast Pilot for the Pacific (California-Washington), 1926, we find the remark pertaining to this general region, that "strong currents are sometimes experienced when the local winds are light." Several investigators, *e.g.*, Dawson, 1909, have pointed out that the current is sometimes found to run more strongly before the heavy wind comes on.

The water which the forces noted above caused to pass through the Bay was undoubtedly, at least in most cases, of coastal nature in the strictest sense of that term. However, in addition there was another type of water which intermittently entered the Bay, thus interfering with the regular unfolding of the normal rhythm. This water apparently represented tongues or detached lateral eddies of the inner margin of the California Current, and thus was of truly oceanic origin. On the occasions when this happened, the Bay water was usually characterized by its bluer color, higher transparency, and by its relative poverty of microplankton. Also, macroplanktonic forms, for instance Coelenterata, which otherwise were not present within the confines of the Bay, made their appearance on those occasions. Sometimes, when this type of water arrived in the Bay, it was evidently more or less decidedly mixed with water of coastal origin. The causes of these irregular entries of water of the California Current are by no means clear. However, there is at least the possibility that, in addition to the primary factor of strong westerly winds, fluctuations in the bulk and rate of flow of the California Current may be implied as causative agents of this phenomenon, but no definite evidence along this line is as yet available; see section 3 of this chapter.

It should perhaps also be noted that some of the minor irregularities call to one's mind the great, tide-like submarine waves which have been described and analyzed by O. Pettersson in a number of publications. Submarine waves of this kind may, of course, exist in our waters, although they have not been discovered so far, due to lack of investigations along this line. However, the chances that this phenomenon was behind most of the minor irregularities, or even of a great number of them, are very remote, on account of the fact that a pronounced discontinuity layer was never found, and, naturally, this type of water movement depends on the presence of water layers of very different densities superimposed upon each other. It should be called to attention, however, that this region apparently is characterized by strata moving in different directions more or less independently of each other, and it may well be that the submarine wave phenomenon occurs between these strata.

To summarize the discussion up to this point: The water of Monterey Bay is in a state of constant flux. Prominent features of its circulation are that the coastal water coming either from the north or from the south tends to pass through the Bay, to a very large extent under the influence of the winds, and that water from the inner margin of the California Current intermittently enters the Bay.

When taken by themselves, these facts would not of necessity suffice to account for the constant thermal changes characteristic of the Bay waters. In order to understand the thermal fluctuations, we must take into consideration also the circumstance that the coastal as well as the offshore waters of the superficial strata are very "spotty," in other words, very heterogeneous from the point of view of their physico-chemical qualities.   The decided heterogeneity of the surface waters along and off the coast of California can readily be demonstrated.   Even within the short distance of ten to twenty miles, the investigator may find quite decided thermal differences in the surface waters.   As a consequence, the "spottiness" is a matter of general knowledge.

We thus reach the result that the frequent passage of these thermally heterogeneous waters through the Bay was the cause of the irregularities in the unfolding of the fundamental rhythm of the superficial strata described in the previous chapters.

As a concluding remark to this section, I may draw the reader's attention to the fact that the intertwining, if such a term may be used, of these water movements and those that underlie the fundamental rhythm creates an enormously complex situation of flow.   For instance, when the cold and highly saline, i.e., very heavy, deep water is lifted towards the surface during the "cold water phase," into the realm of lighter strata, a dynamical instability is created which must be released largely through horizontal water displacements.

C. Water Movements Underlying the Annual Rhythm: As we have seen above, the annual thermal rhythm during 1929–33 embodied three major features, *viz.*, the cold water phase, the warm water phase, and the low thermal gradient phase.   Each of these will be treated separately, and since the cold water phase was by far the most dominant, it will be discussed first.

1. The Cold Water Phase: The fact that the water temperatures in this general region are abnormally low, considering the latitude, has been well known for a comparatively long time.   The relatively early discovery of this abnormality is not surprising, in view of the magnitude of the deviation.   According to Fig. 23 in Bigelow and Leslie (1930), which was based on calculations by G. F. McEwen, of Scripps Institution, the normal surface temperature of the Pacific for the latitude of Monterey, in August, during the period 1919–28, was not less than about 21°, while the corresponding normal for Monterey Bay was about 13°.   Following its discovery, the cold water phenomenon has been subjected repeatedly to discussion, as well as to theoretical analysis.

Whenever abnormal temperatures exist in an oceanic area, it is an absolute indicator of the fact that water from an outside source is coming into, or rather passing through, this area.   A thermal dislocation in a region is thus always a proof of either horizontal or vertical transportation of water.   This follows irrefutably from the fact that if the water were stationary, it must of necessity assume the thermal characteristics of the latitude of the area.   On this point then, all the authors agree; but in respect to the nature and causes of the water movements, they are more or less at variance.   On account of the broad application and the general interest of the cold water phenomenon, a brief review of the ideas proposed by some of these authors may be desirable at this place.

According to Dall (1882, 1904), who was the first to discuss this question, the cold water along the California coast comes from the north and is derived from the originally warm Japan Current which is cooled during its passage through the high northern latitudes. Richter (1887), on the other hand, maintained that the subnormal temperatures along this coast were due to an arctic current, running inside (east) of the Japan (California) Current, which was kept away from the coast by the strength of the flow of the arctic water.   Richter did not attempt to trace the original source of the arctic water beyond denying that it at any time had belonged to the Japan current system. As early as 1899, Andrees proposed a deep local origin of this water. The further tracing of this water was attempted by Bishop (1904) who suggested that the cold water was of an antarctic origin, water which had sunk towards the bottom in the high southern latitudes, afterwards flowing towards the north and east until it reached the coast of North America, between the southern part of Alaska and the northern part of Washington, where it was deflected towards the surface. Holway (1905) introduced a somewhat different idea into the discussion. He, too, supposed the cold water to be derived from the depth. However, he proposed a different place for the upwelling. On account of the fact that that region had the lowest temperatures, he proposed, as the principal focus for the upwelling, the neighborhood of Cape Mendocino, in California, lat. 40° N., thus a place considerably farther to the south than the one suggested by Bishop. Holway's opinion was accepted by Thorade (1909), who had made the most extensive compilation of surface temperatures along the California coast up till that time.

At present the idea of an upwelling of deep water is generally held, and its truth may be considered to be beyond doubt. It may also be interesting to note that the latest investigations strongly support the supposition made by Bishop that the abyssal water of the Pacific is of an antarctic origin (south of the Indian Ocean; see Sverdrup [1931]), even though there is very justifiable doubt as to whether the water that wells up along the California coast is of a truly abyssal origin. It should furthermore be noted that the upwelling process nowadays is considered to take place along a fairly extended part of the coast and that its local variations are connected with the presence of submarine valleys and other irregular features of the continental shelf (Holway, 1905). Also, J. Murray (1899) demonstrated that the temperatures are lower somewhat away from the coast than over the continental shelf, indicating that the upwelling may be most intense at some distance from the coast line.

However, even though the local upwelling phenomenon is largely responsible for the subnormal temperatures at any one point along the California coast, it is not the only causative factor. There is, as we know, during the larger part of the year, a southerly movement of the coastal water sometimes traceable at least as far north as to Vancouver, B. C. This also contributes towards the same trend as the local upwelling in Central California, partly, of course, on account of its upwelling origin, but partly also due to its coming from higher latitudes.

In regard to the causes of the upwelling phenomenon, different views have also been presented. Andrees (1899) maintained that prevailing easterly winds drove the surface water away from the coast and that, in compensation, the deeper waters were raised towards the surface. Hann (1903) added to this idea the suggestion that an offshore deflection of a coastal current would bring forth the same results. The argument of Andrees, however, must be entirely superseded by that of Hann, in view of the fact that the prevailing winds in the coastal belt of California are not easterly, as assumed by Andrees, but northwesterly, as was first emphasized by Thorade (1909). The offshore deflection of the current, suggested by Hann, must occur since the movement of the coastal water is, generally speaking, southerly and the deflection of the currents on the northern hemisphere under the influence of the rotation of the earth is to the right (Ekman, 1906). The deflection thus causes a general lowering of the surface near the coast which must be compensated for in the manner suggested by Andrees. Bishop's (1904) suggestion, referred to in the preceding paragraph, should be remembered in this connection, namely that the rise of the deep water against the North American coast was due to a subsurface pressure from the south; the rise to the surface was the only outlet for the abyssal, antarctic water on its arrival in the closed northern part of the basin of the Pacific.

Thus the abnormally low temperatures along the California coast are supposed to be due largely to upwelling, and this phenomenon is assumed to be caused either by the offshore deflection of the coastal currents or to a subsurface pressure. The question now arises: how do the data gathered in the course of the Hydrobiological Survey agree with these ideas?

First of all, even a cursory inspection of the annual rhythm, as represented by several graphs in this report, particularly by Figs. 13 and 29, will suffice to convince nearly anyone that an upwelling actually takes place. For instance, as will be seen from Fig. 29, the 8° water, symbolized in black, has evidently in the Monterey region its "normal" level at a depth of about 200 meters. During the height of the upwelling season, the mean level of this water may be found within less than 100 meters of the surface. In other words, the 8° water evidently is lifted during this time not less than approximately 100 meters. The rate of this uplift is variable; the average rate may be given as 1.5 meters per day, but distinctly higher rates were observed by us. McEwen has calculated the rate as varying from 27 to 55 meters per month and the total movement to approximate 210 meters. Thus, as already remarked above, this aspect of the generally accepted explanation of the cooling of the coastal water may be considered to be beyond reasonable doubt.

On the other hand, the problem of the cause or causes of the upwelling along the California coast is somewhat less clear and certain. Of the two theories proposed to account for this uplift, the one implying the deflection principle of Ekman appears to be nearly universally accepted at the present time. There can be no doubt that the lack of hydrodynamic equilibrium created by the offshore deflection of the prevailing wind-driven southerly current along the California coast is at least a very essential factor in the vertical circulation of our waters. The reasons for the acceptance of this explanation, which are very excellently presented by McEwen (1912), are too strong for such a doubt. As a matter of fact, the problem at issue is not whether the wind-currents are active in this upwelling process, but rather whether they are the only forces implied.

According to the Ekman deflection principle, the magnitude of the velocity of a current produced by wind decreases as the distance below the surface increases, and the current is deflected more and more to the right with increasing depth. At the depth

$$D = \pi \sqrt{\frac{\mu}{q\omega \sin \phi}}$$

where (μ) is the coefficient of viscosity, (q) equals the density, (ω) is the angular velocity of the earth, and (ϕ) is the latitude of the place, the magnitude of the velocity is only 1/20 of that at the surface, and the movement is in the opposite direction to that at the surface. This distance (D) is called the "depth of the wind-current," and the motion at greater depths is assumed to be negligible. From this statement, which is largely a quotation from McEwen (1912, p. 254), it will be seen that there is a depth, varying under different conditions and in different latitudes, beyond which the deflection principle is not any longer applicable, unless modifying circumstances occur. In the latitude of Monterey, the "depth of the wind-current" should probably not be set below about 75 meters, and the upwelling should be limited to less than the upper 300 meters.

However, this generalization is applicable only in case we are concerned with an unlimited ocean. In case there were a coast running more or less from north to south along a deep sea, then the situation would be fundamentally changed. In that case the entire body of water, from the surface to the bottom would be affected, and there would originate three distinct currents, viz., a bottom current, a mid-water or "deep" current, and a surface current. The bottom current would have a depth D, and it would move approximately in the direction of the slope, thus at right angles to the coast, with a right hand deflection of 45° at the bottom and 90° at the top. The mid-water or "deep" current would extend from a distance D from the bottom to a distance D below the surface, its depth thus varying with the depth of the sea. Its velocity would be proportional to the wind component parallel with the coast and almost uniform, and its direction would be parallel with the coast. The surface current would have the depth D, its velocity would be dependent on the winds and superimposed on the velocity of the mid-water current, and the deflection would be to the right on the northern hemisphere, according to the general principles applicable to wind currents.

These conclusions and generalizations are based on the assumption of steady, uniformly directed winds. In the region with which we are concerned, the winds are quite variable, even though those from the northerly quarters predominate. This fact, however, is of comparatively minor importance on account of the circumstance that, according to the calculations of Ekman, the time required to set up a stationary state of motion is quite short, viz., a few days for the surface current, and some (3-4) months for the "deep" current (which is also very little affected by brief changes). Consequently, the result of this state of affairs would not be the total inhibition of the upwelling, but only that the upwelling process would be quite variable in intensity, and that it would cease intermittently during the period, from December to the middle of February, when strong winds from the southerly quarters are common and the winds from the northerly quarters are correspondingly less frequent. As a matter of fact, and as will be further described later on, in the last period, water of southern origin tends to amass against the coast of the Monterey region, thus creating a situation which is the very opposite to the one under which upwelling will take place, according to the principle of the offshore deflection of the southerly current.

The best available check of the wind theory against actual conditions is obtainable from Fig. 29. As Bigelow and Leslie (1930, p. 475) wrote: "It is generally recognized that upwelling along the California coast is an intermittent process. To gain any reliable picture of active and inactive periods, and to determine the regularity, or reverse, of its seasonal schedule would obviously require frequent periodic record of the temperature of the central part of the bay, as well as of its margin." Fig. 29 represents the changes of temperature occurring in the central part of the Bay in the course of two consecutive years. The great similarity between it and the graphs showing the same development at the margin of the Bay has already been discussed in chapter VII.

Fig. 29 conclusively demonstrates that the upwelling process is intermittent, rhythmic, having its maximum during the summer months when the winds from the northerly quarters are relatively strong and regular. It also shows clearly that there are minor, as well as moderate, irregularities in the upwelling rhythm of each year.   Both of these features are to be expected according to the theory of the winds as the causative factors.

In addition, however, the graph exhibits certain characteristics of this process which do less readily, or even not at all, conform with the wind theory.   These are as follows:

a. The upwelling showed extraordinary differences in magnitude from one year to another, in spite of the fact that commensurate differences in the strength and persistency of the winds from the northerly quarters apparently did not exist; at least, such differences have not been observed.   According to our records, 1929 and 1933 were years of unusually strong upwelling, 1930 and 1932 were in this respect intermediate, and 1931 was very light. Indeed, the differences in the whole configuration, if I may use such an expression, between the upwelling of 1933 and that of 1932 were of such a magnitude that quite conspicuous variations in the underlying forces must, apparently, be postulated.   (Compare also Fig. 18.)

b. While in years of moderate upwelling, such as 1932, the process appears to have been limited, at least largely, to the upper 300 meters, thus within the depth affected by the superficial wind-current, in years of strong vertical shifts, such as 1933, the rise of the deep water can be traced quite clearly to much greater depths.   As was described in chapter VII, the lower isotherm of the 7° water was found, by and large, near the 400 meter level from January until the middle of August, 1932.   Although with some irregularities, described above, this water rose from August, 1932, until the middle of June, 1933, when its lower isotherm reached approximately the 300 meter level; finally, concomitant with the fall of the 8° water, the 7° water subsided to its previous level during the late summer and early autumn months.   In addition, the 6° and 5° waters mirrored, even though on an increasingly reduced scale, the behavior of the 7° water.   In other words, in 1933, the upwelling process could be traced down to a depth of around 700 meters or more, thus far below the limit of the realm of influence of the superficial wind-currents in an open sea.   An emphasis to this description of the difference between the upwelling processes of 1932 and 1933 will be obtained through a study of Fig. 26 and its description in chapter VII.

The fact that the upwelling water along the California coast may be raised from considerable depths was not unknown before these data were collected by the Hydrobiological Survey.   Indeed, even the observations gathered by H. M. S. "Challenger" showed, as demonstrated by Thorade (1909, p. 31), that in this region upwelling occurred at depths between 530 and 730 meters.   Thorade utilized this fact as a proof of the existence of a deep current system, such as described above.   He stressed that on the west coast of north Africa, where the coast runs at right angles to the prevailing winds and a deep circulation of the described type therefore does not exist, the upwelling is limited to the upper 300 meters, while in California where the directions of the prevailing winds are favorable to the creation of this type of deep circulation, the upwelling is not limited to the sphere of influence of the superficial wind-currents.   Thus this deep upwelling would, according to Thorade, actually confirm the wind theory.   However, according to this theory, when upwelling takes place below the limits of influence of the superficial currents, the upwelling water should come from the bottom current; in other words it should be, in this region, truly abyssal water.   As far as available evidence goes, this is, however, not the case. According to our data, the upwelling occurred in the intermediate strata, above the bottom current and below, or somewhat above, the 300 meter level.   Of course, by this I do not intend to state that truly abyssal upwelling does not occur along this coast; indeed, it may very well do so.   But I want to emphasize that the rise, evident between 300 and 700 meters in Fig. 29, probably did not originate close to the bottom which is not less than 1800 meters below the surface a short distance outside station C.

c. The upwelling did not start each year at the same time.   In 1932, it may be said to have begun in the middle of February, *i.e.*, at a time that may be considered to be fairly well in accordance with the wind theory.   On the other hand, in 1933, it began as early as in January, that is, during the month when, as noted above, the winds from the southerly quarters tend to set up a northerly drift (Fig. 45) through which the water level next to the coast of the Monterey region rises rather than subsides, due to the right-hand deflection of the water and to the local trend of the coast line.   As a matter of fact, strictly speaking, the rise of the deep waters began even earlier, *viz.*, during the latter part of December, 1932, when the upper level of the 8° water rose from approximately the 250 to the 150 meter level. This early beginning of the upwelling appears to be utterly incompatible with the deductions from the wind theory.   Indeed, it may be said, probably with full justification, that the rise of the deep water during the period from the middle of December, 1932, to the end of January, 1933, was caused at least very largely, by forces other than those created by the winds off the coast of California.   Furthermore, according to the wind theory, the maximum upwelling should take place in August, since at that time the wind conditions are the most favorable (McEwen, 1915).   In reality, the maximum was distinctly earlier in the years investigated by us (Fig. 29); indeed, the decline of this process began, according to our data, not later than in July.

d. In spite of the fact that, from September to the end of November, winds from the northerly quarters are quite common, even predominating off San Francisco and Monterey, it happens, at least some times, as in 1933, that the upwelling process is, by and large, quiescent during that period.   It should be noted that in 1932, a slight rise and subsidence of the deeper waters took place in that interval.   However, to the facts brought forth in this point, only a minor significance is assigned in this connection.

From the reasons enumerated above, the conclusion appears to be plausible that the upwelling phenomenon along the California coast, when fully developed, is due not only to the offshore deflection of a southerly, coastal current, but also to some other force or forces. As to the relative importance of the former and the latter factors, little can be said with any degree of certainty beyond that the latter factor must play a very fundamental part in the process, and that some features, particularly the vertical dislocation of the deepest strata investigated by us, must be related exclusively to it.   In regard to the nature of the additional force or forces, we can only speculate at present.   It concerns apparently some form of subsurface pressure of variable intensity, probably to be located at an intermediate but variable depth, and not truly abyssal as suggested by Bishop (1904).   This pressure presumably is caused by a very slow and variable current directed nearly at right angles to the continental slope of the California coast.

In the last paragraph, the conditions established in Monterey Bay were deliberately applied to the upwelling along the California coast in general.   That the Monterey situation should be regarded as typical of the full development of the upwelling movement, rather than exceptional, at least until evidence to the contrary be obtained, is suggested by the following fact.   The latitudinal depression of the surface temperatures in Monterey

Bay, according to McEwen (see Bigelow and Leslie, 1930, p. 476, 478), is of the same magnitude, about 7°–8°, as in the region off Cape Mendocino which is considered to be the main focus of this phenomenon.   This fact strongly indicates that the Monterey area is, at least approximately, in the same category as the Cape Mendocino area in respect to the rate of upwelling.   That is, we may justifiably expect that the process reaches its fullest, as well as typical, development in our waters.   It should be noted, however, that the upwelling in southern California may be of a nature somewhat different from the one at Monterey and to the north of that locality.   The southern upwelling may well be explainable exclusively by the Ekman principle presented above.

Another peculiarity of the cold water rhythm is the irregular swelling and shrinking of the various thermal strata, changes which were described in chapter VII.   This process exhibited variations not only from one year to another but also changes of shorter duration. In the summer of 1932, the 8° water showed a pronounced swelling, while in the summer of 1933 the 7° water presented this peculiarity.

The significance of this phenomenon can not be said to be quite clear.   However, it at least appears to suggest that the actual upwelling had taken place on these occasions at some distance outside the Bay rather than against the continental slope.   There is unquestionably a deep southerly current running parallel to the California coast.   This deep current, and not the continental slope, probably acted as the boundary against which the upwelling took place.   After the deep water had been raised in this manner, it entered the area investigated by us as a slow flow, the horizontal shift across the coastwise current evidently taking place at different depths on different occasions, the level presumably being determined, at least partly, by the relative magnitude of the upwelling and of the characteristics of the offshore pressure, as well as by the strength of the coastwise current at different depths.   This suggestion is in harmony with the fact, established by J. Murrey (1899) and noted above, *viz.*, that the temperatures are lower somewhat away from the coast than over the continental shelf (see also Belknap, 1874, p. 38; and Bigelow and Leslie, 1930, p. 466). Thus, in 1932, when the upwelling was of an intermediate magnitude, the inflow of deep water should have occurred principally at the 150 meter level, while in 1933, characterized by strong upwelling, it should have taken place farther down, below the depth of 200 meters. Of course, it should be noticed that the inflow was by no means limited to these depths, but the examples given must suffice in this connection.   A study of the description of the changes in the relative volumes of the various thermal strata, given in chapter VII, should be read from the point of view suggested above.   Finally, it may be worth calling attention to the fact that concomitant with the rise and fall of the upwelling water, there occurred a corresponding outflow of superficial water.   Thus, we gain the impression that over the continental shelf, there is a constant onshore and offshore shifting of the water, the movement being different and, at least partly, opposite in the various strata.

The inward, horizontal shift of the deeper strata from the offshore areas to the outer region of the Bay suggested above may be conjectural, but a similar flow on a smaller scale, *viz.*, from the outer parts of the Bay, outside the continental shelf, to the inner and shallower parts, unquestionably occurs.   It is clearly demonstrable from our data.   The question is: how does this flow take place?   Does the flow occur only horizontally, or are the deeper waters deflected upwards against the continental shelf?

This aspect of the upwelling problem formed one of the main issues in the discussion of Bigelow and Leslie (1930).   For its solution, these authors utilized not only the common principle of hydrographic profiles, but also hydrodynamical calculations and the construction of isothermobaths. The most important part of their work was based on observations collected in the course of two weeks, July 11–24, 1928, while the collection of their entire material covered three weeks. Of course, Bigelow and Leslie were keenly aware of the possibility of misleading and erroneous results by consequence of their observations not having been made simultaneously, nor even nearly so, but, in spite of this, they carried through their calculations, mainly on account of the apparent consistency of the results and due to the fact that nobody at that time realized how extremely unstable the hydrographic conditions of the Bay were. Now we know that their method of lumping data collected over several days is not justifiable, but, as is more fully described in chapter III, this was not established until systematic work to test this method had been carried out for some time under the Hydrobiological Survey. Through our work it was revealed that the summer months, when Bigelow and Leslie were investigating the Bay, usually are among the most variable, and this, of course, makes the procedure of these authors even more precarious. However, the variability is subject to fluctuations from year to year, and it may be, as later suggested by Dr. Bigelow in letters, that the 1928 summer was unusually stable.

Their most outstanding and suggestive results were obtained from the construction of the isothermobath for 9° (Bigelow and Leslie, 1930, Fig. 17). This showed a slope of not less than 65 meters, "with a much more definite valley overlying the entrance to the submarine canōn, and rising thence over the northern and southern slopes of the latter, as well as shoreward along its axis, to flatten out over the more gentle submarine slopes above." Such a configuration of the thermal distribution "points unmistakably to an intensification of the updraft on all sides of the trough contrasted with its axis, as the surface is approached and with expansion of the area included within the picture." The updraft was found to be "active enough to effect considerable thermal distortion upward to within 30–40 meters of the surface, over a large proportion of the shoaler parts of the bay." The dynamical calculations indicated that there was an anticyclonic movement in the center of the Bay, water entering the Bay along the northern side of the deep central channel, and leaving it along the southern side. This eddy would either be closed, or it would form a part of the general north-south drift. The northern part of the Bay was found to be dynamically "dead" at the time of investigation, while the southern part indicated a southerly drift. Nothing could be said in regard to the circulation in the inner parts of the Bay, on account of the small gradients, the strong tidal and wind currents, and the directing effect of an almost straight coast line.

The results of these authors have been reproduced here, in their main outline, first, because they are so far the only ones dealing with the circulation in the Bay, and secondly, since they are very consistent and, in all probability, in part represent a circulatory situation which at times does exist. On the other hand, even though such a situation may at times occur, it is by no means the only one, a condition of which Bigelow and Leslie evidently were fully aware. Indeed, as suggested earlier in this report, the circulation in the Bay is a very changeable phenomenon, and furthermore, at least as a rule, much more complex than as indicated by the results cited above.

Before Bigelow and Leslie undertook their investigation of the Bay, there was a commonly held opinion to the effect that the upwelling water was forced into the central channel of the Bay, and that, near the head of this channel, it was forced up at a considerable rate, thus causing a local turbulence, after which it was spread out over the shallows of the Bay in a fanlike manner.   In order to test this idea, under all states of upwelling, a station (E) was located at the very head of the channel.   Throughout two years (1932, 1933), temperature observations were made at this place.   During this entire interval, there was never even the slightest indication of turbulence at this place, the thermal stratification always being of the same type as at the remaining stations, *i.e.*, quite pronounced during the entire year except in the winter.   This fact, in conjunction with the circumstance that Galliher (1932) found the bottom sediments in the inner portion of the central channel to be fine silt that would be readily removed even by a moderate flow of water, may be considered to justify fully the conclusion that even during the very height of a very strong upwelling period, such as the one of 1933, the region around the head of the central channel does not serve as the outlet for the upwelling water, at least not to a greater extent than does the rest of the edge of the channel.   This conclusion is in agreement with the findings of Bigelow and Leslie, as cited above.

The problem now is: to what an extent does the upwelling water flow over the edge of the channel in general?   In order to answer this question, the temperatures at the 50 and 100 meter levels of station C, on the one hand, and of stations B, D, and E, on the other, collected throughout 1932 and 1933, were compared and the deviations of the latter stations from the former were calculated for each observational day.   The results of the comparison of stations C and B, presented in the form of monthly deviations, are to be found in chapter VII.   From this will be seen that station B, by and large, was somewhat cooler at these levels than station C, and that this condition was somewhat more pronounced during the first six months of the year when the upwelling was, generally speaking, more active.   The same condition was found, when stations C and D were compared.   In regard to station E, the situation was less convincing.   For instance, in 1933, the year of strong upwelling, 26 observational days out of 46 showed the 50 meter level at station E to be somewhat colder, but these days did not exhibit any seasonal concentration, but they were scattered throughout the year in an apparently random manner.   The same was true in regard to the 100 meter level, at which the noted proportion was 24 : 46.   The differences recorded were never large; usually they ranged from 0.1° to 0.3°.   It should be noted, however, that on several days of apparently strong upwelling, neither station B nor D was colder than station C at the mentioned levels.

These observations clearly suggest that the upwelling water, on impinging against the continental shelf, on the edge of which stations B and D were located, was frequently deflected somewhat, although not very much, towards the surface, afterwards to move along the bottom of the shallows of the northern and southern portions of the Bay.   Other data show that on these flats, parts of this cold water reached at least as far in as to stations F and 4.   At the head of the central channel, the upwelling was less frequent and consistent; most of the deeper water was evidently often deflected upwards somewhat farther out from the inner coast line.   This last conclusion is in full harmony with the main results obtained by Bigelow and Leslie.   However, although this situation evidently prevailed frequently, it should be emphasized that the upward deflection did by no means always occur.   Fairly often, the stations on the edge of the continental shelf were either of the same temperatures or slightly warmer at the 50 and 100 meter levels than station C.   When the same temperatures occurred, then, presumably, the upwelling water moved in towards the Bay, according to the suggestion proposed in one of the previous paragraphs, mainly along the 100 meter level. (Observe the swelling of the 8° water, between the 100 and 150 meter levels, in the summer of 1932.) On the other hand, the presence of warmer waters over the continental shelf, at the 50 and 100 meter depths, apparently can be explained, in terms of upwelling, only by the assumption that on those occasions the upwelling water had not quite reached the edge of the shelf, but had moved in only as far as to station C.

To summarize the discussion in this section: The cold water phase of the annual rhythm in Monterey Bay, from 1929 to 1933, was due largely to upwelling. The updraft of cold water took place at an unknown distance outside the Bay, and thus not against the continental slope, and at an intermediate but variable depth. From the offshore focus or foci of upwelling, the cold water moved onshore, towards the Bay, and it was frequently but not always deflected upwards to a moderate or slight extent on reaching the edge of the continental shelf. The presence or absence of this upward deflection evidently depended on the level at which the upwelling water approached the coast. The phenomenon exhibited striking fluctuations both of seasonal and of annual character. The cause of the upwelling should, according to my opinion, be sought not only in the offshore deflection of the superficial, wind driven, southerly current along the coast, as is done exclusively at the present time, but also in a pressure from below caused by an unknown, deep water movement, directed approximately at right angles to the coast and not connected with the water movement caused by the offshore deflection of the surface current. The fact that the upwelling took place in a region to the west of the continental slope would presumably be due to the presence of a coastwise current of intermediate depth, a current possibly of the nature suggested by the Ekman theory discussed above. Against the outside of this current, the upwelling water would have risen. In regard to the relative effect of the surface deflection and the deep pressure on the more superficial upwelling, nothing can be said with any degree of certainty; but the latter factor apparently was either the main or the sole cause of the uplift of the waters from below the 300 meter level, as well as of the pronounced variations in the magnitude of the upwelling from one year to another. Attention should also be drawn to the fact that there is a southerly flow of considerable latitudinal extension along the California coast. This northerly type of water also contributed to depress the temperatures of Monterey Bay. However, it must be remembered that a considerable portion of this northerly water must be assumed to have been of deep origin.

Finally, it should be noted that even though in the preceding discussion, for the sake of simplicity, the wind alone has been mentioned as the cause of the more superficial upwelling, nevertheless, several factors are implied, according to McEwen, such as the rate of evaporation, turbulence, solar radiation, as well as barometric pressure, and others. It is an interesting fact, mentioned here only in passing as a suggestion for future research along this line, that the monthly temperatures in Monterey Bay at a depth of 50 meters moved, in 1930, nearly parallel to the monthly barometric pressure, as recorded at San Francisco.

2. The Warm Water Phase: Of the three phases which characterized the annual, thermal rhythm in Monterey Bay, from 1929 to 1933, the cold water phase, as noted in the beginning of this section, was unquestionably the most important, both in respect to magnitude and to consequences. From the same points of view, the warm water phase should, with equal justification, be allotted the second place. This phase, which occurred in the autumn of each of the five investigated years was of somewhat less certain origin. Most of the available evidence indicates that it was due to the approach to the coast of water of truly oceanic origin derived either from the inner margin or from the middle of the California Current, a branch of the Japan Current. As a consequence, this phase may appropriately be named the "Oceanic Period," as has been done in Fig. 13.

The Japan Current represents a continuation of the warm Pacific North Equatorial Current. That is to say, it is a part of the gigantic clock-wise eddy which encompasses most of the nothern half of the Pacific Ocean, and it is the counterpart of the Gulf Stream and the Atlantic Drift and its branches along the European coasts. After its passage along the eastern coasts of Japan, it moves into the high latitudes of the northern basin of the Pacific, where it is cooled to a very considerable extent. As it approaches the North American continent, it splits. The point of splitting of the current changes in the course of the year; in the winter it takes place farthest to the south, in February off Cape Mendocino, and in the summer farthest to the north, in August off Vancouver, B.C. One of the main branches continues into the Gulf of Alaska, giving to this general region considering the high latitude, its relatively warm climate. Another main branch flows south along the west coast of North America. This southern branch, which often, and justly so, is distinguished under a special name, viz., the California Current, is nearly always removed from the coast line, and as it travels south, it trends towards the west, under the influence of the rotation of the earth.

In spite of its great importance, especially from the point of view of its climatic effects, the Japan Current is comparatively little known. Very deficient is also our knowledge of the American part of this system, the California Current. This deplorable condition is, of course, in a large measure due to the relatively undeveloped state of the shipping business in the northernmost part of the Pacific, when compared with the corresponding region of the Atlantic. In regard to the California Current, it may be said to have a fairly small significance both to the shipping industry and to the climate of California.

A characteristic feature of the Japan Current system is the variability of its bulk and rate of flow. Not only do the bulk and flow decrease from west to east, a fact discovered even by some of the early scientific expeditions and undoubtedly well known to the practical seafaring men for a considerable time, but there are also annual fluctuations. As an example of the west-east differences, it may be mentioned that the "Challenger" Expedition, in 1875, found a speed of about 18 miles per day in the westerly, and about 10 miles per day in the easterly portion of this current. This variability of flow is due to the following circumstances. The Japan Current system is primarily a superficial circulatory phenomenon, caused mainly by the winds, viz., by the north-east trades which blow steadily during the larger part of the year, by the less constant south-east trades, and by the winds which blow out from the northern and north-western sides of the north Pacific high pressure area, in the so-called Horse Latitudes. On the other hand, it is impeded in its flow off the coast of Asia by the northwesterly winds which intermittently blow from that continent, from the end of September to the end of February. There are thus two main opposing forces, and in addition other intervening factors, which regulate the bulk and rate of flow of this current system. As a consequence, on some occasions the current runs strongly, while on others it is more or less obliterated. Due to the variations in the bulk and rate of speed of the current, as well as to the seasonal thermal changes in the high latitudes, the waters of the Japan Current are cooled and mixed to varying degrees on the way across the Pacific towards the coast of California.   Thermal variations in this current in the high latitudes were recorded as early as in 1837 by the "Venus" Expedition.   In regard to the California Current, we should thus expect it to be seasonally variable in temperature, bulk, and rate of flow.   It should also be noticed that this current apparently often is split into what may be termed stripes, just as is the North Atlantic Drift.   Thus, when one travels across these waters, he frequently finds alternating higher and lower temperatures; indeed, there is little evidence of unity and consistency.   Moreover, caused by the friction against neighboring waters, *etc.*, marginal eddies are formed and apparently sometimes detached from the main body of the current, another feature which recalls the situation characteristic of the North Atlantic Drift.   Of course, it is quite obvious that under these extremely complex conditions, the opinions in regard to the origin and nature of the offshore waters of California have been, and indeed still are, more or less at variance.

The main reasons for associating the California Current with the warm water phase are twofold: first, this current is characterized by variable temperatures lying within the ranges recorded during the warm water phase; and, secondly, its chlorinity is relatively low, and the same is true in respect to the superficial waters in Monterey Bay during the autumn.   The California Current is classified as a cold current, and this classification unquestionably is correct.   By the time the water of the warm Japan Current has arrived off the coast of California, it has lost so much of its heat that its temperatures are subnormal from the point of view of the latitudes.   However, even so it is relatively warm when compared with the coastal waters in the strict sense of that term.   Thus we are faced by the paradoxical situation that the warm water phase in Monterey Bay is caused by the influx of a "cold" current.

Another criterion in favor of this interpretation may also be noted.   If the California Current were connected with the warm water phase, then we should expect this phase to be characterized by pronounced fluctuations in magnitude.   This was, as a matter of fact, the case.   In 1932 this phase was extraordinarily well developed, and in the remaining four years it exhibited various gradations; in 1933, for example, it was extremely weak (Fig. 18).   In years of subnormal development of this phase, the incoming water apparently was derived from the inner margin of the California Current, while in years of maximum development, the original source of the warm water must have been located at a considerable distance from the coast.

In regard to the force or forces which caused the waters of the California Current to enter Monterey Bay, we can only conjecture.   Winds undoubtedly may have contributed towards this end, but they can not be considered to have been the sole cause on account of the fact that westerly winds were not only rare but also light during September when this phase reached its maximum development.   As far as I can see, it is necessary to take into account variations in the bulk and rate of flow of the Japan Current, as well as an onshore pressure from outside this current and exercised approximately at right angles to the coast. In regard to the last factor, attention should be called to the fact that a subsurface pressure of a similar direction and source was assumed above as one of the causes of the upwelling phenomenon.   A decrease in the upwelling process could hardly be considered as an explanation.   It should be remembered that the waters of the Japan Current tend away from the coast under the influence of the rotation of the earth.   In 1932, when the warm water phase reached an unusual magnitude, the upwelling water apparently was forced out by the oceanic water, judging by the fact that it came back as soon as the latter type of water departed.   Moreover, in 1931, when the upwelling was weaker than in any of the other years under investigation, the warm water phase was also poorly developed within the confines of Monterey Bay (Fig. 18).

As far as I know, the only available reason that appears to speak against the assumption that the California Current was connected with the warm water phase is that, according to the Pilot Charts of the North Pacific Ocean issued by the Hydrographic Office of the U. S. Navy, the superficial waters off Monterey Bay move offshore rather than onshore during the early autumn months (Fig. 45).   It should be noted, however, that the information concerning the currents embodied in these charts is based on shipping records and not on the results of special hydrographic investigations.   There appears to be no possibility of bringing harmony between the type of flow indicated in the mentioned charts and the hydrographic data collected by us in the autumns of 1929–33.

On the occasions when the California Current, or detached eddies thereof, entered Monterey Bay, the water in the Bay was more or less deep blue, depending on the relative purity and admixture with marginal coastal waters, of more or less high transparency, and characterized by relative scarcity of microplankton and the presence of macroplanktonic forms which otherwise were not found within the Bay.·   Frequently, perhaps even usually, the oceanic water did not penetrate into the innermost regions of the Bay but was found principally at or just outside stations B, C, and D.   On one occasion, the transition between the coastal and the oceanic waters was so sharp that a difference of a couple of degrees was observed between the temperature of the water under the bow and that of the water under the stern of our boat.   In other words, there was a distinct rip at this place, and such rips were seen quite frequently, although we did not stop to examine them.   These rips were often marked by the accumulation of drifting kelp and other debris.

Between 1929 and 1933, the warm water phase, as described above, was always in the autumn, its maximum occurring in September.   This, however, does by no means imply that the oceanic water did not come into the Bay except on the mentioned occasions.   As will be seen from Fig. 16, the curves of the surface temperatures in the two unusually warm years 1930 and 1931 were characterized by great irregularities which in the latter year could be traced to a depth of 50 meters.   These irregularities are probably to be referred to what may be termed unseasonal fluctuations in the California Current.   Particularly worthy of mention in this connection is the fact that during the abnormally warm period from the middle of 1929 to the latter part of 1931, certain plankton forms of oceanic distribution were frequently recorded within the Bay, forms which during the rest of the five year period were either not seen at all, or were recorded only during a short interval in the autumn.

Moreover, a study of Fig. 23 will show, first, that the warm water phase in 1932 was by far the best developed in sixteen years, 1919–34, and, secondly, that this phase is apparently not always to be found in the autumn.   In 1926, for instance, the autumn water was quite cold, while high temperatures prevailed in the spring, April being the warmest month.   In the spring of that year, abnormally warm water swept a large portion of the coast of central and northern California, causing an unusual northerly migration of various forms of sea life.   Some examples of this migration will be found in Hubbs and Schultz (1929).

Finally, it is of interest to note that in the first week of July, 1922, Monterey Bay suddenly was filled with oceanic waters in which occurred enormous quantities of giant salpae, huge pteropods, *etc.*, all strangers to this general region.   This episode was of short duration, but unfortunately no special records were made, except the collecting of a number of these exotic forms.

Some additional observations about this interesting phase will be found in chapter IX, in which fisheries problems are discussed.

3. The Low Thermal Gradient Phase: As will be seen from Fig. 45, representing the surface currents off the west coast of the United States, according to the current charts of the Eastern North Pacific Ocean published in conjunction with the Pilot Charts of the U. S. Hydrographic Office, the movement of the surface waters along the California coast is predominantly southerly in all months except December, January, and February, when a northerly coastal drift prevails.   This northerly surface movement, which usually goes under the name of the Davidson Current, was evidently the main cause of the low thermal gradient phase of the annual rhythm in Monterey Bay, from 1929 to 1933.

The Davidson Current is a wind current.   It is caused by the winds from the southerly quarters which prevail during the winter months and which also are related to the rainy season of this general region.   These winds do by no means blow constantly and their strength is subject to very pronounced variations.   (In regard to the short time required for the establishment of a drift of this kind, see the discussion of the Ekman principle given above in connection with the treatment of the upwelling phenomenon.)   From the variability of the southerly winds it follows that the Davidson Current should also exhibit decided fluctuations in strength as well as in bulk.   That such fluctuations do occur is a matter well known to seafaring men who often utilize this drift in the coastwise shipping business, just as they profit by the opposite current in other seasons.   Another fact to be considered in this connection is that the rotation of the earth causes the northerly drift to tend towards the right, thus towards the coast.   Through the southerly winds and the rotation of the earth the offshore surface waters are thus amassed against the coast, where they are forced down, and from where they move out as a subsurface flow.   Thus the surface of the sea close to the coast is somewhat raised during the periods of southerly winds, a fact which is of interest, as we have seen from the previous discussion, in connection with the upwelling process.   As a consequence of its superficial origin and of its strong vertical circulation, the water amassed against the coast in this manner is thermally very uniform. The vertical circulation and the thermal uniformity often extend to a depth of about 50 meters and on occasions may penetrate as far down as to the 100 meter level.

An analysis of our records of winds and of water temperatures during 1929–33 reveals a very close agreement with the generalizations presented in the previous paragraph.   The presence in Monterey Bay during the low thermal gradient phase of plankton of distinctly southerly facies also supports the assumption that the Bay water during this phase was of an offshore and southerly origin.   For example, the dinoflagellates, which were specially recorded, contained during this period a number of southerly species which were not observed in other seasons.

However, the low thermal gradients were not due exclusively to the phenomena described above.   It should be remembered that during the winter months the superficial waters were subject to local cooling, due to the thermal interaction between the air and the

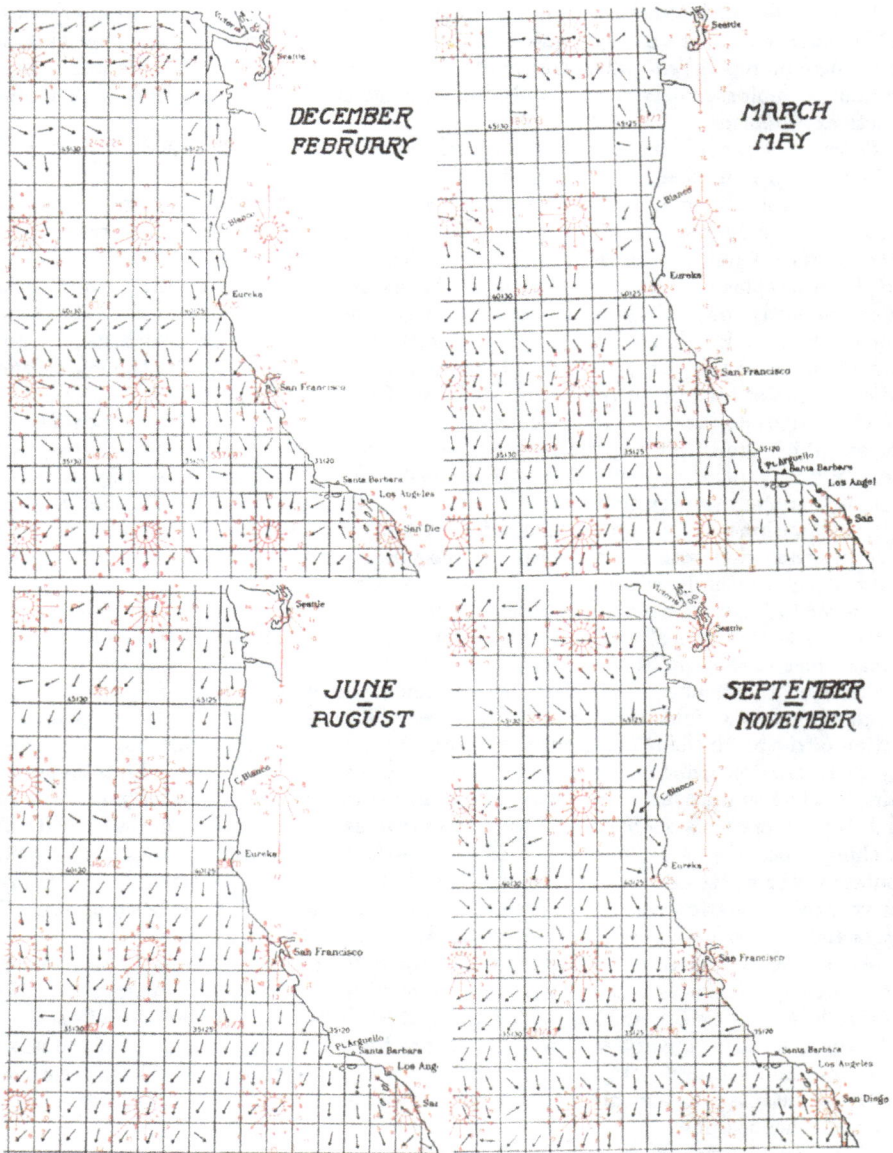

FIG. 45.   Movements of the surface waters off the west coast of the United States, according to the current charts, published in conjunction with the Pilot Charts of the U. S. Hydrographic Office.

water. To a considerable extent, this process contributed to the development of the low thermal gradients. The exact amount of mixing, following the local cooling, was, however, not determined by us. Our observations along this line are described in chapter IV. On the other hand, the relatively high temperatures during this phase were, as far as I can judge, exclusively due to the Davidson Current phenomenon.

It should finally be noted that the development of the Davidson Current from one year to another should show very pronounced variations. In the winters of 1929–30 and 1930–31, this current evidently was unusually strong, while in the following three winters it was either weak or of medium strength (compare Fig. 23).

D. Causes of Other Thermal Characteristics: Of course, some of the conditions described and discussed in the previous sections have a decided bearing on those of the twelve points, given in the beginning of this chapter, which so far have not been treated. I now propose to make a few brief remarks about each of these points, from the angle of the results presented in the previous sections.

The fact that the thermal patterns at the 50 and 100 meter levels were reproduced more closely from one year to another than was the pattern of the surface waters (2) [1] was evidently due to the circumstance that the former strata were dominated by the upwelling, a process which, at least partly, was underlain by forces of strong and persistent nature, while the surface temperatures were under the influence of the vicissitudes of the winds and other phenomena causing great irregularities in the superficial circulation.

The upwelling exercised quite a pronounced influence on the temperatures of the stratum between 25 and 50 meters. This caused the average temperatures for the body of water between the surface and 50 meters to follow the temperatures at 50 meters more closely than they did the superficial ones (3); see Fig. 16.

The shifts in the general levels of the thermal curves of the 50 and 100 meter depths (Fig. 16) were connected with the variations in the intensity of the upwelling from year to year and to a lesser extent with the variations in the California and Davidson Currents (4); see Figs. 16 and 18.

The narrow amplitude of thermal variation in the Monterey region (6) depended on two conditions, *viz.*, that the maximum of the upwelling occurred in the summer and that the relatively warm Davidson Current developed in the winter. It may also be observed that the summer months, generally speaking, were characterized by a high frequency of fog, while the winter months were relatively clear, a fact which, of course, tended to equalize the effects of insolation. The relatively frequent thermal changes, on the other hand, were caused by the instability of the underlying water movements. Even the upwelling process, which was the most stable of these, proved to be characterized by irregular pulsations.

The months in which the thermal fluctuations were greater below than at the surface (7) corresponded to those in which the Davidson Current brought southerly surface water into the Bay. When this current slackened or ceased, under the influence of decreasing winds from the southerly quarters or unfavorable winds from the northerly quarters, the amassed superficial water of this current left the Bay to a varying extent, while simultaneously it was replaced from below. In that manner, the subsurface strata, to a depth of about 100 meters, were characterized alternatingly by cold water from below and com-

[1] Refers, like the following numbers in parentheses, to the corresponding point of the list given in the beginning of this chapter.

paratively warm water forced down from above, in a manner described in the last section, while the superficial layers always were composed of the latter kind of water.

The seasonal progression of the thermal gradients of the upper 100 meters in Monterey Bay was brought forth in a manner partly opposite to that characterizing the similar progression in typical temperate waters (8). This was, of course, due to two of the fundamental water movements in the Monterey region. The raising of the temperatures at the lower levels in the winter was caused by the amassing of relatively warm surface waters against the coast, in conjunction with the Davidson Current, while the lowering of the temperatures at these levels in the spring and summer was a consequence of the upwelling. It should also be noticed that the high gradients in the autumn were accentuated by the ingress of California Current water into the Bay.

The peculiarity that the thermal gradients of the upper 100 meters and those of the stratum between 100 and 250 meters were, in a restricted manner, opposite to each other (9) is also to be explained by the presence of the Davidson Current and the upwelling. The Davidson Current did not affect the waters below 100 meters, or at least not very appreciably, and, as a consequence, the waters below that level maintained their normal gradients during the winter. On the other hand, the upwelling brought comparatively cold water from below towards the upper levels of the 100–250 meter stratum, thus decreasing the gradients of this layer. The upwelling process was at its lowest in the winter, hence the greater gradients, and at its maximum in the summer.

Observations show that annual, thermal changes, under what may be termed normal conditions, can not be traced below 200 meters (daily changes, due to causes other than circulatory, not beyond 20 meters). In Monterey Bay changes were observed as far down as measurements were made (10). This, again was in part due to the upwelling phenomenon, which, as we know, is seasonal, and possibly also in part to the type of water movement noted in the next paragraph.

In regard to the recorded irregular, non-seasonal, fluctuations in the vertical distribution of the temperatures (11), only those of the deeper layers need to be treated here, since those of the upper strata were discussed in section B of this chapter. The irregularities in the temperatures of the deeper strata were apparently due to two causes, viz., the pulsating nature of the upwelling process and the presence of a coastwise, probably southerly, mid-water or "deep" current. The existence of a deep current is, as we have seen above, to be postulated according to the Ekman principles of wind-currents. This current is evidently not confined to the upwelling period but persists all the year around, although in all probability its rate of flow varies seasonally. This persistence is in accord with Ekman's calculations. As was mentioned above, the establishment of the deep current should, according to these calculations, require several months; conversely, the time needed for the complete cessation should also be relatively long. An important fact to be taken into account in this connection is that the winds counteracting this current are of comparatively short duration and that they are interspersed with favorable winds. The deep current apparently causes intermittent elevation and subsidence of the thermal levels in the deeper strata, changes which possibly are connected with variations in the rate of flow, bulk, and original source of the water of the current.

The fact that the thermal conditions, even of the deepest investigated strata, varied from year to year (12) depended on large sized variations in the magnitude of the upwelling, evidently connected with variations in the relative strength of the two main forces under-lying this phenomenon, *viz.*, the offshore deflection of the surface waters, due to winds, and the subsurface pressure. When the former force dominated, the upwelling was con-fined largely to the upper strata, while, when the latter force was the main spring, the deeper waters were much affected.

E. Special Circulatory Conditions: When the circulation in Monterey Bay is considered, we must differentiate between the open regions and the more or less sheltered bights and coves of the bay. In the latter areas, there frequently occur local eddies in which the water is retained for periods of varying lengths. Since these bights and coves are shallow and the insolation frequently intense, their water often is warmer than that in the open parts of the Bay. The presence of very small eddies of this kind was recently established by divers in Carmel Bay, located only a few miles south of Monterey. Apart from differences in the water movements, observed by the divers, these small eddies are marked by the accumulation of a large variety of objects lost by bathers. Each little cove apparently represents a miniature Davy Jones' locker. Examples of local eddies of this kind, but on larger scales, are presented by the sheltered areas around Monterey Harbor and in Soquel Cove, the former in the southern, the latter in the northern part of Monterey Bay. The Soquel Cove evidently is the warmest part of Monterey Bay, a fact already recorded by Bigelow and Leslie (1930, p. 436) who utilized the term "warm pool" to describe this peculi-arity of the thermal distribution; they recorded 15°–16° for this "pool," while in the rest of the Bay 14° was rare. At the same time, these authors erroneously indicated a similar "pool" in the southeastern portion of Monterey Bay, south of the mouth of Salinas River. This region participated very actively in the general circulation, from 1929 to 1933; indeed, a large part of our work was carried out in these waters, and they thus furnished a great deal of the material for the preceding discussions of the circulation in the Bay. The local eddy in this part of the Bay, referred to above, is of much smaller dimensions, occupying only a small area in the close vicinity of the harbor. In these eddies, the water moves clock-wise or anticlock-wise, depending on the direction of the currents just outside them. They undoubtedly receive and lose surface water by the sea breezes and land breezes which frequently blow in this region during the day and the night respectively.

Another type of impediments to the constant renewal of the water in the Bay is to be found in the large beds of kelp (Nereocystis and Macrocystis) which are common in this neighborhood.

The presence of the local shore eddies and the effect of the kelp beds are of importance from a practical standpoint, partly in relation to the disposal of city sewage, partly on ac-count of the large quantities of fish offal emptied into the sea by the Monterey canneries.

A pronounced contrast to these warm and sheltered places is found around the southern headland of Monterey Bay, off Point Pinos. At this place, tidal and other circulatory phenomena are extremely active; in other words, it is a place where turbulence, caused by conflicting currents as well as by heavy waves, prevails. The turbulent condition is frequently made evident by very choppy and irregular wave action. The presence of turbulence does not imply that the entire body of water, to some distance from the point, is churned down to the bottom (at only one mile from the point, the depth is about 50 m.); in most cases, only the upper 15–25 meters, or even less, are affected. Neither does it imply that turbulence always occurs at this place; indeed, often the current flows by the point in a distinctly stratified condition.

Bigelow and Leslie (1930, p. 436) suggested that local upwelling prevailed at Point Pinos in 1928. This was however, hardly the case. The area off this point may be said to form a most frequented gateway through which large quantities of water of the southern part of the Bay enter and leave. The U. S. Coast and Geodetic Survey give a rate of flow of 1.1 knot off Point Pinos at the height of the tide (about one knot, at the corresponding place on the north side of the Bay) and on many occasions the currents run, at least to all appearance, even at greater speeds. That somewhat lower temperatures frequently are found off Point Pinos than to the west of this point or in the southern part of the Bay is undoubtedly due to the fact that cold bottom water from the southern flats of the Bay is forced out of the Bay at this point; in this process, the warmer surface waters are forced somewhat aside, and thus the colder bottom water comes closer to the surface. For example, on May 15, 1933, at stations A, off Point Pinos, 8° water was found at a depth of only 9 meters, while the corresponding thermal level at stations B and 4 occurred at the 29 and 31 meter depths, respectively. Station B was located near the edge of the continental shelf, just outside station A, and station 4 was placed inside station A, on the southern flats of the Bay. Differences of this magnitude were, however, rare.

It may also be mentioned in passing that our knowledge of the circulation over the flats north of the central channel is decidedly less than that of the circulation over the southern flats. Bigelow and Leslie (1930) found the former region dynamically "dead." This, however, is by no means the normal state. The northern region is also characterized by an intensive and complicated circulation, even though this naturally differs somewhat from the one of the southern flats, due to differences in topography, particularly in regard to the position relative to the main headland and to the shape of the bottom. The temperatures at stations F and 4 showed only comparatively slight, and not consistent, differences in the course of 1932 and 1933.

F. Circulation and Bottom Sediments: It may be fitting to finish this chapter with a few remarks in regard to the bottom deposits of the Bay, for the purpose of establishing to what an extent these support or contradict the conclusions concerning the circulation presented above. An excellent chart of the sediments is to be found in Galliher (1932; pl. 3), a chart constructed from data obtained in conjunction with the work carried out by the Hydrobiological Survey.

According to this chart, the bottom just outside the shore line of the inner portions of the Bay consists of coarse sand. The sand gradually becomes finer towards the depth and is replaced by fine silt and clay (usually called mud) on the flats. The center of the southern flat, which appears also to be the center of a more or less permanent eddy-like water movement, is characterized by silt. Generally speaking, the bottom of the northern flat has somewhat coarser material, indicating that the circulation in that locality is somewhat more intensive, a possibility for which we do not have as yet direct hydrographic support. Of the northern flat, only the region next to the central channel has fine silt possibly suggesting the center of another eddy-like water movement.

Along the southern shore of the Bay, where the currents are known to be fairly strong, the bottom material is also comparatively coarse (fine sand, interspersed with broken shells). To the west of the southern headland, the bottom is rocky, with little loose material. This is in good agreement with the known fact that this area is the most turbulent in the Bay and that it forms the main southern gateway for the water entering and leaving the southern part of the Bay.

The edge of the continental shelf along the outer portions of the central channel, against which the upwelling water impinges on entering the Bay, is covered with rock gravel interspersed with finer material.   On the other hand, in the inner regions of this channel where upwelling was found to be comparatively light, fine silt prevails, just as in the center of the southern flat where circulation also is relatively slight.

Finally, in the deep and the outer parts of the channel, thus in the regions of the proposed mid-water or deep current, we find coarse sediments, gravel and sand, overlying silt and clay.   The depth of the coarse material was found to be 1–4 cm., while the depth of the silt and clay is still unknown.   The co-existence of water movement and coarse top material on the bottom is very interesting and certainly suggests a causative connection. However, the possibility of slides from the steep and rocky sides of the channel must be considered when an evaluation of the significance of this type of bottom material at such a great depth is attempted.

Thus, on the whole, the various types of bottom sediments in Monterey Bay are well in harmony with the conclusions relative to the circulation in the Bay presented in this report.

IX. WATER TEMPERATURES AND THE FISHERIES

Of course, it falls largely upon the fishery experts to trace the correlations between the various aspects of the fisheries and the conditions in the sea, with the ultimate goal of obtaining a scientific foundation for predictions of commercial significance.   The main duty of the oceanographer to this industry is to furnish the requisite hydrographic and general biological information, against which the fishery experts may project their own findings. In spite of this natural division of labor, I shall at least attempt to present some examples, largely of local nature, pertaining to the relations which exist between the fishes and the medium in which they live.

The most important effects of the water on the fishes, from the point of view of the practical fishery problems, may be divided into three categories: (a) direct temperature effects; (b) effects of circulation; and (c) effects of fluctuations in the available food supply, caused by the changes in the chemistry of the water.   Naturally, the degree of saltiness of the water is also a very important factor, but this may be disregarded in this connection since most of the fishing in California is carried out in the open sea and the number of rivers of appreciable size is very small in this state.

In respect to the direct effects of temperature, it should be observed that some water organisms are eurythermal, i.e., more or less tolerant to thermal changes, while others are stenothermal, i.e., sensitive, to a more or less high degree, to such changes.   Furthermore, among the latter forms, some are to be classified as cold water stenothermal, and some as warm water stenothermal.   For example, several of our many species of rockfish are limited either to the waters south of San Francisco or to those south of Point Conception, while others extend to the north from these points.   Other examples of this kind are furnished by the California Halibut, which extends south from San Francisco, and the Northern Halibut, the southern limit of which lies in the neighborhood of that city.   Indeed, most of the fishes of California, both those of importance to the fishery industries and those which are commercially more or less insignificant, are, to varying degrees, stenothermal, a fact which can be readily verified by comparing the species in the different fish markets along our extended coast line.   On the other hand, a most important member of the eurythermal class is the California sardine, even though this statement, as we shall see presently, must be made with some reservation. Another, and much more extreme, example of this category is the sablefish (*Anaplopoma fimbria*) which is spread from southern California to the waters of Alaska, with the greatest California landings in San Francisco.

In attempting to evaluate the direct effects of temperature on a species of the higher organisms, differentiation should be made between the effects on eggs and larvae, on the one hand, and the thermal reactions of the post-larval stages, on the other. Generally speaking, the former developmental stages are much more sensitive than the latter. An extremely instructive example of this generalization was furnished by Huntsman (1923, p. 6). In the Bay of Fundy, the American lobster thrives excellently, growing to very large sizes. However, the species can not reproduce there, on account of the prevailing low temperatures; berried females occur, but the young fry can not survive. As a consequence, as a measure of conservation, berried females are now gathered and transported to the warm waters of the Gulf of St. Lawrence, where hatching takes place normally, while at the same time young specimens are taken from the latter region and planted in the Bay of Fundy, where an ample food supply causes them to grow rapidly to commercial size. A well known example from California is the "eastern oyster" which grows excellently in our waters but which must be planted here as spats, imported from other places, since spawning can not take place normally under the relatively low temperatures characterizing the localities favorable for the culturing of this species.

Another case of the difference between the thermal reaction of the spawn and the older stages was recently discovered and investigated by E. C. Scofield (1934). It pertains to the California sardine, and its establishment is of the greatest importance to a scientific understanding of the problem of conservation in relation to the huge industry which centers around this species. The California sardine migrates apparently along the entire west coast of the United States; in other words, it exhibits, as indicated above, a considerable thermal tolerance. However, in spite of the fact that the species is taken commercially throughout this extensive range, its principal spawning area, as demonstrated by Scofield, lies near the southern distributional limit, *viz.*, between San Diego on the south and Point Conception on the north; that is, the spawning occurs in waters of relatively high temperatures. Moreover, the northern limits are more or less flexible. The mentioned boundary holds only during those years which are characterized by "normal" or "subnormal" temperatures in the waters just north of Point Conception. In years of unusually warm water, spawning, although on a small scale, takes place much farther to the north. For instance, limited spawning occurred in Monterey Bay in 1930, 1931, and 1934, in which years the water temperatures in the spring months were above "normal" (the height of the sardine spawning was found to be in April and May). On the other hand, in the cold springs of 1929, 1932, and 1933 (compare Fig. 24), no sardine spawn was recorded in this bay.

If Scofield's findings are substantiated, which they probably will be, then it follows that conservation measures pertaining to the sardine should not be of a local nature, if we wish them to become effective. The amount of sardine taken in southern California then becomes a concern of the industry in central California, and vice versa. Indeed, the problem then assumes an interstate and even an international aspect, since the Canadian catch of large-sized and heavy-spawning sardines in the far north will affect the success of the spawning in the southern California waters. Furthermore, it becomes a task of practical significance to establish whether a concentrated or scattered spawning is the more advantageous from the point of view of a rich replenishment of the natural supply.   There are in this connection two possibilities.   In the northern spawning places, the food supply is richer than in the south, and the supply also is steadier, as far as we know.   This would imply a better chance for the survival of the youngest stages which are incapable of searching extensively for their food.   A relation of this kind has been investigated in European waters, where it was found that the availability of food, just following the hatching, is one of the factors which determine the success or failure of a spawning.   On the other hand, the thermal vicissitudes in the north are more severe than in the south, and the chances of temperatures destructive to the young and tender stages are proportionally greater.   These are problems which should be investigated for the sake of predictions in the field of fluctuations in the natural supply.   The evidence must be gathered both from the field of the biology of the young stages and from hydrography.   In spite of the practical significance of these problems, only the foundation to their solution has been laid.

The statement that the eggs and larvae are, as a rule, more sensitive to temperature than the older stages does, of course, not imply that they always prefer relatively warm waters; it simply signifies that their amplitude of thermal tolerance is narrower.   Thus there are species which spawn when the water is coldest.   It may be of interest to mention in this connection that the suggestion has been made that the forms spawning in the cold season originated in colder regions, later migrating to and establishing themselves in a warmer habitat, but still revealing their early history in the thermal preference of the spawn.   Conversely, the warm season spawners would have originated in warmer climes.   Many extremely interesting examples of this generalization are to be found among the so-called glacial relics, *i.e.*, cold-preferring forms which lagged behind when the glaciers receded during the final phase of the last great ice age.

An important aspect of the problem of the water temperatures in relation to the fisheries is the effect that temperature evidently has on the rate of growth of most of our fishes, *viz.*, low temperatures decrease, or even inhibit, while relatively high temperatures increase the rate of growth.   Instructive examples of this phenomenon were obtained in the course of our work.   For instance, one species of rockfish which lives in the subtidal region along the rocky parts of our shores was found not to have a regularly rhythmic growth, evidently due to the fact that the thermal conditions of its habitat are very variable, relatively low and high temperatures alternating irregularly, but apparently always within the range which is suitable for the growth of the species.   On the other hand, the members of this group of fishes which live between about 50 and 100 meters have a decidedly rhythmic growth expressed, for instance, in the pronounced stratification of their otoliths.   Another example is the sand dab (*Orthopsetta sordida*) which lives on sandy bottom in Monterey Bay at a depth of around 50 meters.   This species has, as was demonstrated by Mr. W. Dill, a distinctly periodic growth.   During the spring months, when the temperature of the water over the flats is sinking (Fig. 16), the growth is very slow, nearly inhibited; in May, when the warmer season begins, growth is resumed.

Only a few of our fishes react differently.   These represent southern stragglers of species, the main centers of distribution of which are to the north and which are so adjusted to lower thermal ranges that relatively high temperatures cause them to become more or less stunted.

The circulation in the sea affects the fishes mainly in the following ways.   Most of the fishes have planktonic eggs and larvae; that is, these stages drift around, suspended in the water.   Therefore, the currents carry the spawn away from the place where reproduction occurred.   This transportation may or may not be favorable to the optimum replenishment of the species, and it varies from year to year, due to the always present variability of the superficial circulation.   Scofield (1934) demonstrated that the sardine spawn in the principal spawning area was carried by the prevailing southerly currents towards the coast of the southern part of California and the northern parts of Lower California where the main nurseries were to be found.   No attempt was made to estimate the percentage of transportation into the offshore regions which are less favorable on account of their lower productivity in food suitable to the early larval stages.   It may also be noted that the rockfishes are very vulnerable, even though they are viviparous.   This is due to the fact that the type of habitat which they require, *viz.*, rocky bottom at intermediate depths, is by no means abundantly represented in our waters.   If, for instance, offshore currents carry the young and fairly helpless stages away from the continental shelf, which is very narrow along the larger part of our coast, out over the great depths, the probability is that these specimens will perish.   I may mention that in the course of our expeditional work, we frequently observed these stages in such situations.   From these few remarks, it will be seen that a proper knowledge of the circulation is of great importance in the study of the biology of the early stages of our commercially exploited species.

Through circulation, irregularities and fluctuations in the distribution of the water in the sea are created, and these affect the active migrations of some of our commercial fish species.   For instance, the Japan Current carries relatively warm water towards the west coast of North America.   The inner, marginal water of the southern branch of this current, the California Current, appears to be a favored feeding place for the albacore, presumably on account of the fact that this water combines tolerably high temperatures (the albacore is a warm water species) with a rich food supply.   It is a well known fact that wherever warm and cold waters mix, there is, generally speaking, abundance of food; examples of this are found around New Foundland and off the coast of Japan.   Some years ago, the albacore fishery of southern California decreased very abruptly.   This sudden decrease was probably only in part due to the very intensive fishing to which the albacore had been subjected during the previous years; at least to a very considerable extent, it must have been caused by the partial abandonment by the species of these waters, an abandonment brought about by unknown hydrographic changes.   At present, as noted above, the albacore appears to roam quite extensively in the offshore waters of California, sometimes continuing its northerly migrations as far as to the waters of Alaska, where, as we know, a branch of the Japan Current is found.   This current is subject to great fluctuations, and the same is probably also true in respect to the migrations of the albacore.   A heavy flow of the current, for instance, will presumably result in a migration farther to the north, and vice versa.   During these migratory movements, the albacore is at present taken only in relatively small quantities.   The failure of the fishermen to locate and catch the fish is in all probability, at least to a very great extent, due to the fact that the waters preferred by the albacore usually are to be found at a fairly long distance from the coast. On rare occasions, however, it does happen that this type of water comes more or less close to land.   The most extreme case of this kind on record, between 1919 and 1934, occurred in the autumn of 1932.   At that time, the oceanic water penetrated into Monterey Bay on a scale vastly surpassing anything observed in that season of other years, as far as available records go.   Furthermore, it is a fact of extreme interest that the albacore fishing began and ended with this hydrographic episode.   During this short period, the heaviest fishing for albacore in the annals of Monterey took place.   The extent to which the oceanic water came into the Bay on this occasion may be seen from Figs. 10 and 29, and Fig. 23 will demonstrate that on that occasion the temperatures at the surface were higher than at any other time since 1919.   Considering the great importance of the albacore to the fish trade of California, and since the distribution of this species appears to be linked, to a considerable degree, with the fluctuations in the Japan Current and its branches along the coast of North America, these fluctuations should be submitted to a careful investigation.   The present indications are that we should not expect a regular albacore fishery in the Monterey area; great irregularities will probably always characterize this fishery.

Moreover, the albacore is not the only species which apparently has changed its habitat in California waters in historic times.   The following statement is taken from Skogsberg (1925, p. 62).   "Nowadays, almost all the California white sea bass of local origin are taken in southern California waters.   For instance, in 1921, out of an annual total of 2,069,544 pounds, only 55,836 pounds were caught north of San Luis Obispo County.   However, in earlier years the situation was quite different.   In 1895, out of an annual total of 669,780 pounds, only 128,980 pounds were recorded for southern California. For 1904 the corresponding figures were 1,056,534 pounds and 293,145 pounds."   These changes can not be referred exclusively to the shift in the center of human population which of late years has taken place in California.   In part, at least, they must be connected with changes in the center of abundance of the white sea bass.   Other examples could also be given.   These shifts in the fishing grounds undoubtedly were brought forth by changes in the hydrographic conditions along our coast, but there is, of course, no possibility at the present time to trace their actual causes, since sufficient hydrographic records are not available from the period concerned.   The main reason for mentioning them here is that they suggest the necessity of being prepared for similar changes in the future and that for this reason we ought to secure records from the field of oceanography for their proper interpretation.   It goes without saying, that in our consideration of problems of conservation, it is of the utmost importance for us to be able to decide whether the decline of a local fishery was due to overfishing or whether it was caused by a shift in the distribution of the species concerned.

Several examples of a northerly migration of commercial fish species during the unusually warm period in the beginning of 1926 were reported by Hubbs and Schultz (1929); among these was the albacore which was taken off the Oregon coast.

In the previous discussion, only horizontal water movements have been mentioned. However, as we know, vertical shifts on a large scale are also characteristic of our region. The most important of these, from the point of view of practical consequences, is evidently the updraft of cold, deep water towards the surface.   Through the upwelling process, the superficial, coastal waters are cooled.   On account of the variability of the phenomenon, the degree of cooling differs from year to year.   The variation in the cooling evidently has important consequences to the fisheries.   For instance, it appears to affect the extent of the northward migration of the sardines in the following way.   When warm waters prevail during the latter part of the year, this migration is more extensive, and the arrival of the sardine in Monterey Bay and neighboring waters on the return migration is correspondingly late. When the thermal conditions are reversed, the behavior of the sardine is also reversed. This rule is illustrated by the following observations, made by the California Fish and Game Commission, showing two of the most important dates pertaining to the "winter fish" in the Monterey catch. By "winter fish" is meant the year classes of the sardine which participate in the northward migration. The first date of each year refers to the time when the first winter fish appeared in the Monterey catch, and the second refers to the time when the entire sardine population in the Monterey waters apparently was composed of winter fish.

1929–1930	Dec. 26, 1929	Jan. 23, 1930
1930–1931	Dec. 29, 1930	Jan. 26, 1931
1931–1932	Dec. 7, 1931	Jan. 26, 1932
1932–1933	Nov. 25, 1932	Jan. 9, 1933
1933–1934	Nov. 27, 1933	Jan. 6, 1934

The years 1929 and 1930, in which the sardine arrived very late, were unusually warm during the period of northerly migration; the autumn of 1931 was subnormal, as was 1933, and the sardine arrived early, particularly in the latter year which was unusually cold. The year 1932 is somewhat more difficult to interpret; it was decidedly subnormal in the early months and normal in the remaining months, except in August and September, during which the Monterey region was characterized by an unusually large influx of water of oceanic (California Current) origin. In all probability, the more northerly districts did not have this influx but were subnormal, hence the early arrival of the sardine (Fig. 19). In order to obtain fully reliable information in respect to this problem, observations should be made at some distance from the coast and at several points from Monterey towards the north. If the correlation proves to be correct, then there would be a possibility of predicting before the beginning of the canning season whether the winter fish should be expected to be early or late, an information of great significance to the industry.

In respect to the effect on the fisheries of the fluctuations in the available food supply, caused by variations in the chemistry of the sea, little can be said as yet concerning our local waters. Of course, it is a well known fact that the water animals, just like the land animals, depend ultimately on the plants for their food. The most important plants in the sea are microscopic and float around in the upper 100 meters. They constitute what is termed the phytoplankton. The growth and reproduction of these plants depend on the availability of certain dissolved nutrients which occur in quantities so small that they are very easily depleted. As a consequence, the supply must be renewed, or plant life will cease. The new supply comes largely from two sources, viz., from run-off water and from the great depths. In regions depending on the former source, there are two principal periods of plant production, corresponding to the two main periods of run-off. In California, the rivers play a subordinate rôle in this process; most of the nutrients come from below through the upwelling. To what an extent this peculiarity in the cycle of the nutrients affects our fisheries remains for future investigations to establish. Quantitative plankton work must be correlated on the one hand with the variations in the upwelling, and on the other with the phenomena of reproduction and growth in the fishes, as well as with the fat and oil content of the fishes.

As stated in the beginning of this chapter, the correlation between the oceanographic information and the fisheries rests largely with the fishery experts.   The preceding remarks and facts are given simply in order to call attention to some phases of interest and to indicate problems that should be investigated.

X. SUMMARY AND CONCLUSIONS

The hydrography, as well as the oceanography in general, of the sea off the coast of California is to a very large extent still an unexplored realm.   This statement refers especially to the waters north of Point Conception.   Our knowledge of the region to the south of this point is still very deficient indeed, but the Scripps Institution of Oceanography, at La Jolla, has made considerable progress towards a scientific understanding of those waters.

This lack of knowledge is especially deplorable on account of the fact that the California fisheries are very highly developed.   Several of our most important fish species are at present under a very heavy strain of exploitation.   Sound measures of conservation depend on a sound knowledge of the biology of the species involved.   In the case of marine forms, such a knowledge depends in its turn to a considerable degree on an advanced state in the knowledge of the hydrography and general biology of the waters in which the species occur.

Due to the extreme difficulties inherent in the investigation of the complex problems of the sea, the present report, even though its material is relatively extensive, does not presume to have accomplished anything beyond the laying of a foundation for future investigations.   The region investigated is very small, hence many of the results undoubtedly are local in nature.   On the other hand, due to the geographical and topographical peculiarities of the investigated area, many of the results appear to be of a more or less broad applicability.

The investigation, out of which the present report grew, is not finished; it still goes on, and we hope that it will continue as a permanent, sustained and growing part of the program of the scientific exploration and investigation of California and its natural resources. It is the task of the oceanographer to keep the ocean under steady surveillance, just as it is the duty of the meteorologist to give to the atmosphere a constant and continual attention.

Most oceanographic reports pertain to brief and disconnected investigations at sea. Only a few oceanographers have carried on continued research over long periods.   As a consequence, our knowledge of seasonal and long range phenomena in the oceans is very small.   However, a scientific understanding of fisheries problems depends on a knowledge of the dynamics of the seas, and this knowledge can not be obtained except through continuity of research.   One of the most outstanding characteristics of our work is its continuity. Our expeditions during the five years covered by this report, 1929–33, were either semi-weekly or weekly, and they were carried on throughout the entire year.

The investigated area, Monterey Bay, is located in central California, between latitudes 36° 36′ N. and 36° 59′ N.   It represents a comparatively shallow indentation in the coast, nearly semicircular in outline, about 23 miles wide between the northern and southern headlands, and measuring only about 11–12 miles from its mouth to its innermost point. It is characterized by a deep central valley which divides the bay into nearly equal parts and which is not separated from the open ocean by means of a threshold.   The continental shelf is very narrow in the region of Monterey, and the continental slope descends rapidly into the great abysses of the sea. As a consequence of these features, Monterey Bay constitutes a bowl through which the oceanic waters can flow in a fairly unimpeded manner, and it receives both superficial and deep waters within its confines. Indeed, this bay reflects, on a small scale, large-featured oceanic water movements, and thus represents an ideal place for a study of these movements by an oceanographer of limited means of investigation.

In the course of five years, 1929–33, the following number of hydrographic data was gathered:

Temperatures	Chlorinity	Phosphate	Silicate
17707	11046	10661	11175

This report represents an analysis and synthesis of the temperature data.

The thermal characteristics of the water in the Bay were found to exhibit a remarkable consistency from year to year; in other words, there was a distinct annual rhythm, a rhythm presenting pronounced and peculiar traits. These traits are given in a summarized form in the beginning of chapter VIII.

There were three hydrographic seasons: *a*. Above the 100 meter isobath there was a cold water season extending from the middle of February to the end of November. *b*. In the upper 25–50 meters, relatively warm conditions prevailed during the latter part of August, through September and the larger part of October. *c*. In December, January, and the first part of February, waters of very low thermal gradients and of relatively high temperatures were found down to a depth of from 50 to 100 meters.

Besides by the annual rhythm, the waters were distinguished by the following fundamental features. *a*. There was a very narrow amplitude of thermal variation, the total amplitude for the five year period being only 6.5° at the surface. *b*. In spite of this constancy, the rate of thermal changes was very great; indeed, in most instances it probably would have been impossible to duplicate a set of readings after the lapse of as short an interval as one hour. *c*. The annual rhythm did not progress in a regular and uninterrupted manner, but was broken up by irregular features distributed throughout the year in a random manner. ·

The cold water season was caused by the upwelling phenomenon. Thus, while the "normal" level of the 8° water was found at a depth of about 200 meters, at the height of the cold water season this level was raised as far as to about 40–60 meters below the surface. The rate of elevation of the deep waters varied from one year to another; as an example it may be mentioned that the 8° water rose in 1933 at an average rate of 1.5 meters a day. The later subsidence of that water took place at a rate of about three meters a day. The upwelling apparently took place against a deep coastwise current, not against the continental slope. It was evidently due to two causes: first, to the offshore deflection of a southerly coastal current generated by winds from the northerly quarters, and secondly, by an onshore pressure of unknown origin.

The warm water phase of the annual rhythm was interpreted as having been caused by the approximation to the coast of water from the California Current, a southerly branch of the Japan Current.

The low thermal gradient phase was evidently caused by the Davidson Current, a northerly coastal current produced by the winds from the southerly quarters which prevailed during the winter months. Through the rotation of the earth, this current is banked against the coast; hence the surface waters are forced down and low thermal gradients are brought forth.

The narrow amplitude of thermal variation was caused by the circumstance that the upwelling phenomenon prevailed during the summer and the Davidson Current occurred during the winter. The fact that the summer months were foggy and the winter months clear also equalized the effects of insolation.

The high rate of thermal change was caused by the intensity of the circulation and the thermal heterogeneity of the water passing through the Bay.

The breaks in the annual rhythm were due to irregular changes in the superficial circulation largely caused by the winds. Sometimes the coastal waters moved through the Bay in a northerly direction, sometimes in a southerly direction, and in addition unseasonal influx of waters from the California Current complicated the picture.

XI. MONTHLY AVERAGE TEMPERATURES STATIONS 4 AND B

	Surf.	5	10	15	20	25	30	35	40	45	50	55	60	65	70	75	80	85	90	95	100
January																					
1929	12.3	12.0	12.0	12.0	12.0	11.8	11.8	11.7	11.6	11.4	11.2										
1930	12.7	12.8	12.8	12.8	12.7	12.7	12.7	12.6	12.6	12.5	12.5										
1931	13.2	13.2	13.2	13.2	13.2	13.2	13.2	13.2	13.2	13.2	13.1	13.1	13.1	13.1	13.0	13.0	12.9	12.8	12.6	12.3	12.1
1932	11.7	11.6	11.7	11.7	11.7	11.6	11.6	11.6	11.5	11.4	11.3	11.2	11.2	11.1	11.1	10.9	10.7	10.5	10.4	10.1	10.1
1933	11.4	11.4	11.4	11.3	11.3	11.2	11.1	11.0	11.0	10.9	10.9	10.7	10.6	10.5	10.5	10.5	10.5	10.4	10.3	10.3	10.2
1929–33	12.3	12.2	12.2	12.2	12.2	12.1	12.1	12.0	12.0	11.9	11.8										
1931–33	12.1	12.1	12.1	12.1	12.1	12.0	12.0	11.9	11.9	11.8	11.8	11.7	11.6	11.6	11.5	11.5	11.4	11.2	11.1	10.9	10.8
February																					
1929	12.1	11.9	11.8	11.7	11.6	11.5	11.4	11.3	11.3	11.1	11.0										
1930	13.0	13.0	12.9	12.8	12.7	12.7	12.6	12.5	12.4	12.1	11.9										
1931	13.9	13.9	13.9	13.9	13.9	13.9	13.8	13.8	13.6	13.6	13.5	13.4	13.2	13.0	12.8	12.8	12.8	12.7	12.6	12.4	12.4
1932	11.1	11.1	11.2	11.1	11.0	11.0	11.0	11.0	10.9	10.8	10.7	10.5	10.4	10.4	10.3	10.2	10.1	10.0	9.9	9.9	9.9
1933	10.9	10.8	10.8	10.7	10.6	10.4	10.4	10.2	10.2	10.1	10.1	10.0	10.0	9.9	9.7	9.6	9.5	9.4	9.2		
1929–33	12.2	12.1	12.1	12.0	12.0	11.9	11.8	11.8	11.7	11.6	11.5										
1931–33	12.0	11.9	11.9	11.9	11.8	11.8	11.7	11.7	11.6	11.5	11.5	11.3	11.2	11.1	11.0	11.0	10.9	10.8	10.7	10.6	10.5
March																					
1929	11.8	11.7	11.3	10.8	10.6	10.4	10.2	10.0	9.8	9.5	9.4										
1930	13.3	12.8	12.5	12.1	11.9	11.8	11.6	11.5	11.3	11.1	10.9										
1931	13.2	13.0	12.9	12.6	12.3	12.2	12.0	11.6	11.4	11.0	10.9	10.7	10.6	10.4	10.2	10.1	9.9	9.8	9.8	9.6	9.6
1932	11.3	11.3	11.0	10.9	10.7	10.6	10.5	10.3	10.1	10.0	9.9	9.8	9.7	9.7	9.6	9.5	9.3	9.2	9.2	9.1	9.1
1933	11.1	10.8	10.7	10.4	10.3	10.1	10.0	9.8	9.7	9.6	9.5	9.5	9.4	9.4	9.3	9.3	9.2	9.1	9.1	9.0	
1929–33	12.1	12.0	11.7	11.4	11.2	11.1	10.9	10.7	10.5	10.3	10.1										
1931–33	11.9	11.8	11.6	11.4	11.1	11.0	10.9	10.6	10.4	10.2	10.1	10.0	9.9	9.8	9.7	9.6	9.5	9.4	9.4	9.3	9.2
April																					
1929	11.5	11.2	10.8	10.4	9.9	9.5	9.3	9.2	9.1	9.1	9.0										
1930	14.1	13.9	13.4	12.6	11.9	11.5	11.2	10.9	10.6	10.4	10.2										
1931	12.2	11.8	11.6	11.4	11.2	10.9	10.7	10.5	10.4	10.3	10.2	10.2	10.1	10.1	10.0	10.0	9.9	9.8	9.7	9.4	9.4
1932	11.2	11.1	10.9	10.8	10.3	10.2	9.9	9.8	9.7	9.6	9.4	9.3	9.1	9.1	8.9	8.8	8.8	8.8	8.8	8.7	8.7
1933	11.2	11.1	11.0	10.5	10.1	9.6	9.4	9.0	9.0	9.0	8.9	8.9	8.8	8.7	8.7	8.6	8.6	8.6	8.5	8.4	8.4
1929–33	12.0	11.8	11.5	11.1	10.7	10.3	10.1	9.9	9.8	9.7	9.5										
1931–33	11.5	11.3	11.2	10.9	10.5	10.2	10.0	9.8	9.7	9.6	9.5	9.5	9.3	9.3	9.2	9.1	9.1	9.1	9.0	8.8	8.8

XI. MONTHLY AVERAGE TEMPERATURES STATIONS 4 AND B—

(Continued)

Depth in Meters

May

	Surf.	5	10	15	20	25	30	35	40	45	50	55	60	65	70	75	80	85	90	95	100
1929	11.8	11.6	11.1	10.2	9.7	9.4	9.1	9.0	8.8	8.8	8.8										
1930	13.7	13.1	12.4	10.4	9.7	9.5	9.3	9.2	9.2	9.0	8.9										
1931	14.2	13.9	13.1	12.7	12.3	12.0	11.7	11.6	11.4	11.1	10.9	10.7	10.4	10.1	10.0	9.9	9.6	9.6	9.5	9.3	9.3
1932	12.8	12.7	12.4	11.5	10.9	10.3	10.0	9.8	9.6	9.6	9.4	9.3	9.2	9.1	9.1	9.0	8.9	8.9	8.8	8.8	8.8
1933	10.6	10.5	10.4	10.3	10.1	9.8	9.5	9.1	8.9	8.8	8.7	8.5	8.4	8.3	8.3	8.3	8.2	8.1	8.1	8.0	8.0
1929-33	12.6	12.4	11.9	11.0	10.5	10.2	9.9	9.7	9.6	9.5	9.3										
1931-33	12.5	12.4	12.0	11.5	11.1	10.7	10.4	10.2	10.0	9.8	9.7	9.5	9.3	9.2	9.1	9.1	8.9	8.9	8.8	8.7	8.7

June

	Surf.	5	10	15	20	25	30	35	40	45	50	55	60	65	70	75	80	85	90	95	100
1929	13.3	12.5	11.1	10.6	10.4	10.2	10.0	9.9	9.8	9.6	9.6										
1930	13.5	13.1	12.4	11.3	10.7	10.4	10.3	10.1	10.1	9.9	9.8										
1931	14.4	14.2	13.6	13.2	12.8	12.1	11.5	10.7	10.2	10.0	9.8	9.8	9.6	9.5	9.4	9.3	9.2	9.2	9.2	9.0	9.0
1932	12.9	12.6	12.0	11.7	11.1	10.5	10.2	10.0	9.8	9.7	9.5	9.3	9.1	9.0	9.0	8.9	8.8	8.8	8.7	8.7	8.6
1933	11.0	10.9	10.7	10.3	9.9	9.8	9.7	9.5	9.4	9.1	8.8	8.7	8.6	8.5	8.5	8.4	8.3	8.2	8.2	8.2	8.1
1929-33	13.0	12.7	12.0	11.4	11.0	10.6	10.3	10.0	9.9	9.7	9.5										
1931-33	12.8	12.6	12.1	11.7	11.3	10.8	10.5	10.1	9.8	9.6	9.4	9.3	9.1	9.0	9.0	8.9	8.8	8.7	8.7	8.6	8.6

July

	Surf.	5	10	15	20	25	30	35	40	45	50	55	60	65	70	75	80	85	90	95	100
1929	14.1	13.7	13.0	12.2	11.7	11.5	11.2	11.1	10.9	10.6	10.5										
1930	13.3	13.1	11.9	11.1	10.8	10.5	10.3	10.2	10.0	9.8	9.7										
1931	12.9	12.5	12.1	11.1	10.6	10.5	10.3	10.1	10.1	9.9	9.8	9.7	9.6	9.6	9.5	9.5	9.4	9.4	9.3	9.3	9.2
1932	13.2	12.8	12.0	11.4	10.9	10.3	10.2	10.0	9.6	9.4	9.4	9.3	9.3	9.2	9.2	9.2	9.1	9.1	9.0	9.0	
1933	12.2	11.8	11.3	10.8	10.2	9.8	9.6	9.5	9.3	9.2	9.2	9.1	9.0	8.9	8.8	8.8	8.8	8.7	8.7	8.7	8.7
1929-33	13.1	12.8	12.1	11.3	10.8	10.5	10.3	10.2	10.0	9.8	9.7										
1931-33	12.8	12.4	11.8	11.1	10.6	10.2	10.0	9.9	9.7	9.5	9.5	9.4	9.3	9.2	9.2	9.2	9.1	9.1	9.0	9.0	9.0

August

	Surf.	5	10	15	20	25	30	35	40	45	50	55	60	65	70	75	80	85	90	95	100
1929	14.0	13.4	11.8	11.3	11.0	10.8	10.7	10.6	10.5	10.3	10.2										
1930	13.4	13.1	12.1	11.4	11.0	10.8	10.6	10.5	10.5	10.4	10.3										
1931	14.1	13.7	13.2	12.4	11.9	11.6	11.4	11.2	11.0	10.8	10.6	10.4	10.3	10.2	10.1	10.0	9.8	9.8	9.7	9.6	9.6
1932	13.6	13.4	12.9	12.1	11.4	11.0	10.9	10.5	10.4	10.5	10.4	10.3	10.2	10.1	10.0	10.0	9.9	9.9	9.8	9.7	9.7
1933	13.0	12.9	12.2	11.6	11.0	10.2	9.9	9.8	9.7	9.7	9.6	9.5	9.4	9.4	9.3	9.3	9.2	9.1	9.0	9.0	9.0
1929-33	13.6	13.3	12.4	11.8	11.3	10.9	10.7	10.5	10.4	10.3	10.2										
1931-33	13.6	13.3	12.8	12.0	11.4	10.9	10.7	10.5	10.4	10.3	10.2	10.1	10.0	9.9	9.8	9.8	9.6	9.6	9.5	9.4	9.4

September

	Surf.	5	10	15	20	25	30	35	40	45	50	55	60	65	70	75	80	85	90	95	100
1929	14.7	14.2	13.4	12.5	12.0	11.7	11.5	11.3	11.0	10.8	10.6										
1930	15.0	14.2	13.3	12.8	12.4	12.0	11.9	11.7	11.5	11.2	11.0										
1931	14.6	14.1	13.7	13.0	12.5	11.9	11.7	11.5	11.3	11.1	10.9	10.8	10.7	10.5	10.4	10.3	10.2	10.1	10.0	9.9	9.7
1932	15.9	15.7	15.5	15.4	15.1	14.2	13.7	12.7	11.9	11.4	10.9	10.7	10.6	10.5	10.5	10.5	10.5	10.5	10.4	10.3	10.2
1933	13.5	13.4	12.8	12.1	11.9	10.9	10.6	10.4	10.1	10.0	10.0	9.9	9.8	9.8	9.8	9.6	9.6	9.5	9.4	9.2	9.1
1929-33	14.7	14.3	13.7	13.2	12.8	12.1	11.9	11.5	11.2	10.9	10.7										
1931-33	14.7	14.4	14.0	13.5	13.2	12.3	12.0	11.5	11.1	10.8	10.6	10.5	10.4	10.3	10.2	10.1	10.1	10.0	9.9	9.8	9.7

XI. MONTHLY AVERAGE TEMPERATURES STATIONS 4 AND B—

(Continued)

Depth in Meters

October

	Surf.	5	10	15	20	25	30	35	40	45	50	55	60	65	70	75	80	85	90	95	100
1929	14.0	13.7	13.1	12.7	12.3	12.0	11.8	11.6	11.5	11.2	11.1										
1930	14.2	13.7	12.8	12.2	11.8	11.7	11.5	11.4	11.4	11.2	11.1										
1931	13.3	13.0	12.7	12.4	12.1	11.9	11.7	11.5	11.2	11.0	10.9	10.8	10.7	10.6	10.4	10.4	10.3	10.2	10.1	10.0	9.9
1932	14.0	13.8	13.6	13.0	12.4	12.0	11.7	11.3	11.0	10.6	10.5	10.3	10.2	10.2	10.1	10.1	10.0	10.0	9.9	9.8	9.7
1933	12.2	11.9	11.6	11.2	10.8	10.5	10.4	10.3	10.2	10.2	10.1	10.1	10.1	10.0	9.9	9.9	9.8	9.8	9.7	9.6	9.5
1929–33	13.5	13.2	12.8	12.3	11.9	11.6	11.4	11.2	11.1	10.8	10.7										
1931–33	13.2	12.9	12.6	12.2	11.8	11.5	11.3	11.0	10.8	10.6	10.5	10.4	10.3	10.3	10.1	10.1	10.0	10.0	9.9	9.8	9.7

November

	Surf.	5	10	15	20	25	30	35	40	45	50	55	60	65	70	75	80	85	90	95	100
1929	13.0	13.0	12.9	12.8	12.6	12.5	12.4	12.3	12.3	12.1	11.8										
1930	12.6	12.6	12.5	12.4	12.3	12.2	12.1	11.9	11.7	11.5	11.5										
1931	12.3	12.2	12.2	12.0	11.9	11.8	11.6	11.4	11.3	11.2	11.1	11.0	11.0	10.8	10.6	10.5	10.5	10.4	10.3	10.1	10.0
1932	12.5	12.5	12.3	11.7	11.4	11.1	10.6	10.6	10.6	10.5	10.5	10.5	10.4	10.3	10.3	10.3	10.2	10.0	10.0	9.9	9.8
1933	12.0	11.9	11.6	11.4	11.0	10.9	10.7	10.6	10.5	10.4	10.3	10.1	10.0	9.9	9.9	9.9	9.8	9.7	9.6	9.5	9.4
1929–33	12.5	12.4	12.3	12.1	11.8	11.7	11.5	11.4	11.3	11.1	11.0										
1931–33	12.3	12.2	12.0	11.7	11.4	11.3	11.0	10.9	10.8	10.7	10.6	10.5	10.5	10.3	10.3	10.2	10.2	10.0	10.0	9.8	9.7

December

	Surf.	5	10	15	20	25	30	35	40	45	50	55	60	65	70	75	80	85	90	95	100
1929	13.6	13.6	13.4	13.4	13.2	13.1	12.9	12.8	12.7	12.5	12.4										
1930	13.5	13.5	13.5	13.5	13.4	13.4	13.3	13.2	13.2	13.1	13.0										
1931	11.8	11.9	11.9	11.9	11.8	11.8	11.8	11.8	11.8	11.7	11.7	11.6	11.6	11.6	11.5	11.5	11.4	11.3	11.3	11.1	11.0
1932	12.4	12.4	12.4	12.4	12.3	12.3	12.2	12.1	12.0	11.8	11.7	11.6	11.5	11.4	11.3	11.1	11.0	10.9	10.7	10.6	
1933	11.3	11.2	11.2	11.1	11.0	10.9	10.8	10.7	10.7	10.6	10.6	10.5	10.4	10.3	10.3	10.2	10.0	9.9	9.8	9.8	9.7
1929–33	12.5	12.5	12.5	12.5	12.4	12.3	12.2	12.1	12.1	12.0	11.9										
1931–33	11.8	11.8	11.8	11.8	11.7	11.7	11.6	11.6	11.5	11.4	11.4	11.3	11.2	11.1	11.1	11.0	10.8	10.7	10.7	10.5	10.4

XII. BIBLIOGRAPHY

ANDREES
  1899   Geographisches Handbuch zu Andrees Handatlas.
  1900   Allgemeiner Handatlas.
BELKNAP, G. E.
  1874   Deep Sea Soundings in the North Pacific Ocean Obtained in the United States Steamer "Tuscarora."   U. S. Hydrographic Office, No. 54; 51 pp.; 12 pls.; 9 profiles; 1 chart.
BIGELOW, H. B.
  1927   Physical Oceanography of the Gulf of Maine.   Bull. U. S. Bur. Fish., Vol. 40: 2; 207 figs. in text; pp. 511–1027.
BIGELOW, H. B., AND M. LESLIE
  1930   Reconnaissance of the Waters and Plankton of Monterey Bay, July, 1928.   Bull. Mus. Comp. Zool. Harvard Coll., Vol. 70: 5; 43 figs. in text; pp. 427–581.
BISHOP, S. E.
  1904   The Cold-Current System of the Pacific, and the Source of the Pacific Coast Current.   Science, Vol. 20; pp. 338–41.

CHAMBERS, S. W.
    1929 Vertical Sections of One Thousand Meters and over in the Northeast Pacific Ocean.   Scripps Inst. of Oceanogr. (mimeogr.), 23 pp.

CLARK, A. H.
    1916 On the Temperature of the Water Below the 500-Fathom Line on the West Coast of South and North America.   Jour. Washington Acad. Sci., Vol. 6: 13; pp. 413–7.
    1916 On the Temperature of the Water Below the 1000-Fathom Line Between California and the Hawaiian Islands.   Jour. Washington Acad. Sci., Vol. 6: 7; pp. 175–7; 1 fig. in text.

DALL, W. H.
    1882 Report on the Currents and Temperature of Bering Sea and Adjacent Waters.   U. S. Coast and Geodetic Surv.; App. No. 16.
    1904 Currents of the North Pacific.   Science, Vol. 20; pp. 436–7.

DAVIDSON, G.
    1885 The Temperature of the Water of the Golden Gate.   Bull. Calif. Acad. Sci., Vol. 1; pp. 354–7.
    1897 The Submerged Valleys of the Coast of California, U. S. A., and of Lower California, Mexico.   Proc. Calif. Acad. Sci., Geol. (3), Vol. 1: 2; pp. 73–103; pls. 4–12.

DAWSON, W. B.
    1909 Effect of the Wind on Currents and Tidal Streams.   Trans. R. Soc. Canada, Ser. 3, Vol. 3: 3; 20 pp.

DORMAN, H. P.
    1927 Quantitative Studies on Marine Diatoms and Dinoflagellates at Four Inshore Stations on the Coast of California in 1923.   Bull. Scripps Inst. Oceanogr., Tech. Ser., Vol. 1: 7; 4 figs. in text; pp. 73–89.

EKMAN, V. W.
    1905–06 On the Influence of the Earth's Rotation on Ocean Currents.   Arch. f. Mat. Astr., Fysik., Vol. 2; 53 pp.; 1 pl.; 10 figs. in text.
    1906 Beiträge zur Theorie der Meeresströmungen.   Ann. d. Hydrogr. u. Marit. Meteorol., Vol. 34; pp. 423–30, 472–84, 527–40, 566–83; 38 figs. in text.
    1931 Meeresströmungen.   Handb. Phys. Techn. Mech., F. Auerbach and W. Hort.   Vol. 5, pp. 177–206; 10 figs. in text.   Leipzig.

GALLIHER, E. W.
    1932 Sediments of Monterey Bay, California.   Mining in Calif., Jan., 1932; pp. 42–79; 17 figs. in text; 1 chart.

HAMLIN, H.
    1904 Water Resources of the Salinas Valley, California.   Dept. Inter., U. S. Geol. Surv.   Ser. J. Water Storage, 9.   Water Suppl. and Irrig. Pap. No. 89; 91 pp.; 12 pls.

HARVEY, H. W.
    1928 Biological Chemistry and Physics of Sea Water.   Cambridge Univ. Press, 194 pp.; 65 figs. in text.

HELLAND-HANSEN, B.
    1911–12 The Ocean Waters, an Introduction to Physical Oceanography.   1.   General Part (Methods).   Internat. Rev. ges. Hydrobiol. u. Hydrogr. Suppl., pp. 1–84; 46 figs. in text.

HOLWAY, R. S.
    1905 Cold-water Belt along the West Coast of the United States.   Univ. Calif. Publ. Geol., Vol. 4: 13; pp. 263–86; pls. 31–7.

HUBBS, C. L., AND L. P. SCHULTZ
    1929 The Northward Occurrence of Southern Forms of Marine Life Along the Pacific Coast in 1926.   Calif. Fish Game, Vol. 15: 3; pp. 234–40; figs. 80, 81 in text.

HUMPHREYS, W. J.
    1911 Ocean Currents—Their Relation to One Another.   Meteorol. Chart of the North Pacific Ocean, Jan., 1911.   U. S. Dept. Agricult.

HUNTSMAN, A. G.
    1923 Natural Lobster Breeding.   Bull. Biol. Board Canada, 11 pp.; 2 pls.; 1 fig. in text.
    1924 Oceanography.   Handbook of Canada; pp. 274–90; figs. 32–6 in text.   Toronto.

KRÜMMEL, O.
  1907 Handbuch d. Ozeanographie. Bd. 1; xvi + 526 pp.; 69 figs. in text.
  1911 Handbuch d. Ozeanographie. Bd. 2; xvi + 766 pp.; 182 figs. in text.
MAKAROFF, S.
  1894 Le "Vitiaz" et l'Océan Pacifique. (Kummar, St. Petersburg), 337 pp.
McEWEN, G. F.
  1910 Preliminary Report on the Hydrographic Work Carried on by the Marine Biological Station
         of San Diego. Univ. Calif. Publ. Zoöl., Vol. 6: 9; pp. 189–204; 1 fig.; 1 pl.
  1912 The Distribution of Ocean Temperatures along the West Coast of North America Deduced
         from Ekman's Theory of the Upwelling of Cold Water from the Adjacent Ocean Depths.
         Internat. Rev. d. ges. Hydrobiol. Hydrogr., Vol. 5; pp. 243–86; 21 figs. in text.
  1914 Peculiarities of the California Climate Explained on the Basis of General Principles of Atmos-
         pheric and Oceanic Circulation. Monthly Weather Rev., Vol. 42; pp. 14–23 (Washington).
  1915 Oceanic Circulation and Temperature off the Pacific Coast. Nature and Science on the
         Pacific Coast, pp. 133–40; figs. 14–7 in text.
  1916 Summary and Interpretation of the Hydrographic Observations Made by the Scripps Institu-
         tion for Biological Research of the University of California, 1908–15. Univ. Calif. Publ.
         Zoöl., Vol. 15: 3; pp. 255–356; 38 pls.
  1919 Ocean Temperatures, their Relation to Solar Radiation and Oceanic Circulation. Semicent.
         Publ. Univ. Calif., Misc. Stud.; pp. 334–421; 19 figs. in text.
MICHAEL, E. L., AND G. F. McEWEN
  1915 Hydrographic, Plankton and Dredging Records of the Scripps Institution for Biological Re-
         search of the University of California, 1901–1912, etc. Univ. Calif. Publ. Zoöl., Vol. 15;
         pp. 1–206; 4 figs. and 1 map in text.
  1916 Continuation of Hydrographic, Plankton, and Dredging Records of the University of Cali-
         fornia, 1913–1915. Univ. Calif. Publ. Zoöl., Vol. 15; pp. 207–54; 4 figs. in text.
MILLER, R. C., W. D. RAMAGE, AND E. L. LAZIER
  1928 A Study of Physical and Chemical Conditions in San Francisco Bay Especially in Relation
         to the Tides. Univ. Calif. Publ. Zoöl., Vol. 31: 11; pp. 201–67; 5 figs. in text; 5 charts.
MURREY, J.
  1898 Effects of Winds on Distribution of Temperature, etc. Scot. Geogr. Mag., 1898; p. 345.
  1899 On the Temperature of the Floor of the Ocean. Geogr. Jour., Vol. 14.
PAGE, J.
  1902 Ocean Currents. Nat. Geogr. Mag., April, 1902; pp. 135–42.
PETTERSSON, O.
  1909 Gezeitenaehnliche Bewegungen des Teifenwassers. Cons. Perm. Internat. Expl. Mer. Publ.,
         Circ. No. 47; 21 pp.; 1 pl.; 3 figs. in text.
RICHTER, C. M.
  1887 Ocean Currents Contiguous to the Coast of California. Bull. Calif. Acad. Sci., Vol. 2: 7; pp.
         337–50; 8 pls.
RITTER, W. E.
  1903 Marine Biological Survey Work of the University of California. Science, Vol. 19; pp. 360–6.
SCOFIELD, E. C.
  1934 Early Life-History of the California Sardine (Sardina cærulea), with Special Reference to the
         Distribution of the Eggs and Larvæ. Div. Fish and Game, Calif. Bur. Comm. Fish.—
         Fish Bull. No. 41; 48 pp.; 24 figs. in text.
SKOGSBERG, T.
  1925 Preliminary Investigation of the Purse Seine Industry of Southern California. Calif. Fish
         Game Comm., Fish Bull. No. 9; 95 pp., 23 figs. in text.
  1930 Hydrobiological Survey of the Hopkins Marine Station. Calif. Fish Game, Vol. 16: 1; pp.
         35–9.
SUMNER, F. B., G. D. LOUDERBACK, W. L. SCHMITT, AND E. C. JOHNSTON.
  1914 A Report upon the Physical Conditions in San Francisco Bay, Based upon the Operations of

the United States Fisheries Steamer "Albatross" during the Years 1912 and 1913.   Univ. Calif. Publ. Zoöl., Vol. 14: 1; pp. 1–198; 13 pls.; 20 figs. in text.

SVERDRUP, H. U.
   1931   The Origin of the Deep-Water of the Pacific Ocean as Indicated by the Oceanographic Work of the Carnegie.   Gerlands Beitr. z. Geophys., Vol. 29; pp. 95–105; 6 figs. in text.

THORADE, H.
   1909   Über die Kalifornische Meeresströmung.   Oberflächentemperaturen und Strömungen an der Westküste Nordamerikas.   Inaug. Diss.; 3 pls.; 31 pp.   Göttingen.

TOWNSEND, C. H.
   1901   Dredging and Other Records of the United States Fish Commission Steamer "Albatross" with Bibliography Relative to the Work of the Vessel.   U. S. Comm. Fish Fisheries Rep., 1900; 7 pls.; pp. 387–562.

United States Coast and Geodetic Survey
   1926   United States Coast Pilot, Pacific Coast: California, Oregon, and Washington.   330 pp.; 1 chart; 3 figs. in text.

WÜST, G.
   1929   Schichtung und Tiefenzirkulation des Pazifischen Ozeans auf Grund zweier Längsschnitte. Veröff. d. Inst. f. Meeresk.   N.F. A.   Geogr.-naturwiss. R. Heft 20; 63 pp.; 4 pls.; 14 figs. in text.

ZÖPPRITZ, K.
   1878   Hydrodynamic Problems in Reference to the Theory of Ocean Currents.   Phil. Mag., Ser. 5, Vol. 6; pp. 192–211.

TRANSACTIONS

OF THE

AMERICAN PHILOSOPHICAL SOCIETY

HELD AT PHILADELPHIA

FOR PROMOTING USEFUL KNOWLEDGE

NEW SERIES—VOLUME XXIX, PART II
NOVEMBER, 1937

ARTICLE II

The Variation in the Silicate Content of the Water in Monterey Bay, California, During 1932, 1933 and 1934

AUSTIN PHELPS

PHILADELPHIA:

THE AMERICAN PHILOSOPHICAL SOCIETY

104 SOUTH FIFTH STREET

1937

THE VARIATION IN THE SILICATE CONTENT OF THE WATER IN MONTEREY BAY, CALIFORNIA, DURING 1932, 1933, AND 1934

By Austin Phelps

Hopkins Marine Station, Pacific Grove, Calif.

INTRODUCTION

The present paper deals with a portion of the results of the chemical investigations of the waters of Monterey Bay, California, obtained by the Hydrobiological Survey during the years 1932, 1933, and 1934. The Hydrobiological Survey is a joint project of the California Department of Natural Resources, Division of Fish and Game, and of Stanford University. It was originated at the initiative of Professor Tage Skogsberg of the Hopkins Marine Station of Stanford University, and has been operated under his direction since 1928. A full account of the history and aims of the Survey is to be found elsewhere (Skogsberg, '36).

The chemical investigations of the Survey commenced in 1929, when Miss Lucina Stanford, after spending some time at the Scripps Institution of Oceanography, began an analysis of sea water collected at a number of stations in Monterey Bay, at which temperature readings were also taken. In 1931 the direction of the chemical investigations was taken over by Dr. Danella Cope, who studied the methods employed and made a few alterations in them. From 1932 to 1935 the analysis of sea water has been entirely in the hands of one of the laboratory's technicians, Miss Pearl Murray. The present writer became associated with the Hydrobiological Survey at the end of 1934, and accordingly has played no part in the collection of any of the original data which are reported in this paper. His sole rôle has been to tabulate, interpret, and report on the data previously obtained.

Acknowledgment should be made to the following persons who have been responsible for the collection of the data: Professor Tage Skogsberg, who not only conceived of and directed the work, but also took a major part in the actual collection of water samples; Dr. Rolf Bolin, also of the Hopkins Marine Station, who assisted in the handling of the water bottles, often under the most trying conditions; to the crew of the California Fish and Game Patrol Launch, "Albacore," and especially to Captain Lars Weseth for his unfailingly cheerful and efficient cooperation; and to Miss Pearl Murray, who, for the three years comprising this report, conducted the chemical analyses with unvarying conscienciousness. We are also indebted to the Scripps Institution of Oceanography for providing the outline of methods of water analysis which have been largely followed in this work. The author wishes to thank Professor Tage Skogsberg for many helpful suggestions in regard to the arrangement of the material, and for a critical study of the manuscript; he also wishes to thank Professor Frank W. Weymouth for suggestions concerning the statistical treatment of the data, and to Mr. Herbert W. Graham for reading and correcting the manuscript.

153

MATERIALS AND METHODS

All data to be reported were obtained from Station "C," which is located in Long. 122° 01' W., and Lat. 36° 44' N., seven miles NNW. of Point Pinos Lighthouse. The reason for the choice of this particular station from which to collect chemical data lay in the fact that Station "C" lies on the intersection of a line going straight across the mouth of the bay and a line following the deep submarine valley which bisects the bay. It was felt that this station combined the advantages of possessing water which was as representative of the bay as a whole as could be found at any one station; also of permitting observations to be made as deep as 900 meters; and finally of constituting the best position from which to observe the upwelling of deep water which so fundamentally influences the hydrography of Monterey Bay.

Water samples during 1932 were collected once a week from the launch "Albacore" at 10 meter intervals from the surface to 100 meters, at 50 meter intervals from 100 meters to 500 meters, and at 100 meter intervals from 500 meters to 900 meters.[1] During 1933 and 1934 the collections below 500 meters were made every two weeks. The water samples were brought up in Nansen reversing water bottles. As soon as the water arrived on deck about 100 cc. were used to rinse thoroughly a numbered storage bottle, after which the latter was filled, tightly stoppered, and placed in a covered box. The samples of sea water were taken at about 11 A.M., and were analysed for silicate 24 hours later. The assumption was made that the silicate content of the samples did not change appreciably during this period of storage. Citrate of magnesia bottles were used for this purpose. Following their purchase they were leached in the dark for over three months with sea water which was changed every few days during this time. The bottles had been in constant use for over two years before the present work was started, so that it was felt that they had been as thoroughly conditioned as possible. In order to determine whether or not the silicate content changed during the 24 hour storage period, the following experiment was performed: Ten pairs of the regular storage bottles were filled with sea water containing varying amounts of silicate from 0.47 to 5.35 mg./liter in such a way that each pair contained sea water of an identical concentration of silicate. The water in one of each of the pairs of bottles was tested immediately, and that in the second was tested after 48 hours of storage in the dark. The average increase of silicate due to the storage in the dark was found to be 1.2%. From this experiment it was assumed that the silicate content changed less than 1% during 24 hours of storage.

The colorimetric method of Dienert and Waldenbulke ('23) was employed for the determination of the concentrations of silicate in the sea water samples. Thirty to 50 cc. of the water sample were placed in a 100 cc. volumetric flask, and 2.4 cc. of a mixture containing 25% of 7N H_2SO_4 and 75% of 10% ammonium molybdate were added. More of the sample was added to make the total just 100 cc. After 10 minutes the sample was compared with picric acid standards. These latter were made as follows: Commercial

---

[1] All depths mentioned in this paper refer solely to the amount of cable which was let out into the water, as registered by a meter wheel. Owing to the fact that the cable never falls perfectly vertically from the boat, the true depth is slightly less than the depth registered. Whenever the boat drifted considerably owing to wind or heavy weather it was backed into the cable to counteract the drift and keep the cable as nearly vertical as possible. Recent comparisons of protected thermometers with unprotected ones indicate that at 900 meters in the majority of cases the true depth is 0–10 meters less than the observed depth.

picric acid was twice recrystallized from distilled water, and then dried *in vacuo* over fresh concentrated H_2SO_4 for some time.   To 1000 cc. of distilled water was added 0.256 gm. of the dried picric acid, and the resulting solution was held to contain a color equivalent to that produced by 500 mg./liter of SiO_2.   Appropriate dilutions were made in order to obtain a set of 16 permanent color standards, representing 0.2 to 10.0 mg./liter of SiO_2. The colorimeter employed consisted of two 100 cc. Hehner tubes which stood on a glass plate, and were illuminated by daylight (north exposure) reflected up through them by a frosted glass plate which stood underneath at an angle of 45°.   After one tube had been filled with the sample to be tested, the other was filled with the standard which appeared most nearly to match the sample.   Liquid was withdrawn by means of the stopcocks from either one tube or the other, until the strength of color appeared to be exactly equal in each case.   The amount of SiO_2 in the sample was then found by a table which was prepared in accordance with the proportion:

$$X = Y \cdot \frac{A}{B} \cdot \frac{100}{97.5}$$

where X is the amount of SiO_2 in the sample, Y the amount of SiO_2 represented by the standard, and A and B represent the depths of the standard and sample respectively.

In order to determine the relative degree of accuracy of the method of testing, paired samples of sea water were taken as follows: In the course of the regular collection of samples two reversing bottles were placed about one meter apart on the cable, and these were sent down to a given depth, usually 500 meters.   The contents of each bottle were then transferred to a storage bottle in the usual way, and brought in with the regular samples; it was assumed that the silicate content was the same in both samples, and hence that a comparison of the observed content of one sample with that of the other would give an index of the accuracy of determination.   In order to avoid the personal element, the technician was kept unaware of the purpose of the experiment or of which bottles were pairs. In all, 30 such pairs were sampled over a period of about 6 months.   A statistical study of the results showed a standard deviation of 3.4% and a probable error of 2.1% in the determinations.

RESULTS

The results of the present study will be presented under the following headings: General Observations on the Silicate Content; Month by Month Descriptions of the Silicate Content; The Annual Cycle; The Gradient of Silicate; A Correlation Between the Temperature Variation and the Silicate Variation in the Annual Cycle.

GENERAL OBSERVATIONS ON THE SILICATE CONTENT

In Table I will be found the tabulation of the silicate content at various depths for each month, averaged over the 3 year period under consideration, and also the total average for the various depths.   The average variation of silicate with respect to depth for the 3 year period is shown in Fig. 1.   The ordinate in this graph represents depth, and the abscissa represents concentration of silicate expressed as mg./liter of SiO_2.   It will be seen first, that the silicate content increases consistently with depth, down to 900 meters.   This increase with depth has been almost universally observed wherever silicate in the ocean

TABLE 1

MONTHLY AVERAGES AT GIVEN DEPTHS AS MG./LITER

AVERAGE 1932, 1933, AND 1934.

Figures in parentheses refer to the number of observations

Depth	Jan.	Feb.	Mar.	Apr.	May	June	July	Aug.	Sept.	Oct.	Nov.	Dec.	Average
Surf.	0.676 (13)	0.757 (13)	0.598 (12)	0.784 (11)	0.684 (12)	0.934 (10)	0.705 (11)	0.671 (11)	0.648 (5)	0.679 (10)	0.592 (9)	0.527 (6)	0.688 (123)
20 M.	0.817 (13)	0.865 (13)	0.699 (12)	1.03 (11)	0.878 (12)	1.16 (10)	1.20 (11)	1.14 (11)	0.686 (5)	0.956 (10)	0.689 (9)	0.533 (6)	0.888 (122)
40 M.	0.980 (13)	0.951 (12)	1.07 (12)	1.49 (11)	1.41 (12)	1.57 (10)	1.66 (11)	1.47 (11)	0.893 (5)	1.19 (10)	0.958 (9)	0.745 (6)	1.20 (122)
60 M.	1.12 (13)	1.15 (12)	1.36 (12)	1.80 (11)	1.80 (12)	1.83 (10)	1.91 (11)	1.75 (11)	1.00 (5)	1.31 (10)	1.27 (9)	0.894 (6)	1.44 (122)
80 M.	1.36 (13)	1.38 (12)	1.52 (12)	2.01 (11)	2.04 (12)	1.99 (10)	2.12 (11)	1.87 (11)	1.18 (5)	1.47 (10)	1.45 (9)	1.26 (6)	1.65 (122)
100 M.	1.64 (13)	1.68 (12)	1.83 (12)	2.14 (11)	2.13 (12)	2.29 (10)	2.23 (11)	1.94 (11)	1.38 (5)	1.69 (10)	1.61 (9)	1.49 (6)	1.92 (122)
200 M.	2.34 (12)	2.34 (12)	2.39 (12)	2.56 (11)	2.53 (12)	2.64 (10)	2.62 (11)	2.31 (9)	2.13 (5)	2.23 (10)	2.13 (8)	1.88 (6)	2.36 (115)
300 M.	3.28 (12)	3.15 (13)	2.98 (12)	3.03 (11)	3.04 (12)	3.17 (10)	3.01 (8)	2.86 (9)	2.79 (5)	2.89 (10)	2.76 (8)	2.67 (6)	2.94 (116)
400 M.	3.78 (12)	3.71 (13)	3.56 (12)	3.55 (11)	3.50 (12)	3.71 (10)	3.55 (8)	3.30 (9)	3.42 (5)	3.55 (10)	3.50 (8)	3.41 (6)	3.55 (116)
500 M.	4.67 (12)	4.42 (13)	4.13 (12)	4.34 (11)	4.24 (12)	4.34 (10)	4.10 (8)	3.62 (8)	4.05 (5)	4.27 (10)	4.24 (8)	4.10 (6)	4.21 (116)
600 M.	5.54 (9)	5.34 (11)	4.83 (10)	5.14 (9)	4.66 (11)	4.89 (9)	4.93 (7)	4.87 (7)	4.84 (3)	5.03 (8)	5.12 (7)	4.55 (5)	4.98 (96)
700 M.	6.50 (9)	5.66 (11)	5.46 (10)	5.57 (9)	5.53 (11)	5.70 (9)	5.47 (6)	5.38 (7)	5.41 (3)	5.62 (8)	5.45 (7)	5.58 (5)	5.64 (95)
800 M.	6.55 (8)	6.31 (11)	5.87 (10)	6.19 (9)	5.80 (11)	6.12 (6)	5.84 (6)	5.79 (7)	6.07 (3)	6.21 (8)	5.69 (7)	6.06 (5)	6.07 (91)
900 M.	7.25 (8)	6.73 (11)	6.39 (10)	6.59 (9)	6.35 (11)	6.55 (6)	6.40 (6)	6.26 (6)	6.28 (3)	6.69 (6)	6.48 (7)	6.54 (4)	6.54 (87)

has been studied. The only exceptions seem to occur near the mouths of rivers, owing to the addition to the surface layer, of fresh water with a high silicate content (Lucas, Hutchinson, and King, '28), and in some regions where a heavy plankton population depletes the silicate content a few meters under the surface to a greater extent than at the surface, thus disturbing the gradient in the upper 15 meters of water (Hutchinson, Lucas, and McPhail, '29; Thompson and Johnson, '30). The nature of this gradient has not heretofore been adequately studied, however, because too few observations have been made in deep water, and in the same location, to give significant results in this regard.

The present data demonstrate that in Monterey Bay the average silicate gradient takes the following course: From the surface to 100 meters the gradient is regular and much higher than in the deeper waters; from 100–200 meters it is low, being only 36% of that in the upper 100 meters; below 200 meters the gradient increases, until a deep water maximum is reached between the 500 and 600 meter levels, where it is 63% of that in the

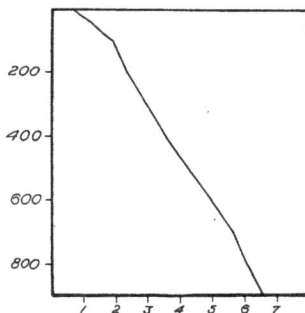

Fig. 1. Average Fluctuation of Silicate With Respect to Depth. Ordinate represents depth expressed in meters and abscissa concentration of silicate as mg./liter.

upper 100 meters. Between 600 and 700 meters the gradient is almost as great as between 500 and 600 meters, but from 700 to 800 meters it is markedly less, being about the same as the gradient from 100 to 200 meters. Finally, the gradient from 800 to 900 meters is insignificantly different from that of 700–800 meters. A more extended analysis of the gradient will be found in another part of this paper, on page 175. It is not believed to be profitable to speculate as to the causes of this variation of gradient with depth, because too little is at present known in regard to the details of the silicate cycle in the ocean. It is of interest to note, however, that the gradient of silicate is very different from that of temperature. The latter, averaged over several years in Monterey Bay, shows a maximum value in the top 100 meters and decreases very regularly at a slowly declining rate from 200 to 900 meters.

Additional features of general interest in regard to the data shown in Table I may be derived from a comparison of the silicate content of Monterey Bay with the silicate values which have been found elsewhere. In order to make this comparison, a very brief review of typical observations on the silicate content of the ocean is desirable.

The question of the silicate content of sea water has proven of particular interest to oceanographers because of the rôle which this constituent plays in the biological economy of the sea.   Since the major part of the photosynthetic microplankton consists of diatoms, and since these forms utilize silicate as the predominating element in the formation of their tests, it has long been realized that silicon must be one of the important ecological factors in the ocean.   It has for many years been maintained that lack of silicon was one of the limiting factors which eventually brought the growth of diatom populations to a halt. Recently there has been evidence presented which indicates that reduction in the concentration of other nutrient salts, particularly nitrates and phosphates, is the real limiting factor to the indefinite increase of the diatom population, inasmuch as these elements at times approach complete disappearance from the surface waters of the ocean far more closely than does silicate.   However, in view of the experiments of Harvey ('33) which show that even in the presence of an excess of phosphate and nitrate a population of *Nitschia* will grow better in a high concentration of silicate than when a concentration as low as is frequently found in the ocean is present, the question of the relative amount of silicate as a limiting factor in diatom growth must be considered to be still open.

The earliest reliable estimates of the amount of silicate in sea water were made by Raben ('05) on water from the Baltic and the North Seas.   He found distinct evidences of a seasonal cycle, the silicate content being high in the late autumn and winter and low in the late spring and summer; the values from the surface ranged from 0.40 to 1.10 mg./liter.   In two further publications ('10 and '13) he reported further data from the Baltic Sea and the North Sea.   These observations for the most part confirmed the seasonal cycle reported in the earlier paper, and the values found were of the same order of magnitude.   In the majority of cases, the silicon content in the upper layers of the North Sea was somewhat lower than that in the Baltic.   In the course of his observations from 1902 to 1912 Raben made a number of observations on the silicate content of subsurface water, the majority of samples coming from the upper 100 meters.   It is interesting to note that in many cases, the deeper waters showed a lower silicate value than did the surface water.

Brandt ('19) found much the same situation in the Baltic where observations taken in 1911 showed silicate values which ran from 0.9 mg./liter in February to 0.6 mg./liter in April and May with an increase to 0.9 mg./liter in June at the surface.   During this same period values found at a depth of 19 meters lay between 0.6 mg./liter and 1.15 mg./liter. Wells ('22) reported some silicate values found a short distance off shore from Gloucester, Massachusetts.   The samples were obtained from the surface, and the silicate content was found to vary from 0.3 mg./liter to 2.9 mg./liter.

An extended series of observations of silicate was made by Atkins ('23; '26; '28; '30), largely at stations in the North Sea, but also in other parts of the Atlantic.   This author found a decided seasonal cycle which followed very closely the plankton cycle.   There was a maximum of silicate in the winter, followed by a strong depletion in the spring associated with the outburst of plankton at that time.   In the summer, when the growth of plankton was less marked, there was a regeneration of silicate.   In the late autumn the silicate again was depleted, this depletion being accompanied by the second seasonal peak in the plankton population.   These findings of Atkins confirmed forcefully the earlier theories of Raben, Johnstone ('08), and Brandt that the cycle of silicate in several regions of the ocean is intimately bound up with the annual cycle of populations of silicious plankton.   Typical values found by Atkins in the North Sea range from 0.05 to 0.35 mg./liter* [2] at the surface, and about the same range at the bottom, at 70 meters.

The first report in regard to the silicate content of water in the stratosphere was provided by Atkins and Harvey ('25). At two stations between Lisbon and the Canary Islands they found an increase of silicate with depth, the values being 0.22 mg./liter* at the surface, 0.28* at 500 meters, 0.45* at 1000 meters, 0.48* at 2000 meters, and 1.20* at 3000 meters. From samples taken from seven stations in the Faroe-Iceland and Faroe-Shetland Channels, off the south coast of Ireland, in the Bay of Biscay, and off Portugal, Atkins ('26) found silicate values ranging from 0.136–0.216 mg./liter* at the surface, 0.26–0.36* at 100 meters, 0.40–0.45* at 500 meters, 0.52–0.65* at 1000 meters, 0.70* at 2000 meters, and 1.60* at 3000 meters.

The first work on the silicate content of the Pacific Ocean is that of Moberg ('26), who obtained samples at two stations, five miles and ten miles off shore from LaJolla, California. The values at both stations were essentially the same. At the 5 mile station 0.20 mg./liter* were found at the surface, 0.65* at 50 meters, 1.12* at 100 meters, and 1.19* at 150 meters. At the 10 mile station 0.23 mg./liter* were found at the surface, 0.68* at 50 meters, and 1.24* at 100 meters. Two years later the same author ('28) reported data from the 10 mile station taken in the summer. These showed values of 0.46 mg./liter* at the surface, 0.91* at 50 meters, 1.32* at 100 meters, and 1.68* at 150 meters. In the same year Hutchinson ('28) reported finding approximately 4.0 mg./liter* at the surface, 1.0* at a depth of 4 yards, and 2.0* at both 10 and 20 yards depths in the Straits of Georgia near Vancouver Island. Thompson and Johnson ('30), making weekly observations from September 1928 to September 1929 in Puget Sound, found an average of 2.8 mg./liter at the surface, with a maximum of 3.3 and a minimum of 2.1. At 12.8 meters the average was 2.7 mg./liter with a maximum of 3.3 and a minimum of 2.1.

Bigelow and Leslie ('30) reported observations of the water at 31 stations all located in Monterey Bay, during July of 1928. These authors tested the silicate content at 18 stations, going down to a maximum of 600 meters at 3 stations. A summary of their maximum, minimum, and average values at various depths is given in Table 2. Thompson, Thomas, and Barnes ('34) reported on the silicate content of the water at stations about 100 miles off the coast of Vancouver Island in July 1933.[3] They found an average of 0.805 mg./liter at the surface, 2.29 at 100 meters, 5.67 at 500 meters, and 9.59 at 1000 meters. Uda ('34), making investigations in the Sea of Japan, found typical values of 0.40–1.5 mg./liter at the surface, 0.73–2.2 at 100 meters, 2.2–3.0 at 500 meters, 2.8–3.7 at 1000 meters, and 3.6–4.0 at 2000 meters. In samples collected in Bering Strait in July 1934, Barnes, Thompson, and Zeusler ('35) found surface values of 0.124–1.24 mg./liter and at 25 meters 0.31–1.73. The same authors found, at two stations located in the region of the tip of the Aleutian Islands, and visited in August of the same year, the following con-

[2] It has been pointed out by King and Lucas ('28) that in performing colorimetric analysis of silicate, the picric acid employed should first be purified several times and then thoroughly dehydrated. Since this precaution has been commonly neglected in work prior to 1928, it has been found necessary to multiply the results of such work by the factor 1.44. Data which have been so corrected in this paper are marked with an asterisk.

[3] In this paper and in the paper of Barnes, Thompson, and Zeusler ('35) the values are given in terms of mg. atoms per kilo. For the sake of uniformity this notation has been changed to mg./liter.

TABLE 2

SILICATE VALUES IN MONTEREY BAY IN 1928 AS MG./LITER

(After Bigelow and Leslie)

Depth meters	Maximum	Minimum	Mean
0	1.12*	0.206*	0.57*
50	2.73*	1.63*	2.17*
100	3.47*	1.99*	2.55*
200	3.47*	2.32*	2.70*
400	4.19*	3.44*	3.59*
600	5.14*	4.41*	4.71*

Note: These authors have given a value of 1.43 mg./liter for the minimum at the surface. Obviously 0.143 mg./liter was intended, and this correction has been made in the above table.

centrations: Surface, 1.85–2.04 mg./liter; 50 meters, 3.10–3.71; 100 meters, 4.03–4.34; 200 meters, 4.64–4.95; 500 meters, 6.19–7.12; and 1000 meters, 7.43–8.05.

The observations made by Graham and Moberg on the cruise of the "Carnegie" in 1929 are of particular significance to the advance of our understanding of the distribution of silicate in the ocean in general and especially in the north and central Pacific.[4] It was found that the silicate content of the Pacific Ocean was considerably higher than that of the Atlantic as reported by Atkins. An average of all the stations at which silicate was determined showed a content of 0.42 mg./liter at the surface, 2.79 at 500 meters, 5.10 at 1000 meters, 7.07 at 2000 meters, and 7.33 at 3000 meters. In addition it was found that there was a definite reduction in the silicate content of the water at given levels in the more southerly latitudes. An average of 16 stations between Lats. 20° N. and 34° N. showed 0.51 mg./liter at the surface, 0.60 at 100 meters, 3.22 at 500 meters and 5.43 at 1000 meters; an average of 12 stations between Lats. 20° N. and 14° S. showed 0.31 mg./liter at the surface, 0.62 at 100 meters, 1.96 at 500 meters, and 3.78 at 1000 meters. In general an increase in silicate was found with increase in depth down to 3000 meters at which level there was a maximum in both the northern and the southern stations.

From the very brief review just given a number of general statements in regard to the distribution of silicate in the ocean may be made. First, it is apparent that the concentration differs very widely at the surface and at given depths in different regions of both the Atlantic and Pacific Oceans. Also, in many localities silicate concentration varies seasonally in conjunction with the waxing and waning of the population of surface plankton. The northeastern portion of the Pacific Ocean shows a concentration of silicate which is greater than that of most of the regions of the Atlantic which have been studied and is considerably larger than that of the west coast of the Pacific. This unusually strong concentration in the northeastern Pacific is particularly marked in the deeper water. Finally, it is evident that the silicate content of the waters of the northeastern coast of the Pacific decreases regularly from north to south at all depths which have been studied.

[4] Grateful acknowledgment is made for permission by the Carnegie Institution of Washington through the Director of its Department of Terrestrial Magnetism to use the silicate data obtained by the "Carnegie," and to examine before printing the manuscript of the valuable investigation of these data by H. W. Graham and E. G. Moberg.

In comparing the average silicate values shown in Table I with the values found elsewhere by other observers, two features should be noted. First, the surface water of the Atlantic Ocean appears in most cases to contain only about 30% as much silicate as is found in Monterey Bay. Notable exceptions to this generalization are the findings of Raben and of Brandt in the Baltic and North Seas and of Wells off the coast of Massachusetts; in both these regions the silicate content was of the same order of magnitude as that reported in the present work. It should be noted, however, that the results of the three authors just mentioned were obtained by means of gravimetric analysis and a comparison of them with results obtained by colorimetric analysis may not be entirely safe. This is indicated by the fact that Raben, using the former method, found surface values of silicate in the North Sea which were significantly higher than those found by Atkins who applied the latter method of analysis to samples taken in the same region. A second point to be noted is that the values found below the surface are lower than those found by Thompson, Thomas, and Barnes ('34) in the open ocean off the coast of Vancouver, and higher than those found by Moberg ('26, '28) off the California coast in the region of La Jolla. This furnishes further evidence in support of the statement by Thompson and Robinson ('32) that the silicate content along the eastern coast of the Pacific Ocean increases as the latitude increases.

THE MONTHLY CONDITION OF THE WATER

The monthly condition of the water is shown in Tables 3, 4, and 5 in which are given the monthly averages of the silicate content at various depths. In Figures 2, 3, 4, and 5 these monthly averages are depicted graphically, the ordinates representing depth, and the abscissae concentration of silicate. It will be seen that these tables and figures show data from only about half of the depths at which samples were taken. This condensation was decided upon for reasons of economy and because the omitted data did not contribute essentially to the clarity of the presentation of the hydrographic picture. The omitted data were employed, however, to this extent: In every case where the value at a certain depth was suspected of being faulty, either through error of collecting or of testing, the values which were found just below and above the level in question were employed to estimate the true silicate content.

Although the data for only 3 years are at present under consideration, and it is thus dangerous to generalize to too great an extent, it seems profitable to give a month by month description of the conditions obtaining in the bay in order that this information may be quickly available for reference.

January: During this month in 1932, the silicate content was above average in the upper 20 meters of water, below average from 40 to 80 meters, and above average from 100 to 900 meters. The highest values of the year occurred during this month from 400 to 600 meters. In 1933, the content was above average at the surface, below average from 20 to 200 meters, and above average from 300 to 900 meters. The highest annual values were found from 700 to 900 meters. In 1934, the silicate was subnormal from the surface to 200 meters, and above average from 300 to 900 meters. The three year average shows the content to be below average from the surface to 200 meters and above average from 300 to 900 meters.

TABLE 3

MONTHLY AVERAGES AT GIVEN DEPTHS AS MG./LITER

AVERAGE 1932

Figures in parentheses refer to the number of observations

Depth	Jan.	Feb.	Mar.	Apr.	May	June	July	Aug.	Sept.	Oct.	Nov.	Dec.	Average
Surf.	0.772 (4)	0.758 (5)	0.618 (4)	0.510 (4)	0.560 (5)	0.851 (4)	0.902 (4)	0.973 (4)	0.613 (1)	0.887 (3)	0.614 (3)	0.693 (3)	0.729 (44)
20 M.	0.964 (4)	0.853 (4)	0.774 (4)	0.635 (4)	0.837 (5)	1.18 (4)	1.68 (4)	1.29 (4)	0.615 (1)	0.938 (3)	0.690 (3)	0.696 (3)	0.930 (43)
40 M.	1.05 (4)	0.926 (4)	1.15 (4)	1.21 (4)	1.32 (5)	1.57 (4)	1.85 (4)	1.49 (4)	0.615 (1)	1.23 (3)	1.03 (3)	0.769 (3)	1.18 (43)
60 M.	1.08 (4)	1.16 (4)	1.42 (4)	1.66 (4)	1.68 (5)	1.83 (4)	2.05 (4)	1.81 (4)	0.628 (1)	1.39 (3)	1.29 (3)	0.887 (3)	1.41 (43)
80 M.	1.42 (4)	1.24 (4)	1.67 (4)	1.97 (4)	1.94 (5)	1.87 (4)	2.28 (4)	1.89 (4)	0.803 (1)	1.54 (3)	1.45 (3)	1.09 (3)	1.59 (43)
100 M.	1.90 (4)	1.33 (4)	1.85 (4)	1.88 (4)	2.13 (5)	2.24 (4)	2.51 (4)	1.96 (4)	1.08 (1)	1.69 (3)	1.65 (3)	1.43 (3)	1.80 (43)
200 M.	2.50 (3)	2.01 (4)	2.30 (4)	2.46 (4)	2.48 (5)	2.64 (4)	2.78 (4)	2.26 (2)	2.05 (1)	2.19 (3)	2.09 (2)	2.08 (3)	2.32 (36)
300 M.	3.03 (3)	2.75 (5)	2.98 (4)	2.98 (4)	2.99 (5)	3.19 (4)	3.07 (1)	2.93 (2)	2.87 (1)	2.87 (3)	2.51 (2)	2.73 (3)	2.91 (36)
400 M.	3.91 (3)	3.44 (5)	3.53 (4)	3.26 (4)	3.42 (5)	3.79 (4)	3.82 (1)	3.55 (2)	3.49 (1)	3.53 (3)	3.51 (2)	3.60 (3)	3.57 (36)
500 M.	4.95 (3)	4.05 (5)	4.02 (4)	4.26 (4)	4.07 (5)	4.32 (4)	4.30 (1)	4.15 (2)	4.10 (1)	4.61 (3)	4.60 (2)	4.66 (3)	4.36 (36)
600 M.	5.81 (3)	4.90 (5)	4.54 (4)	5.00 (4)	4.83 (5)	4.36 (4)	5.25 (1)	5.42 (2)	5.12 (1)	5.50 (3)	5.32 (2)	5.46 (3)	5.13 (36)
700 M.	6.49 (3)	5.29 (5)	5.28 (4)	5.44 (4)	5.33 (5)	5.68 (4)		5.63 (2)	6.03 (1)	5.84 (3)	5.87 (2)	6.50 (3)	5.76 (35)
800 M.	6.15 (3)	6.00 (5)	5.59 (4)	6.05 (4)	5.82 (5)			6.11 (2)	6.15 (1)	6.21 (3)	6.35 (2)	7.03 (3)	6.15 (34)
900 M.	6.72 (3)	6.55 (5)	6.15 (4)	6.21 (4)	6.33 (5)			6.15 (2)	6.21 (1)	7.05 (3)		7.82 (2)	6.58 (32)

TABLE 4

MONTHLY AVERAGES AT GIVEN DEPTHS AS MG./LITER

AVERAGE 1933

Figures in parentheses refer to the number of observations.

Depth	Jan.	Feb.	Mar.	Apr.	May	June	July	Aug.	Sept.	Oct.	Nov.	Dec.	Average
Surf.	0.780 (5)	0.942 (4)	0.615 (4)	0.973 (3)	0.742 (5)	0.963 (3)	0.787 (5)	0.629 (4)		0.630 (3)	0.683 (3)	0.410 (1)	0.741 (40)
20 M.	0.860 (5)	1.07 (4)	0.672 (4)	1.16 (3)	0.977 (5)	1.09 (3)	1.10 (5)	1.28 (4)		1.09 (3)	0.818 (3)	0.410 (1)	1.05 (40)
40 M.	1.01 (5)	1.24 (4)	1.14 (4)	1.77 (3)	1.75 (5)	1.66 (3)	1.65 (5)	1.60 (4)		1.29 (3)	1.00 (3)	0.668 (1)	1.34 (40)
60 M.	1.23 (5)	1.46 (4)	1.37 (4)	2.04 (3)	2.23 (5)	1.91 (3)	2.03 (5)	1.84 (4)		1.40 (3)	1.37 (3)	1.02 (1)	1.63 (40)
80 M.	1.39 (5)	1.73 (4)	1.70 (4)	2.17 (3)	2.43 (5)	2.12 (3)	2.21 (5)	1.98 (4)		1.51 (3)	1.52 (3)	1.32 (1)	1.83 (40)
100 M.	1.60 (5)	2.10 (4)	1.84 (4)	2.46 (3)	2.38 (5)	2.45 (3)	2.42 (5)	2.09 (4)		1.78 (3)	1.66 (3)	1.56 (1)	2.03 (40)
200 M.	2.51 (5)	2.63 (4)	2.43 (4)	2.65 (3)	2.77 (5)	2.79 (3)	2.72 (5)	2.48 (4)		2.38 (3)	2.18 (3)	2.21 (1)	2.52 (40)
300 M.	3.25 (5)	3.63 (4)	3.02 (4)	3.10 (3)	3.20 (5)	3.44 (3)	3.08 (5)	2.93 (4)		2.91 (3)	2.76 (3)	2.70 (1)	3.09 (40)
400 M.	4.10 (5)	4.15 (4)	3.79 (4)	3.83 (3)	3.85 (5)	3.81 (3)	3.64 (5)	3.54 (4)		3.61 (3)	3.56 (3)	3.28 (1)	3.74 (40)
500 M.	5.09 (5)	5.22 (4)	4.23 (4)	4.72 (3)	4.63 (5)	4.62 (3)	4.10 (5)	4.09 (4)		4.10 (3)	4.07 (3)	3.84 (1)	4.43 (40)
600 M.	5.70 (5)	5.99 (4)	5.20 (4)	5.31 (3)	4.88 (5)	5.19 (3)	5.00 (5)	4.70 (4)		5.12 (3)	4.93 (3)	4.10 (1)	5.10 (40)
700 M.	6.87 (5)	6.38 (5)	5.86 (4)	5.81 (3)	5.74 (5)	6.12 (3)	5.79 (5)	5.12 (4)		5.90 (3)	5.12 (3)	5.16 (1)	5.80 (40)
800 M.	7.34 (4)	6.79 (4)	6.48 (4)	6.60 (3)	6.06 (5)	6.23 (3)	6.29 (5)	6.15 (4)		6.26 (3)	5.61 (3)	5.45 (1)	6.31 (39)
900 M.	7.78 (4)	7.48 (4)	7.02 (4)	7.18 (3)	6.56 (5)	6.71 (3)	6.64 (5)	6.40 (4)		6.69 (2)	6.23 (3)	6.15 (1)	6.80 (38)

TABLE 5

MONTHLY AVERAGES AT GIVEN DEPTHS AS MG./LITER

AVERAGE 1934

Figures in parentheses refer to the number of observations

Depth	Jan.	Feb.	Mar.	Apr.	May	June	July	Aug.	Sept.	Oct.	Nov.	Dec.	Average
Surf.	0.475 (4)	0.569 (4)	0.564 (4)	0.871 (4)	0.750 (2)	0.988 (3)	0.427 (2)	0.410 (3)	0.681 (4)	0.519 (4)	0.478 (3)	0.478 (2)	0.601 (39)
20 M.	0.627 (4)	0.674 (4)	0.653 (4)	1.31 (4)	0.820 (2)	1.20 (3)	0.828 (2)	0.846 (3)	0.756 (4)	0.819 (4)	0.558 (3)	0.494 (2)	0.798 (39)
40 M.	0.944 (4)	0.686 (4)	0.913 (4)	1.49 (4)	1.15 (2)	1.47 (3)	1.47 (2)	1.31 (3)	1.17 (4)	1.05 (4)	0.845 (3)	0.787 (2)	1.11 (39)
60 M.	1.05 (4)	0.839 (4)	1.30 (4)	1.70 (4)	1.49 (2)	1.77 (3)	1.65 (2)	1.59 (3)	1.38 (4)	1.16 (4)	1.12 (3)	1.14 (2)	1.33 (39)
80 M.	1.27 (4)	1.17 (4)	1.59 (4)	1.90 (4)	1.75 (2)	1.97 (3)	1.88 (2)	1.73 (3)	1.55 (4)	1.35 (4)	1.38 (3)	1.38 (2)	1.58 (39)
100 M.	1.43 (4)	1.62 (4)	1.70 (4)	2.08 (4)	1.88 (2)	2.17 (3)	1.76 (2)	1.76 (3)	1.67 (4)	1.59 (4)	1.51 (3)	1.49 (2)	1.73 (39)
200 M.	2.00 (4)	2.30 (4)	2.45 (4)	2.57 (4)	2.35 (2)	2.49 (3)	2.35 (2)	2.18 (3)	2.21 (4)	2.13 (4)	2.09 (3)	2.06 (2)	2.27 (39)
300 M.	2.55 (4)	3.06 (4)	2.93 (4)	3.01 (4)	2.93 (2)	2.89 (3)	2.87 (2)	2.73 (3)	2.70 (4)	2.89 (4)	3.01 (3)	2.59 (2)	2.78 (39)
400 M.	3.32 (4)	3.53 (4)	3.36 (4)	3.56 (4)	3.24 (2)	3.52 (3)	3.20 (2)	3.27 (3)	3.34 (4)	3.48 (4)	3.44 (3)	3.35 (2)	3.35 (39)
500 M.	3.97 (4)	3.99 (4)	4.14 (2)	4.03 (4)	4.01 (2)	4.08 (3)	3.89 (2)	3.72 (3)	3.99 (4)	4.09 (4)	4.06 (3)	3.81 (2)	3.93 (39)
600 M.	5.12 (1)	5.15 (2)	4.76 (2)	5.09 (2)	4.27 (1)	5.11 (2)	4.55 (1)	4.50 (1)	4.55 (2)	4.46 (2)	5.12 (2)	4.10 (1)	4.73 (19)
700 M.	6.15 (1)	5.31 (2)	5.23 (2)	5.45 (2)		5.28 (2)	5.18 (1)	5.39 (1)	5.39 (2)	5.12 (2)	5.41 (2)	5.10 (1)	5.36 (18)
800 M.	6.15 (1)	6.08 (2)	5.59 (2)	5.92 (2)	5.53 (1)	5.91 (2)	5.39 (1)	5.12 (1)	5.98 (2)	6.17 (2)	6.00 (2)	5.69 (1)	5.68 (19)
900 M.		6.15 (2)	5.99 (2)	6.27 (2)	6.19 (1)	6.50 (2)	6.11 (1)		6.41 (2)	6.34 (1)	6.73 (2)	5.57 (1)	6.22 (16)

February: In 1932 the silicate content was subnormal from 20 to 900 meters, but somewhat above average at the surface. In 1933 the water was above average in the upper 20 meters, below average from 40 to 80 meters, and distinctly above average from 100 to 900 meters; from 300 to 600 meters occurred the highest values of the year. In 1934 the silicate content was low between the surface and 100 meters, and showed the

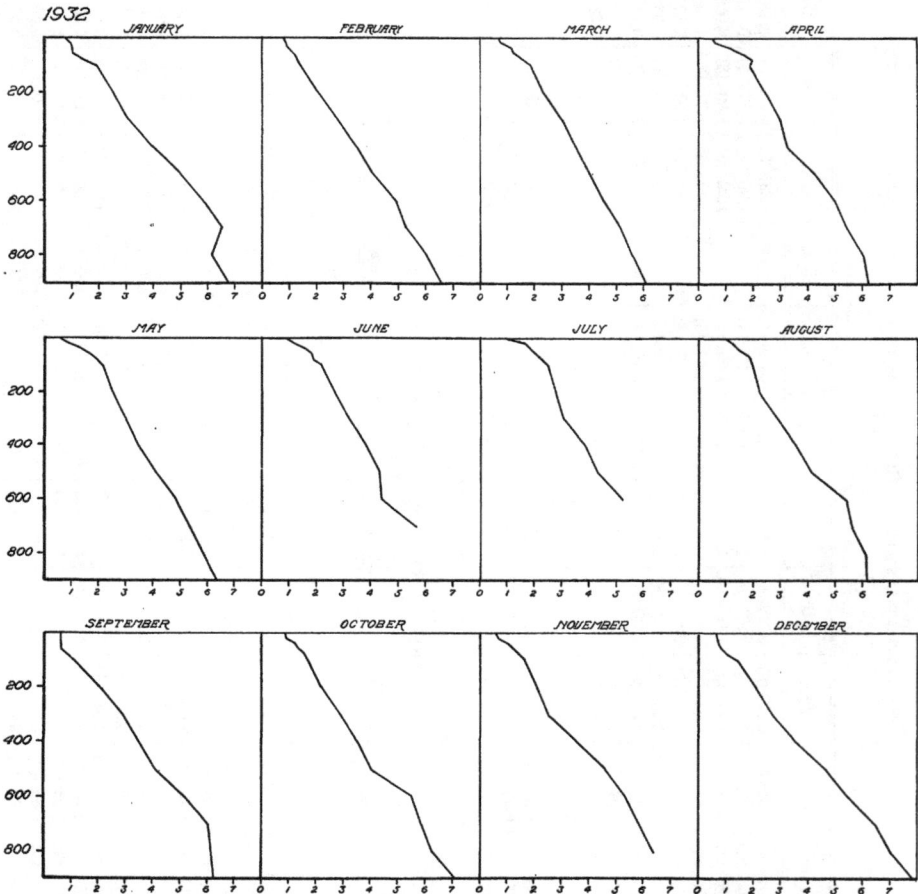

FIG. 2.   Fluctuation of Silicate With Respect to Depth During Each Month of 1932.

lowest values of the year from 40 to 80 meters. From 200 to 600 meters the values were above average, and at 300 meters was found the highest content of the year; at 700 and 900 meters the values were below average and at 800 meters above average. The three year average showed in general the same conditions which held in January, except that the average at the surface was above normal.

March: In all three years the water from the surface to 100 meters showed average to slightly low values.   In 1932 the silicate content was average from 100 to 400 meters, at 500 and 600 meters it was low, and from 700 to 900 meters it showed the lowest values of the year.   In 1933, the values from 100 to 500 meters were likewise about average, but

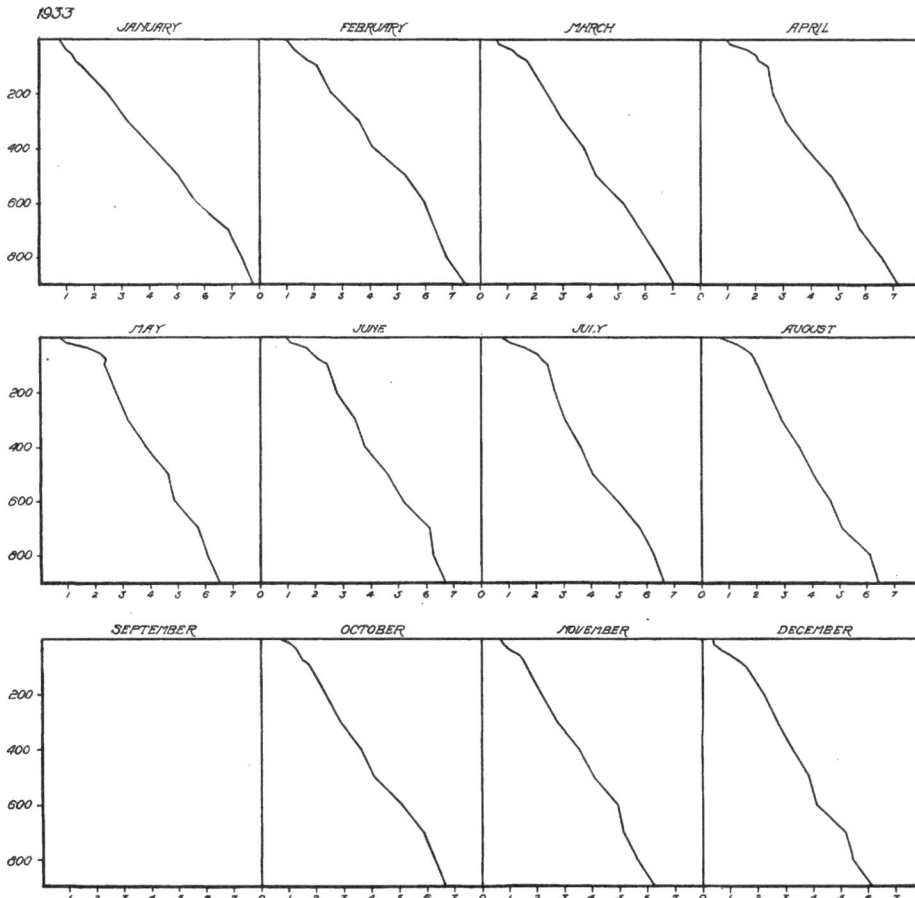

FIG. 3.   Fluctuation of Silicate With Respect to Depth During Each Month of 1933.

from 600 to 900 meters they were somewhat above normal.   In 1934, on the other hand, they were high from 100 to 400 meters, and showed the highest value for the year at 500 meters.   From 500 to 800 meters the silicate content was about average for the year, but at 900 meters it was distinctly low.   The averages were all subnormal with the exception of the 200–400 meter levels which are about normal.

April: During April of 1932 the silicate content was distinctly low from the surface to 20 meters. From 40 to 300 meters the content was above average and from 400 to 900 meters the values were low. In 1933 the content was decidedly high from the surface to 200 meters, and showed the highest values of the year at the surface, at 40 meters, and at 100 meters. From 300 to 900 meters it was high to average. In 1934 it was distinctly

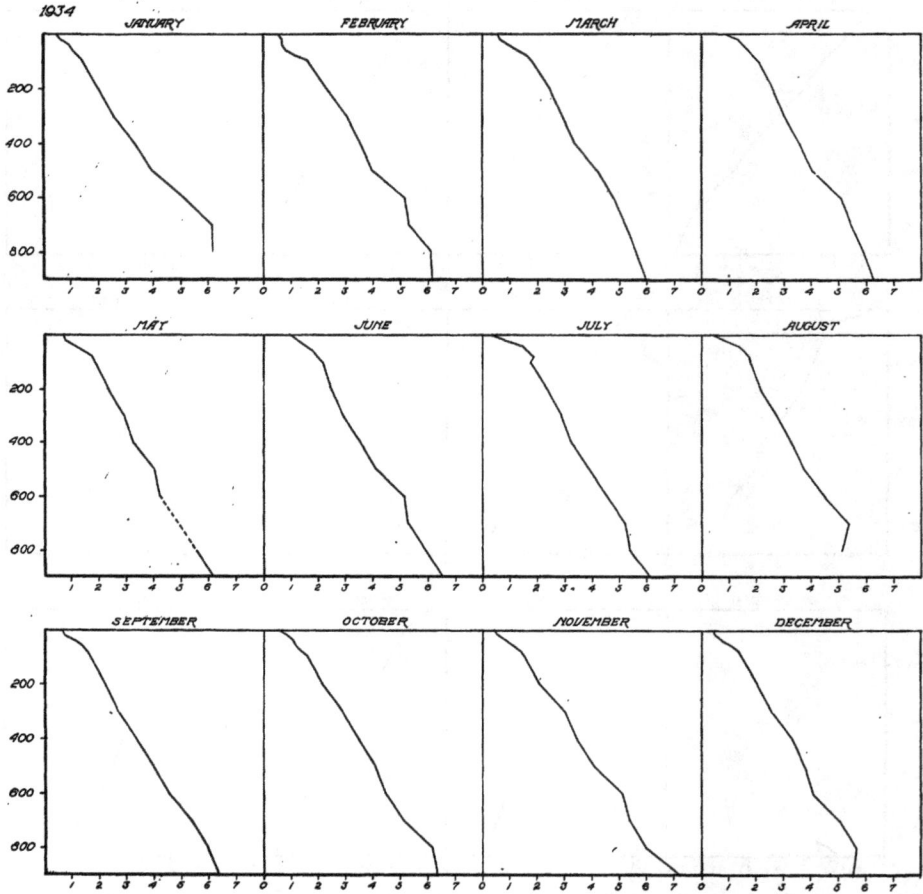

FIG. 4.   Fluctuation of Silicate With Respect to Depth During Each Month of 1934.

above normal from the surface to 900 meters, and the 20 to 40 meter levels of water, as well as the 200 and 400 meter levels, showed the highest values of the year. The three year average indicates that the values from the surface to 300 meters were very considerably above normal and that from 400 to 900 meters they were average to somewhat high, with the exception of the 700 meter level which was a little subnormal.

May: In 1932, the condition of the water almost exactly parallelled that in April, except that in general, the variations from the normal were a little more strongly marked, except in the upper 20 meters. In 1933, conditions from the surface to 20 meters were about average, but from 40 to 500 meters the silicate content was greatly above normal.

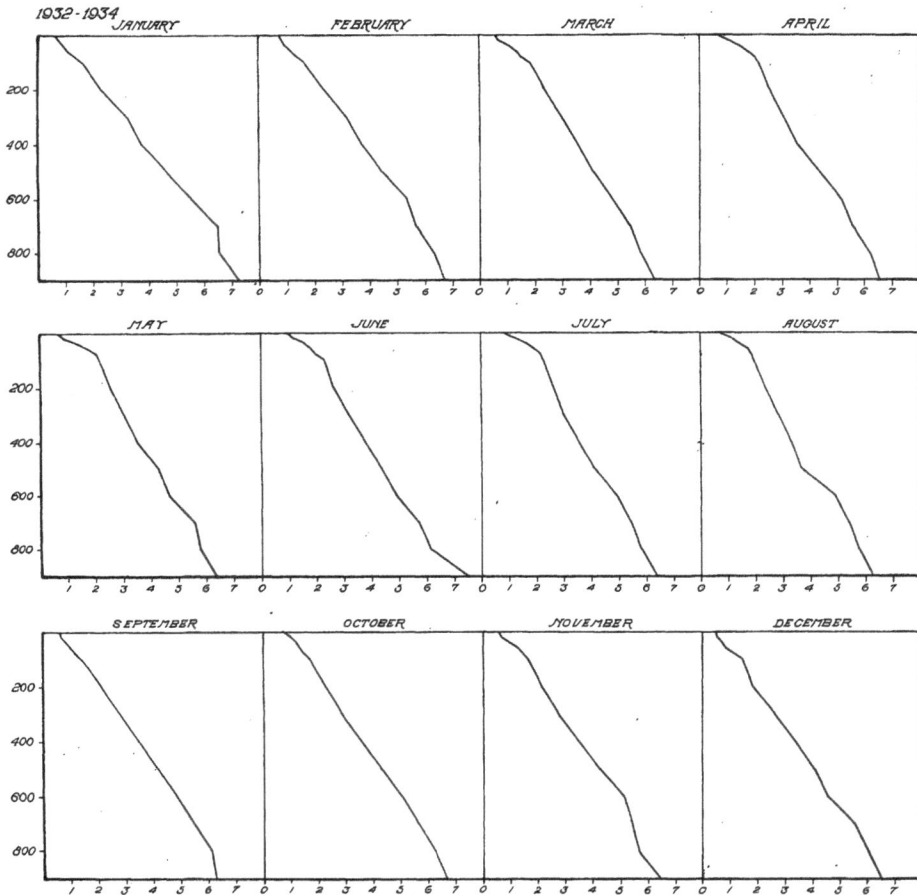

FIG. 5.   Fluctuation of Silicate With Respect to Depth During Each Month, Averaged Over the Three Year Period.

showing the highest values of the year from 60 to 80 meters.   Below 600 meters the content was subnormal.   In 1934, all the values from the surface to 300 meters were somewhat above average and below 500 meters the content was in general a little subnormal.   The three year average indicates that from the surface to 20 meters the silicate content was subnormal; from 40 to 300 meters it was above average; at 400 and 500 meters it was normal; and from 600 to 800 meters it was subnormal.

June: From the surface to 400 meters the content was above average, and from 500 to 700 meters it was below average; at 300 meters it showed the highest values of the year. In 1933 the values were above average at all depths, except at 800 and 900 meters, this effect being most marked in the upper 300 meters; at 200 meters occurred the highest value of the year. In 1934 the values at all depths except at 700 meters were distinctly above average; with the exception of the 20 and 40 meter levels, the content from the surface to 200 meters was the highest of the year. The three years average showed the values at all depths except 600 meters to be above average.

July: During this month, in 1932, the silicate content was very greatly above average down to 400 meters, and showed the highest values for the year from 20 to 300 meters; at 500 and 600 meters, however, the values were about average. In 1933 the values were considerably above normal from the surface to 200 meters; at 300 meters the content was about average and from 400 to 900 meters it was subnormal. In 1934 the upper 20 meters were subnormal, the interval from 40 to 300 meters was above normal, and from 400 to 900 meters the silicate content was below average. The three years average showed values above normal from the surface to 300 meters, average conditions at 400 meters, and a content below average from 500 to 900 meters.

August: In 1932 the values during August were above normal from the surface to 100 meters, and about average from 200 meters on down, except that they were above normal at 600 meters, and subnormal at 900 meters; the content at the surface was the highest of the year. In 1933 the values at the surface were below average, those from 20 to 100 meters were above average, and those from 200 to 900 were increasingly below average; at 20 meters the content was the highest of the year. In 1934 the content was below normal in the upper 10 meters of water, above normal from 20 to 100 meters, and below normal from 200 to 800 meters, with the exception of the 700 meter level, which was about normal; at the surface and at 500 meters occurred the lowest values of the year. The three year average showed values which are subnormal at the surface, above normal from 20 to 100 meters, and below normal from 200 to 900 meters.

September: In 1932 the silicate content was below average from the surface to 900 meters, with the exception of 700 meters, which was above average, and 800 meters which was average; the values were low for the year from 20 to 200 meters. In 1933 there are no records for September. In 1934, the values were consistently subnormal from the surface to 900 meters. The average for the two years showed all values to be below the general average, except for the 900 meter level which was normal.

October: In October of 1932 the silicate values were above average in the upper 40 meters of water, average at 60 meters, below average from 80 to 400 meters, and above average from 500 to 900 meters. In 1933 the content was subnormal from the surface to 500 meters, except for the 20 meter level, which was above average; it was above average at 600 and 700 meters, and below average at 800 and 900 meters. In 1934 the values were below normal from the surface to 200 meters, with the exception of the 20 meter level; from 300 to 500 meters they were above average, from 600 to 700 meters they were below average, and at 800 and 900 meters they were again above average. The three years average showed subnormal values from the surface to 300 meters, with the exception of the 20 meter level; at 400 meters they were average and from 500 to 900 meters they were above normal, except for the 700 meter level, which was below average.

November: In 1932 the content was below average from the surface to 400 meters, and above average from 500 to 800 meters; the lowest values of the year occurred at 300 meters. In 1933 all values were below average from the surface to 900 meters.   In 1934 they were below normal from the surface to 200 meters, and above normal from 300 to 900 meters; the highest content of the year occurred at 900 meters.   The three years average showed subnormal values from the surface to 400 meters; at 500 and 600 meters they were about average, and from 700 to 900 meters they were again subnormal.

December: In December of 1932 the values were below average from the surface to 300 meters, and above average from 400 to 900 meters; from 700 to 900 meters occurred the highest values of the year.   In 1933 the silicate content was very subnormal, showing the lowest values of the year from the surface to 900 meters; too much significance cannot be placed in these data, since only one set of observations was made during this month in 1933.   In 1934 the content was subnormal from the surface to 900 meters, except for the 400 and 800 meter levels, which were about average.   The lowest values of the year occurred at 20, 600, 700 and 900 meters.   The three year average showed subnormal values from the surface to 700 meters, and average values at 800 and 900 meters.

THE ANNUAL CYCLE

The annual cycle of the silicate content of the water of Monterey Bay is shown in Figs. 6 and 7.   These graphs are constructed on the same principle as that employed by Skogsberg ('36) in depicting the seasonal cycle of the temperature of the bay, and they are designed to show the weekly fluctuations in the concentration of silicon, with respect to depth.   The ordinates represent depths expressed in meters, and the abscissae represent time.   The silicate content of the water is indicated in each graph by means of different areas, the limits of which represent units of concentration.   When certain units of concentration were found to occur between two of the levels which were actually sampled, the depth at which these units occurred was found by interpolation, assuming that the silicate content increased at a uniform rate with depth.   This assumption was not always perfectly justified, but nevertheless it is felt that the errors so introduced were unimportant.   The limits of the various areas are 1.00 mg./liter $(16.19 \times 10^{-3}$ mg. atom/kilo)[5] apart beginning with 1.50 mg./liter (24.29), and each area may be distinguished by a characteristic design. The blank area at the top of the figures represents a silicate content of less than 1.5 (24.29), the black area, a content of 1.5–2.5 mg./liter (24.29–40.48), the diagonally crosshatched area, a content of 2.5–3.5 mg./liter (40.48–56.67), the area marked by vertical crosshatching with dots, 3.5–4.5 mg./liter (56.67–72.86), the area marked by vertical crosshatching without dots, 4.5–5.5 mg./liter (72.86–89.05), the area marked by diagonal lining sloping upward to the right, 5.5–6.5 mg./liter (89.05–105.2), the area marked by diagonal lining sloping upward to the left, 6.5–7.5 mg./liter (105.2–121.4), and finally the blank area at the bottom

[5] All the data reported in the present work have been collected and tabulated in units of mg./liter of SiO_2.   Inasmuch as there is at present a strong tendency among oceanographers toward expressing most of their chemical data in terms of mg. atoms/kilo, and since many investigators have come to think in these terms, both units are given in this paper, the units in parentheses referring in every case to mg. atoms/kilo $\times 10^{-3}$ of Si.   In all future papers on the chemistry of Monterey Bay, the data will be reported in accordance with the suggestions of Carter, Moberg, Skogsberg, and Thompson ('34).

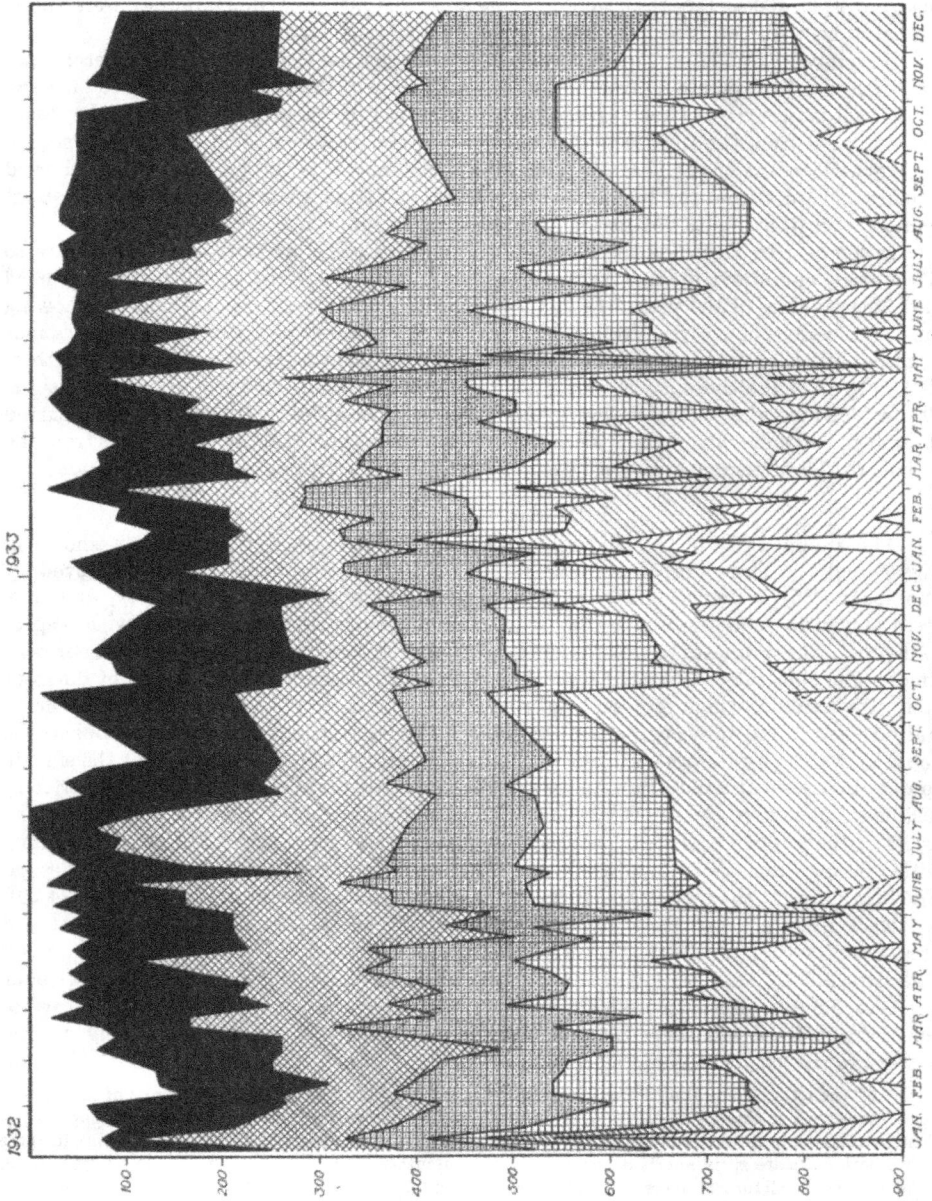

Fig. 6.   Weekly Fluctuation of Silicate During 1932 and 1933.   (For explanation, see text.)

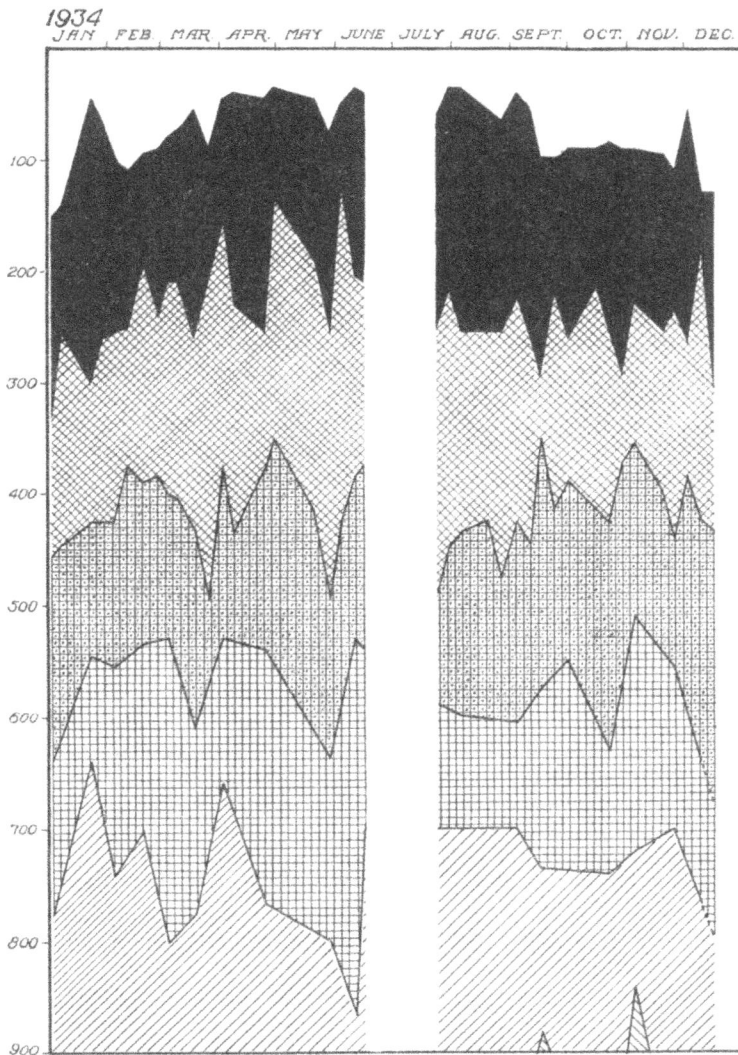

FIG. 7. Weekly Fluctuation of Silicate During 1934. (For explanation, see text.)

of the figures represents a content in excess of 7.5 mg./liter (121.4).   In the ensuing description, the silicate content will be referred to in terms of blocks differing from each
other by 1.00 mg./liter (16.19), and all the water having a silicate content between 0.50
mg./liter (8.09) and 1.50 mg./liter (24.29) will be termed "0.50 (8.09) water," etc.

It is only in the upper 200 meters of water that a distinct annual rhythm is evident, so the detailed description will be confined to this region.   Beginning with January, 1932 it will be seen that during this month the upper 75 meters was occupied by 0.50 (8.09) water, and the remainder by 1.50 (24.29) water, except for a single pulse of 2.50 (40.47) water which came up almost to the 100 meter level in the second week.   During February the silicate content was decidedly lower, the upper 130 meters being occupied largely by 0.50 (8.09) water, and the deeper layer by 1.50 (24.29) water.   It is the last time until the following September that 0.50 (8.09) water invaded the region below 100 meters.   In March the silicate content of the upper 200 meters became noticeably higher.   The 1.50 (24.29) water climbed up as high as to the 25 meter level, and for the second time during the year the lower layers were occupied by 2.50 (40.47) water.   During April conditions were very stable, the upper 50 meters of water being occupied by 0.50 (8.09) water, and the remainder by 1.50 (24.29) water.   There was an increase in the silicate content in May; the 50 meter level was occupied by 1.50 (24.29) water during the latter part of the month, and both at the beginning and again at the end of May the deeper portions of the upper 200 meters were invaded by 2.50 (40.47) water.   In June the silicate continued to increase, and the 1.50 (24.29) water came up as high as the 20 meter level; at the same time the 2.50 (40.47) water extended on one occasion as high as the 100 meter level.   The climax of this continuing increase in silicate content came in July.   It is seen that during the latter part of this month 0.50 (8.09) water was entirely driven out of the bay, the upper layers being occupied up to the surface by 1.50 (24.29) water.   The deeper portion of the upper 200 meters was occupied entirely by 2.50 (40.47) water, and this latter came up to the 65 meter level, this being the highest point reached by it during the whole year.   In August there was a sharp and strongly marked falling off in concentration.   The 0.50 (8.09) water returned to the bay, and descended to the 50 meter level.   The region from 100 to 200 meters was occupied entirely by 1.50 (24.29) water for almost the entire month.   During the last week in August and the first week in September, the falling off in silicate content was more strongly marked than at any other time during the entire year.   By the end of the first week in September the 0.50 (8.09) water had descended to the 130 meter level, and the remainder of the upper 200 meters was occupied by 1.50 (24.29) water.   Unfortunately no observations were made between the first week in September and the middle of October, so we know nothing concerning the silicate content of the water during that time.   In the middle of October the content had risen again somewhat, but it fell off immediately, so that by the end of October the values were almost as low as they were in the first part of September.   During November the silicate content rose a little in the first two weeks of the month, and fell off a little during the latter part; during the whole month, however, it was decidedly low.   In the beginning of December the values fell off very markedly, the 0.50 (8.09) water descended as low as 130 meters, and the next 70 meters were occupied by 1.50 (24.29) water; but during the latter part of the month there was a recovery.

In 1933 the silicate content of the upper 200 meters of water was found to be on the whole distinctly higher than in 1932.   This may not at first be apparent upon examination of Fig. 6, because of the fact that the highest points reached by the 1.50 (24.29) water and by the 2.50 (40.47) water were achieved in 1932, when the former water reached the surface, and the latter water came up to the 65 meter level, an occurrence unparallelled in 1933.   That the silicate content was actually higher in 1933 is shown by the fact that the 1.50

(24.29) water invaded the upper 50 meters during a total of 182 days in this year, whereas in 1932 it was present for only 108 days in that region.   Similarly, in 1933 the 2.50 (40.47) water was found in the upper 200 meters during 188 days, while in 1932 it was present for only 110 days.

Starting in January 1933, the water was found to contain an average amount of silicate; the upper 50 meters was occupied by 0.50 (8.09) water and the remainder by 1.50 (24.29).   Toward the latter part of the month, however, there was a strong falling off in the content of the upper layers and the 0.50 (8.09) water descended as far as the 125 meter level, a depth which was achieved only once again during the entire year, in the latter part of October.   During February there was a tremendously strong sudden surge upward of the silicate content, so that by the end of the month the 1.50 (24.29) water had reached the 20 meter level, one of the three highest points of the year.   Similarly, the 2.50 (40.47) water came up above the 100 meter level at that time.   During March there was another falling off in the silicate, the upper 50 meters of water being occupied by 0.50 (8.09) water, and the remainder containing almost entirely 1.50 (24.29) water.   In April came the increase in silicate which was to last through the rest of the spring, and through the summer; the 1.50 (24.29) water came up to the 30 meter level and the 2.50 (40.47) water climbed steadily until by the end of the month it had reached the 100 meter level.   During May the depth of the 0.50 (8.09) water remained quite constant at 30–40 meters, but during the middle of this month the 2.50 (40.47) water left the upper 200 meters for a short time.   In June the upper portion of the water varied very little and showed a slightly lower silicate content than in May; the lower portion was irregular, the 2.50 (40.47) water coming up almost to the 100 meter level in the middle of the month, and leaving the upper 200 meters altogether at the end of the month.   In July the upper layers were just about the same as they were in May, but in the lower levels the silicate showed a sharp decrease during the month.   In August the upper layers remained practically unchanged, with the 0.50 (8.09) water extending down to the region of the 30 meter level; in the deeper layers, however, the silicate content was reduced so that during most of the month the 2.50 (40.47) water was entirely pushed out by 1.50 (24.29) water.   No data were collected during September or the first part of October.   During the latter part of October there was a very sharp falling off in the silicate, the 1.50 (24.29) water descended to the 125 meter level, and the 2.50 (40.47) water left the upper 200 meters for the rest of the year.   In the upper layers, in November, the 1.50 (24.29) water at first rose sharply and then gradually settled back until by the end of December it lay at a depth of 95–100 meters.   The condition of the water at that time was very similar to that found in the first part of January 1932 and at the end of December of the same year.

January of 1934 was marked by a steady encroachment of the upper layers by 1.50 (24.29) water, which started with its upper limits at 150 meters and by the third week reached as high as the 45 meter level.   During the last week of January, however, the silicate content fell rapidly, and by the middle of February the 0.50 (8.09) water had descended to the 110 meter level, the last time that it reached to such a depth until November.   From the middle of February until the first of May there was an almost unbroken increase in the silicate content of the water, at the end of which time the 1.50 (24.29) water had come up to the 35 meter level.   During this period the 2.50 (40.47) water made its first appearance of the year and invaded the upper 200 meters three times, each time rising higher and remaining longer.   In the first three weeks of May there was a gradual decrease in silicate which was followed by an increase so that by the middle of June the conditions were about the same as they were in the beginning of May.   In the beginning of June occurred the greatest upwelling during the year of the 2.50 (40.47) water, which reached the 125 meter level and then promptly receded.   This water did not appear again in the upper 200 meters until the end of December.   There are no data available for the last half of June and the first three weeks of July.   At the end of July the conditions were the same as they were early in June, except that the 1.50 (24.29) water extended deeper. During the first three weeks of August there was a slight decrease in the silicate, followed by an increase during the last week.   In September there was a sharp and persistent decrease in the silicate, and the 0.50 (8.09) water descended from the upper 30 to 60 meters where it was during the spring and summer, to the 100 meter level.   Through October and November the condition of the water was about the same as it was at the end of September, the upper 100 meters being occupied by 1.50 (24.29) water.   The first few days in December were marked by a brief but pronounced upwelling of 1.50 (24.29) water which came as high as the 55 meter level.   This was followed by an equally sharp falling off, so that by the middle of the month the 1.50 (24.29) water was not found above 130 meters.   During this last sharp decline of the silicate content of the upper 150 meters of water there was a brief upwelling of deeper water which brought the 2.50 (40.47) water into the upper 200 meters for a few days.

Combining and summing up briefly the data of the three years, we find evidence of a distinct annual cycle.   In January the water was of about average silicate content but the latter was decreasing so that by the end of January and in February it was definitely subnormal.   Following this there was a general rise in silicate content which continued during most of the spring and then evened off, leaving a high content during the latter part of the spring and summer.   Toward the end of July and in August the silicate fell off more or less sharply and from October through December this decline continued, but at a slower rate.

In regard to the water below 200 meters it is impossible to draw general conclusions from the limited data at hand.   It will be seen that in 1932 there was no general trend observable below the 200–300 meter level, except possibly for an increase in the silicate content during the last three months of the year in the deeper layers.   In 1933 on the other hand, there was a marked increase in the silicate content during January and February, followed by a very gradual decrease during the remainder of the year, so that during almost all of that year the content was higher than in 1932.   This feature was particularly marked in the deep water; thus, 6.5 (105.23) water entered the bay only six times from January 1932 to December of that year, and each time it remained for not more than two to three weeks,[6] while from December 1932 to May 1933 6.5 (105.23) water was found in the bay without interruption, and from May through August it was present about half the time.   Likewise, prior to December 1932, 7.5 (121.41) water was not found in the bay, but during that month, and January and February of 1933 it was found on three occasions.

[6] It should be noted, however, that there was a possible exception to this statement, namely the period following June 5th.   Unfortunately, owing to a broken cable, no observations were made below 700 meters during the summer of 1932.   Likewise, the length of time that 6.5 (105.23) water remained in the bay is uncertain in two other cases, for about six weeks in September and October of 1932 and 1933 respectively, owing to missing data.

Despite the fact that only half as many observations were made in the deeper water in 1934 as during the two preceding years, there is sufficient data to indicate that there was no definite seasonal trend or cycle during that year.   Throughout 1934 the conditions were just about the same as they were during the first nine months of 1932, except that from 800 to 900 meters the water showed a slightly lower concentration.

If the seasonal variations in the silicate content of the water be looked at from another angle, the annual cycle in the upper layers will be made increasingly evident and the situation in the deeper water will be seen even more clearly.   In Fig. 8 the water from the surface to 900 meters has been divided somewhat arbitrarily into three layers, an upper layer from the surface to 100 meters inclusive, an intermediate layer from 200 to 500 meters inclusive, and a deep layer from 600 to 900 meters inclusive.   The average values of the silicate content of each layer during the three years under consideration were taken as a base line and the average values during each month of each year were plotted against this base line.   In the upper layer during 1932 it is seen that the first three months were subnormal, from May until July the content was increasingly above normal, and from August until December the values decreased regularly.   In 1933 the values with the exception of those in March increased until a maximum was reached in May, then there was a steady decrease until December, the averages probably becoming subnormal in September. During the last year under consideration the values increased until a maximum was reached in June, and from that point on there was a steady monthly decrease until December.   In each of the three years the values during the first three months were subnormal with the exception of those of February 1932; the values were above normal during the summer; and during the last three months were below normal with the exception of those of October 1932.   In the intermediate layer there was no real evidence of a seasonal cycle; during 1932 and 1934 the values were distributed more or less at random, but in 1933 there was evidence of a somewhat relative and uniform decrease in the values during the entire year. In the deep layer there was again no evidence of a seasonal cycle.   In 1932 there appeared to be a cycle which was the reverse of that in the upper layer; in 1933 the relatively consistent decrease in the silicate content during the entire year, which was noted in the intermediate layer, was strongly marked, and in 1934 the values occurred apparently at random. Particularly striking was the evidence of a non-seasonal cycle in the deep layer during 1932 and 1933.   Starting with January of 1932, which showed values distinctly above average, the silicate content decreased, coming to a minimum in March of that year.   Following this, there was a relatively consistent increase until maximum values for the three year period were reached in January of 1933.   During 1933 there was a remarkably consistent fall in the silicate content, and the values were decidedly subnormal during 1934.   This cycle was reflected, though to a less marked extent in the intermediate layer.

THE GRADIENT OF SILICATE

The gradient of all the values of silicate averaged together, are shown in Fig. 1, which has already been briefly discussed.   In Fig. 5 is shown the silicate gradient for each month, averaged over the three year period under consideration.   In Figs. 2, 3, and 4, will be seen the gradient which was found during each of the 36 months comprising this report.   A study of Fig. 5, showing the three year average for each month, will reveal a distinct indication of a regular seasonal cycle in the superficial water.   In the upper 50 meters of water

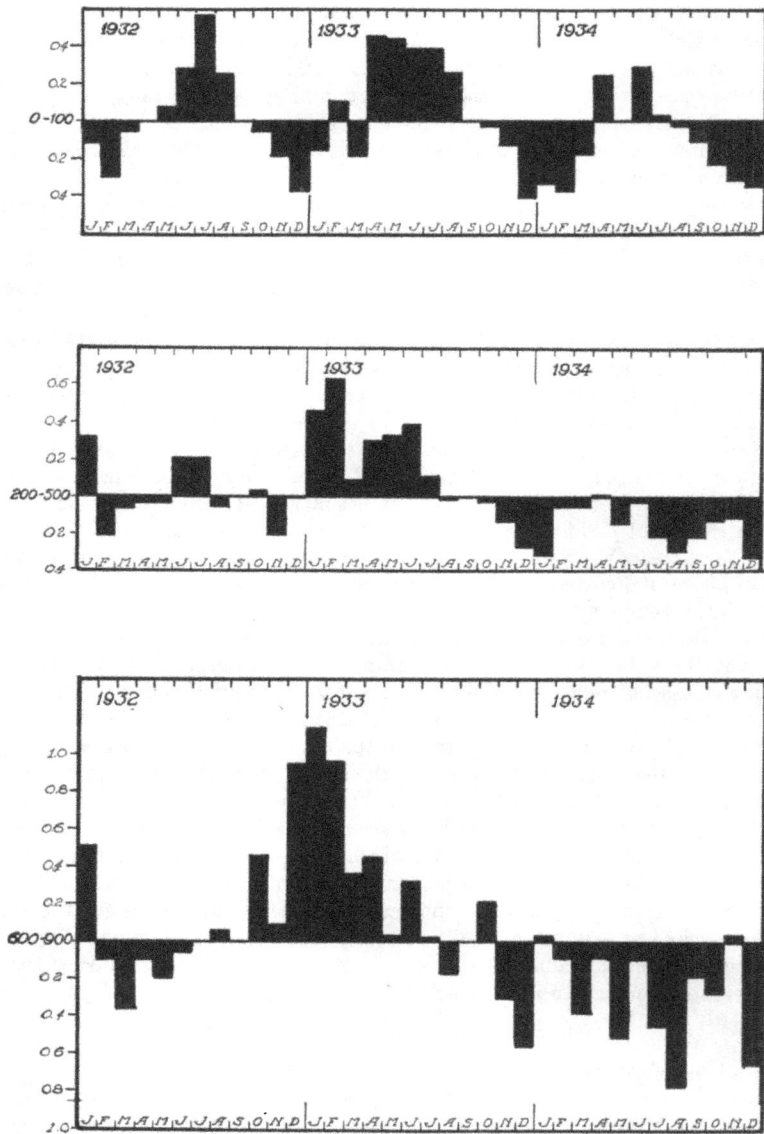

FIG. 8.   Deviation of Silicate Values From the Three Year Average For Each Month in Three Different Regions.

the gradient was low in the winter, rose through the spring, reached a maximum in the summer, and fell off during the autumn. The gradient from 80 to 200 meters was much lower during the 4 months from May to August than it was during the remainder of the year. Below 200 meters there was no evident seasonal cycle; however, there was one phenomenon in the deeper water which was so definitely marked that it is deserving of special mention. This was the striking consistency of the deep water maximum gradient which occurred during 8 months between the 500 meter and the 600 meter levels and during the other four months (January, May, June and December) just below this, in the region between 600 meters and 700 meters. A comparison of the silicate gradient found in Monterey Bay with that found by Graham during the cruise of the "Carnegie" is of particular interest. At 16 stations lying between Lats. 20° N. and 34° N. in the Pacific Ocean this investigator found that the silicate gradient formed a smooth sigmoid curve; the gradient was very low in the upper 100 meters of water and increased until it reached a maximum between 400 and 500 meters, following which it decreased steadily with increasing depth. In the southern group, comprising 12 stations between Lats. 20° N. and 14° S., the silicate gradient was much more constant from the surface to a very considerable depth; indeed, the increase in silicate with depth was practically linear from the surface to 300 meters and from 700 meters to 1500 meters. There was, however, a clearly marked maximum in the silicate gradient between the 500 meter and the 600 meter levels. The northern group of "Carnegie" stations showed a much lower silicate gradient in the upper 100 meters than occurs in Monterey Bay; from 100 meters to 300 meters the silicate gradient was about the same as that found in the bay; and from 300 to 900 meters the gradient was very considerably greater. Curiously enough the gradient found in the "Carnegie's" tropical group of stations resembled that found in Monterey Bay more closely than did the gradient found in the Temperate Zone. The silicate increased with depth far less in the upper 100 meters at the southern stations than in Monterey Bay, but from 100 to 900 meters the gradient followed very much the same course as that in the bay, except that it was slightly less steep. It is of particular interest to note that the deep water maximum of gradient occurring in Monterey Bay between 500 and 700 meters was clearly reflected in the "Carnegie" data. In the more northerly group of stations this maximum occurred between 400 and 500 meters, whereas in the tropical group of stations it lay between the 500 and 600 meter levels. The great difference in gradient in the upper 100 meters between Monterey Bay and the open waters of the Pacific Ocean is undoubtedly due to the strong upwelling of deep water which occurs in the former region. This upwelling furnishes the surface layers with a high concentration of nutrient salts, including silicate, and thus permits the flourishing of a much heavier plankton population than is found in the open ocean. The heavier plankton population tends to create a higher gradient in the upper 100 meters of water by using up the nutrient salts in the photosynthetic zone and by increasing the concentration of these salts below the photosynthetic zone as a result of the dissolution of their sinking dead bodies.

The gradient between the surface and 50 meters and between 50 and 100 meters in Monterey Bay for each month of the three years, is shown in Fig. 9; the average gradients for each month are also shown. In the interval from the surface to 50 meters the gradient was low during the first 2 months of 1932, and increased during the spring to come to a maximum in July. There was a decrease until October and an increase during the rest of the year.   In 1933 there were again low values at the beginning of the year which increased during the spring to reach a maximum in May; the values which were high during the late summer and early autumn decreased regularly during the autumn and early winter.   During the first 6 months of 1934 the gradient was irregular and differed from that observed during the similar period in 1932 and 1933.   Like its predecessors, however, this year showed a maximum gradient in the summer and regularly declining values during the autumn and early winter.   It is believed that the assumption is justified that there is a regular annual cycle in the gradient of the upper 50 meters, the gradient being low in the winter, increasing during the spring, high in the summer and decreasing during the autumn.

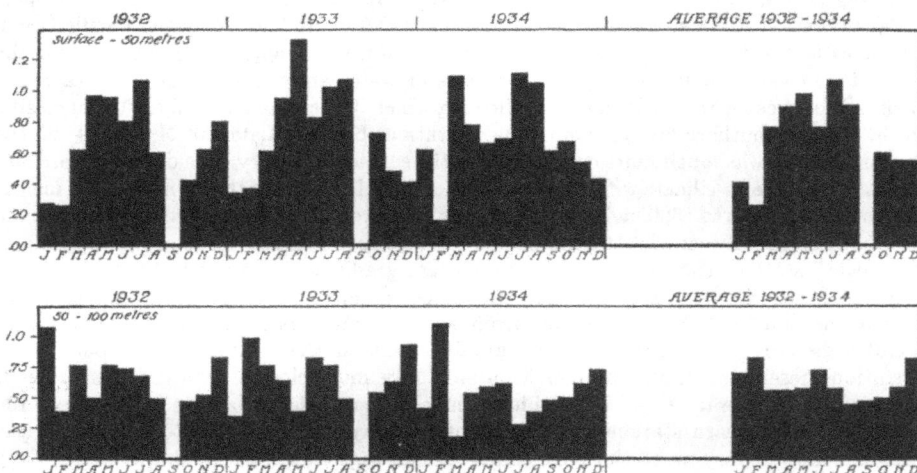

FIG. 9.   Average Gradient From the Surface to 50 Meters, and From 50 to 100 Meters During Each Month From 1932–1934.   Also the Gradients for Each Month Averaged Over the Three Year Period.

Between 50 and 100 meters the monthly variation in gradient was far less regular than in the upper layer.   During 1932 the gradient in the deeper water appeared in most cases to parallel that of the upper 50 meters, especially during the last half of the year.   In 1933 and 1934, however, the reverse seemed to be true, there being a low gradient between 50 and 100 meters when there was a high gradient in the upper 50 meters, and *vice versa*.

COMPARISON OF SILICATE WITH TEMPERATURE

Although our knowledge of the details of the cycle of silicate in the ocean, that is of the mechanism responsible for its peculiar distribution horizontally, vertically, and through time, is still incomplete, it has been definitely established that the dominant rôle in this cycle is played by biological factors and by ocean currents.   Despite the fact that the rivers which empty into the ocean frequently contain several hundred times as much silicate as is found in sea water, the vast majority of this silicate occurs in a suspended form, and is precipitated upon reaching the sea (Thompson and Robinson, '32).   As a result, the amount of soluble silicate which is added in this way effects the distribution in the ocean as a whole only to a negligible extent.   Similarly, although it has been suggested that contact with the shore (Bigelow and Leslie '30) or with the ocean floor may very possibly increase the amount of dissolved silicate in the water, actual observation does not indicate that this is a significant distributional factor.   It has been shown by Moberg ('26) that water taken from the end of a pier along a sandy beach has about the same silicate content as that found 5 and 10 miles out at sea.   Our own data tend to confirm the latter observation; water taken at the surf line at the Hopkins Marine Station has a somewhat higher silicate content than is found at Station "C" (this partially confirms the suggestion of Bigelow and Leslie in regard to the very local influence of the shore upon the silicate content of the water).   Inasmuch, however, as the water at Station "L," 50 miles due west of Station "C," shows approximately the same silicate concentration at all depths down to 900 meters as is found at Station "C," it is felt that the silicate at the latter station is not affected by proximity to the shore.   In considering the possibility that contact with the ocean floor may have an important effect upon the distribution of silicate, it is interesting to note that Graham on the cruise of the "Carnegie" found that where observations were made at a sufficient depth, a definite maximum of silicate was found, not next to the bottom, but frequently at a considerable distance above it.   It should be clearly understood that it is not being maintained that contact with rivers, with the shore, and with the bottom has absolutely no effect upon the silicate content of the ocean, but it is felt that these factors are negligible when the cycle of silicate in the ocean as a whole is being considered.   Since the silicate cycle is determined primarily by biological phenomena and by currents which distribute the results of biological phenomena, a comparison of the variation of silicate content with the variation of temperature should be of particular interest, since the latter is controlled exclusively by physical factors.

The temperature conditions in Monterey Bay have been fully described by Skogsberg ('36), and a comparison between the weekly thermal conditions and the silicate content can be made by reference to his Figs. 7–12 and 29, and Figs. 6 and 7 of the present work. In order to extend the comparison through the year 1934 the weekly thermal conditions are shown in Fig. 10.   In this figure, water having a temperature over 13° is indicated by an area marked by lines which slope upward and to the left, water between 12° and 13° by lines which slope upward and to the right, water between 11° and 12° by vertical crosshatching, water between 10° and 11° by vertical crosshatching with dots, water between 9° and 10° by diagonal crosshatching, water between 8° and 9° by black shading.   Water having temperatures which lie between 7° and 8°, 6° and 7°, 5° and 6°, respectively, and water having a temperature below 5°, are indicated by the four blank areas in the lower portion of the figure.

All figures are on the same scale both with respect to depth and to time.   If the temperature and silicate figures be compared, one of the first points which will be noticed is the striking similarity between the general seasonal cycles in each case.   The marked upwelling of deep water in the spring and early summer of each year which is reflected in the temperature conditions is closely paralleled by the conditions with respect to silicate content; likewise the same close similarity is apparent in the decline of values in the late summer and the autumn, and the low values in the winter.   This of course is to be expected since both silicate and temperature vary consistently with depth, and the seasonal cycle is determined by the upwelling of deep water in the spring and summer and by special current phenomena

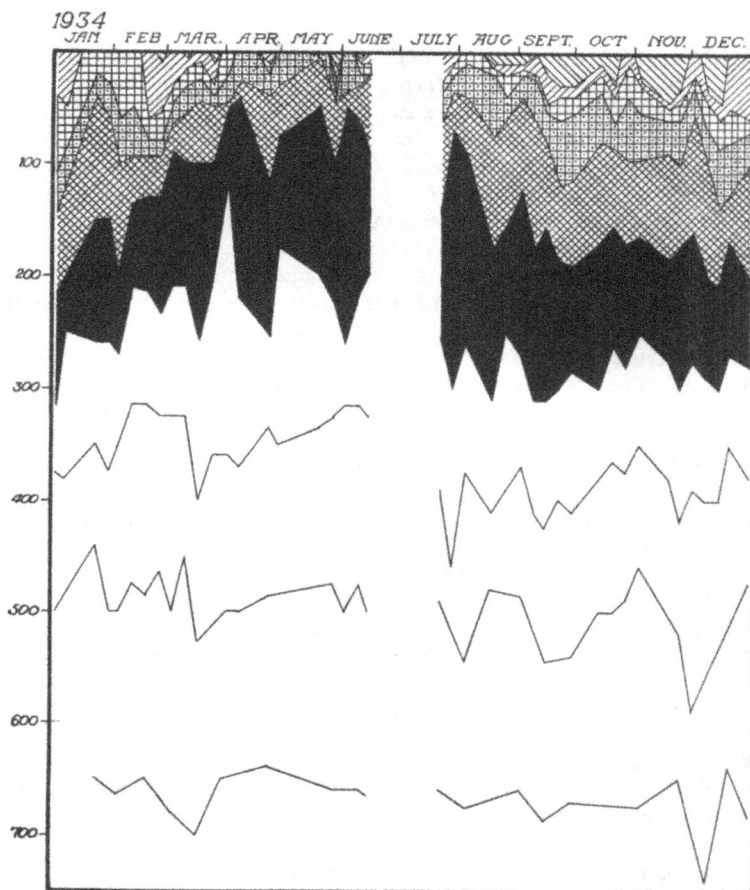

FIG. 10. Weekly Fluctuation of Temperature During 1934. (For explanation, see text.)

in the autumn and winter. The similarity between temperature and silicate is also clearly shown by a comparison of Fig. 8 with Fig. 11. This latter figure is based on exactly the same principle as the former, and shows for each of three layers of water variation of the average temperature during each month from the average for the entire three year period. It is seen that, while the correlation between silicate and temperature is not exact, the same general trends are found in both cases in each of the three layers. It will be realized that the variation of silicate above and below the average is the reverse of that of the temperature, which is to be expected inasmuch as temperature decreases with depth, whereas silicate increases with depth.

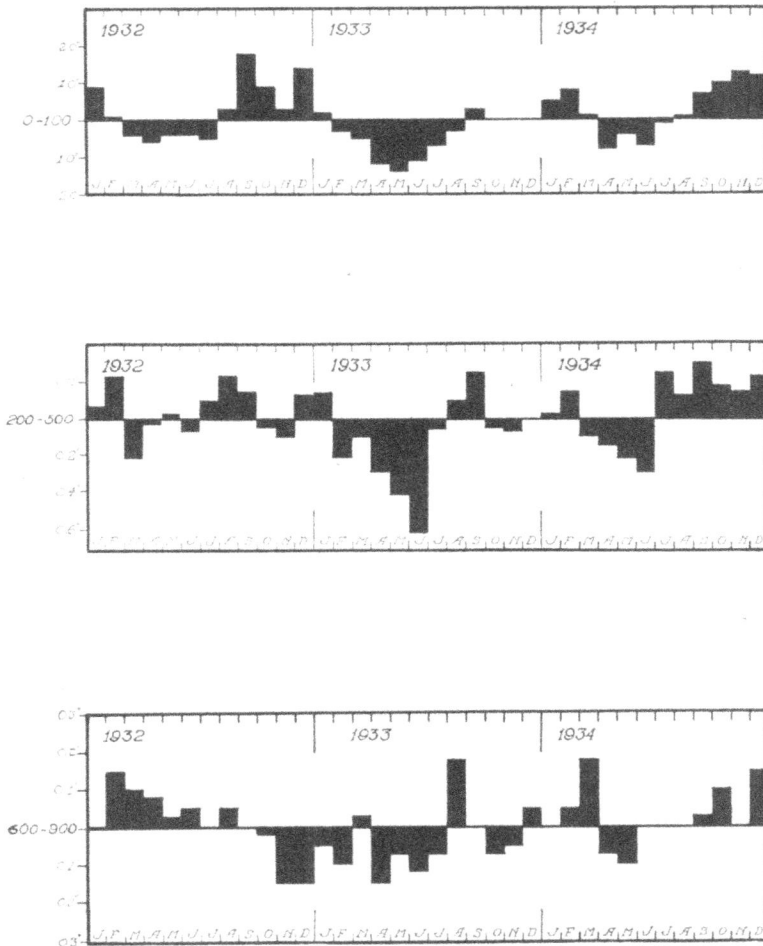

FIG. 11.   Deviation of Temperature Values From the Three Year Average for Each Month in Three Different Regions.

The seasonal silicate cycle, while paralleling the temperature cycle qualitatively shows a distinct quantitative difference in the upper layers.   This is shown by Figs. 12 and 13 in which the silicate content is plotted against temperature in the same way in which it was plotted against depth in Figs. 6 and 7.   It will be noted that in each of the three years in the upper layers of water which contained a given quantity of silicate it was decidedly warmer during the late summer and the autumn and was colder in the late winter and the spring.   Since this seasonal difference of silicate with respect to temperature is

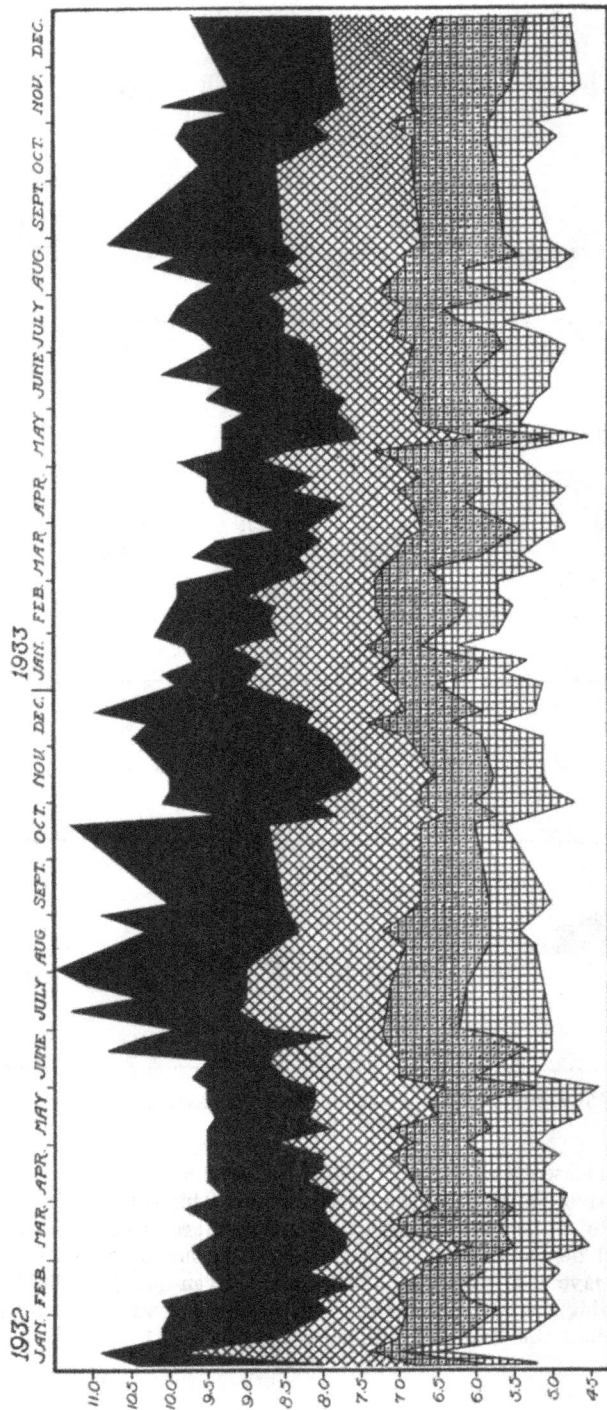

FIG. 12. Fluctuation of Silicate with Respect to Temperature During 1932 and 1933.

largely limited to the upper layer of water it seems reasonable to assume that it is due partly to a greater surface warming of the water in the late summer and autumn than occurs in the spring and early summer; that is, that the temperature is influenced by local physical factors which do not influence the silicate content.   On the other hand, it should be observed that probably there are strictly biological factors which influence the silicate content but have no effect upon the temperature.   It may be mentioned at this point that there is no particular seasonal variation of silicate which can be correlated with the seasonal fluctuation of plankton, such as has been observed in the North Sea, in the Baltic, and elsewhere.   Indeed, no regular fluctuation in the plankton of Monterey Bay has been observed, although systematic weekly estimates have been made over a period of two years.

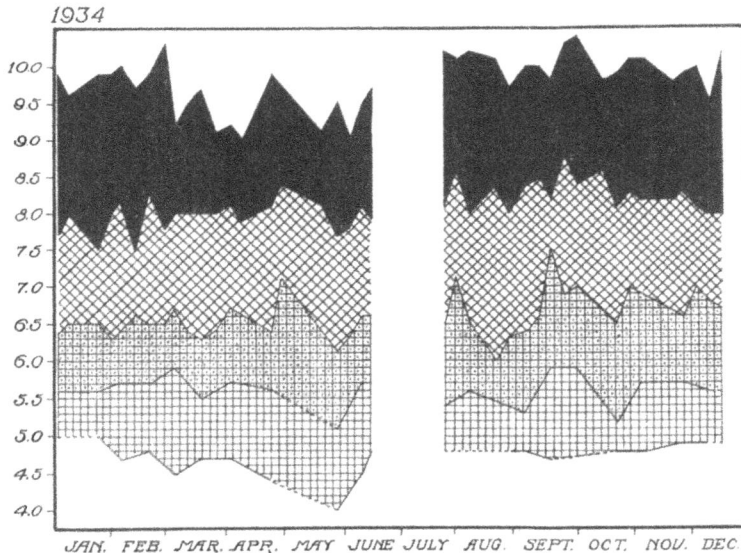

FIG. 13.   Fluctuation of Silicate with Respect to Temperature During 1934.

A comparison of the individual variations of silicate from week to week with the corresponding variations of temperature brings to light a number of interesting points. Rough inspection shows that in many instances the fluctuations of silicate from one week to the next are paralleled by the fluctuations of temperature.   This is most particularly true in the upper 200 meters and is undoubtedly largely due to vertical movements of the water.   In order to determine the exact amount of correlation between the silicate and temperature fluctuations, the coefficients of correlation between the two have been calculated for three periods of four months each, in each of the three years, at four representative levels, the surface, 100 meters, 500 meters, and 900 meters.   The results are shown in Table 6.   The figures in the column marked "r" show the coefficient of correlation between the temperature and the silicate and the figures in the column marked "p"

TABLE 6

	Dec., Jan., Feb., Mar.		Apr., May, June, July		Aug., Sept., Oct., Nov.	
	r.	p.	r.	p.	r.	p.
Surf. 1932	−0.041	10.1	−0.231	64.0	−0.449	89.4
" 1933	−0.029	8.0	−0.412	87.6	−0.709	96.7
" 1934	−0.022	5.4	−0.493	88.1	−0.240	61.0
100 M. 1932	−0.540	92.6	+0.249	68.0	−0.644	98.0
" 1933	−0.883	99.8	−0.437	90.9	+0.009	2.2
" 1934	−0.636	97.3	−0.110	27.7	−0.150	41.1
500 M. 1932	+0.051	13.5	−0.841	99.8	+0.001	0.3
" 1933	+0.029	8.0	−0.663	99.0	−0.314	65.2
" 1934	−0.012	3.0	+0.082	20.5	+0.036	4.0
900 M. 1932	+0.665	95.3	+0.167	31.8	−0.025	5.5
" 1933	+0.066	17.0	−0.206	55.9	−0.117	25.8
" 1934	−0.153	24.3	+0.307	50.6		

indicate the probability expressed as the number of chances in one hundred, that there is a true correlation and that the results are not due to chance. It will be seen that at the surface the correlation is very poor indeed during the four month period from December through March in all three years; from April through July, which represents the period of strong upwelling of subsurface water there is a slight correlation in 1932, and a good one in 1933 and 1934; from August through November the correlation is good in 1932, very strong in 1933 and fair in 1934.   The correlation at 100 meters is very strong from December through March in each of the three years; during the upwelling period it is fair in 1932, strong in 1933 and very poor in 1934; from August through November the correlation is pronounced in 1932 and insignificant in 1933 and 1934.   During each period of each year the correlation with but one exception is negative, as would be expected from the fact that the temperature decreases and the silicate increases with depth, and that much of the fluctuation in both temperature and silicate is occasioned by pulses of deeper water surging toward the surface.   At 500 meters the correlation is insignificant except in three cases; from April through July it is very strong in 1932 and 1933 and during the period from August through November it is doubtful.   At 900 meters the correlation is insignificant throughout, except in one case; from December through April the correlation is strong and positive.   There is every reason to think that this is without much significance inasmuch as the seven other periods show such a very low probability of any correlation being present.

Because of the extreme variation from one year to another which occurred in so many instances it is difficult to see the significance of the correlations just outlined.   There are, however, two points which should be noted.   One is that there is a much greater degree of negative correlation at the surface and at 100 meters than at the two deeper strata. A second feature to be observed is the lack of correlation of any sort in the week to week fluctuation of silicate and of temperature which occurs so frequently in the upper layers and almost without exception in the deeper water. Temperature is generally considered to be one of the more reliable indices of the origin of a given body of water. Fluctuations in temperature in Monterey Bay are particularly significant in this respect and are due almost wholly to varying current phenomena, with the exception of course of the superficial stratum which is influenced by local action of air, and warming by the sun. This has been clearly brought out by Skogsberg ('36) in his study of the temperature conditions of Monterey Bay. Since the fluctuations of silicate values show on such frequent occasions lack of correlation with the fluctuations of temperature the conclusion seems evident that the silicate fluctuation is not due solely to the same factors which influence the temperature variation, *viz.*, differences in origin. A factor which may suggest itself as an explanation of the variation from week to week is the error inherent in the method of testing. In order to determine the extent to which the fluctuations observed might be due to this error the average fluctuation from one week to the next was calculated at four representative levels, the surface, 100 meters, 500 meters, and 900 meters, for the three years under consideration. The fluctuation was calculated in terms of the percentage of deviation from the mean of each of the pairs of consecutive weeks. The average values obtained were as follows: For the surface, 18.9%; for 100 meters, 7.1%; for 500 meters 4.3%; and for 900 meters, 3.4%. It has been shown earlier that the method of testing gives a probable error of 2.05%; if the fluctuations from one week to the next which exceed three times the probable error of sampling be taken as significant the surface then shows a significant fluctuation from one week to the next 71% of the time, the 100 meter level fluctuates 42% of the time, the 500 meter level fluctuates 18% of the time, and the 900 meter level fluctuates 16% of the time.

Since neither the fluctuations of currents in the bay as indicated by temperature nor the errors inherent in the method of sampling are sufficient to account for much of the weekly fluctuation in silicate content, the question arises as to the source of this variation. The most reasonable explanation would seem to be that it is due to biological causes. It has been observed for a long time and by many investigators that the plankton which utilize a considerable quantity of silicate in the formation of their tests, especially the diatoms, frequently occur in large numbers in relatively local patches in the ocean; statistical evidence of this "swarming" has been furnished by McEwen ('29). The reasons for this tendency to aggregate are still obscure; the fact remains that where it does occur the silicate must be taken from the surface water at a greater rate than it is elsewhere. It has been pointed out (Skogsberg, '36) that the water in Monterey Bay is in a continual state of flux, and is constantly being renewed by water from outside. The fluctuations in silicate content which are observed at the surface may thus be assumed to be frequently due to the arrival in the bay of water which contains or has contained a swarm of silicious plankton.

In considering the deeper water, again it seems probable that on the many occasions when the weekly fluctuation in silicate content does not show a good negative correlation with the fluctuation of temperature, the cause is biological. If we consider the local patches of silicious plankton which occur at the surface in the open ocean, it will be evident that upon the death of these forms they will sink and in the deeper water there will begin a slow process of dissolution. The deeper water which has lain beneath a local swarm of silicious plankton will thus contain a higher concentration of soluble silicate than does the surrounding water.   As the dead plankton organisms sink deeper and deeper they will tend to spread out horizontally more and more owing to the differences in direction and velocity of the deep currents, and the increased silicate content for which they are responsible will become more diffuse, due both to this spreading out of the dead plankton and to diffusion of soluble silicate.   Thus the deeper the water the less heterogeneous it will be with respect to silicate.   The observations reported above are thoroughly consistent with this hypothesis. It is seen that the percentage of significant fluctuation from week to week in the deeper water falls rapidly with increasing depth.   On the other hand, much of this decrease in fluctuation with depth is of course undoubtedly due to the fact that the ocean currents have a considerably lower velocity in the deep water, so that the water in the bay is changed less rapidly at these levels.

While considering the fluctuations of silicate from one week to the next, and the frequent lack of correlation between such fluctuations and the variations in temperature, one must not lose sight of the fact which has been previously stressed, that the general annual cycle of silicate follows the cycle of temperature with but few and minor exceptions.   Skogsberg ('36), following an analysis of the temperature conditions in Monterey Bay for the five years from 1928 to 1933, arrived at these conclusions in regard to the origins of the water in the bay: There is a period of strong upwelling of deep water which is most apparent during the four months from March to July.   This period is marked by cold water in the upper 100 meters.   Following this there is a period extending from August through October when the bay is dominated by water from the edge of the California Current.   This condition is indicated by warm water in the upper layers.   Lastly, during December, January and the first part of February the water is strongly influenced by the intermittent northerly flowing Davidson Current, and is characterized by a very low thermal gradient in the upper 50 to 100 meters of water.   A consideration of the silicate data shows clear evidence of two of these three annual current phenomena.   Most strongly marked is the effect of the upwelling period extending from the latter part of February to the end of July.   During these months the silicate content in the upper 100 meters of water is very much higher than it is during the rest of the year; it begins to increase in February and continues steadily until it reaches a maximum in July.   This undoubtedly is due largely to the fact that the upwelling phenomenon brings deeper water having a higher silicate content to the surface.

The second annual current phenomenon which is reflected in the silicate data is the encroaching of the water of the Davidson Current during the winter months.   This current tends to flow in a northwest direction along the California coast, and because of the deflection to the right which is characteristic of all currents in the northern hemisphere and which is a resultant of the forces set up by the earth's rotation, the water is banked up against the coast.   As a result of this superficial current against the coast the upper 50 to 100 meters are occupied largely by water which is of surface origin and therefore, as mentioned above, this period is marked by a low thermal gradient in the upper layers.   It has been pointed out that this condition of low gradient is also found in the silicate distribution, the months of December, January and February showing a much lower gradient of silicate in the upper 50 meters than is found at any other time during the year.

The third period in the annual water cycle, the "oceanic period," which occurs in the latter part of August, September, and October, is less definitely marked in the silicate content than in the temperature.   There is a regular, uniform decrease in the silicate content from August through December in the upper 100 meters of water, and there is no definite evidence of the minimal values during September which would be expected if the silicate situation were completely dominated by the close approach of the California Current at this time.   It is particularly unfortunate in this respect that data for September are available in only one of the three years under consideration.   When a further study of the silicate content during this month is undertaken it may very possibly appear that the one year for which the September data are available, 1934, was irregular.   That this was probably the case is indicated by the temperature data obtained during September of 1934, which, unlike those of most of the other years since 1928, do not show the usual maximum September values.

At present it would appear that the most probable explanation of the decrease in silicate content, during the autumn in the upper 100 meters of water, lies in the fact that the heavy upwelling of water which occurs during the spring and early summer comes to an end in August, leaving behind it water which is exceptionally rich in the nutrient salts, including silicate.   During the autumn there is no further replenishing of these salts, and as a result they are steadily depleted by the plankton, this depletion continuing until the first signs of upwelling in the late winter and early spring.

CITATIONS

ATKINS, W. R. G.   1923   The silica content of some natural waters and of culture media.   Journ. Mar. Biol. Assoc. N. S. 13, pp. 151–159.

———   1926   Seasonal changes in the silica content of natural waters in relation to the phytoplankton. Journ. Mar. Biol. Assoc. N. S. 14, pp. 89–99.

———   1928   Seasonal variations in the phosphate and silicate content of sea-water during 1926 and 1927 in relation to the phytoplankton crop.   Journ. Mar. Biol. Assoc. N. S. 15, pp. 191–205.

———   1930   Seasonal variations in the phosphate and silicate content of sea-water in relation to the phytoplankton crop.   Pt. V.   November 1927 to April 1929, compared with earlier years from 1923. Journ. Mar. Biol. Assoc. N. S. 16, pp. 821–852.

———, AND H. W. HARVEY.   1925   The variation with depth of certain salts utilized in plant growth in the sea.   Nature 116, p. 784.

BIGELOW, H. B., AND MAURINE LESLIE.   1930   Reconnaissance of the waters and plankton of Monterey Bay, July, 1928.   Bull. Mus. Comp. Zool. Harvard College 70, pp. 430–581.

BRANDT, K.   1919   Über den Stoffwechsel im Meere.   Dritte Abhandlung. Wiss. Meeresunter. Abt. Kiel 18, s. 187–429.

CARTER, N. M., E. G. MOBERG, TAGE SKOGSBERG, AND T. G. THOMPSON.   1933   The reporting of data in oceanographical chemistry.   Fifth Pac. Sci. Cong., pp. 2123–2127.

DIÉNERT, F., AND F. WALDENBULCKE.   1923   Sur le dosage de la silice dans les eaux.   C. R. Acad. des Sci., Paris 176, pp. 1478–1480.

GRAHAM, H. W., AND E. G. MOBERG.   The distribution of silicate in the sea.   Reports on the chemical results of cruise VII of the Carnegie during 1928–9 (unpublished).

HARVEY, H. W.   1933   On the rate of diatom growth.   Journ. Mar. Biol. Assoc. N. S. 19, pp. 253–276.

HUTCHINSON, A. H.   1928   A bio-hydrographical investigation of the sea adjacent to the Fraser River mouth.   Trans. Roy. Soc. Canada 3d series, 22, pp. 293–311.

———, C. C. LUCAS, AND M. McPHAIL.   1929   Seasonal variations in the chemical and physical proper-ties of the waters of the Straight of Georgia in relation to phytoplankton.   Trans. Roy. Soc. Canada 3d series, 23, pp. 177–183.

JOHNSTONE, JAMES.   1908   Conditions of life in the sea.   Cambridge University Press.

KING, E. J., AND C. C. LUCAS.   1928   The use of picric acid as an artificial standard in the colorimetric estimation of silica.   Journ. Amer. Chem. Soc. 50, pp. 2395–2397.

MOBERG, E. G.   1926   The phosphate, silica, and fixed nitrogen content of seawater.   Proc. 3d Pan-Pacific Sci. Cong., Tokyo, pp. 229–232.

———   1928   The interrelation between diatomes, their chemical environment, and upwelling water in the sea, off the coast of Southern California.   Proc. Nat. Acad. Sci. 14, pp. 511–518.

RABEN, E.   1905   Weitere Mitteilungen über quantitative Bestimmungen von Stickstoffverbindungen und von gelöster Kieselsäure im Meerwasser.   Wiss. Meeresunter. Abt. Kiel 8, s. 279–287.

———   1910   Dritte Mitteilung über quantitative Bestimmungen von Stickstoffverbindungen und von gelöster Kieselsäure im Meerwasser.   Wiss. Meeresunter. Abt. Kiel 11, s. 305–319.

———   1913   Vierte Mitteilung über quantitative Bestimmungen von Stickstoffverbindungen im Meerwasser und Boden, sowie von gelöster Kieselsäure im Meerwasser.   Wiss. Meeresunter. Abt. Kiel 16, s. 209–229.

SKOGSBERG, TAGE.   1936   Hydrography of Monterey Bay, California.

THOMPSON, T. G., AND M. W. JOHNSON.   1930   The sea water at the Puget Sound Biological Station from September 1928 to September 1929.   Publ. Puget Sound Biol. Station, 7.

THOMPSON, T. G., AND R. J. ROBINSON.   1932   Chemistry of the sea.   Physics of the Earth—V Oceanography.   Bull. Nat. Res. Council 85.   Nat. Res. Council of Nat. Acad. of Sciences, Washington, D. C.

THOMPSON, T. G., B. D. THOMAS, AND C. A. BARNES.   1934   Distribution of oxygen in the North Pacific Ocean.   James Johnstone Memorial Volume, pp. 203–234, University Press of Liverpool.

THOMPSON, T. G., AND F. A. ZEUSLER.   1935   Summary of the oceanographic investigations of Bering Sea and Bering Strait.   Trans. Amer. Geophys. Union, Sixteenth Meeting, pp. 258–264.

UDA, MITITAKE.   1934   Hydrographical studies based on simultaneous oceanographical surveys made in the Japan Sea and in its adjacent waters during May and June, 1932.   Rec. Ocean. Works in Japan 6, pp. 19–107.

WELLS, R. C.   1922   Determination of silica in filtered sea-water.   Journ. Am. Chem. Soc. 44, pp. 2187–2193.

TRANSACTIONS

OF THE

AMERICAN PHILOSOPHICAL SOCIETY

HELD AT PHILADELPHIA
FOR PROMOTING USEFUL KNOWLEDGE

NEW SERIES—VOLUME XXIX, PART II
MARCH, 1938

ARTICLE III

The Old Stone Age in European Russia
EUGENE A. GOLOMSHTOK

PHILADELPHIA:

THE AMERICAN PHILOSOPHICAL SOCIETY

104 SOUTH FIFTH STREET

1938

LANCASTER PRESS, INC., LANCASTER, PA.

THE OLD STONE AGE IN EUROPEAN RUSSIA

By Eugene A. Golomshtok

PREFACE

At the suggestion of the Director of the University Museum, Mr. Horace H. F. Jayne, the author has attempted to gather together all the available data concerning the Palaeolithic Period in European Russia.

Judging by the extent of information existing in non-Russian literature, this did not, at first, seem to be a large or a difficult task. Aside from the few sites discussed in short reviews of Russian Archaeological Congresses before 1917, and in occasional notes and scientific news, practically the only site to be thoroughly described is that of Ilskaya, published by Zamiatnin in l'Anthropologie. The standard summaries give us very little information. Thus, M.˙C. Burkitt lists nine sites, with some description; G. G. MacCurdy lists fourteen, with no details except the dating; Max Ebert indicates ten, only seven of which are of the Palaeolithic Period; O. Menghin discusses six with some interpretation.

A cursory study of the information available in Russian, however, showed the real dimensions of the task. It was soon realized that there exists a considerable amount of literature on the Palaeolithic period.[1] This literature is scattered through various small publications and reports, and the difficulty of assembling the information is increased by the fact that a certain portion of the literature is written in the Ukranian and Bielorussian dialects, which differ sufficiently from the main Russian language to make the careful reading of scientific works rather a difficult task.

Being an anthropologist of American training with little experience in Quaternary archaeology, the author found himself at times perplexed by the diversity of existing opinions, which he was not always able to reconcile. Some of these confusing statements and contradictory reports, as well as many omissions, might have been corrected by an actual study of collections and sites. This, however, should form a separate task for the future.

The present work also suffers from the fact that a certain number of publications, known by title, were impossible to obtain. Others may have existed, but for some reason may have escaped the attention of the author.

Therefore, this paper should not be considered as a final survey of extant material but, it is hoped that it may prove to be useful to those who do not know the Russian language, and perhaps even to some Russian students.

The author takes this opportunity to express his gratitude to the Board of Managers of the University Museum for the support they have given this project. Thanks are also due

---

[1] The amount of archaeological literature existing in the Russian language may be judged from the fact that S. A. Dubinsky's "Bibliography of the Archeology of Bielorussia and Adjacent Regions," Minsk, 1933, gives 4031 titles.

to the National Research Council for a Grant in Aid in 1934, which enabled the author to finish this manuscript.

The opportunity to publish this work in its present form was provided by The American Philosophical Society at Philadelphia, to whom the author wishes to express his sincere appreciation.

The author is very grateful to many Russian colleagues, who very kindly placed information at his disposal.   He thanks especially P. P. Ephimenko, S. N. Zamiatnin, G. A. Bonch-Osmolovsky, V. I. Gromov, and V. A. Gorodzov.   He is also extremely grateful to all those authors whose specific works have been used and whose illustrations have been reproduced.

The critical advice of M. L'Abbe Breuil, M. R. Vaufrey, Mr. H. Kelly and Mrs. Alice Kelly has helped greatly to clear up many obscure points in the evaluation of the industries. Thanks are due to M. L'Abbe Breuil for his kind permission to use some of his photographs as well as drawings of the industry of Kiik-Koba.

The author wishes to thank Miss Vivian Falk, Miss Dorothy Spencer, and Miss Lucy Cykman for the editorial work in connection with the preparation of this manuscript.

Last, but not least, the author wishes to thank Mr. Horace H. F. Jayne, Director of the University Museum, without whose enthusiastic help and encouragement this work would never have been accomplished.

EUGENE A. GOLOMSHTOK,
University Museum, Philadelphia.

TABLE OF CONTENTS

LIST OF PLATES

191

TEXT FIGURES

ABBREVIATIONS

B—in Bielorussian.
E—in English.
F—in French.
G—in German.

Gg—in Georgian.
P—in Polish.
R—in Russian.
U—in Ukrainian.

AA—American Anthropologist.
AC—Archaeological Congress.
ACBA—Archaeological Committee of the Bielorussian Academy of Sciences.
ALV—Annual Report of the Anthropological Laboratory of Theodore Vovk
    (T. Volkov).
BQC—Bulletin of the Quaternary Committee.
ESA—Eurasia Septentrionalis Antiqua.
GAIMK—The State Academy for the History of Material Culture.
PHPS—Problems of the History of Pre-Clan Society.
Q—Die Quartarperiode, Ukrainian Academy of Sciences.
RAJ—Russian Anthropological Journal.
RANION—Russian Association of the Social Science Research Institutes.
RAS—Russian Anthropological Society.

I. INTRODUCTION

SCOPE AND PURPOSE

Our knowledge of the Palaeolithic Period in European Russia is far from satisfactory. The sum total of information which has appeared from time to time in English, French and German, in the shape of small notices and miscellaneous notes, is quite negligible in comparison with the size of the country.

Yet, almost every specialist, particularly when writing on the general subject of the development of the culture of Prehistoric man, invariably points out the great importance of obtaining more precise and detailed information about the actual conditions of the cultural stages in Eastern Europe, the greatest part of which is European Russia.

It was decided, consequently, to gather as much data as possible about this area regardless of the language of the publication, continually endeavoring to present the first-hand information as given by the excavator, with the additional material of evaluation and correction when available or warranted.

The borders of the present European part of the U.S.S.R. were taken as geographical limits and the Palaeolithic Period proper (until the beginning of microlithic industries of Azilian and Tardenoisian times) as approximate temporal limits.

A short chapter on Quaternary Geology and Palaeontology was thought to be desirable and was prepared in collaboration with Mr. A. W. Postel of the Department of Geology in the University of Pennsylvania.

A SHORT HISTORY OF THE STUDY OF PALAEOLITHIC PERIOD IN RUSSIA
BEFORE 1917

The first site of Palaeolithic age discovered in Russia was Gontzi, in the Poltava region. The find of fossil bones associated with flint implements, in 1874, attracted the attention of a local teacher, Kaminsky, who started the first investigation of the site. In 1877, Count A. S. Uvarov discovered mammoth bones with some stone implements on his estate near the village of Karacharovo. The resulting excavations proved the existence there of a Palaeolithic site. Two years later, Poliakov, who was participating with Uvarov in the excavation of Karacharovo, discovered one of the most important Russian sites, that of Kostenki I.

The trip of K. S. Merezkovsky, a professor of botany, to Crimea, resulted in the discovery of a series of cave sites, such as Suren I, Cherkess-Kermen, etc. About the year 1800, Antonovitch pointed out several Palaeolithic locations in the Dnieper region. This was followed in the next decade by a series of important discoveries, such as the work of N. I. Kristaphovitch in Novo-Alexandria, Kaschenko in Tomsk, V. V. Khvoiko in Cyrill Street, Kiev, Baron de Baye in Ilskaya, etc.

The work of the father of Russian archaeology, Count A. S. Uvarov, with his first attempts to summarize existing material and especially the studies of Theodor Volkov (later referred to in Ukrainian spelling as Th. Vovk) were the most important stimuli for the study of the cultures of the Quaternary period.   The knowledge of Western European material enabled Volkov to correct many naive assertions of amateur investigators, and thus place Palaeolithic researches in Russia on a truly scientific basis.

The discovery of the late Palaeolithic site of Mezine in 1908, gave Volkov the opportunity to train in methods of excavation and interpretation of material a number of younger students, the most able of whom is P. P. Ephimenko, who today occupies the foremost place among the Russian specialists.

After an interval of almost forty years, the investigations in Gontzi were continued by Scherbakovsky, whose work has contributed to the understanding of this important site.   In 1915, the Polish investigator, Krukovsky, made very extensive excavations in Kostenki.   At about the same time the investigations by the foreign specialists, R. R. Schmidt, L. Kozlovsky and S. A. Krukovsky of several caves in the Caucasus yielded interesting results.

PALAEOLITHIC RESEARCH IN SOVIET RUSSIA

After the revolution, the interest in the study of early cultures of the U.S.S.R. re-awakened with a tremendous impetus.   This resulted in a series of very important new discoveries and the checking and following up of older sites.   Thus, in 1922, Zamiatnin discovered and studied Borshevo near Kostenki, Zhukov and Gorodzov worked in Suponevo, Voevodsky and Gorodzov excavated Timonovka.[1]   A series of sites was discovered and studied by G. A. Bonch-Osmolovsky and others, in Crimea; Berdizh, Gagarino [2] and new sites in Kostenki were excavated by Zamiatnin.

In Siberia, extensive studies were made in the Palaeolithic Period of Yenisei by Gero von Merhart, in Afontova Gora by Sosnovsky and Auerbach, and in the region of Irkutsk by Petri, Gerasimov (Malta) [3] and others.

The archaeological research was supplemented by a more detailed study of Quaternary geology (Pavlov, Mirchink, Krokos) and fauna (Gromov and Gromova), which afforded better dating and correlation.

A tendency to check the old works and study the material available in collections resulted in the fine description of Karacharovo and Ilskaya by Zamiatnin, and a series of old sites in Crimea by Bonch-Osmolovsky.   The first attempt to summarize existing material by Uvarov in 1881 (Archaeology, The Stone Age), was followed by a special article of Spitzin in 1915 (The Russian Palaeolithic Period).   Vishnevsky in 1924, gives a short discussion of the Stone Age in Russia in a special chapter following his translation of Osborn's work.   It was followed by the only existing summary of the Russian Palaeolithic in foreign language by L. Sawicki.[4]   It is a true foundation work for further researches from which a long list of illustrations has been reproduced in this survey.   It is unfortunate

---

[1] Golomshtok, E. A., "Trois Gizements du Paléolithique Supérieur, Russe et Sibérien," L'Anthropologie, Vol. XLIII, 1933, pp. 301–27.

[2] *Ibid.*

[3] *Ibid.*

[4] Ephimenko in his general work on prehistory, "Pre-Clan Society," Leningrad, 1936, gives considerable material pertaining to the most important Russian sites.

that the work is in Polish,[1] and consequently, like Russian scientific literature, not easily available to Western European students.

The materialistic interpretation of history influenced the general attitude of Russian specialists toward the study of the culture demanding the investigation of every factor which may have influenced it.   This was manifested by a number of so-called "complex" expeditions in which, alongside the ethnographical and anthropological data, information on local geology, fauna, flora, climate, etc., was obtained.

The same scheme, perhaps to a lesser degree, has been pursued of late in the realm of archaeology.   A special Committee for the Study of the Quaternary Period was organized by the Russian Academy of Sciences in Leningrad for the purpose of studying Quaternary man in relation to his surroundings, as represented by climate, fauna, flora, etc.[2]   The lead was followed by the Ukranian Academy which started a special periodical called "Die Quartarperiode," devoted to the same problems.   Here, with the same aim in mind, geologists and palaeontologists work side by side with archaeologists, checking each other's work, and supplying necessary information.[3]

Several institutions such as the Institute of Archaeological Technology of the State Academy for the History of Material Culture, in Leningrad, and the Technological section of the Russian Association of Scientific Institutions in Moscow (RANION), devote their time to research in the technological processes encountered in archaeological studies.

Methods of excavation, correlation of material, and other theoretical subjects are often discussed in special conferences.   The accepted decisions are published for guidance in future work.

It should also be borne in mind that modern students of the development of culture in the U.S.S.R. are firm believers in the evolutionary materialistic explanation of cultural changes.   They are sternly opposed to the borrowing and migration theories [4] which, one must admit, are brought up much too often by the Western European scientists in their effort to explain puzzling cultural phenomena.

The same desire to obtain the clearest possible picture of culture with social, economic, and other factors outlined, coupled with the evolutionary theories of Morgan and Tylor, resulted in a series of attempts to reconstruct Palaeolithic society in all its phases.[5]   While some of the results may be startling from the point of view of western scientists, and others may prove to be erroneous, the Russian archaeologists deserve definite commendation for their struggle with the purely mechanistic attitude which forgets the peoples who made artifacts and is interested in the objects *per se*.

[1] Sawicki, L., "Materials for the Study of Russian Archeology," Przeklad Archeologicznii, Poznan, Vol. III, Pt. 2–3, 1926–28.

[2] Bonch-Osmolovsky, G. A., "The Problems of the Complex Study of the Quaternary Period," Soobschenia of Gaimk, No. 3–4, Leningrad, 1932, pp. 44–49.

[3] Golomshtok, E. A., "Anthropological Activities in Soviet Russia," A. A., Vol. 36, No. 2, pp. 301–327.

[4] Meschaninov, I. I., "Palaeontology and Homo Sapiens," Izvestia of Gaimk, Vol. VI, Pt. VII, Leningrad, 1930, pp. 5–36; "The Theory of Migrations in Archeology," Soobschenia of Gaimk, No. 9–10, Leningrad, 1931, pp. 33–39.

[5] Ravdonikas, V. I., "The Marxian History of Material Culture," Izvestia of Gaimk, Vol. VII, Pt. III, IV, Leningrad, 1930, pp. 5–94; id., "To the Question of the Sociological Periodization of the Old Stone Age in Connection with the Views of Marx and Engels on Primitive Society," Izvestia of Gaimk, Vol. IX, Pt. 1–2, pp. 1–33.

In their effort to obtain the most complete information about the people whose remains they are excavating, Soviet archaeologists [1] are stressing the advisability of quantitative as well as qualitative analysis of the industry, thus quite often correcting the erroneous impression which results from overstressing the so-called "leading" forms, regardless of the comparatively small part those forms may have played in the culture of studied groups. In conformity with the materialistic philosophy of science, Russian archaeologists strive for functional interpretation of their discoveries. It is perhaps due to this hyper-sensitiveness to functionalism that Ephimenko, Zamiatnin and Gorodzov were able to recognize the remains of the semi-subterranean dwellings of Palaeolithic man. According to Ephimenko, similar habitations were found in Europe on several occasions, but passed totally unnoticed by the investigators.

The recently adopted method of microscopic analysis of charcoal and ashes found in the open hearths and fireplaces of Palaeolithic sites, supplies floral data and thus gives additional information on the climate.

The tremendous interest on the part of the Government and general public toward archaeology, coupled with their more systematic methods of excavation and study, resulted during the past seventeen years in the accumulation of a considerable body of material, which, taken with what was done before the revolution, permits us to form a comprehensive picture of the Old Stone Age in Russia.

SPELLING, TRANSLITERATION, ETC.

Because of technical reasons, the ideal method of giving all references in the alphabets of their corresponding countries cannot be used. Therefore, all original titles of works written in languages other than English, French, or German, are given in English translation. In those cases where several series are published by the same scientific organization under different headings, the name of the institution is translated into English, but the specific terms designating the particular series are given in transliteration. Thus, such terms as "Izvestia" (memoirs or reports) of the State Academy in Leningrad, or "Zapiski" (memoirs or reports) of the Imperial Geological Society, are retained.

The regional scientific institutions, such as the Ukrainian and Bielorussian Academies, publish their works in the Ukrainian and Bielorussian dialects respectively, with corresponding differences of pronunciation and spelling of proper names. Thus, the Ukrainian dialect has the tendency to change the G sound (as in good) to H (as in horn), e.g., "Hontzi" instead of "Gontzi"; the sound E (as in yet) and O (as in pole) to I (as in kill), e.g., "Kiiv" instead of "Kiev" and "Voronii" instead of "Voronoi." The Bielorussian dialect changes O to A (as in far), T to Tz. I to E; e.g., "Sazh" instead of "Sozh," "Tzimanavka" instead of "Timonovka," "Mensk" instead of "Minsk," etc.

Names of authors and place names are, therefore, given uniformly in a simple transliteration, preserving each time the "Great Russian" pronunciation; except in cases where a name has attained a standard English form.

The following transcriptions have been adopted in this work to represent some of the Russian sounds not found in English:

[1] Bonch-Osmolovsky, G. A., "The Question of the Evolution of Lower Palaeolithic Industries," "Chelovek," 1928, No. 2–4.

Tch equal to ch as in chalk—Tchernigov
ia equal to ya as in German ja (yes)—Viatka
ie equal to ye as in yes—Dnieper
zh equal to French je (I)—Sozh
kh equal to ch as in German ach—Khvalinsk
tz equal to Z in German Zeit—Donetz

RECENT CHANGES IN PLACE NAMES AND BOUNDARIES

After the revolution, the old system of dividing Russia into Governments (Gubernia) and Counties (Uyezd) was changed.   European Russia now consists of the Russian Federal Socialistic Soviet Republic, the Bielorussian Socialistic Soviet Republic, the Ukrainian Socialistic Soviet Republic, and a number of semi-independent and independent territories.

The larger geographical units (smaller than one of the above political divisions) are designated in this paper as *regions* and the smaller ones as *districts*.   For example Gontzi, Lubni district,. Poltava region.

The following changes of names, which went into effect after the revolution, should be noted:

Former Name	*Present Name*
St. Petersburg, later Petrograd	Leningrad
Nizhnii-Novgorod	Gorkii
Tzaritzin	Stalingrad
Simbirsk	Ulianovsk
Ekaterinodar	Krasnodar
Ekaterinburg	Sverdlovsk

II. THE PALAEOLITHIC PERIOD IN EUROPEAN RUSSIA

A. QUATERNARY GEOLOGY AND PALAEONTOLOGY OF EUROPEAN RUSSIA [1]

1. PLEISTOCENE GLACIATIONS

Geologists have shown a great divergence of opinion concerning the relative number of glaciations that Russia has undergone, as well as their correlation with the glacial periods of the rest of Europe.   S. N. Nikitin [2] recognizes only one glaciation of this part of the world, whereas A. Girmounsky believes that the central U.S.S.R. underwent two glaciations, an opinion upheld for the southern part of the U.S.S.R. by D. Sobolev.   L. Sawicki, A. Girmounsky, H. Hausen, H. Martensen, E. Kraus, and A. A. Grigoriev have also put themselves on record as believing that the western and northeastern regions of the U.S.S.R., respectively, were subjected to two periods of glaciation.   Girmounsky correlates his two periods of glaciation with Mindel and Riss, while Sawicki establishes his correlation to Riss and Würm.[3]

[1] Prepared in collaboration with Mr. A. W. Postel of the Geological Department of the University of Pennsylvania.

[2] Mirchink, G. F., "Number of Glaciations of Russian Plains," " Priroda," 1928, No. 7–8.

[3] Antevs, E., "Maps of the Pleistocene Glaciation," Bulletin of the Geological Society of America, 1929, Vol. 40.

The following writers have shown evidence to warrant the support of three great advances of the ice sheet in Russia: G. F. Mirchink, A. P. Pavlov, E. Antevs, N. Florov, B. Vishnevsky, and A. Pravoslavlev; later, in 1927, Girmounsky also came to hold this opinion. Mirchink, Pavlov and Laskarev, place a fourth and earliest period of glaciation in the late Pliocene, correlating it with Günz; the three Pleistocene glaciations they link with Mindel, Riss and Würm. Laskarev differs from Pravoslavlev in his sea parallels; he attributes the layers containing the shells of *Cardium edule* to Würm, thus placing the Khvalinsk and Hazar transgressions farther back. With this Mirchink disagrees, because the finds of *Cardium edule* in the highest horizons of the loess contradict it. W. I. Krokos believes, however, that all four epochs of glaciation may be placed in Quaternary time, citing as evidence four horizons of loess separated by three horizons of fossil soils of "tchernozem" (humus) type [1] in Ukraina. Mirchink criticizes this opinion on the basis that Krokos failed to take into consideration the general history of the region in which he was working, and that his fourth loess horizon was established on the basis of twelve cross-sections situated far apart, and not continuous. [2]

Considerable disagreement of opinion is shown concerning the great peripheral moraine border in Russia. It is dated as Mindel by Pavlov, whereas Krokos and Mirchink place it in Riss. (Mirchink places the Mindel drift inside, running irregularly from lat. 52° N. long. 27° E.—Garyn River—over Mozir, Mogiliev, Moscow and Narechta to the Volga, lat. 57° N.) Florov agrees with the Polish geologists that in Eastern Europe the most extensive drift is Würm, and that it reaches far south into the U.S.S.R. The view that this moraine belongs to Würm, however, seems to be most inadequately supported. In Germany, the outer limit of the topographically young lake region is regarded also as the limit of the last ice sheet. This lake region runs northeast through northern Poland, the eastern Baltic area and across the northwestern part of the U.S.S.R. to the southern shore of the White Sea, and it may here, as in Germany, mark the extreme limit of the last glaciation. If the last ice sheet reached as far south in the U.S.S.R. as the drift, the accepted border in Germany would be wrong. The fact that the drift border, extending east from Khelm, Poland, cannot very well be Würm, may make it Riss. The older drift border running through Khelm, Przemysl, and Krakow, may be Mindel, which agrees with the determination made by G. Graham for Germany in 1928. As the probable Riss drift east of Khelm and the Boug River forms the outermost drift, it may perhaps connect with the outermost drift in the U.S.S.R., which, according to Mirchink, is of Riss age. [3]

In the U.S.S.R. this peripheral drift runs E.N.E. from lat. 51° N., long. 26° E. to lat. 61° N., long. 59° E. in the Urals; it has two long lobes that project south in the valleys of the Dnieper and the Don (see Map, Fig. 100). The youngest drift border may run from Vilna southeast to a point north of Minsk (according to Vollosovitch and Koslovsky), then northeast for a short distance to change east and again northeast to a point at lat. 56°45′ N., long. 33° E. (according to Hausen, Kraus and Mirchink). Mirchink's [4] latest map then carries it east to Yaroslav on the Volga, whence the border may run northeast to Osserdok

[1] Antevs, E., "Maps of the Pleistocene Glaciation," Bulletin of the Geological Society of America, 1929, Vol. 40.

[2] Mirchink, G. F., "Number of Glaciations of Russian Plains," Priroda, 1928, No. 7–8.

[3] Antevs, E., *op. cit.*

[4] Zhirmunsky, A. M., "The Question of the Limits of Glaciation on the Russian Plains," Bulletin of Quaternary Committee; 1929; Vol. I, pp. 21–26; is inclined to disagree with the work of Mirchink.

on the Dvina, whence it probably goes north across the Gulf of Mezen to the Kanin Peninsula. Since Antev's maps have been published, Bobnovo marks the Brandenburg (Weichael), and Vishnii Volochok stage differently, extending the Flaming (Warthe) moraines in a fairly smooth curve over Moscow to beyond the Volga.[1]

From the foregoing it may be seen that there is a great diversity of opinion concerning the designation, relationship, and nomenclature of the various glacial periods. That these relationships may be more readily seen and understood, the following table of the systems drawn up by different authors has been added. In the main, this outline has been based on the glacial time-sequence determined by Mirchink.

Sobolev,[2] describing the glacial deposits of the Ukraine, recognizes general but unequally distributed elevations of the earth's crust as contributing factors in the formation of centers from which the continental glaciation spread. These elevations caused an increase of erosion in the river valleys in the regions free from ice. Glaciation continued until crustal elevation ceased, and a reverse movement set in. Consequently, in the second half of each period of glaciation, sedimentation predominated over erosion in the river valleys, as the rivers lost their powers of erosion due to a decrease in their elevations. The deep erosion cuts were filled with fluvio-glacial formations. By covering, changing, and dispersing moraine materials the drainage systems formed terraces.

2. Mindel Glaciation

In the beginning of the Quaternary period, the northern part of Russia was covered by an ice sheet which, according to Pavlov, was the Mindel glaciation. In the eastern part of Russia at this time the familiar outlines of Volga could not be found; the large Volga valley did not exist. The tongues of the glacier extended to the contemporary bed of the Volga between Tzaritzin and Kamishin and penetrated farther east in the present Trans-Volga region. One of these tongues spread from Tzaritzin toward Manich.[3]

Pavlov[4] states that the deposits are of a highly altered red-brown material and extend farther than the line usually considered as the maximum extent of glaciation. The extent of this glaciation is hard to determine exactly, as much of the terminal material has been destroyed by subsequent glaciation to the south and southwest. In the district of Khorol, near Ostapia, beneath the moraine of the second glaciation is a lacustrine deposit thirty feet thick containing fresh-water shells, below which is found material of the first glaciation: glacial sands, gravel and erratics. In other places the second glacial moraine is separated from that of the first by a bed of black earth which indicates a mild interglacial period. In this region the first and second glaciations seem to have been of equal extent.

In southwestern Russia and Galicia it is difficult to set a precise limit to the Mindel glaciation, because, in this region the moraines of the first glaciation were also removed by

[1] Antevs, E., "Maps of the Pleistocene Glaciation," Bulletin of the Geological Society of America, 1929, Vol. 40.

[2] Sobolev, D., "Quaternary Morphogenesis in the Ukraine," Second International Conference of the Association for the Study of the Quaternary Period of Europe, Vol. II, Moscow, 1933; pp. 71–101.

[3] Vishnevsky, B. N., "Prehistoric Man in Russia," appendix to the translation of Osborn's "Men of the Old Stone Age," Leningrad, 1924, pp. 441–508.

[4] Pavlov, A. P., "Époques glaciaires et interglaciaires de l'Europe et leur rapport à l'histoire de l'homme fossil," Bulletin de la Société des Amis des Sciences Naturelles, d'Anthropologie et d'Ethnographie, 1925.

TABLE 1

COMPARISON OF THE GLACIAL NOMENCLATURE OF VARIOUS AUTHORS

Penck and Bruckner (1909)	P. Range 1926	German Designations 1925	Krokos[1]	G. F. Mirchink[2]	H. Breuil[3] N. France S. England 1934	A. L. Reingard[4] 1933 Caucasus	A. Vardaniantz[5] 1933 Central Caucasus	G. Graham 1928 Germany	L. von Waweke 1927
Feman							7 Period XVII–XIX Cent		
Egessen							6th Period Historical		
Daun							5th Period		
Geschnitz			Post Glacial		1 Post Würm I	Retreat	4th Period		
Buhle	Baltic Stage	Third or Last Glaciation (Wisla)		Würm II	Würm III		3 Period 2 Period 1 Period		
Würm	4th Glaciation		Würm II Würm Interglacial Würm I	Würm I	Würm II Würm I	a-Main Period b-Second Period	Maximum	Warthe	Last Glaciation
Riss-Würm			Riss-Würm	Riss-Würm	Riss-Würm				
Riss	3rd Glaciation	2nd Glaciation (Saale)	Riss	Riss	Riss	a-2 Period b-Main Period		Saale	Glaciation Before the Last
Mindel-Riss			Mindel-Riss	Mindel-Riss	Mindel-Riss	Mindel-Riss (of Long Duration)			
Mindel	2nd Glaciation	1st Glaciation (Elster)	Mindel	Mindel	Mindel			Elster	First Glaciation
Gunz-Mindel			Gunz-Mindel	Gunz-Mindel	Gunz-Mindel				
Gunz	1st Glaciation		Gunz	Gunz	Gunz	Suspected		Elbe	Oldest Glaciation (Elbe)

[1] Krokos, V. I., "Materials for the Study of Quaternary Deposits in the Ukraine," Materials for the study of Ukranian Soils, Vol. 5, pp. 1–326, Kharkov, 1927.

[2] Mirchink, G. F., "Number of Glaciations of Russian Plains," "Priroda," No. 7–8, 1928.

[3] Breuil, H., unpublished chart.

[4] Reingard, A. L., Second International Conference of the Association for the Study of the Quaternary Period of Europe, Vol. II, pp. 3–14, Moscow, 1933.

[5] Vardaniantz, L. A., Second International Conference of the Association for the Study of the Quaternary Period of Europe, Vol. II, pp. 15–20.

the second advance of the great ice sheet, as shown in the central part of the Volyn region near Ostrog and Doubno, and in Galicia near Brody. The glacial erratics here are not of the country rock, and the moraine material is highly altered and wind-polished. That these materials belong to the first glaciation is confirmed by Gagel and Karn, because decomposed materials are found beneath the Mindel-Riss interglacial beds containing typical time markers (botanical remains and fresh water molluscs) at three points on the borders of the Boug, descending towards Voldava. Thus, in the southwest also, the Mindel and Riss glaciations seem to have nearly the same distribution, and are separated by an interglacial bed, which indicates a long time interval.[1]

As in the rest of Europe, there are abundant beds of river alluvium in the regions of the lower Dnieper, in Bessarabia, and the Don. A striking example is the Tiraspol gravel on the left bank of the Dniester near Tiraspol, which was probably laid down at the end of Mindel.

The work of Sobolev[2] in the Ukraine shows how the elevation accompanying the Mindel glaciation accentuated erosion to form wide valleys. As a lowering of the ground occurred, erosion changed to sedimentation. This is illustrated on the middle Dnieper terrace by the deposition of fluvial glacial sands on the erosion base; in places these sands are displaced by dark bluish loams; on the drainage divides these are in turn displaced by loess-like clays. At the end of Mindel (or the beginning of Mindel-Riss) these beds were covered by loams containing fresh water fauna (*Paludina* were absent). On the main Dnieper terrace the typical Mindel-Riss terrace sands occur with *Paludina diluviana* and *Elephas trogontherii;* these sands cover the Mindel fluvio-glacial formation.

3. MINDEL-RISS INTERGLACIAL EPOCH

The retreat of the ice in the interglacial period left traces in the form of a moraine in the northwestern part of the Saratov region, and in the neighboring terrain of the Tambov and Penza regions. The boulders of local and more northern formations are the markers of its route. These traces are in general almost absent in the south and in the southwest.

With the retreat of the glacier, a regional subsidence occurred. The northern regions of Russia were inundated by the sea in the vicinity of the basins of the Northern Dvina and Pechora. This wide sea was of considerable depth and was inhabited by a polar fauna in its deeper parts, while along the shores molluscs of a more temperate habitat have been found. In the south there was a somewhat different picture, with lakes and streams dominating. The powerful rivers washed out wide vallies and deposited sands and gravels. In the northern part of the Trans-Volga area, slower streams, lakes and swamps prevailed.

In the districts of Khorol and Ostapia the lacustrine deposits of this period cover the erratics of the Mindel glaciation and are in turn covered by the sands and erratics of the Riss advance.[3]

---

[1] Pavlov, A. P., "Époques glaciaires et interglaciaires de l'Europe et leur rapport à l'histoire de l'homme fossil," Bulletin de la Société Naturaliste de Moscow, 1922.

[2] Sobolev, D., "Quaternary Morphogenesis in the Ukraine," Second International Conference of the Association for the Study of the Quaternary Period of Europe, Vol. II, Moscow, 1933, pp. 71–101.

[3] Pavlov, A. P., "Dépôts Néogènes et Quaternaires de l'Europe Mériodionale et Orientale," Mémoir de la Section Géologique de la Société des Amis des Sciences Naturelles, d'Anthropologie et d'Ethnographie, Vol. V, Moscow, 1925, pp. 1–215.

In the south, the climate became temperate and moist. The elevated areas were covered by vegetation and the stony soil was mantled by humus. In the eastern part of the Saratov region, as well as in Penza, the Don region, Poltava, and to the south of Moscow, there were large lakes, the remains of which still exist to-day. The southern mammoth rhinoceros, *Cervus latifrons, Bos,* and horse constituted the fauna.[1]

4. RISS GLACIATION

The Riss glaciation, according to Mirchink, represents the epoch of greatest glaciation in Russia. The new advance of the ice brought with it the debris of northern formations. Northern Russia was subjected to a land elevation which caused the seas occupying the basins of the Northern Dvina and Pechora to retreat. The glaciers of Scandinavia spread southward filling the Baltic sea, while in the Caucasus masses of ice descended slowly, send-ing tongues downward and outward to fill the valleys. The glaciation covered the valleys of the Dnieper and the Don in the depression of the Russian ravine. (See map, Fig. 100.)

Pavlov says that this second advance of the ice was of long duration, and, like Mirchink, he correlates it with the Riss of the Alps. Its southern limit can be more or less definitely traced. The boulders carried by this glaciation are from the Volyn region, north of Lutzk and Rovno. In the Kiev region, the southern border of their distribution descended along the valley of the river Dnieper, reaching the northern part of the Ekaterinoslavl region, thus forming one of the southernmost tongues. The ice sheet then receded northward to form another tongue which, turning again in a southern direction, reached to the head waters of the Don, almost to the village Ust-Medveditza. From there it turned north again through the Saratov, Penza and Simbirsk, regions to Nijnii-Novgorod, and then by a broken line reached across Viatka and Perm to the Ural Mountains.[2]

In the south of Russia the waters from the melting glacier formed rivers depositing clays and sands. This enormous territory was not cut by river beds at that time. The higher and dryer places show the results of wind and water action, and form the most ancient loess layer in the Kursk, Tchernigov and Volyn regions. The southeastern part of Russia had a continental climate during this period. Strong rains filled the valley with debris, the higher regions received a deposition of sandy-clay formations over which at the end of this epoch the layer of loess reached its maximum thickness.

According to Sobolev,[3] the ice advanced on Poliesie and Ukraine along two routes, from the south through Poland along the Poliesie depression, and from the north along the Dnieper depression. Due to specific topographical conditions and differences in the country rock, the fluvial deposits of the two regions are different. Fluvio-glacial sands were deposited on the Poliesie terrace, while fluvio-glacial loams were deposited on the main terrace of the middle Dnieper. These formations are nevertheless synchronous.

Erratics left by the ice sheet have been found in Poltava, Kalouga, and Tchernigov. In the southwestern parts of the great European plain the glacier blocked the streams flow-ing south until they overflowed, the swifter currents depositing sands and the slower ones

[1] Vishnevsky, B. N., "Prehistoric Man in Russia," appendix to translation of Osborn's "Men of the Old Stone Age," Leningrad, 1924, pp. 441–508.

[2] Vishnevsky, B. N., *loc. cit.*

[3] Sobolev, D. N., "The Quaternary Morphogenesis in the Ukraine," Second International Conference of the Association for the Study of the Quaternary Period of Europe, Vol. II, Moscow, 1933, p. 85.

fine loess-like muds.   The sandy beds along the valleys of the Mokcha, the d'Alatyr and the Soura rivers belong also to this period.[1]

In the Caspian region the brackish sea diminished in size, and there appeared lakes, swamps, and areas rich in fine grass and vegetations where large herds of camels, wild horses, antelops, and numerous steppe rodents found their habitat.[2]

The complex of the glacial formations of this time consist of moraines which are usually expressed in the Moscow region by red-gray pebble clays, often superimposed on fluvio-glacial formations of cross bedded sands, containing lenses of gravel.   Forming a part of the glacial complex loess-like loams and clays covering the moraines are found; very likely they represent the products of a washing out of the moraine *in situ* during the melting of the ice.   Farther south the moraine becomes thinner, and contains more material from the local formations.   In the Poltava tongue this is noticeable in the loess-like character of the moraine, the result of a re-working of the underlying loess; in the Don tongue, this is evident from the gray coloring and a considerable quantity of chalk pebbles that are present.

The peculiarity of this glacial complex lies in the fact that the moraines of both tongues participate in the formation of the upper river terraces.   This is seen in the descent of these moraines toward the valleys of the rivers which have these terraces.   Thus, it can be traced along the Dnieper from Orsha up to the southern border of its spread.   In nonglaciated regions, in the river valleys, such as along the river Vorskl, the lower Poltava, and along the Dnieper above the Kremenchug, it can be seen that stratigraphically the moraine is supplanted by the ancient alluvial sands, which form the upper terrace.   These sands in their turn are covered by a layer of the upper horizon of the loess.   The position corresponding to the sands of the upper terrace is taken by the second (counting from above) horizon of the loess, which is separated from the upper by a layer of fossil humus.[3]   In reference to this same complex Mirchink states that V. I. Krokos confused the normal loess with the loess-like loams which are intimately connected with this moraine, and attributes them to Riss glaciation.   Mirchink feels that he has failed to trace the further development of these loams to the point where they turn into fine grained sands of fluvio-glacial origin.

In places, the erosion accompanying the Riss has cut through the main terrace of the middle Dnieper to the Jurassic clays.   This terrace is superimposed, after an erosion interval, on the Mindel-Riss terrace which is composed of sands and loams.   The main terrace here was changed during the Riss glaciation both by erosion and deposition and on its surface fresh water loam deposits and moraine material were laid down.   This Riss terrace is equivalent to the moraine terrace of Poliesie, the fluvial-glacial deposits of which are contemporaneous with the loams of the Middle Dnieper.   Part of the Riss terrace was formed by the filling of the pre-Rissian erosion depressions, first by Riss fluvial-glacial deposits, and then by valley glacier sands.   (This process may have continued through Riss-Würm.)   The terrace has been considered Würm by some investigators, as it has only

[1] Pavlov, A. P., "Dépôts Néogenes et Quaternaires de l'Europe Meriodionale et Orientale," Mémoir de la Section Géologique de la Société des Amis des Sciences Naturelles, d'Anthropologie et d'Ethnographie, 1925.

[2] Vishnevsky, B. N., "Prehistoric Man in Russia," appendix to translation of Osborn's "Men of the Old Stone Age," Leningrad, 1924, pp. 441–508.

[3] Mirchink, G. F., "Number of Glaciations of Russian Plains," "Priroda," No. 7–8, 1928.

one layer of loess; the argument, though, is not convincing as only one layer of loess is likewise to be found on the main terrace in the region of Tzibli.[1]

Under these deposits no formations are found that can be attributed to the glacial formations proper. Thus in the Dnieper region, we have only a layer of loess separated by a layer of fossil soil from the covering fluvio-glacial formations.

Moving in a northern direction we encounter, along the Dnieper and south from Riechitza, the horizon of the red-gray and gray pebble moraine and sandy soils, systematically traceable in cross sections, which stratigraphically occupy the same place that in more southern regions is taken by the horizon of loess under the glacial formations of the Riss period.

In the ice-free region of the Dnieper, the glacial formations of Riechitza are contemporaneous with the loess layer which underlies the Riss glacial deposits. North from Riechitza this horizon of the glacial formations is more easily traced, and allows an opportunity to follow the southern limits of the Mindel glaciation through Mozir, Riechitza, Roslavl, and the central part of the Moscow and Vladimir regions.[2]

The advance of the Riss ice sheet caused great changes in climate. During the height of this glaciation the climate was intensly cold; but farther to the south in Russia there still existed rich prairies which supported a tundra fauna.[3] Many animals migrated to these surroundings, and either accommodated themselves to the new environment or died out. In the tundra region herds of wild horses existed, while closer to the ice belt lived the woolly mammoth, the reindeer and the musk-ox.[4]

The topographical changes of this epoch were also pronounced. The eastern part of the Volga region was depressed, forming the present region of the trans-Volga meadow lands; this sunken area was occupied by the waters of the enlarged Caspian Sea, which spread far north during this epoch as well as in the following interglacial period. Toward the end of this glaciation, the waters from the retreating glacier cut the present valleys of rivers, making beds in the deposits of loess and in underlying red clays. The final freshwater conversion of the enormous inland sea, which was receiving the rivers of southern Russia and filling the Black Sea depression, took place during this epoch. The water level of this inland sea fell considerably. In other regions quite different events took place. First in the southwest, then also in the west, in the regions of Simbirsk and Nijnii-Novgorod, the glacier obstructed the rivers flowing northward, causing them to overflow and fill with sand the lower regions where there now exist wide belts of sand and small rivers.[5]

5. RISS-WÜRM INTERGLACIAL EPOCH

After the final retreat of the Riss ice sheet the northern regions of Russia were left covered by large erratic boulders and stony materials that had been carried down from Scandinavia and Finland with the advancing ice. Lakes were formed in the lower area, while large swamps occupied the ice-freed land.

[1] Sobolev, D., "Quaternary Morphogenesis in the Ukraine," Second International Conference of the Association for the Study of the Quaternary Period of Europe, Vol. II, Moscow, 1933, p. 95.

[2] Mirchink, G. F., "Number of Glaciations of Russian Plains," "Priroda," No. 7–8, 1928.

[3] Gorodzov, V. A., "Archaeology, The Stone Age," Vol. I, Moscow, 1925.

[4] Vishnevsky, B. N., "Prehistoric Man in Russia," appendix to translation of Osborn's "Men of the Old Stone Age," Leningrad, 1924, pp. 441–508.

[5] Vishnevsky, B. N., *op. cit.*

The fresh water marls and peats along the Boog near Vlodava belong to this period, as probably do the lacustrine deposits of Troitskoye near Moscow; below these deposits are gravels, while over them lie moraine materials (sands and gravels). The Riss-Würm is also represented by the lake formations in the Drutskoi ravine near Grodno on the Nieman; similar deposits occur near the village of Pepelevo on the river Chuya in the region of Kostroma. These lakes probably existed in the earlier half of Riss-Würm.

In the middle portions of the Russian plain (regions of Poltava, Chernigov, and Koorsk), in valleys cut by the Riss glaciation, are sand deposits which, with the deepening of the valleys in the following epoch, form terraces above the alluvial terrain.

In the more northern regions of the Russian plain this period is represented by deposits laid down by marine transgressions. Deposits of this type are to be seen near Petrosavodsk and along the northern Dvina.[1]

6. WÜRM GLACIATION

The limit of Würm glaciation is difficult to determine because the fluvio-glacial outwash of the Baltic pause masked or destroyed the true moraine materials. In central Russia it is also hard to determine whether the moraine materials belong to Riss or Würm, which also makes it difficult to ascertain whether the adjacent interglacial deposits are Riss-Würm or Mindel-Riss.[2]

Gorodzov describes the ice sheet of the fourth glaciation at its greatest extent as spreading southward from its Scandinavian center into Germany and Poland, and from there into southeastern Russia. In the region of Germany and Poland its border was less extensive than the third glaciation, while in Russia the southeastern limit of the ice lay some 500 kilometers northwest of the border line of the third glaciation. It lies on the line of Suvalki-Vilno-Vitebsk-Novgorod from about lat. 54° N., long. 23° E. to lat. 51° N., long. 31° E. and further on to the beginning of the Onega River. In these regions the remains of the glacial deposits are especially fresh, the moraines are well preserved as are also the various fluvio-glacial formations. The glacial deposits consist of reddish-gray boulder clays. The end moraines have the appearance of wide elevated bands of boulders and pebbles, or of sharply outlined chains of boulder-pebble hills which are well preserved in the Vitebsk and Pskov regions. The end moraines usually are accompanied by sands deposited by streams flowing from under the ice.

Contemporaneously with the formation of the topography of the glaciated zone, gully erosion became active in the area south and west of the zone and moved the loess and other clay soils from the heights to redeposit them at lower elevations. During the same time sands were deposited in the basins of the Pripiat, Desna, Oka, Viatka, Pechora, and Northern Dvina, which later developed into dunes.[3]

Increased erosion preceded the Würm glaciation. Its action can be traced northward along the valleys of the rivers of the Ukraine, though it did not proceed far enough to

---

[1] Pavlov, A. P., "Dépôts néogènes et quaternaires de l'Europe, mériodionale et orientale," Mémoir de la Section Géologique de la Société des Amis des Sciences Naturelles, d'Anthropologie et d'Ethnographie, Moscow, 1925.

[2] Pavlov, A. P., "Époques glaciaires et interglaciaires de l'Europe et leur rapport á l'histoire de l'homme fossil," Bulletin de la Société Naturalists de Moscow, 1922.

[3] Gorodzov, V. A., "Archaeology, The Stone Age," Vol. I, Moscow, 1925.

destroy the gullies of the tributaries. The outwash from the melting ice covered the Poliesie terrace with Würmian sands. This terrace rests on Riss foramtions. In the Ukraine the Würm terraces are composed of sandy deposits. Here the deposition of the Würm loess was completed before the formation of the terrace. (There is no loess capping the terrace.) The upper loess covers the Riss moraine and is separated from it by a layer of fossil soil. Consequently the covering sands of Poliesie are of the same age as the sands of the second terrace and the upper loess of the middle Dnieper. In southeastern Poliesie there is a zone of aeolian deposits which were formed and transported during Würm.

Sobolev considers the Dnieper moraine with its corresponding fluvio-glacial formations to be Riss. He dates the moraine of the east Poliesie tongue as the Poliesie stage of Würm; to the same age belong the covering sands, the valley glacial deposits, and the terrace. These formations were later incorporated in the sands of the second (Borovaya) Dnieper terrace, though this terrace may also contain materials of the Bielorussia stage, and perhaps even later stages of Würm.[1]

Vishnevsky in his work describes the Würm glaciation as being less extensive than the preceding glacial epochs. According to him, the ice sheet covered only the Baltic Sea, spreading along its eastern shores but not penetrating far inland. The nearby ice sheet covering northern Germany caused an arid and cool climate to prevail. In the areas neighboring the glacier a tundra developed with characteristic fauna and flora. The mammoth, so abundant in previous epochs, became rare and its place was taken by the reindeer.

The land in the far south, free from ice, gradually sank. This allowed the waters of the Mediterranean to flow into the Black Sea basin, which formerly had been comparatively small. This newly formed Black Sea had at this time a higher level, and covered the mouths of most of the south Russian rivers. It was at the end of these events that the south of Russia acquired its contemporary topography.

Toward the end of the Würm epoch the climate became warmer and dryer; the ice commenced to thaw, and the formation of diluvial loess began.

The upper horizon of loess in the Dnieper region is separated in the ice-free areas from the underlying loess horizon by a layer of fossil soil; this corresponds to the Würm glacial complex. In the regions of the Dnieper which were affected by this glaciation, this horizon is separated by the fossil soil from the underlying fluvio-glacial and alluvial formations of the Riss glaciation which are represented by thin and sometimes cross-bedded sandy soils that pass into true sands further to the north. The upper non-inundated terraces of the loess of this horizon lie either on the ancient alluvial formations of the Riss epoch (in non-glaciated regions), or on the horizon of the Riss moraine which descends on the upper non-inundated terraces. In a northern direction the ancient alluvial deposits of these terraces turn into sands which encircle the end moraines along the line: Slutzk-Minsk-Lukoml-Charea-Orsha-Smolensk-Tvier-Kostroma-Lake Kubanskoe [2] (lat. 53° N., long. 67° E. to lat. 59° N., long. 40° E.).

In summarizing the glaciations undergone by Russia in Pleistocene time, Mirchink says: "The three complexes of the glacial formations in the regions affected by the ice are associated with three groups of continental formations in the form of three horizons of

---

[1] Sobolev, D., "Quaternary Morphogenesis in the Ukraine," Second International Conference of the Association for the Study of the Quaternary Period of Europe, Vol. II, Moscow, 1933, pp. 71–101.

[2] Mirchink, G. F., "Number of Glaciations of Russian Plains," " Priroda," No. 7–8, 1928.

loess. The upper one (the first) covers the glacial formations of the maximum activity of the Riss glaciation.

"The upper non-inundated terrace corresponds to the epoch of the Riss glaciation. This formation covers the third horizon of the loess and underlies the formation of the minimum glaciation. This is the third terrace of Lichkov. The second non-inundated terrace (the second of Lichkov) which covers the glacial formations of the Dnieper tongue corresponds to the Würm glaciation. The Mindel terrace is evidently covered so completely that it cannot be well traced." [1]

Mirchink also points out the importance of the fluctuations of the level of the Caspian Sea as an indication of ice movements and climatic changes. He says: "The fluctuations of the Caspian basin, which furnish a very sensitive geological barometer, corresponded to the three periods of glaciations. This occurred in each case when the increased flow of water from the melting of the receding glacier took place at the end of an epoch of glaciation. Three of these transgressions are known: Khvalinsk, Hazar and Babinsk. The more ancient transgressions of the Caspian such as Apsheron and Akchalig took place in Pliocene time, and if they could be connected with glaciations, as was indicated by A. P. Pavlov, we could deal with the Pliocene glaciations, in which we will have put the problematical Günz glaciation."

It is more difficult to draw similar parallels for the Black Sea basin, which previous to its joining with the Mediterranean should have been reacting in the same manner as the Caspian to the climatic changes. However, to trace these changes as accurately as in the case of the Caspian is impossible, because the joining of the Black Sea with the Mediterranean, which took place in post-Würmian time, caused such a rise of the sea level as to obliterate all traces of the previous transgressions.[2]

7. FAUNA OF THE WÜRM GLACIATION

V. A. Gorodzov says of the fauna of this epoch that the mammoth and rhinoceros were commencing to die out here as in Western Europe, and therefore did not penetrate into the zone occupied by the last glaciation of Russia, because they had perished before the land was sufficiently freed from ice to give them a habitat.[3] Pavlov also states that *Elephas antiquus* and *Rhinoceros merckii* are no longer found during this epoch, and *Rhinoceros tichorhinus* is also extremely rare, being found only in the lower layers of the horizon.[4]

Speaking of *Elephas antiquus*, Gorodzov says that this animal (which is not known as far east as Moscow-Odessa) died out with the approach of the Riss ice sheet. *Elephas trogontherii* was more adaptable than *Elephas antiquus*, and outlived the latter. *Rhinoceros merckii* spread across Europe and Siberia, but had died out by the end of the Quaternary period.[5]

[1] Mirchink, G. F., "Number of Glaciations of Russian Plains," "Priroda," No. 7–8, 1928.
    Mirchink, G. F., *ibid.*
[3] Gorodzov, V. A., "Archaeology, The Stone Age," Vol. I, Moscow, 1925.
[4] Pavlov, A. P., "Époques glaciaires et interglaciaires de l'Europe et leur rapport à l'histoire de l'homme fossil," Bulletin de la Société Naturaliste de Moscow, 1922.
[5] Gorodzov, V. A., "Archaeology, The Stone Age," Vol. I, Moscow, 1925.
Further notes on the fauna of this period will be found in the concluding section of this outline.

8. Post-Glacial Epoch

The opening of this period is marked by the deposition of sand and silt on the valley floors and was accompanied by accumulations of alluvial loess in the valley declivities. These diluvial deposits often contain archaeological sites (Cyrill Street at Kiev, Mezine in the region of Chernigov, and Gontzi on the River Udai in Poltava). *Elephas antiquus* and *Rhinoceros merckii* are lacking in this epoch. *Rhinoceros tichorhinus* is also absent in this period—or extremely rare—in the lower horizon of the post-glacial formations. The mammoth still survived, but conditions were becoming increasingly unfavorable for it as it was becoming an easy prey for man. Thus mammoth bones are already rare in the upper layers of the Cyrill Street station,[1] which are of Magdalenian age.[2]

According to Pavlov, the first glacial pause of the final retreat, the Buhl, was marked by a lowering of temperature throughout Europe, and by an extension of the tundra environment. After the termination of the Buhl stage, the climate became milder and more humid; the reindeer disappeared from the middle latitudes of Europe, and forests supporting deer appeared. As the ice continued to retreat to the north and west, the Russian plain took on steppe characteristics with such fauna as wild horses and antelope. When the retreating ice again paused in southern Finland and Scandinavia, the climate of the plain became more severe and humid.

In northern Europe the retreating ice of the last glaciation left shallow depressions behind it. These were filled by the waters of the North Sea, which then connected with the Baltic to form the Yoldic Sea. The climate was cold with a corresponding cold fauna of tundra type, while steppe conditions still prevailed farther south. With time the climate became warmer; the sea which developed after the glacier became smaller, and finally became separated by dry land from the White Sea and the North Sea on the northeast, and from the German Sea on the southwest. This resulted in the formation of an enormous fresh-water lake with an area of some 570,000 sq. km., called the Ancyl Sea after the mollusk most commonly found there. The polar fauna of the Yoldic Sea changed to one corresponding to more temperate and more humid conditions. The epoch of Ancylus was characterized by a land elevation followed by a renewed sinking that connected this lake with the German Sea. The climate became milder and warmer than that of the present, and in the salt waters of the Baltic *Littorina litorea* and a temperate zone vegetation existed. The beginning of this Littorine Sea is in the contemporary geological epoch, the Holocene.[3]

Mirchink describes the geographical distribution of the end moraines of the glacial pauses during post-glacial times. North from the line of distribution of the Würm terminal moraines, and connected with them, we find another series of end moraines definitely expressed. The southernmost of them are situated along the line of continuation of the Baltic end moraines of Germany. Those which border on the so-called lake region show better form than the others. They run along the line of Vilno-Lepel-Senno-Vitebsk-Toropetz-Ostashkov-Borovichi—that is roughly lat. 54° N., long. 25° E. to lat. 57° N.,

---

[1] Pavlov, A. P., "Dépôts néogènes et quaternaires de l'Europe mériodionale et orientale," Mémoir de la Section Géologique de la Société des Amis des Sciences Naturelles, d'Anthropologie et d'Ethnographie, 1925.

[2] According to P. P. Ephimenko, "Pre-Clan Society," Leningrad, 1936.

[3] Vishnevsky, B. N., "Prehistoric Man in Russia," Appendix to the translation of H. F. Osborn's "Men of the Old Stone Age," Leningrad, 1924.

long. 34° E.—and the northern shore of the Oniega Lake and Niandoma.   Behind this boundary is situated another line of end moraines in the direction of Tukkuma-Riga-Valk-Pskov-Luga-Petrozavodsk-Oniega-Archangelsk from lat. 57° N., long. 22° E. to lat. 65° N., long. 41° E.   Finally, the last row is formed by the southern Finnish terminal moraines.   No other formations or independent terraces are found which can be placed in actual connection with the time of formation of these end moraines.   The only exception is the southernmost moraine.   During its formation occurred the accumulation of the alluvial deposits of the lower terrace, the top of which was formed during the maximum of the Würm glaciation.   Consequently, it is reasonable to consider these end moraines as marking the final retreat of the glacier, and to see in them an analogy to the Buhl, Geshnitz, and Daun stages of the Alps.[1]

9. LOESS DEPOSITS OF EUROPEAN RUSSIA

The following description of the loess formations of Russia has been condensed from a recent paper by Vera Malycheff.[2]

The origin of the loess deposits is still a matter of dispute; therefore the question will not be gone into here.   It is sufficient to say that various authors have ascribed the origin of these deposits to fluvio-glacial streams, to normal stream deposition, and to wind action.

The loess deposits of Russia have their principal development south or southeast of the Baltic stage of the Würm moraines.   In their distribution, it is possible to distinguish four regions: first, the northern region including the upper basins of the Dnieper and the Volga; second, the eastern region which includes the middle and lower basins of the Dnieper and the Crimean steppes; third, the central region which takes in the basin of the Don, the middle basin of the Volga, and the steppes to the north of the Caucasus; and fourth, the southeast region around the Caspian depression.

The northern region runs south from the Baltic moraines and the central part of the Arctic marine transgressions.   The loess blanket is not continuous but forms a series of disconnected islands, which, in the basin of the Dnieper, according to Mirchink, are composed of aeolian loess.   By other authors, as Berg, they are considered to be sandy loess-like muds of aqueous origin; the beds are often pebbly, sometimes stratified, and always less rich in lime than true loess.   Throughout this region the loess is localized along declivities; it is lacking on the high elevations.   The base of the formation is often composed of coarser, more pebbly material which sometimes grades into deposits of a lacustrine type.   Under the moraine deposits that this loess covers is found in several places a second level of loess-like silt which separates these deposits from the next underlying older moraine.   In the eastern part of this region some argillaceous loess-like formations have been reported.

The other three regions, lying to the south of the last region described, possess a certain number of traits in common.   The loamy blanket attains a considerable thickness, and a more or less perfect continuity; in places the formation enclosed beds of fossil soil.   In the parts of this region that were covered by the Riss glaciation, the loamy formation is superimposed on the Riss glacial deposits, from which it is ordinarily separated by a

[1] Mirchink, G. F., "Number of Glaciations of Russian Plains," "Priroda," No. 7–8, 1928.

[2] Malycheff, Vera, "Le Loess," Revue de Géographie physique et de Géologie dynamique, Vol. VIII. No. 4, 1930.

zone of outwash sediments or fossil soil.   Below the moraine deposits there is frequently found a second and older layer of loess.

The western region, which includes a large part of Ukraine, is the locality of the typical loess.   To the north it is localized on the plateaus, while in the valleys it gives place to aqueous deposits.   In the south, where the loess is better developed, attaining here, in fact its maximum development, it frequently covers the country with a continuous mantle, showing a thickness of twenty meters in some places.   With the exception of the beds of fossil soil that it encloses, the loess is homogeneous throughout its mass.

The loess of the central region is often replaced by aqueous deposits.

The region to the southeast, the Caspian depression, does not show a true loess; its place is occupied by argillaceous silts associated with sand deposits of the Quaternary transgression of the Caspian.

According to Krokos,[1] the loess formations in Russia are divided into two complexes: the recent loess and the ancient loess.   In the north, the recent loess is represented by at least two levels of loess-like formations.   The upper level covers the deposits of the Würm moraines, and a lower level which in several places lies under these formations.   In the region of Smolensk, the lower level of the loess is separated from the underlying Riss sands by a peat formation.   The loess in its upper parts contains *Carpinus betulus*, *Corylus betulus*, etc., while the bottom parts contain *Brasenia purpurea*, *Trapa natans*, etc.   In the neighborhood of Moscow the lower level of the recent loess contains *Elephas primigenius*, *Bos* sp., *Equus* sp., and in the region of Kaluga, near Lichvine, *Elephas primigenius* and *Rhinoceros tichorhinus* have been found.

Two levels of the recent loess are also found in the west of Russia, in the region occupied by the Riss glaciation (the Dnieper tongue).   Each of these levels, but particularly the older level, often pass beneath aqueous deposits, or deposits of lacustrine loess.   These two levels are generally superimposed and separated by a layer of fossil soil (first fossil soil of Krokos).   They rest on the deposits of the Riss moraine but are kept from actual contact with these deposits by another layer of fossil soil (second fossil soil of Krokos).   These relationships may be observed on the high terrace of the Dnieper.

These two levels of the recent loess along the Dnieper probably correspond to two separate advances of the Würm ice sheet—the lower level of the recent loess corresponding to Würm I, and the upper level to Würm II.

The typical loess of the central region is not as well known as the loess of the Dnieper area, but the geological and morphological characteristics allow one to suppose that the recent loess with two levels exists here also.

The following faunal types have been found in the upper level of the recent loess:

Elephas primigenius,*	*Rangifer tarandus*,*
Rhinoceros tichorhinus,*	*Saiga tartarica*,
Equus sp.,	*Ovibos* sp.,*
Sus scrofa,	*Bos* sp.,
Cervus elaphus,	*Canis lupus*,*
Cervus megaceros,	*Vulpes vulpes*,*
Cervus sp.,*	*Vulpes lagopus*,*

[1] Krokos, V., "Stratigraphie der quartaren Ablagerungen der Ukraine," Die Quartarperiode, Vol. 4, 1922, Kiev, 1932, pp. 1–4.

Vulpes corsac,	*Lepus* sp.,*
Gulo sp.,*	*Ochotona* sp.,
Hyaena spelaea,	*Citellus rufescens,**
Ursus sp.,*	*Alactaya saliens,*
*Felis leo,**	*Arvicola terrestris,*
Felis lynx,	*Marmota bobac,**
Castor fiber,	*Elobuis talpinus.*

The asterisks indicate species which have been described in the loess itself, and which are encountered in the loams of the region of Voronezh. The other species, with the exception of the lynx, are also encountered in this last region, and have been found in the beds of the covering rock. The horse, deer, bear, polar fox, and rabbit are common to the loess and the covering beds.

The complex of the ancient loess is often separated into levels, the upper level being contemporaneous with the Riss glaciation and the lower level being pre-Riss. These two levels, frequently separated by a layer of fossil soil (third fossil soil at Krokos) pass laterally into loess-like fresh-water formations. Together with the Riss moraine they form the high terrace of the Dnieper.

The fresh water portion of the upper level contains the molluscs *Carbicula fluminalis* and *Paludina diluviana.* The vertebrates are represented by a single worn tooth of *Elephas trogontherii.*

No archaeological sites can be connected with this ancient loess complex.

10. River Terraces

It has been demonstrated that the formation of river terraces is closely connected with the sea level which controls the base of erosion of the rivers. Thus a lowering of the sea level (or an increase of continental elevation) would lower the base of erosion and cause the rivers to cut their valleys deeper and to start building new terraces. That this process was operative during Quaternary time is shown by numerous river terraces occupying different levels.

Pavlov [1] describes the results obtained by various investigators in correlating the river terraces with the Pleistocene glaciations. Thus, C. Depéret in the region of the Rhone has connected the four terraces found there with the four main glacial periods. These terraces have, starting at the top, been named Sicilian, Milazzian, Tyrrhenian and Monastirian. The Sicilian terrace lies 90-100 meters above the river level and is correlated with the Günz glaciation of the Alps. The Milazzian lies at an elevation of 55-60 meters and is connected with the Mindel glaciation of the Alps. The Tyrrhenian occurs at 28-30 meters and can be correlated with the Mindel-Riss inter-glacial period. The Monastirian lies at an elevation of 18-20 meters and corresponds to the moraines of the Würm glaciation.

The high terrace of the Dnieper, lying at an elevation of 20-30 meters, is described as Riss by Malycheff [2]; the low terrace, 14-15 meters above the river level, can be corre-

[1] Pavlov, A. P., "Dépôts néogènes et quaternaires de l'Europe mériodionale et orientale," Mémoir de la Section Géologique de la Société des Amis des Sciences Naturelles, d'Anthropologie et d'Ethnographie, 1925.

[2] Malycheff, Vera, "Le Loess," Revue de Géographie physique et de Géologie dynamique, Vol. VIII, No. 4, 1930.

lated with Würm, for in the upper valley of the Dnieper it passes laterally into the Würm fluvio-glacial deposits. Mirchink [1] describes three terraces on the Dnieper near Orsha: an upper terrace 40 meters above the present level of the Dnieper, a middle terrace at 20 meters, and a lower terrace at 10–12 meters.

11. LACUSTRINE DEPOSITS OF EUROPEAN RUSSIA

The best known of the lake deposits in Russia are those of Lichvinsk, investigated by the geologist N. N. Bogolubov. [2] There it was proved definitely that the moraines of the fourth (Würm) glaciation did not reach the Kaluga region. The lake deposits were found under the moraine of the third (Riss) glaciation, the deposits themselves belonging to the second glaciation. Under the Riss moraine is, first of all, a layer of loess indicating the interglacial period. Beneath it are the lake deposits with numerous remains of buried fauna and flora; lower still are pebble sands of the second (Mindel) glaciation, the typical moraine of which was discovered near Kaluga, where two moraines divided by a layer of loess can be observed. According to the opinion of the investigator the lake deposits merged gradually and changed in an upward direction into loess, forming with it an organic whole. This shows that the lake deposits and the loess belong to the same epoch as the second interglacial period, being only its most ancient phase of deposition.

The flora of Lichvinsk lake bears witness to two climatic periods in the life of the lake: the earlier moderate, and the second warmer. During the first period the remains of *Picea excelsa* Lk., *Parix* sp., and *Salix* sp. were deposited in large quantity. Later appeared *Carpinus betulus*, *Taxus* sp., *Fagus* sp., and finally *Euryale europaea weberi*, the species of semi-desert regions. It is similar to *Euryale ferux* growing at the present time in Bengal, China, Japan, and on the shores of the river Ussúry in Siberia. The mean temperature of the latter locality is −18° C. in January and 21° C. in July, averaging 4° C. for the year. This indicates that the climate was becoming warmer, surpassing the present day climate until it began to turn dry with a steppe character, returning to loess forming conditions. The fauna of the lake consisted for the most part of molluscs, fish, and the otter (*Lutra vulgaris*). But the most interesting species of fauna found in the loess were the remains of the mammoth (*Elephas trogontherii*) and rhinoceros (*Rhinoceros merckii*).

In central and southern Russia, many other lake deposits of this time exist; such are the lake deposits on the banks of the Moscow River, in the Studeniniy Gully, and the Troitzki Gully (both near Moscow) and other regions. In the Troitzk lake deposits, among the rich forest and lake fauna and flora, the whole skeleton of a mammoth was found which evidently had become bogged while watering and had died in the deep lake mud, being unable to extricate itself. [3]

In the region of the Caspian transgression, the lacustrine deposits, according to Chernishev, [4] may be described as follows: These deposits are found principally in the region

[1] Mirchink, G. F., "Number of Glaciations of Russian Plains," " Priroda," No. 7–8, 1918.

[2] Bogolubov, N. N., "Materials for the Geological History of the Kaluga Region in the Glacial Period." "The Yearbook of Geology and Mineralogy of Russia," by N. N. Krishtafovich, Vol. VII, pp. 111–119.

[3] Gorodzov, V. A., "Archaeology, The Stone Age," Vol. I, Moscow, 1925.

[4] Chernishev, Th., "Aperçu sur les dépôts postérieurs en connection avec trouvailles des restes de la culture préhistorique au nord et à l'est de la Russie d'Europe," Congrés International d'Archéologie Préhistorique et d'Anthropologie, Moscow, 1892.

of Viatka and are petrologically identical with the Caspian deposits.   They represent fresh water deposits formed between the limits of Caspian transgression and the northern marine transgression.   To these may be related the peat deposits.   They are associated with clays and sands, with sphaerosiderite, and occupy a considerable expanse in the districts of Kotelnitch, Glagov and Slebodsk in the region of Viatka.   The presence of peaty deposits of considerable thickness (60 to 70 meters) in these formations, and the fact that their extension does not coincide with the direction of the Viatka and other rivers leads one to believe that these formations do not correspond with the present river system, but that they represent sediments laid down by a system of lakes connected by a series of straits. Though petrologically these lacustrine deposits are similar to the deposits of the Caspian transgression they contain a fresh water fish fauna (*Alosa, Perca fluviatilis, Abremts brauma*) and molluscs (*Dreissena, Cyclas, Anodonta,* and *Paludina*), and a few rare remains of *Elephas primigenius* and rhinoceros.[1]

Nikitine [2] says of the lacustrine deposits of the Baltic region: "Lacustrine basins are found filled with stratified marly clay, sometimes arenaceous, in which are found a polar fauna; consequently these deposits were laid down at the beginning of the post-glacial epoch.   These clays are usually covered by fresh water calcareous earth, passing in the upper part into carbonized vegetable matter."   The lacustrine deposits of central Russia he cites as being particularly important because it is in them that many of the mammoth remains are found.   These formations he believes can be correlated with the glacial pause or retreat.

12. QUATERNARY FLUVIAL DEPOSITS

Chernishev describes the fluvial deposits of the Caspian region as follows: the fluvial beds, forming terraces in the old river valleys, are very distinct in the valleys of the Urals, where two terraces may be distinguished, the older and upper being post-Pliocene while the lower is composed of recent alluvium.   Farther west the lower terrace divides in two. The more widespread of these formations is made up of a more or less sandy, calcareous, yellowish clay, usually distinctly stratified, both in the region of the Urals and in regions immediately adjacent.   Sometimes this clay does not show an exact stratification; it is porous and contains calcareous concretions.   Breaking up into escarped walls it often appears similar to loess; thus it is often termed "loessiforme."   Related to the above, a bluish-gray and fawn colored clay, as well as a pebble conglomerate, is often found in the structure of the old fluvial terraces.   Although this formation can be regarded as a constant horizon, the yellow clay (loessiforme) forms, in general, the upper horizon of the upper terrace, while the bluish-gray clays and pebble conglomerates form the lower part of the horizon.   This formation attains its greatest development in the regions where the rivers ceased to be torrents on leaving the mountains, while the loessiforme deposits are often found well developed at elevations of 600 meters.   In these beds in the upper terraces are found: *Elephas primigenius, Rhinoceros tichorhinus, Rhinoceros merckii, Bos priscus, Bos taurus,*

---

[1] Chernishev, Th., "Aperçu sur les dépôts postérieurs en connection avec trouvailles des restes de la culture préhistorique au nord et à l'est de la Russie d'Europe," Congrés International d'Archéologie Préhistorique et d'Anthropologie, Moscow, 1892.

[2] Nikitine, S., "Sur la constitution des dépôts quaternaire en Russie et leurs relations aux trouvailles resultant de l'activité de l'homme préhistorique," Congrés International d'Archéologie Préhistorique et d'Anthropologie, Moscow, 1892.

and *Ovibos moschatus*. The period of formation of these deposits is the same as that of the Caspian transgression.

In northern Russia, the fluvial deposits are represented by stratified sands intercalated with gravel and a few rare boulders. These sands show their best development where the rivers flowed into the Arctic Sea; they lie on the post-Pliocene marine beds. The fluvial deposits are clearly set off from the grayish marine formation by their clear yellow color. *Elephas primigenius* and *Rangifer tarandus* have been found in these sands (which testifies that the mammoth inhabited north European Russia at the same time that it inhabited northwestern Siberia, for its remains have been found in fresh water deposits in the region of the Yenisei River).

The alluvial deposits of the valleys which are exposed to the spring floods belong to this group of formations. In contradistinction to the valleys of the rivers of southeastern and southern Russia, the river valleys belonging to the basin of the Arctic Sea show only a weak development of terraces. The river valleys of this region are comparatively narrow and are bordered by abrupt escarpments of post-Pliocene beds.

13. Deposits of the Caspian Transgression

In southeastern Russia, in the west part of the region of Ufa, the southeastern part of the region of Viatka, and the central part of the region of Perm, are found the deposits of the Caspian transgression. The Caspian sediments change into the fluvial terrace sediments as one goes from west to east. The transgression took place by means of deep gulfs, following chiefly the ancient valleys of the Kama and the Bielaya Rivers. These gulfs were a complicated network of coves and bays in the western parts of the districts of Birsk and Menselinsk of the region of Ufa, and in the Kazan and Viatka regions as well as along the length of the Kama River. Through this region the following formations are represented by typical sections, commencing at the top: [1]

A. Fawn-yellow clay, stratified and containing marly concretions.

B. Gray stratified sand passing into gravel and conglomerates.

C. Plastic light gray clay, sandy at the base. In the middle of this clay one can observe layers of peat passing into lignite.

14. Boreal Marine Transgressions

With the freeing of Northern Russia from the great glaciers marine transgressions occurred. The deposits of these transgressions are represented by argillaceous sand beds and sandy clay beds of a brownish gray color, as well as by beds of dark gray clay, the latter is not a distinct horizon but is mutually intercalated with the former. In all these beds a similar fauna is found, resembling the fauna of the present Mourman Sea littoral. These sediments contain boulders of both igneous and sedimentary rocks often marked with striae. The material for these stratified deposits was furnished chiefly by Jurassic and Cretaceous formations. At its greatest extent the Arctic Sea encroached on the region of moraine formations in the form of deep gulfs, and changed the direction of the old valleys of the Dvina, Vaga, and Soukhona river systems.

[1] Chernishev, Th., "Aperçu sur les dépôts posterieurs en connection avec trouvailles des restes de la culture préhistorique au nord et à l'est de la Russie d'Europe," Congrés International d'Archéologie Préhistorique et d'Anthropologie, Moscow, 1892.

PLATE I.

BEREZOVKA MAMMOTH IN PROCESS OF EXCAVATION. Photo courtesy of Zoological Museum of Academy of Sciences.

This transgression reworked and redeposited much of the moraine material of northern Russia. Many of the marine deposits contain striated pebbles and boulders. The fauna of these deposits includes shells of *Cyprina islandica, Astarte borealis, Cardium groenlandicum, Cardium islandicum,* and *Mactra elliptica;* this is a moderate Arctic fauna and is closely related to that of the present Mourman littoral. In the regions of the southwest, in the basin of the Vaga, abundant shells of *Cyrpina islandica* and *Cardium edule* are found. The main epoch of this transgression (as also the transgression of northwestern Siberia) corresponds in general to the epoch of the deposition of clays and sands with *Yoldia arctica* in Sweden.[1]

Chernishev also lists these formations:

A. Eluvial Formations. They are represented by sandy clay deposits developed on the flanks and summits of mountain chains. The non-stratified irregular masses of clay formations found in the broken surfaces of the carboniferous limestone also belong to this type of formation. To present a chronological classification of the eluvial deposits of the Urals and its vicinity is difficult; these deposits are characterized by no sorting of materials, and as they are due to chemical and mechanical disintegration *in situ,* they pass at various depths into the original parent rock. They are therefore represented by clays and sands containing debris from the disintegration of the rocks from which they were formed. Faunal traces are rare in these formations, though *Elephas primigenius* and *Bos priscus* have been found.

B. Aeolian Formations. These are represented by a range of dunes on the littoral of the Arctic Sea and in the large valleys of the great rivers, the Pechora, the Vichegda, the Dvina, and others.

C. The Cave Deposits of the Ural Region. These are of interest to archaeologists. Their floors are composed of clay containing rock debris from the cave roofs. The character of these clays would seem to show that they are the results of inundation; in them have been found *Bos priscus, Cervus alces,* etc.[2]

15. VERTEBRATE FAUNA OF THE QUATERNARY PERIOD

The following description of the Quaternary fauna of the Ukraine has been taken from a paper by J. G. Pidoplichka.[3]

The data on the Quaternary fauna is apt to be poor, because to date insufficient work has been done on them, and also because the loess formations are not a favorable medium for the preservation of skeletal remains, particularly those of small animals.[4]

---

[1] Chernishev, Th., "Aperçu sur les dépôts postérieurs en connection avec trouvailles des restes de la culture préhistorique au nord et à l'est de la Russie d'Europe," Congrés International d'Archéologie Préhistorique et d'Anthropologie.

[2] Further brief petrological description of the Quaternary deposits are given in the "Map of the Quaternary Deposits of the European Part of the U.S.S.R. and the Adjacent Regions," Explanatory Note, Leningrad, 1932.

[3] Pidoplichka, J. G., "Die Fauna der Quartaren Saugetiere der Ukraine," Die Quartarperiode, Ukrainische Akademie der Wissenschaften, No. 4, 1932.

[4] The unusually well preserved mammoth remains from Siberia (Plates I–VI) have been described fully. See also: E. Golomshtok, "Le Trompe du Mammouth Sibérien," L'Anthropologie, t. XLII, no. 5–6, 1932, 548–550.

Fifth loess (Gunz): in this horizon no strictly dated conclusions as to faunal types can be drawn.

Fourth loess (Mindel) and fourth fossil bed (Gunz-Mindel): in 1927 Krokos assigned to this period the mammals of the Tiraspolian gravels; that is, the terrace depositions of the old Dniester. At present, this demands further proof.

Third fossil bed (Mindel-Riss): no mammalian remains have been found in the fossil bed of this period in Ukraine. However, in the sands of this period *Elephas trogontherii* has been found near Melnyk at Kaniev. The remains of this same animal have also been found at the village of Lyssa Gora in the neighborhood of Pervomaiska. The sand in which these finds were made is located beneath the third loess horizon (counting from the top).

Third loess (Riss): the faunal contents of this horizon are inadequately known. Krokos reports finding the remains of *Elephas primigenius* in the sub-moraine sands near the village of Kovali near Lubni. To this stratum may also be assigned remains of *Marmota* sp. found in Dorenburg, near Askania Nova. *Ursus spelaeus rossicus*, Boriss has been reported from this loess at Lubimovka near Chersov.

Second fossil bed (Riss-Würm): this horizon in the Ukraine has not been well investigated. However, to this period belong such steppe animals in western Europe as:

Equus hemionus,	*Saiga tatarica,*
Cricetulus migratorius,	*Alactaga jaculus.*

Second loess (Würm I): At present few finds can be assigned to this period. The remains of *Spalax* and other burrowing animals have been found.

The fauna of the Crimean caves is mixed. The presence of *Vulpes logopus* and *Rangifer tarandus* (lower Aurignacian) indicates the immediate influence of the neighboring tundra fauna (Mezine), while the presence of many steppe animals, among them *Alactaga jaculus* and *Hyaena spelaea*, bears witness to the southern character of the Crimean fauna.

The fauna of the first fossil bed (Würm interstadial) has been subjected to close examination as the result of excavations that have been carried out at the palaeolithic sites of Mezine and Zuravka. Conclusions regarding the distribution of the mammalian types of this horizon were also made possible by the investigations of Bialinitzky-Birulya in the Crimea. The fauna of the Mezine site shows classical tundra forms:

Elephas primigenius,	*Lepus timidus,*
Rhinoceros tichorhinus,	*Citellus rufescens,*
Equus caballus,	*Dicrostonyx torquatus,*
Bos sp.,	*Gulo gulo,*
Rangifer tarandus,	*Ovibus moschatus.*
Vulpes lagopus,	

The fauna of the Palaeolithic site of Zuravka is an arctic tundra fauna, namely:

Elephas primigenius,	*Marmota bobak,*
Rangifer tarandus,	*Citellus rufescens.*
Saiga tatarica,	

PLATE II.

BEREZOVKA MAMMOTH IN THE ZOOLOGICAL MUSEUM OF U.S.S.R. ACADEMY OF SCIENCES. Photo courtesy of Zoological Museum.

Examination has shown the first loess horizon (Würm II) to contain the faunal remains listed below:

Meles meles,	Spalax microphtalmus,
Ochotona pusilla,	Spalax polonicus,
Marmota bobak,	Microtus arvalis,
Citellus citellus,	Cricetulus migratorius,
Citellus suslicus,	Silvimus silvaticus,
Citellus pigmaeus,	Vormela sarmatica,
Cricetus cricetus,	Putorius eversmanni.

With the exception of *Ochotona pusilla* all the forms in the foregoing list have still living representatives. The late Würm loess-free terrace of the Dnieper has also shown *Cervus megaceros hibernicus.*

TABLE II.

	Mousterian		Aurignacian		Proto-Solutrean	Magdalenian		
	Crimea	Caucasus	Crimea	Central Russia		Early	Middle	Late
1. *Elephas primigenius*..............	×	×	×	×	×	×	×	.....
2. *Rhinoceros tichorhinus*............	×	×	×	×	....	traces	.....	.....
3. *Equus caballus*..................	×	×	×	×	×	×	.....	.....
4. *Equus hemionus*.................	×	.....	......	×	.....	.....	.....	×
5. *Equus asinus*....................	×	×	.....	.....	.....	.....	.....	.....
6. *Sus scrofa*......................	×	×	×	.....	×	×	.....	.....
7. *Bos* (species)...................	×	.....	×	×	×	×	×	.....
8. *Cervus elaphus*..................	×	×	×	.....	×	×	.....	×
9. *Cervus megaceros*................	×	×	×	.....	.....	.....	.....	.....
10. *Rangifer tarandus*...............	.....	.....	×	×	×	×	×	×
11. *Capreolus* (species).............	×	.....	.....	.....	.....	.....	.....	.....
12. *Saiga tatarica*.................	×	.....	×	.....	.....	.....	.....	.....
13. *Canis lupus*....................	×	×	×	×	×	×	.....	.....
14. *Canis familiaris*...............	.....	.....	.....	.....	.....	.....	.....	×
15. *Vulpes vulpes*..................	.....	.....	×	.....	.....	.....	×	.....
16. *Vulpes lagopus*.................	×	.....	×	.....	.....	.....	×	.....
17. *Vulpes corsac*..................	×	.....	×	.....	.....	.....	.....	.....
18. *Meles meles*....................	×	.....	.....	.....	.....	.....	.....	.....
19. *Martes foina*...................	.....	.....	.....	.....	.....	.....	.....	.....
20. *Putorius evesmanni*	.....	.....	×	.....	.....	.....	.....	.....
21. *Putorius ermineus*..............	.....	.....	×	.....	.....	.....	.....	.....
22. *Putorius nivalis*...............	.....	.....	×	.....	.....	.....	.....	.....
23. *Ursus arctos*...................	.....	.....	.....	.....	.....	×	.....	.....
24. *Ursus spelaeus*.................	.....	×	×	.....	×	.....	.....	.....
25. *Felis leo*......................	×	.....	.....	.....	.....	.....	.....	.....
26. *Felis lynx*.....................	×	.....	.....	.....	.....	.....	.....	.....
27. *Felis silvestris*................	.....	.....	.....	.....	.....	.....	.....	.....
28. *Hyaena spelaea*.................	.....	×	×	.....	.....	.....	.....	.....
29. *Lepus europeus*.................	×	.....	.....	.....	.....	×	.....	.....

TABLE II (*Continued*)

	Mousterian		Aurignacian		Proto-Solutrean	Magdalenian		
	Cri-mea	Cau-casus	Cri-mea	Central Russia		Early	Middle	Late
30. *Lepus timidus*	×	......	×	....	×	....	....	sp.
31. *Ochotona pusilla*			×					
32. *Marmota bobac*	×		×	×	×	×		
33. *Citellus rufescens*	×		×	×	×			
34. *Castor fiber*			×					
35. *Cricetus cricetus*	×		×	×				
36. *Cricetus (Mesocricetus)* sp.			×					
37. *Cricetus (Cricetulus)* sp.			×					
38. *Mus sylvaticus*	×		×					
39. *Arvicola amphibius*	×		×					
40. *Microtus arvalis*			×	×				
41. *Evotomys glareolus*			×					
42. *Lagurus luteus*	×		×					
43. *Ellobius talpinus*	×		×			×		
44. *Alactaga jaculus*			×					
45. *Alactaga elater*			×					
46. *Scirtopoda telum*			×					
47. *Bison (priscus)*		×		×	×	×		
48. *Spalax microphtalmus*				×	×			×
49. *Vermella perugusna*				×				
50. *Alopex lagopus*					×	×		×
51. *Ovibos moschatus*						×		
52. *Dicrostonyx torquatus*						×		
53. *Alces alces*						×	×	
54. *Antelope* (species)							×	
55. *Ursus* (species)							×	
56. *Vulpes* (species)							×	
57. *Gulo gulo*								

Many animals have become extinct in historical times. These may be listed as follows:

Equus prjewalskii,
Bos primigenius,
Bison bonasus,
Cervus elaphus,
Saiga tatarica,

Ursus arctos,
Gulo gulo,
Vulpes corsak,
Alactagulus acontion.

Further faunal listings may be given in conjunction with the various culture periods found in Russia. The correlation of these different cultures with the glacial and inter-glacial periods may be made as follows: Mousterian covers late Riss-Würm to early Würm; Aurignacian to full Magdalenian may be taken to occupy the rest of Würm. The final stages of Magdalenian may be taken as extending over into post-glacial time.

The faunal remains of the Caucasus are on the whole similar to those of the Crimea, though the latter does show a wider variety of small animals. This may perhaps be ex-

PLATE III.

LEG OF BEREZOVKA MAMMOTH. Photo courtesy of Zoological Museum of U.S.S.R. Academy of Sciences.

plained on the grounds that environmental conditions were more favorable for the preservation of the bones of the small mammals in the Crimea.   If conditions are unfavorable for preservation, small bones tend to be destroyed more rapidly than large ones.   This fact may explain the scarcity of small bones in the Caucasus.

16. Invertebrate Fauna of the Quaternary Period

The following table (III, pp. 224–228 inclusive) of invertebrate fauna for Ukraine have been taken from the work of W. G. Bondartchuk.[1]

17. Flora of the Quaternary Period

J. D. Kleopov [2] gives the following divisions of the Quaternary flora for the Ukraine:

1. Alpine fossil types of Riss age.
2. Mesophile fossils of forest type belonging to the Riss-Würm period.
3. Xerophile fossils of the dry loess-forming period contemporaneous with Würm.
4. Boreal fossils of types belonging to the close of Würm and the post-glacial periods.

Some Alpine types of fossil species of the deep plains are classified (according to the system of Litvinov) with the glacial fossils.   Such species may be listed as follows:

Androsace villosa,	*Potentilla pimpinelloides,*
Daphne cneorum (including *D. Julia* K.-Pol.),	*Woodsia alpina,*
Schiwereckia podolica,	*Polygonum alpinum,*
Anemone narcissiflora,	*Cynanchum minus,*
Bupleurum ranunculoides (*Rasse multinerve*),	*Rhododendron flavum.*
Chrysanthemum sibiricum (*Rasse alaunicum*),	

Some of these forms existed in the Tertiary and have existed down to the present day. The Riss glaciation drove this flora into the Donetz region and the Podol-Volyn region where the ice did not penetrate.

The mesophile flora includes the following scattered forest species in the broad sense:

Orobus variegatus,	*Silaus peucedanoides,*
Coronilla elegans,	*Cynanchum scandens,*
Arabis turrita,	*Laserpitium hispidum,*
Callamintha officinalis,	*Arum orientale,*
Myosotis idaea ucrainica,	*Equisetum maximum,*
Evonymus nana,	*Polystichum braunii,*
Erysimum aureum,	*Geranium phaeum,*
Lysimachia verticillata,	*Anthriscusnitidus,*
Physospermum cornubiense,	*Symphytum.*

The Xerophile species include examples from the loess island-like stands as the *Gypsophila altissima,* and *Kochia prostrata* of the chalk cliffs of the Toltra mountains and the steep banks of the Dniester near the Podolian forest division, as well as the stony steppe

---

[1] Bondartchuk, W. J., "Die Fauna der Quartaren Ablagerungen der Ukraine S. S. R.," Die Quartarperiode, Ukrainische Akademie der Wissenschaften, 1932, No. 4.

[2] Kleopov, J. D., "Über das Alter der Relikte der Ukraine in Konnex mit der Sukzession in ihrer Vegetation im Laufe der Quartarzeit," Die Quartarperiode, Ukrainische Akademie der Wissenschaften, 1932, No. 4.

TABLE III. INVERTEBRATE FAUNAL SPECIES

	1. Fifth loess.	2. Fourth loess.	3. Caspian formations.	4. Alluvium of fourth terrace.	5. Submoraine part of third loess.	6. Fresh water clay contm. with third loess.	7. Moraine and glacial lakes.	8. Upper moraine part of third loess.	9. Alluvium of third terrace.	10. Second loess.	11. Alluvium of second terrace.	12. First loess.	13. Alluvium of first terrace.
1. *Hyalinia cristalina* Sand.				+									
2. *Zonitoides nitidus* Mull.													+
3. *Zonit ammonis* Str.											+		+
4. *Agriolimax agrestis* Lin.													+
5. *Monacha rubiginosa* Schm.				+							+	+	
6. *Euconulus trochiformis*				+									+
7. *Coniodiscus ruderatus* Stud.								+					
8. *Punctum pigmeum* Drap.													
9. *Euolota fruticum* Mull.				+									
10. *Helicella striata* Mull.												+	
11. *H. ericetorum* Mull.				+									
12. *H. arenosa* Kryn.												+	
13. *Xerophilla deyecta* San.												+	
14. *Fruticicola hispida* Brann.	+	+	+				+						
15. *Fruticicola hispida* v. *septentrionalis*	+								+				
16. *Helix strigela* Drap.				+									
17. *Clausilia dubia* Drap.				+									
18. *Succinea putris* Lin.		+		+					+	+	+		
19. *S. elegans* Riss.				+									
20. *S. pfeifferi* Rossm.	+	+	+	+	+			+	+	+	+	+	+
21. *S. oblonga* Drap.	+	+	+	+	+	+	+	+	+	+	+	+	+
22. *S. oblonga* v. *elongata* Sand.	+	+								+		+	
23. *Vallonia pulchella* Mull.												+	+
24. *V. costata* Mull.				+							+	+	+
25. *V. tenuilabris* Braun	+	+			+		+	+	+			+	
26. *V. vall.* aff. *adela* West.					+								
27. *Vertigo autivertigo* Drap.													+
28. *V. substriata*											+		
29. *V. genesii* Gredler.										+			
30. *V. parcedentata* Sand.					+					+			
31. *V. pygmaea* Drap.					+								
32. *Columella edent. col.* Mart.					+					+			
33. *C. edentula* Drap.		+			+							+	
34. *C. columella* Mart.	+							+				+	

PLATE IV.

PARTS OF BEREZOVKA MAMMOTH. Photo courtesy of Zoological Museum of U.S.S.R. Academy of Sciences.

TABLE III (Continued)

	1. Fifth loess.	2. Fourth loess.	3. Caspian formations.	4. Alluvium of fourth terrace.	5. Submoraine part of third loess.	6. Fresh water clay contm. with third loess.	7. Moraine and glacial lakes.	8. Upper moraine part of third loess.	9. Alluvium of third terrace.	10. Second loess.	11. Alluvium of second terrace.	12. First loess.	13. Alluvium of first terrace.
35. *Pupilla muscorum* Mull.	+	+			+			+	+		+	+	
36. *P. muscorum* v. *edentula*					+			+	+		+	+	
37. *P. musc.* v. *unidentata* Pf.					+						+	+	
38. *P. cupa* Voit.												+	
39. *Chondrina avenacea* Brug.					+								
40. *Pupa minutissima* Hal.				+									
41. *Fruncatulina cylindrica* Fer.								+					
42. *Zebrina detrita* Mull.												+	
43. *Chondrula tridens* Mull.					+							+	
44. *Cochlicopa lubrica* Mull.												+	+
45. *Corichium minimum*													+
46. *Lymnaea stagnalis* Mull.													+
47. *Radix auricularia* Lin.						+							
48. *R. ovata* Drap.											+	+	+
49. *R. ovata fontinalis* Geyer								+	+				
50. *Stagnicola palustris* Mull.			+	+	+	+		+				+	+
51. *S.* v. *fusca* Pfeiff.	+	+	+								+	+	
52. *S.* v. *curta* Cless.												+	+
53. *S.* v. *tunicula* Held.												+	+
54. *S.* v. *septentrionalis*	+	+							+		+		
55. *Leptolimnaea glabra* Mull.						+		+				+	
56. *Galba truncatula* Mull.					+	+					+	+	+
57. *Coretus corneus* Lin.			+	+		+					+		
58. *Planorbis carinatus* Mull.					+								
59. *P. planorbis* Lin.	+	+	+	+	+			+	+		+	+	+
60. *P. marginatus* Lin.				+									
61. *P. micromphalus*				+									
62. *Armiger nautileus*									+				
63. *Spiralina vortex* Lin.											+		+
64. *S.* v. *compressa* Mith.												+	
65. *Paraspira spirobis* Lin.			+	+	+	+			+		+	+	
66. *P. bucostoma* Mith.	+	+	+		+			+	+			+	
67. *P. septemgyrata* Zicq.												+	
68. *Gyraulus albus* Mull.					+	+	+				+	+	+

TABLE III (*Continued*)

	1. Fifth loess.	2. Fourth loess.	3. Caspian formations.	4. Alluvium of fourth terrace.	5. Submoraine part of third loess.	6. Fresh water clay contm. with third loess.	7. Moraine and glacial lakes.	8. Upper moraine part of third loess.	9. Alluvium of third terrace.	10. Second loess.	11. Alluvium of second terrace.	12. First loess.	13. Alluvium of first terrace.
69. *G. gudleri* Gredl.	+	+			+		+	+		+		+	
70. *G. rossmaessleri* Auerw.					+		+			+		+	
71. *G. lacvis* Aldlr.						+							
72. *Bethyomph contortus* Lin.	+	+			+			+	+		+	+	+
73. *Segmentina nitida* Mull.			+			+			+	+	+		
74. *Ancylus fluviatilis* Mull.				+									
75. *A. involutus* Pav.				+									
76. *Acrolotus lacustris* Lin.													+
77. *Physa fontinalis* Lin.													+
78. *Plexa hipitorum* Lin.								+				+	
79. *Valvata piscinalis* Mull.									+		+		+
80. *V. v. antiqua* Sow.			+	+					+				
81. *V. pulchella* Stud.	+	+	+					+	+				+
82. *V. cristata* Mull.						+							
83. *V. macrostoma* Sand.				+		+							
84. *V. sulekiana* Bruss.				+									+
85. *Paludina fasciata* Mull.									+				
86. *P. soxolovi* Pav.			+	+					+				
87. *P. zickendrathi* Pav.			+	+						+			
88. *P. istriena* Pav.			+	+									
89. *P. diluviana* v. *crassa* Kunt.			+	+									
90. *P. diluviana* v. *gracilis* Kunt.			+	+									
91. *P. romaloi* Cob.			+	+									+
92. *P. getica* Pav.				+					+				
93. *P. achatinoides* Desh.			+										
94. *P. pseudoachatinoides* Pav.			+	+									
95. *P. achatinoides* Desh.										+			
96. *P. subconcinna* Sinz.			+										
97. *P. sinzovi* Pav.			+										
98. *P. tanaitica* Boudar.			+										
99. *P. bugensis* Boudar.			+										
100. *P. pseudoneumayri* Pav.			+										
101. *P. obtusa*.			+										
102. *P. hungarica* Sabba.			+										

PLATE V.

MOUNTED MAMMOTH FROM BEREZOVKA. After Vishnevky.

TABLE III (*Continued*)

	1. Fifth loess.	2. Fourth loess.	3. Caspian formations.	4. Alluvium of fourth terrace.	5. Submoraine part of third loess.	6. Fresh water clay contm. with third loess.	7. Moraine and glacial lakes.	8. Upper moraine part of third loess.	9. Alluvium of third terrace.	10. Second loess.	11. Alluvium of second terrace.	12. First loess.	13. Alluvium of first terrace.
103. *Paludina dresseli* Tourn.			+										
104. *P. contecta* Pav.			+										
105. *P. cretzestiensis* Pav.			+										+
106. *P. murgescki* Sabba			+										
107. *P. bockhi* Hal.			+	+									
108. *P. aethiops* Pav.			+	+									
109. *P. mammata* Sabba			+	+					+				
110. *P. tiraspolitana* Pav.				+									
111. *P. costae* Held													
112. *P. rodensis* Pav.												+	
113. *Bithynia tentaculata* Lin.				+									
114. *B. leachi* Schep.	+		+			+		+	+		+		
115. *B. spoliata* Sabba			+										
116. *Lythoqlyphus naticoides* Pf.			+	+									+
117. *L. mumayri* Sabba			+	+									+
118. *L. michaeli* Hal.				+									
119. *Melanopsis acicularis* Fer.			+										
120. *M. esperoides* Sabba			+	+									
121. *M. cotrocensis* Cob.				+									
122. *Theodoxus fluviatis* Lin.													+
123. *Neretina serra filiniformis* Gel.				+									
124. *N. transversalis* Pfeuff.				+									
125. *N. scripta* Sabba				+									
126. *N. licherdopoli* Sabba				+									
127. *N. semiplicata* Hal.				+									
128. *Unio pictorum* Lin.													+
129. *U. tumidus* Retz.			+	+									+
130. *U. crassus* Retz.			+	+									+
131. *U. batavus haassica* Haas			+	+									
132. *U. batavus pseudocrassus* Rossm.			+	±									
133. *U. cubranollici* Brus.				+									
134. *U. kungurensis* Rossm.			+	+									
135. *U. rumanus* Tour.			+	+							+		
136. *Unio* sp.						+			+				

TABLE III (*Continued*)

	1. Fifth loess.	2. Fourth loess.	3. Caspian formations.	4. Alluvium of fourth terrace.	5. Submoraine part of third loess.	6. Fresh water clay contm. with third loess.	7. Moraine and glacial lakes.	8. Upper moraine part of third loess.	9. Alluvium of third terrace.	10. Second loess.	11. Alluvium of second terrace.	12. First loess.	13. Alluvium of first terrace.
137. *Anodonta* sp.			+	+									+
138. *Cyclas rivicola* Lam.			+	+		+			+				+
139. *Sphaerium solidum* Nor.				+									
140. *Corbicula fluminalis*			+	+									
141. *Corb. jassiensis* Cob.			+	+									
142. *Dreissensia polymorpha* Pall.			+	+									+
143. *D. v. miusica* Bondar.			±										
144. *D. v. nova* Andr.			+										
145. *D. v. etima* Andr.			+										
146. *D. v. regularis* Pall.			+										
147. *D. v. marina* Pall.			+										
148. *D. v. occidentalis* Sabb.			+										
149. *D. v. fluviatilis* Pall.			+										
150. *D. v. latior* Andr.			+										
151. *D. v. oblonga* Andr.			+										
152. *D. retovskii* Andr.			+										
153. *D. rostriformis* Desh.			+										
154. *D. caspia* Eichw.			+										
155. *D. crassa* Andr.			+										
156. *D. eichwaldi*													
157. *D. semilimaris*				+									
158. *Pisidium amnicum* Mull.				+					+				
159. *P. casertanum* Poll.													+
160. *P. substriatum* Malm.						+			+				
161. *P. nitidum* Mull.									+				
162. *P. obtusale* Pfeiff.				+		+			+		+	+	+
163. *P. hibernicum* West.									+				
164. *P. tenuilincatum* Stef.								+					
165. *Pisidium* sp.	+							+					
166. *Didacna rudis* Naliv.			+										
167. *D. crassa* Eichw.			+										
168. *D. pseudocrassa* Pav.			+										
169. *Monodacna colorata* Eich.			+										

PLATE VI.

MAMMOTH TRUNK FOUND IN SIBERIA. Photo courtesy of Zoological Museum of U.S.S.R. Academy of Sciences.

regions. The majority of the Xerophile species tend to the Mediterranean, especially to the east and in the neighborhood of Bulgaria. Species in this region may be listed as follows:

Dianthus pseudoarmeria,	*Teucrium polium,*
Silene supine,	*Alyssum tortuosum,*
Silene compacta,	*Silene cyri* (including *S. hellmanii*),
Pimpinella tragium s.l.,	*Alsine glomerata,*
Zizyphora teniuor,	· *Arenaria rigida,*
Jurinea stoechadifolia,	*Trinia hispida.*

In the Ural-Caspian region the following species are found:

Silene sibirica,	*Caragana grandiflora* (including *C. scythica*),
Gypsophila altissima,	*Caragana frutex.*
Geranium linearilobum,	

The majority of the Xerophile migrated to the steppe regions during the loess deposition of the various glacial periods, though by far the greatest number of this group are found in the Würm.

The boreal fossils are of north Asiatic origin; they are scattered island-like in the forest steppes, and the steppes along the Samara and the Donetz. For the greatest part they are not connected with the large river terraces. The following plants may be listed as typical: *Pinguicula vulgaris, Saxifraga hirculus, Vaccinium myrtillus,* and *Vaccinium vitisidaea.* The moss series of Podolia are to be considered as belonging to post-glacial times.

18. Conclusions

On the basis of the foregoing data, the existence of only three definitely traceable glaciations, Mindel, Riss and Würm, seems to have been established in European Russia. The Gunz was very weak and either its traces were obliterated by subsequent advances, or it was not effective in European Russia. The other three glacial advances appear to correspond in time, and in general characteristics, with those of Western Europe. Of these, the Riss glaciation had the longest duration, and the widest spread.

The fauna which accompanied the various climatic changes generally corresponds to that of Western Europe, but has on the whole a colder character. The extremely warm species are few and not abundant. Some local variations due to specific physical environment are encountered in the Crimea and the Caucasus.

Further work is needed to tie up the terraces with the ice advances more definitely. The best existing correlation is furnished by the series of loess strata with interlayers of fossil soils.

B. SKELETAL REMAINS IN EUROPEAN RUSSIA

1. PODKUMOK [1]

In the fall of 1918, in Piatigorsk, while excavating sewers on Bazarnaya Street, workmen came across a pottery vessel, which had been broken by spades, and a polished stone implement in the shape of a thick disc perforated in the middle. This was found at a depth of 4–6 meters. Beneath it the remains of a very brittle human skeleton were dug out and picked up by A. N. Voronin, an employee of the University of Piatigorsk.

The bones were given to M. A. Gremiatsky the next day with the details of their discovery. On the following day an attempted investigation was terminated by civil war conditions and the capitulation of the city. Therefore, the only means of dating is, on the basis of the general geological conditions of the country. V. P. Rengarten,[2] who made a subsequent geological survey, gives the following account.

The find was made on the slope of a large terrace, 15 meters above the Podkumok River. This terrace is well pronounced along the left bank of the river, and composed of boulder clay deposits. There are three terraces: (1) fifteen meters high, (2) sixty meters, and (3) one hundred and twenty meters high, the general condition for the North Caucasus, all of them of the Quaternary Age. The lower one is connected with the moraine of the last glaciation, as the deposition of the pebbles comprising its lower part was shown to be contemporary with the Würm glaciation on the basis of the process of the lowering of the snow line. The loess-like clays, which, as it is supposed, contained the remains of Podkumok man, were deposited during a short period between the beginning of the retreat of the glacier and the land raising. On the basis of this, with some corroborating evidence from the Black Sea region, Rengarten definitely dates the deposit of Podkumok as belonging to the Würm glaciation.

Gremiatsky indicates, however, a weak point in the dating due to the absence of faunal material; moreover, the find itself was not made by him, and he was forced to take the evidence of workmen as to the precise location of bones.

According to him the following human bones were found:

1. Part of a cranium, consisting of an almost whole frontal bone, the front parts of both temple bones, and a small part of the nasal bone (Pl. VII, Figs. a, b).

2. A fragment of the right side of the lower jaw with five teeth.

3. Two fragments of both temple bones.

4. A small fragment of the left part of the lower jaw.

5. A fragment of the left shoulder bone.

6. Other small fragments.

The bones in general are rather thin and small, have very weak traces of the areas of muscle attachments.

[1] This account represents the summary of M. A. Gremiatsky's article "The Podkumok Cranium and its Morphological Peculiarities," Russian Anthropological Journal, Vol. XII, Pt. 1–2, Moscow, 1922, pp. 92–110.

[2] Rengarten, V. P., "The Age of Deposits Containing the Remains of Podkumok Man," Russian Anthropological Journal, Vol. XII, Pt. 1–2, Moscow, 1922, pp. 193–195.

PLATE VII.

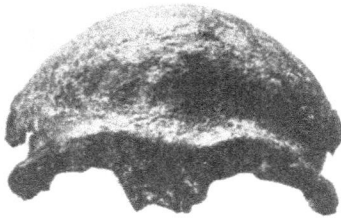

a

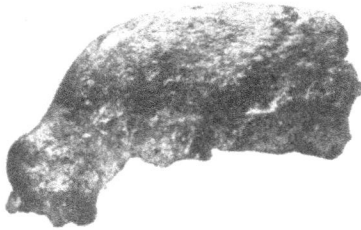

b

PODKUMOK SKULL. *a*, front view; *b*, side view. Photo, courtesy of Prof. Gremiatsky.

The Skull.—The small size of the skull, not too well pronounced *lineae temporalis*, and some other characteristics indicate that this skeleton belongs to a female.

The almost complete closing of the sutures and the thinness of the bones indicate that the individual died between the ages of fifty-five and sixty-five, so that this skull can be called senile, a fact which is also supported by the condition of the teeth.

After an examination of the available material, Professor Gremiatsky arrives at the following conclusions:

1. On the basis of geological evidence, the Podkumok cranium can be considered with a great degree of probability as belonging to the time of the last glaciation of North Caucasus.

2. In spite of the comparatively small size of the recovered fragment of the cranium, it is possible to establish its morphological affinity with the Neanderthal group on the basis of the following:

(a) The presence of uninterrupted *torus supraorbitalis* passing to *processus zygomaticus*, similar in its development to the corresponding formation of Spy II and the Krapina fragments.

(b) The presence, typical for anthropoids and Neanderthal groups, of the "over eye beak" (obtained by drawing the sagittal curve through the middle of the eye socket).

(c) The fact that the sagittal curve, drawn on the border of the inner and middle third of the eye socket, touches the medial curve in the upper glabella region.

3. By using Schwalbe's method for the special orientation of the skull, it is possible to measure the bregma angle and the frontal angle. The results obtained are: bregma angle 42–47, frontal 68–73.

4. The comparison of the Podkumok fragment with other Neanderthaloids, Anthropoids and Hominidae, permits to establish a new diagnostical characteristic for the Neanderthal man, the presence of *depressia supraorbitalis* (*fossa supraorbitalis* of Klaatch). The index of the latter is rarely over seven in the living groups and varies in the Neanderthal specimens from nine to thirteen.

	Angle of Depression	Index of Depression
Neanderthal	159	9.8
Spy II	150	12.8
Podkumok	155	9.8
Alsas, modern [1]	170	7.4
Moscow kurgan, late [1]	165	6.8
Armenian, modern [1]	162	7.8
Chimpanzee	139	16.6
Many modern skulls	180	

The angle of depression is less characteristic than the index. The importance of this method is that it permits a diagnosis on the basis of a small fragment of frontal bone.

5. The glabella index—33.6—and the considerable difference between the bregma angle and the frontal angle places the Podkumok skull somewhat nearer to modern man than other Neanderthaloids.

[1] *Homo sapiens* with the unusual supraorbital ridges.

6. Besides the above diagnostical characteristics, the following morphological peculiarities confirm the primitive character and genetic affiliations:

> (a) The large interorbital space.
> (b) The high interorbital index.
> (c) The very slight development of the *Tubera frontalis*.
> (d) The evidently small size of the cranium.

The Lower Jaw.—Gremiatsky [1] points out that the fragments of the lower jaw have the same degree of preservation as the rest of the find. Morphologically this jaw has a number of characteristics which place it somewhere between the typical Neanderthaloids and Homo sapiens.

In his article, quoted above, Gremiatsky gives a detailed description of the fragments, the degree of preservation, etc. Comparing this jaw with other fossil and modern jaws he points out:

> I. The features analogous to Neanderthaloids:
> > (a) Considerable size of *foramen maxilliare*.
> > (b) Presence of supra-marginal notch of considerable length.
> > (c) Considerable thickness of lower edge.
> > (d) General massiveness.
> > (e) Roundness of *angulus maxillae*.
> > (f) Considerable (reconstructed) width of the upper part.
> II. The features not found in Neanderthaloids:
> > (a) Suggestion of the appearance of chin protrusion.
> > (b) Shortness of the alveolar arc.

The Teeth.—The general length of the alveolar arc is 41.2 mm. The teeth are fairly well preserved and in their absolute dimensions do not exceed the average norm of variations for the living races. The only difference is the reduction of the number of cusps in molars, very probable in the second (4) and indisputable on the third (4).

In general, Gremiatsky considers that the lower jaw with its teeth occupies an intermediate position between that of *Homo neanderthalensis* and *Homo sapiens*.

Conclusions.—Gremiatsky feels that this new and unexpected discovery of the Neanderthaloid skeleton is very important, as it represents the first established find so far removed from the region of Western Europe. It is apparent that Neanderthal man lived in Northern Caucasus in conditions very similar to those of the diluvial period of Western Europe.

It should be pointed out that Sir Arthur Keith [2] does not accept Gremiatsky's assertion; but says that "those familiar with the diagnostic features of Neanderthal man have only to examine these illustrations [3] to be convinced that Podkumok man was altogether of the Neanthropic species."

Lately, in Russia, there has also been a tendency to question the Neanderthaloid affinities of Podkumok. The author does not feel qualified to enter into this controversy, but the data given above may help to clear the situation.

[1] Gremiatsky, M. A., "Remains of the Lower Jaw and Teeth of Podkumok Man" (Appendix to Russian Anthropological Journal, Vol. XIV, Moscow, 1925, pp. 91–99).

[2] Keith, Sir Arthur, "New Discoveries Relating to the Antiquity of Man," London, 1931, p. 365.

[3] Unfortunately the illustrations in the article of Gremiatsky are so bad, that hardly any diagnosis is possible (E. G.). The photos on Table I were furnished to the author by Prof. Gremiatsky.

2. KIIK-KOBA

In the lower layer of the cave Kiik-Koba, Crimea (see details, page 240) remains of two Neanderthaloid skeletons were discovered.

3. DEVIS-HVRELI

Nioradze discovered a human lower jaw together with an industry of Upper Palaeolithic character in the cave Devis-Hvreli, Caucasus (see page 293).

4. HRIAS'CHEVKA

The finds of human remains in the region of the lower Volga are described by A. P. Pavlov.[1]   There were four general localities from which these finds were reported.

A rich and varied fauna of the Richsdorf type (Mammoth) was found in many places on the banks of the Volga, in the northern part of the Samara and Ulianovisk (former Simbirsk) regions.   The southernmost location is five miles below the village of Hrias'chevka on the left bank of the Volga, on the small peninsula of Tungus.

Here, in 1913, on the surface of the slightly colored gravel which lay under a considerable thickness of sands deposited by the river, bones of the following animals were found.

1. *Elephas primigenius,*
2. *Rhinoceros tichorhinus,*
3. *Equus caballus,*
4. *Bos primigenius,*
5. *Bos priscus* var. *latifrontis,*
6. *Alces fossilis,*
7. *Cervus elaphus,*
8. *Rangifer tarandus,*
9. *Camelus fossilis,*
10. *Carnivorae* sp.

All the bones were covered by a characteristic dark-brown patination.

A *human frontal bone* found with these bones, was somewhat lighter in color.   This may indicate that it was not contemporaneous with the animal remains.[2]

5. SOBACHIA PRORVA

The easternmost site of this group is located on the left bank of the Volga, opposite the small channel, "Sobachia Prorva."   The bones of the same fauna as in Hrias'chevka were found in 1913 in the sand bar at the base of the alluvial terrace, in the layer of gravel covered by sands.   The character of the gravel and bones indicate that the animal remains were found in secondary deposition.   Yet, the abundance of bones, as well as the slight degree of rolling, make it apparent that their primary location was not far off.   Pavlov believes that these bones were washed out from the ancient terrace, composed partly of diluvial, partly of marshy-lake formation.   This terrace was underwashed by the Volga, and pieces of Jurassic concretions, Permian marls, and chalk which formed the sand bar, were deposited together with the bones.

[1] Pavlov, A. P., "Fossil Man Contemporaneous with Mammoth in Eastern Europe, and Fossil Men of Western Europe" (Supplement to the Russian Anthropological Journal, Vol. XIV, Pt. 1–2, Moscow, 1925, pp. 5–36).

[2] Osokov, P. A., an article in the Addition to the Minutes of the Meeting of the Moscow Society of Naturalists in 1913.   Quoted by Pavlov.

The dark, almost black, polish of the bones indicates, according to Pavlov, that they were originally deposited beneath the marshy-lake formations. With these bones was found a *human humerus*, which had an almost identical patination (Pl. IX, Fig. *c*).

6. UNDORA I AND II

The third site is situated on the island of Undora, just below the channel of "Sobachia Prorva." The animal bones were found here in a similar horizon: a layer of gravel, covered by sands. The general character of the gravel and bones is the same as in location 5. Here bones of mammoth, *Bos* sp. and *two human crania* of identical color and degree of preservation, lying side by side were found in 1913 (Pls. VIII, IX).

Parlov designates the finds of crania in site 6 as Undora I and Undora II.

Undora I (Pl. VIII, Figs. *a*, *b*).—In the cranium of Undora I only the frontal bone and the right temporal bone are preserved, thus restricting the possible number of measurements. Consequently, Pavlov made an attempt to reconstruct the whole cranium in order to obtain the position of the *inion*.

Using "the cubus Martini" to obtain the *norma lateralis* (Fig. 1, *a*) it was comparatively easy for him to establish the position of *lambda*, since the lambdoid as well as the sagittal sutures were well preserved.

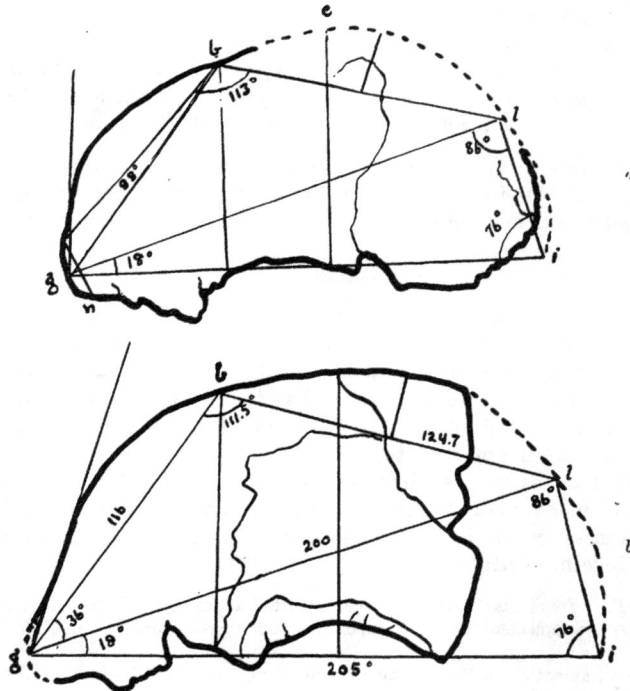

FIG. 1.   *a*, Reconstruction of Undora I.   *b*, Reconstruction of Undora II.   After Pavlov.

PLATE VIII.

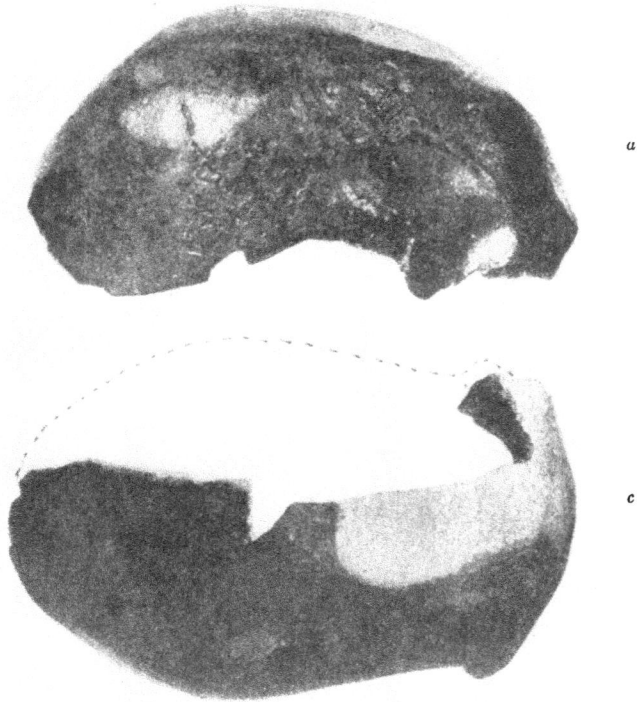

a

c

UNDORA I.   *a*, side view; *b*, top view.   After Pavlov.

To determine the *inion* he used the method of D. G. Schwalbe, with this difference: instead of allowing 20° for the angle LGI (the average for modern races), he used 18°, which is in his opinion more suitable for fossil man.   He estimated then the size of the angle GLI which he considers very important, though it is somewhat neglected by anthropologists.   This angle is for:

Pithecanthropus erectus... 102
Homo neanderthalensis... 94
South Australian.. 88
Homo sapiens (other).. 75–84

Pavlov takes 86° as a conservative size of this angle which is one degree higher than the average for the fossil skulls (non-Neanderthal).   Thus he obtained the *inion*.

Undora II (Pl. IX, Figs. *d, e*).—The second cranium differs markedly from Undora I. Its forehead is more sloping, the supraorbital ridges are more pronounced, and it appears to have been longer.   Unfortunately a smaller part of the cranium is preserved; in the front only the left part of the orbit is preserved, so that the right side had to be reconstructed.

Since the sagittal suture does not reach *lambda*, the positions of both *lambda* and *inion* are unknown.   To determine them, Pavlov used the same method as in the case of Undora I, with some modification.

TABLE IV. COMPARISON OF UNDORA I AND II WITH OTHER SKULLS. After A. P. Pavlov.

	Undora I	Undora II	Galley-Hill	Brünn	Combe-Capelle	Canstatt	Brüx	Egisheim	Homo neanderthalensis	Pithecanthropus erectus
Maximum length......................	177	206	205	203	198	191	206	190	199	181
Maximum breadth.....................	120	132	132	134	130	130	130	150	147	130
Cephalic index........................	67	64	69.4	66	65.7	68	63	80.2	73.9	71.8
Calvarial height......................	88	102	97	103	104	91	101	97	80.5	62.5
Calvarial height index.................	50.3	49.7	48.2	51.2	51.4	48	48	51	40.4	34
Frontal angle........................	80	70	82	75	84	79	80.5	89	62	54
Bregma angle........................	53	54	52	54	58	51	54	58	44	35
Index of position of bregma...........	33.7	33.6	34	33	32	33.8	33.3	32	40.5	50.
Upper cranial angle...................	113	112	113	108.5	110	111.5	112	111.5	120	120
Glabella-lambda arc...................	169	200	190	196	192	184	198	181	—	173
Glabella-inion arc....................	175	205	201	201	191	189	204	190	199	181
Occipital angle.......................	86	86	88	85	79	86	86	86	96	103
Lambdoid angle......................	76	76	74	78	82	76	77	76	65.5	64
Glabella-bregma chord.................	99	116	110	115	111	114	115	113	—	—
Sagitto-temporal chord................	104	125	118	126	124	108	123	116	—	—
Index of frontal curvature.............	19	14.6	20	18.6	18	16.5	17.7	19	13.6	—
Index of frontal parietal curvature......	19	19	18	19	17	12	19.5	16.3	16	—
Frontal curvature angle...............	138	145	135	147	139.5	142	142	138	153	137
Parietal curvature angle...............	136	138	138	138	136	152	130	143	145	136
Lambda-calvarial height...............	56	57	64	68	68	62	67	70	54.5	—
Lambda-calvarial height index..........	33	28.5	33.6	35.6	35.4	33.6	33.8	38.6	—	—

Taking (Fig. 1, *b*) the average length of the line B-L of the known fossil skulls as 124.7, he assumed that they are of the same size as Undora II.   Then he determined the average angle GBL as 115.5, thus establishing the position of the *bregma*.   The angle BGL is consequently equal to 36°, which is the average for the fossil skulls used for the comparison.

The *inion* was determined as in the case of Undora I.   Though realizing the short-comings of measurements based on reconstructions, Pavlov feels that the true values are not very far from those obtained.   Then he proceeded to reconstruct the crania of Brüx, Cannstatt, and Egisheim by his method, feeling that the Schwalbe reconstruction is not accurate, and tends to make the posterior part of the skull too modern.   The results of the measurements of both Undora I and II in comparison with other skulls are given in the table of measurements and indices (Table IV).

Professor Pavlov, analyzing his results, points out that both skulls Undora I and II are dolichocephalic.   Undora I (67) is near Brünn (66); Undora II (64) near Galley Hill (64).   The same relationship exists in the height of the cranium (hauter de la voute) Undora II (102), Brünn (103); Undora I (88), Galley Hill (97).   The frontal angle of Undora I (81) is near Galley Hill (82), while Brünn is less (75) and Undora II is still less (70). The bregma angle of Undora I (53) occupies an intermediate position between the Galley Hill (52) and Brünn (54) and Undora II (54).

Comparing the two Undora skulls, Pavlov remarks that Undora I is that of a male and Undora II of a female.

Concluding, he points out that Undora I and II belong to the series of crania of the Galley Hill type, which should include Brünn and Cannstatt, appearing in Europe soon after the last glaciation.

7. GORODIS'TCHE

The fourth location is on the left bank of the Volga, opposite the village of Gorodis'tche on the island of Mulinov.   Here, in 1912, likewise in a layer of gravel covered by sands, were found bones of *Elephas primigenius, Rhinoceros tichorhinus, Rangifer tarandus, Alces alces, Bison priscus* and a *human lower jaw* with a protruding chin.   The bones were not so deeply colored as in the previous sites, and the fauna was somewhat different.   Pavlov believes that these bones came from the diluvial deposits, and not from the marshy-lake formation.

8. SENGILEI

Professor V. A. Gorodzov [1] mentions the find of another cranium near the town of Sengilei, the date of which he questions, though, according to him, Pavlov considers it of the Galley Hill type.

9. KHVALINSK

Gorodzov also mentions the find of a part of a human cranium 25 kl. south of the town of Khvalinsk.   Only the upper part with the supraorbital ridges was preserved.   Both finds (Sengilei and Khvalinsk) were accompanied by the typical mammoth fauna.   According to Gorodzov, Pavlov considers this skull also of the Galley Hill type.   The condi-

[1] Gorodzov, V. A., "The Results of Archaeological Activities in U.S.S.R. During 1918–1930."   Manuscript.

PLATE IX.

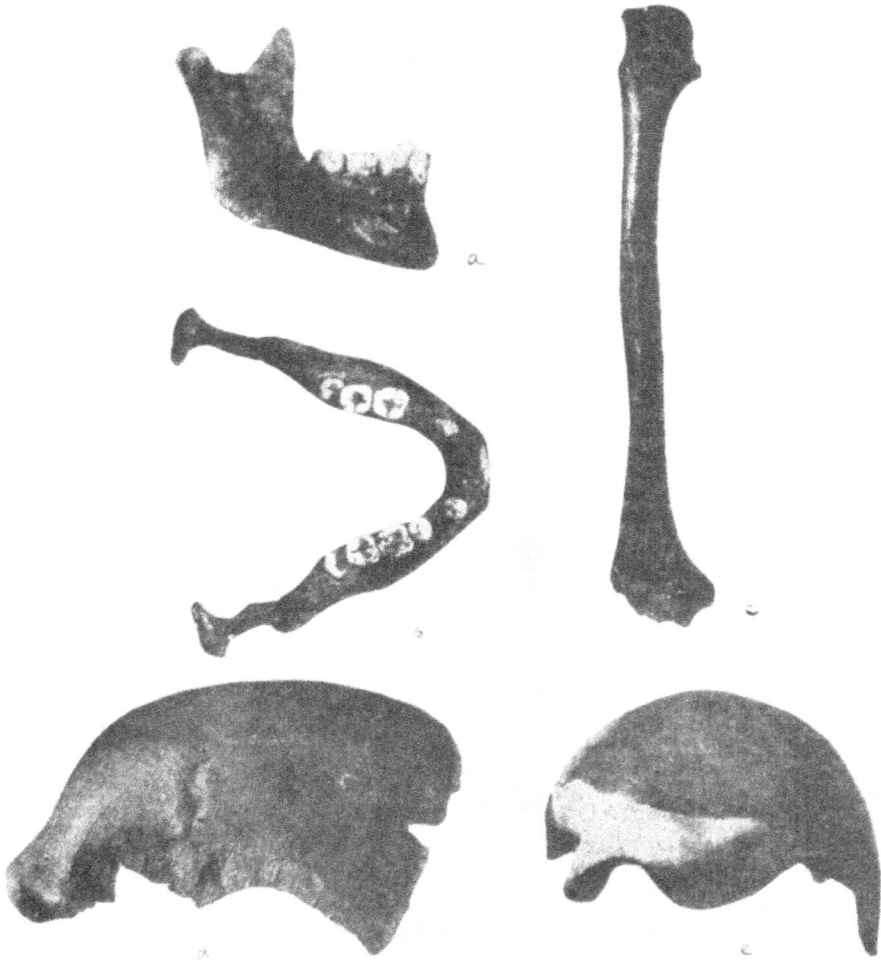

a, b, I ower Jaw found in the Same Location as Undora I and II.   *c,* Humerus found at "Sobachia Prorva."   *d, e,* Undora II, front and side view.   After Pavlcv.

PLATE X.

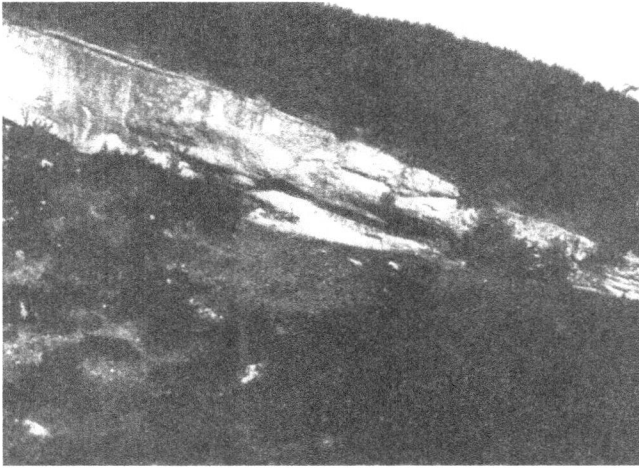

GENERAL VIEW OF THE ROCK-SHELTER FATMA-KOBA. Photo courtesy of G. A. Bonch-Osmolovsky.

PLATE XI.

Tardenosian Burial from the Rock-shelter Fatma-Koba. Photo courtesy of G. A. Bonch-Osmolovsky.

tions of the find are described by Gorodzov in a manuscript [1] to which the author did not have access.

At the present time only the description of the Undora finds, in Pavlov's article referred to above, are available.

10. FATMA-KOBA

In the Tardenoisean layer of the cave Fatma-Koba G. A. Bonch-Osmolovsky discovered a finely preserved skeleton, which was transported in block to Leningrad for study (Plates X, XI).

[1] Gorodzov, V. A., "Investigations of Quaternary Conditions in the Region of the Lower Volga in 1929." Manuscript.

FIG. 2. Map No. 1. Palaeolithic Sites in European Russia.

THE PALAEOLITHIC SITES ON MAP I

1 Podkumok,
2. Kiik-Koba and Koukrek,
3. Devis-Hvreli,
4. Hrias'chevka,
5. Undora I and II, Sobachia Prorva,
6. Sengiley,   6a. Gorodische,
7. Khvalinsk,
8. Fatma-Koba,
9. Rzhev,
10. Wolf Grotto,
11. Chokurcha,
12. Kosh-Koba,
13. Gamkovo,
14. Ilskaya,
15. Derkula,
16. Shaitan Koba,
17. Suren I and II and Cheken-kermen,
18. Borshevo I and II,
19. Kostenki I, II, III, IV, and V,
20. Gagarino,
21. Berdizh,
22. Mezine,
23. Novo-Bobovichi,

24. Urovichi,
25. Karacharovo,
26. Gontzi,
27. Cyrill str, Kiev,
28. Iskorost,
29. Timonovka,
30. Suponevo,
31. Elisevichi,
32. Zhuravka,
33. Dubova Gully,
34. Dovginichi,
35. Khotianovka,
36. Kastrova Gully,
37. Lugansk,
38. Gvardzhilas-Klde,
39. Kolodiazhnoye,
40. Samara, Voskresenskii Spusk,
41. Samara, Postnikov Gully,
42. Kamenetsk-Podolsk,
43. Studenitza,
44. Selis'che,
45. Kiev, Protasov Yar,
46. Kiev.

47. Viazovka,
48. Meltinova,
49. Urkusta,
50. Stenina,
51. Shapovalovka,
52. Kachinsk,
53. Sergeevka,
54. Osokorovka,
55. Kizil-Koba,
56. Schurii Rog,
57. Kolchakovichi,
58. Kosheleva,
59. Chatirzhdag,
60. Skakalka,
61. Babanka,
62. Gai and Uhnovo,
63. Degtiarevo,
64. Umrihino,
65. Virchov's cave,
66. Bartashvilli cave,
67. Uvarov's cave,
68. Novo-Alexandria.

For detailed location of Nos. 2, 3, 8, 10, 11, 12, 16, 17, 55, 59, and other Crimean sites see Map No. 2.   (page 241.)

C. LOWER PALAEOLITHIC SITES OF EUROPEAN RUSSIA

1. RZHEV

Implements of Acheulean type were found by V. M. Lemeshevsky in the Rzhev district of the Tvier region.   As they were for the most part surface material, the investigator could not definitely determine the original position of the find.   The specimen illustrated (Fig. 3) is preserved in the Museum of Anthropology and Ethnography of the Academy of Sciences in Leningrad.

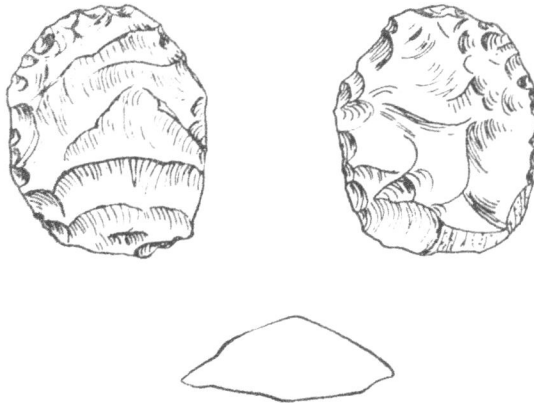

FIG. 3.   Implement of Acheulean type from Rzhev district, Tvier region.   ⅛ natural size.   After Vishnevsky.

P. P. Ephimenko[1] considers these implements, so similar to the upper Volga and Nijni-Novgorod finds, as early Neolithic. A need for large cutting axe-like implements, created by changing natural surroundings and modes of life, changed the flint industry. The fine technical methods of the Upper Palaeolithic period could not be used in the manufacture of these tools, so Neolithic man had to use the same crude technique which was known in the Lower Palaeolithic times. This explains the reappearance of the *coup de poing*, which soon developed into a series of new forms. Further investigations will make clear the real significance of these puzzling finds which are so much like the tools of the Lower Palaeolithic period.

2. Volga Coup de Poing

This implement was found, before the World War, on the banks of the Volga in the territory of the former Samara province. At present it is in the possession of V. A. Gorodzov,[2] who is very much interested in a more accurate determination of the spot of this find which is so important for Eastern Europe. The lower Volga, where the find was made, was never under Quaternary glaciation, remaining always available for the habitation of ancient man and the contemporary fauna. By studying the rich palaeontological collections in the museums of the lower Volga towns, such as Samara, Khvalinsk and Saratov, it is easy to establish the fact that during the Lower Palaeolithic epoch, the vicinity of these towns was inhabited by such fauna as *Elephas trogonterii* and *Elasmotherium*. The bones of these animals, according to recent investigation, are found in the lower layers of the Quaternary deposits of the Volga in association with each other.

The Volga specimen of a *coup de poing* (Plate XII, *a, b*) weighing 300 grams is 11.6 cm. long, 8.4 cm. wide and 3.2 thick. The tool is made of bluish yellow flint. The original surface of the formation is preserved on the lower part, and the entire surface is covered by dendritice markings and patination. In form this specimen is very near the French Acheulean type (*coup de poing Acheuléen*).

3. Kiik-Koba

The site of Kiik-Koba [3] is situated at the base of the cliff overlooking the southern end of the village of Kipchak, in the valley of the Zuiya River, a tributary of the Salgir. It is 25 klm. east of Siempheropol, Crimea and 3 klm. from the village of Niezatz. (See Map No. 2, Fig. 4.)

The narrow canyon of Zuiya is about 200 meters deep. It was formed by the action of the river, cutting through the limestone formations of the plateau. The highest point there is 540 meters above sea level, 200 meters higher than the river.

In the perpendicular slope of the limestone a number of niches, corridors and rock shelters had been formed by the processes of erosion and weathering. The largest of them, Kosh-Koba, was excavated in 1920 and some traces of human industry, with a number of bones of Quaternary animals, were found.

---

[1] According to Vishnevsky, B. N., "Prehistoric Man in Russia," Appendix to translation of Osborn's "Men of the Old Stone Age," Leningrad, 1934, p. 470.

[2] Gorodzov, V. A., "The Results of Archaeological Activities in U.S.S.R. During 1917–30." Manuscript.

[3] See preliminary account, Golomshtok, E. A., "Anthropological Activities in Soviet Russia," American Anthropologist, Vol. 35, No. 2, 1933, p. 313.

PLATE XII.

a   b

a, b. Volga Coup de Poing.   Photo courtesy V. A. Gorodzov.

The grotto, Kiik-Koba, is situated 400 meters south and some 20 meters lower than the first one. It is a rock shelter, facing southward, 16 meters wide, 9 meters long, and 7 meters high. The entrance of the grotto, well protected by overhanging rocks, is very near the edge of the steep forest-covered slope of the cliff. Just in front of it there is an enormous rock, fallen from above. The ceiling slopes sharply inward, forming a very low inside passage, and a sort of spherical niche, which reflects the rays of the sun and heats the floor which is a flat area of 100 square meters (Fig. 5).

Several hardly noticeable trails lead from the village of Kipchak to Kiik-Koba, which recently has been used as a shelter for sheep, as is seen from their excrements and the

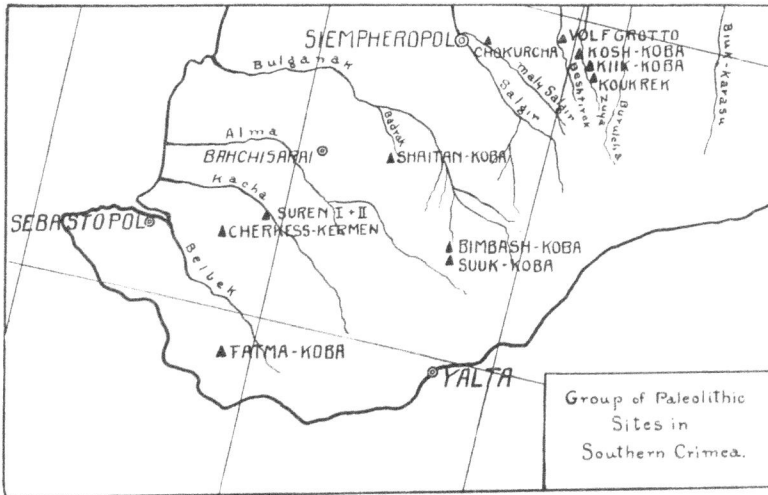

FIG. 4.

remains of the open fires of the shepherds. The position of the cave, almost on the border line of forest and steppe, with favorable sun conditions and the presence of a fresh water spring 100 meters away, made it a very suitable place for human habitat at all times.

The excavations were started in 1924 and continued in the summer of 1925. The work was done by a group of students of Leningrad University, under the leadership of G. A. Bonch-Osmolovsky [1] and Professor N. L. Ernest. In 1926 a special committee from

[1] The material which follows represents a summary and extracts from the following works:

Bonch-Osmolovsky, G. A., "Le Paléolithique de Crimée," Bulletin of the Quaternary Committee, No. 1, 1929.

Bonch-Osmolovsky, G. A., "The Question of Evolution of Old Palaeolithic Industries," "Chelovek," 1928, No. 2–4.

Bonch-Osmolovsky, G. A., an article in the Guide for Crimea, Siemperopol, 1925.

Bonch-Osmolovsky, G. A., "The Prehistoric Cultures of Crimea," Krim, No. 2.

Bonch-Osmolovsky, G. A., an article in "Priroda," 1926, No. 5–6.

Bonch-Osmolovsky, G. A., "A Palaeolithic Site in Crimea," Russian Anthropological Journal, Vol. 14, Pt. 3–4, Moscow, 1926, pp. 81–87.

Glavnauka,[1] consisting of Professors Bunak, Gorodzov, and Zhukov, visited the site, and confirmed the conclusions of the excavators. At present the material is in Leningrad, forming part of the exhibition of the Quaternary Committee of the U.S.S.R. Academy of Sciences.

The stratigraphy of the site, as revealed by the excavations, is as follows (Plate XIII, a):

 I. The contemporary black layer.
 II. The gray compact layer.
 III. Quaternary clay.
 IV. The upper hearth layer.
 V. The intermediate yellow layer.
 VI. The lower hearth layer.
 VII. The rock bottom of the cave.

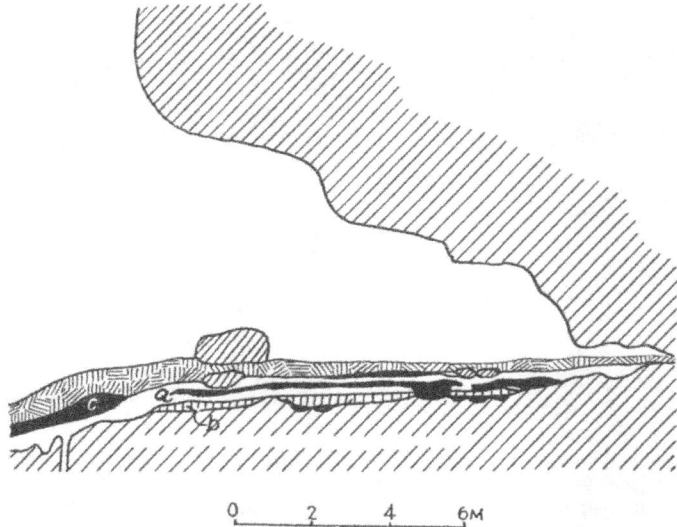

Fig. 5. Section of the Cave Kiik-Koba. a, upper layer; b, lower layer; c, Tardenosian level. After Ephimenko, "Pre-Clan Society."

On the top was (I) a contemporary black layer of ashes and sheep dung with pottery of different epochs, ranging from the Bronze to the Tartar age. This layer is 25 cm. thick and poor in finds, and does not give sufficient basis for further subdivisions.

Under it, separated in places by (II) a thin gray layer, was (III) a loose yellow clay containing stony debris, and (IV and VI) two fire-place layers. The lower one lay immediately on VII, the even floor of the cave.

A careful examination of the floor showed that the weathered limestone formed, in spots, "carst" depressions, which were filled by absolutely dry, pale-green clay, schistous in structure.

[1] The highest official scientific body of U.S.S.R.

Bonch-Osmolovsky gives the following reconstruction of the history of this Grotto: The low entrance indicates that the beginning of the formation of the cave was started by the small spring which found an outlet in the more permeable layers of limestone. The water action washed out "carst" depressions in the rocky bottom of the cave, and, during the period preceding the disappearance of the spring, it filled the depressions with clay brought from the depths of the formation, and thus evened the floor of the cave.

The spring disappeared long ago, previous to the formation of the layers covering the floor of the cave and before weathering became the main factor.[1]

The location of Kiik-Koba, in the very corner of the formation, made it especially open to wind action and weathering processes. The pieces of limestone fell down from the ceiling and disintegrated into loose debris of a brilliant yellow color.

Two later periods of man's occupation of the cave are indicated by the remains of open fires and deposits of organic matter, which colored the ground gray or black, depending on the thickness and duration of occupation. Thus two hearths were formed (layers IV and VI).

They are divided by a well-pronounced yellow layer (V), containing a small quantity of flint and bones. This shows that the periods of occupation were separated by certain intervals. The thickness of the intermediate layer, averaging 15 cm., testifies to the considerable time between the two periods of occupation, which is confirmed by an analysis of the contents of both cultural layers.

The gray compact layer (II) is the upper part of the Quaternary clay (III), with many streaks of later open fires. Finally, the contemporary black layer (I) is really of the same origin as the hearth spots of the Quaternary epoch. Due to the faster accumulation of ashes, and organic remains from the flocks which spend the nights here, it is of a darker color and more loosely formed.

All layers of Kiik-Koba are very thin, considering their antiquity; near the wall they are about 50 cm., increasing toward the entrance to 150 cm. This can be explained by the strong winds blowing off the loose debris, and the character of the canyon's slope in Pleistocene times.

The section cut in 1925 gives reason to suppose that, in the beginning of the formation of the cave's layers, the slope was not covered by humus, and was steep and rocky. Only later the process of accumulation of debris reached the level of the cave. Naturally, a considerable part of the deposits fell down the slope. The excavation of the slope may add, therefore, to the finds of Kiik-Koba.

The contents of both cultural layers differ sharply from one another. The lower hearth layer, 10–20 cm. thick, occupies almost the whole floor, except the innermost southern part of the cave, where elevations of the floor are noticeable. There it is absent, and the intermediate or upper hearth layer lies directly on the rocky floor. (Fig. 5.)

The lower (VI) layer is distinguishable almost always by its dark, almost black color. In it were found a large quantity of flint tools and chips, and a very few animal bones, the latter being in a very poor state of preservation. They are completely devoid of organic matter, though not mineralized, and at the least touch they crumble and require lacquer treatment. Many bones show evidence of fire action. The whole deposit was lying on

---

[1] Malycheff, Vera ("Le Loess," Revue de Géographie physique, 1929–30, Vol. II, III) considers that the layers of Kiik-Koba may be contemporaneous to the recent loess.

the rocky surface of the cave. Preliminary examinations of bones found there, made by A. A. Bilianitsky-Birulia, Director of the Zoological Museum of U.S.S.R. Academy of Sciences, showed the following:

Equus caballus,	*Saiga tatarica,*
Equus hemionus,	*Canis lupus,*
Sus scrofa,[1]	*Vulpes vulpes,*
Bison (priscus),	*Lepus timidus,*
Cervus elaphus,	*Citellus rufescens.*
Cervus megaceros,	

An entirely different picture presents itself in the upper (IV) cultural layer. Both in area and thickness it is less than the lower one, but the quantity of bones exceeds the lower one ten times in weight. The bones are broken, the long bones split lengthwise for extracting marrow; many are charred. The following bones were found:

Elephas primigenius,	*Canis aureus,*
Rhinoceros tichorhinus,	*Canis lupus,*
Equus caballus,	*Vulpes vulpes,*
Equus hemionus,	*Vulpes corsac,*
Equus asinus,	*Ursus spelaeus,*
Sus scrofa ferus,	*Hyaena spelaea,*
Bos sp.,	*Lepus timidus,*
Cervus elaphus,	*Marmota bobac,*
Cervus megaceros,	*Citellus rufescens,*
Saiga tatarica,	*Cricetus cricetus.*

This fauna, though undoubtedly indicating the Quaternary age of the hearth layers, still does not give sufficient basis for more specific indications of the time period, because of the lack of knowledge of the succession of faunal stages in Crimea. Being of a mixed forest and steppe type, it indicates a landscape somewhat similar to that of present times. It should be noted that Kiik-Koba is only 2 klm. from the present southern limit of the forest hill and steppe area.

On the basis of the above lists, it is hard to arrive at any opinion as to differences in both faunae, as all species of the lower (VI) are found also in the upper layer (IV) and the variety of the latter is perhaps due to the quantity of bones found. It is also possible that the marked increase of species may be related to the degree of perfection attained in the flint industry of the second (IV) layer, as compared with the cruder tools of the lower. At any rate, the fauna does not give any basis for the reconstruction of the climate. Some light is shed on this subject by the examination and analysis of the charcoal, which was made by the Institute of Technological Archaeology of the Academy of the History of Material Culture.

This study of charcoal and ashes by Palibin and Hammermann [2] resulted in the identification of the following flora (in both horizons): (Fig. 6).

[1] Those marked with asterisks are added from Gromov, V. I., "The Geology and Fauna of the Palaeolithic of U.S.S.R.," Problems of the History of Material Culture, Parts 1–2, 1933, pp. 22–33.

[2] Palibin, J., and Hammermann, A., "Kohlenreste aus dem Palaolithikum der Krim, Hohle Kiik-Koba," Bulletin of the Quaternary Committee, No. 1, 1929, pp. 35–37.

1. *Acer* (?).   2. *Rhamnus cathartica*.   3. *Juniperus* sp.

Industry.—The flint implements of the upper and lower layers differ sharply in quantity, color, technique, and patination.

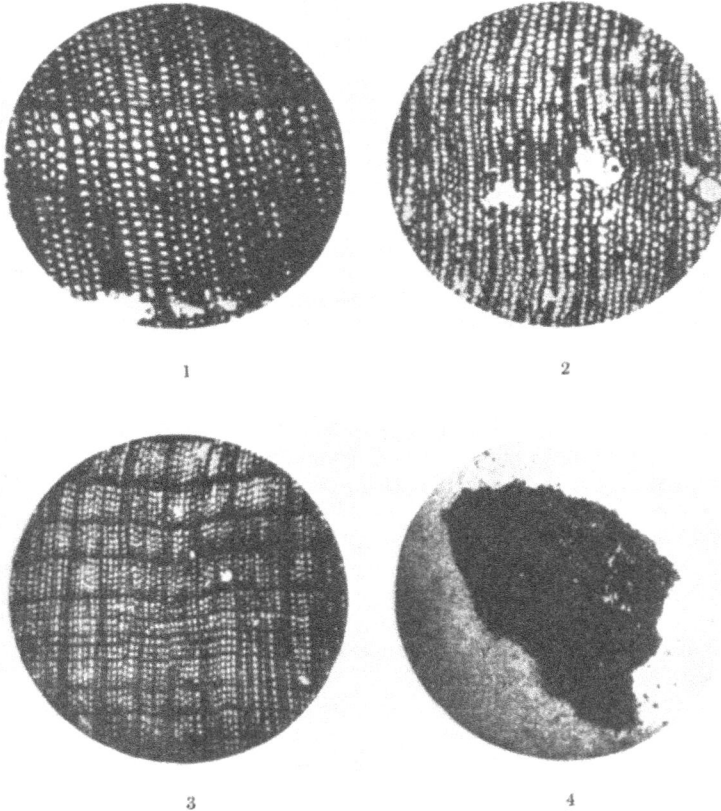

FIG. 6.   Microscopic Analysis of Ashes from Kiik-Koba.   After Palibin and Hammermann.

1. Querschnitt des rezenten Kiefernholzes.
2. Querschliff der Kohle aus demselben Kiefernholzstück.
3. Querschliff von Juniperus sp. Kohle aus Kiik-Koba.
4. Querschliff von Rhamnus cathartica Kohle aus Kiik-Koba.

Lower Layer (Figs. 7, 9, 10).—In the lower layer were found over 10,000 flint tools and chips.   Almost all are of a dark color, covered by gray spots of patination.   The technique of manufacture was very primitive: there is an absence of established forms; the retouch which improves the working edges is rough, as if the unskilled master could not make the desired forms, but followed the accidental outlines of roughly made flakes, only slightly improving them on the edges. The chips found are, in the majority of cases, rejects, due to the unskilled flaking of the nucleus.

According to the excavator, the general impression of the culture of the lower layer leads one to the conclusion that it was the first attempt to master the technique, when in the preparation of each tool, time, work, and material were wasted. Perhaps this imperfection of hunting tools can explain the limited number of animal bones found in this layer.

The typical industrial forms are absent in the lower layer. One can distinguish a certain number of small *coups de poing*, notched scrapers, and rather formless points—perforators. Bonch-Osmolovsky refers to this industry as "atypical."

Upper Layer (Figs. 8, 10, 11, 12, 13).—In the upper (IV) layer the flints are found in much lesser quantity. Altogether 500 tools and some 4000 flakes were found there. They are light brown in color, semi-transparent, with less patination, and differ markedly from the flints of the lower layer. Many well made tools of a definite form, finely retouched, and showing good workmanship, were found. About sixty per cent of them are triangular points up to 7 cm. long, made on a more or less flat flake with one sharp corner, retouched on both sides, of La Micoque type. There are also smaller points, triangular in shape, worked on one side only. About twenty per cent of the implements were scrapers, retouched on one side.

The absence of other forms indicates the universality of the use of the point; it was used for many purposes for which, in upper Palaeolithic times, man developed many different tools. According to the excavator, points were used for cutting, drilling, perforating, and scraping. It is possible that at the same time it served as a point for striking and throwing tools.

The small number of forms represented in this layer, as well as the technique of their making, place this culture in the Middle Palaeolithic or Mousterian. Among the points and scrapers of Kiik-Koba, we find many which are worked on both sides having been made on a core and not on a flake. This links the upper layer with the culture of La Micoque, widely distributed throughout Central Europe and Poland, and which belongs to the end of the Acheulean or the beginning of the Mousterian period.[1]

On the basis of the dating of the upper layer, Bonch-Osmolovsky dates the lower as Lower Palaeolithic, which is confirmed by the primitive character of its technique.

In his opinion, the atypical character of the tools and the rudeness of technique links it with the industries of the lower layers of La Micoque, Le Moustier, Hudenus and perhaps Crapina, Taubach, Ehringsdorf and Wildkirchli, and which are best characterized by the word "amorphic."[2]

Bonch-Osmolovsky illustrates without further discussion the implements from the VI and IV layers reproduced on Figs. 7, 8.

[1] See Cave Okenik in Poland, the site of Golonskaya in Poland, in Kozlowsky, "The Old Stone Age in Poland," Poznan, 1922. It is possible that two artifacts found by Merezhkovsky in the Bieshtirietsk Cave in Crimea belong to the same culture (E. G.).

[2] According to the latest classification of M. L'Abbé Breuil, La Micoque is Acheulian VII. ("Etudes de stratigraphie paléolithique dans le nord de la France, la Belgique et l'Angleterre," L'Anthropologie, Vol. XLI, 1931.)

Fig. 7. (Natural Size.)   Lower layer VI.

1. Side scraper.
2. Side scraper (?).
3. Side scraper.
4. Side scraper with point.
5. Point perforator (?) formed by two notches.
6. Convex side scraper.
7. Small biface.
8. Nucleiform tool (?).
9. Triangular blade with partial retouch on right edge and traces of retouch (?) on the lower face.

Fig. 8. (Natural Size.)   Upper layer IV.

1. Triangular side scraper on thin flake.
2. Triangular point on flake with fine retouch along the edges.
3. Small biface.
4. Finely retouched biface with round base.
5. Triangular point on flake, retouched along the edges.
6. Small biface.
7. Small biface.

The following implements are illustrated here from the drawings made by· H. Breuil from original material from Kiik-Koba exhibited in Paris by Bonch-Osmolovsky:

Lower Horizon or Layer VI.

Fig. 9.

1. Round scraper with lower face partly retouched.
2. Pointed tool with retouched notch.
3. End scraper with partial peripheral retouch.
4. Small bifacial tool.
5. Small bifacial tool (?).
6. Irregular flake retouched along the edges.
7. Crude point with secondary retouch along the left edge.
8. Small bifacial tool.
9. Small bifacial tool.
10. Small *coup de poing* with secondary retouch.
11. Fragment of tanged point (?).
12. Scraper.
13. Side scraper.

Fig. 10.

1–3. Irregular pointed flakes with secondary retouch.
4. Triangular flake with partial bifacial retouch.
5. Scraper (?).
6. Notched flake with secondary retouch.
7. Side scraper.
9. Very small point.
10. Very small scraper.
11. Very small scraper.
12. Oval scrapers with peripheral retouch.
13. End scraper on blade (?).
17. End scraper, retouched along all edges.

Layer V.

Fig. 10.

8. Small scraper with some retouch on the lower face.
14. Side scraper.
15. Round scraper.
16. Scraper of type *carenée atypique*.

Layer IV or Upper Layer.

Fig. 11.

1. Rectangular biface.
2. Triangular point with fine peripheral retouch.
3–4. Small bifaces.
5. Long flake finely retouched along the right edge.
6. Finely retouched triangular biface with flaked base.
7. Small *coup de poing* of La Micoque type.
8. Finely retouched point.
9. Point on a thin flake.

Fig. 12.

1, 2, 4, 7, 8, 12, 14. Small bifaces with more or less pronounced points.
3, 5, 6, 9, 10, 11, 13. Points approaching classical Mousterian type.

Fig. 13.

1, 2, 3. Side scrapers.
4. Irregular tool retouched all along the edges.
5, 6, 7, 8. Points.
9. Round scraper with partial retouch on the lower face.
10, 13, 15. Points.
11, 12, 16. Small pointed bifaces.
14. Point on a narrow small blade.

Layer III.

Fig. 13.

17. Perforator (?) on a thin flake retouched on both faces. The fine retouch on the edges and lower face makes a sharp point.
18. Implement of Upper Palaeolithic type made on a long narrow flake with retouch along the right edge, forming a point of Chatelperron type.

The Human Remains. In the lower hearth layer of the Kiik-Koba cave were found two incomplete human skeletons; an adult and a child. The bones of the adult lay in a grave which, undoubtedly, had been intentionally dug in the rocky floor of the cave." The grave was rectangular in shape, 170 cm. long, 55 cm. wide, and up to 30 cm. deep. In the middle was a small depression for the pelvis, and, at the eastern end, one for the head.

The grave was situated almost in the center of the cave and orientated east-west. In the western end, in normal position, were the right tibia and bones of both feet (Plate XIV). Over these bones the lower hearth layer was 7 cm. thick, being separated here from the upper layer by a yellow intermediate layer.

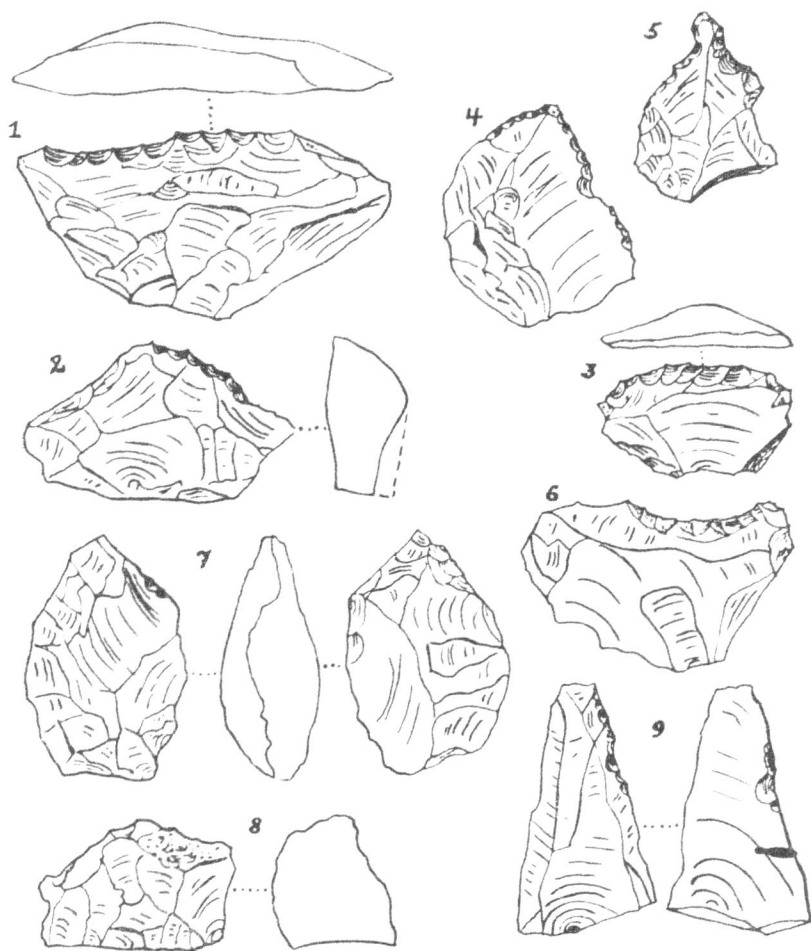

FIG. 7.   The Flint Industry of Kiik-Koba, Lower Layer (VI).   After Bonch-Osmolovsky.

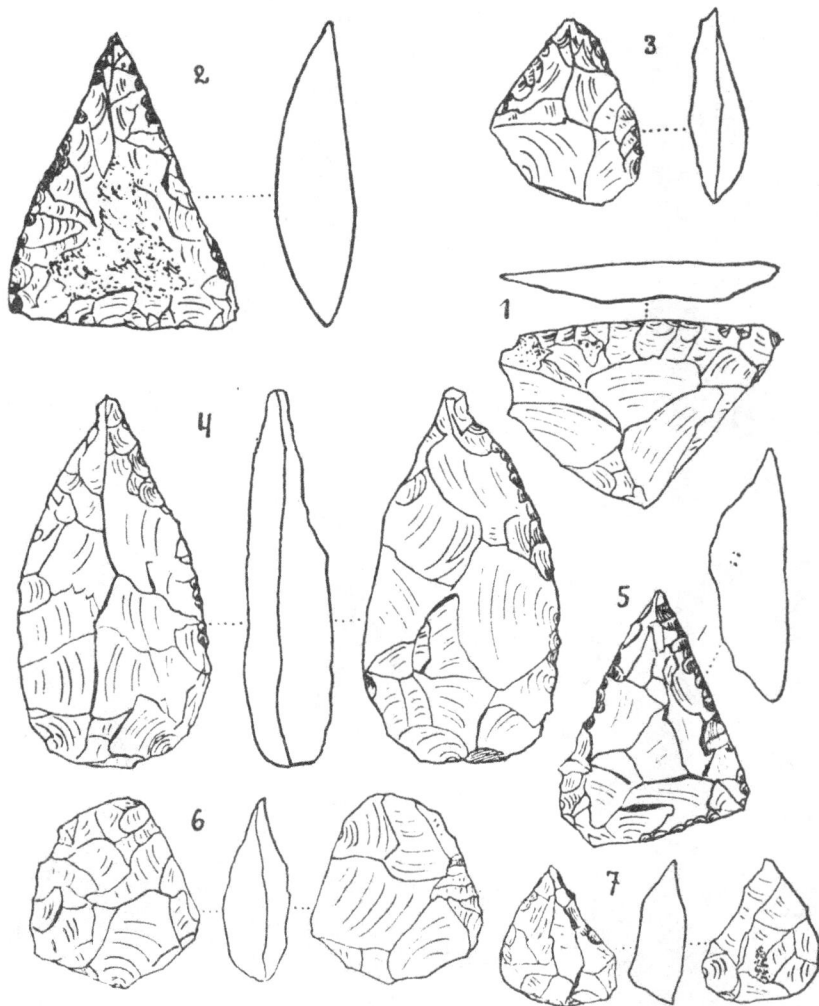

FIG. 8. The Flint Industry of Kiik-Koba, Upper Layer (IV). La Micoque. After Bonch-Osmolovsky.

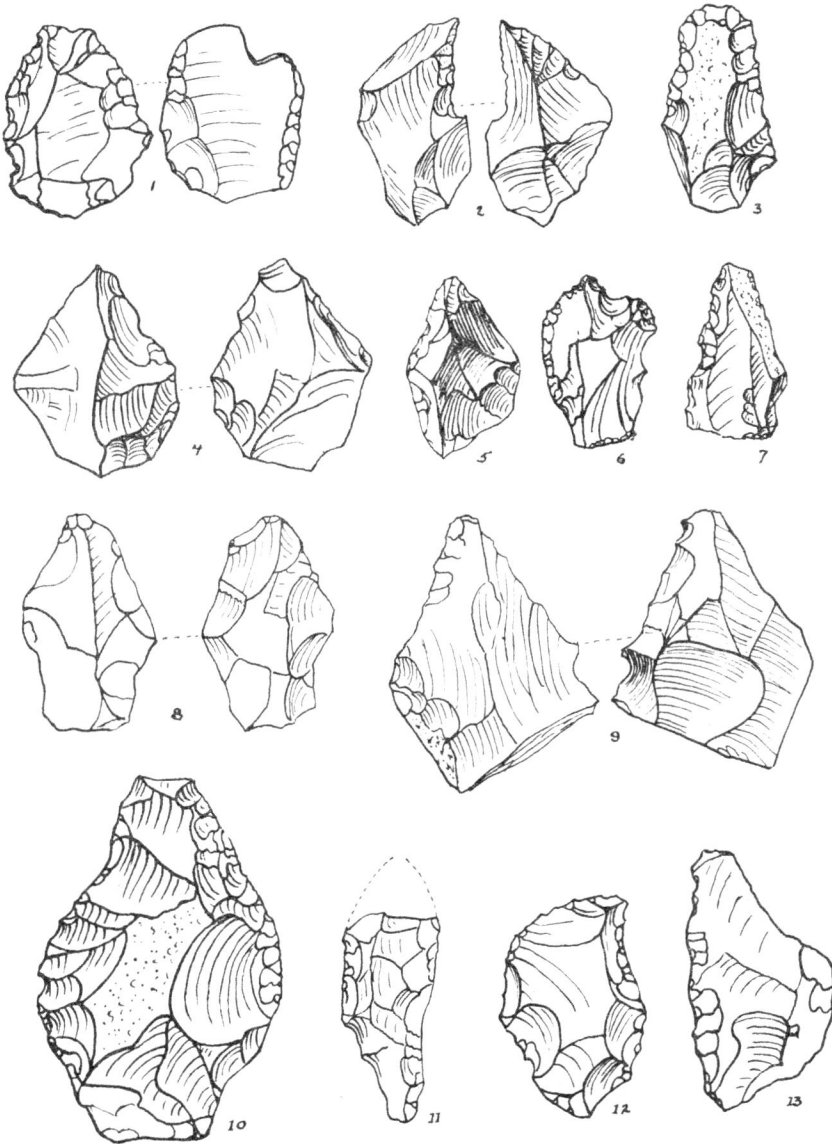

Fig. 9. Lower Layer (VI) of Kiik-Koba. After H. Breuil.

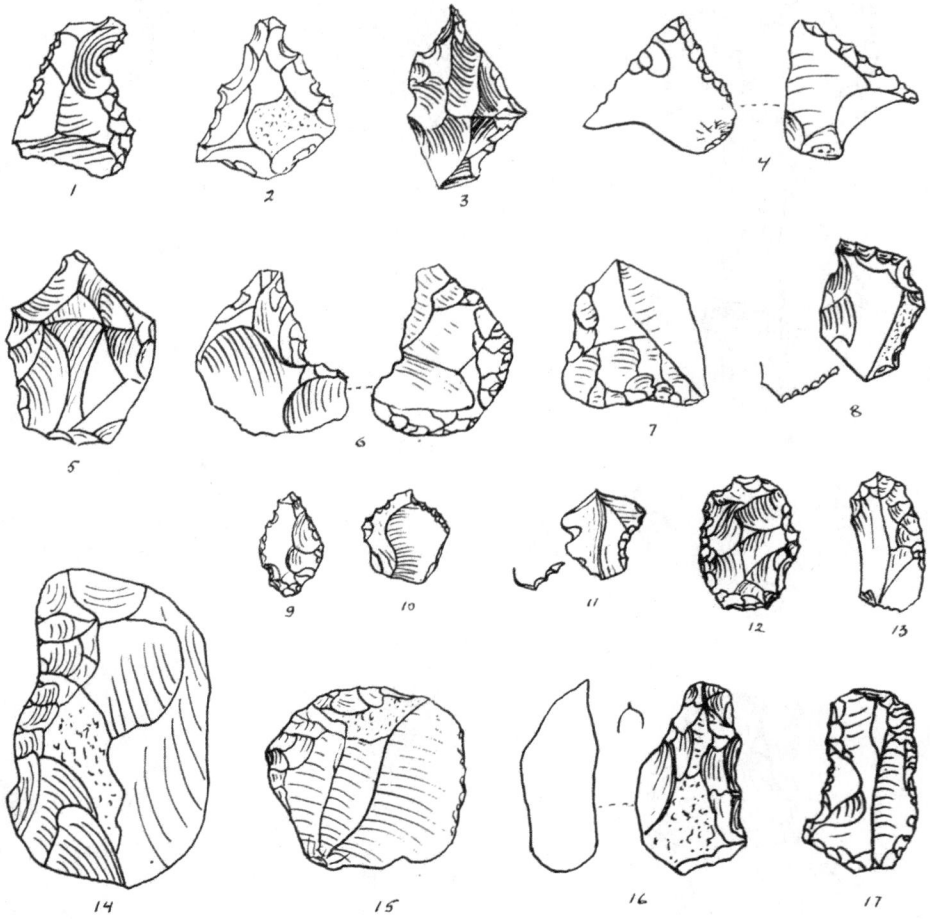

FIG. 10. Lower Layer of Kiik-Koba (VI), Nos. 1–7, 9–13. Next Layer of Kiik-Koba (V), Nos. 8, 14–17. After H. Breuil.

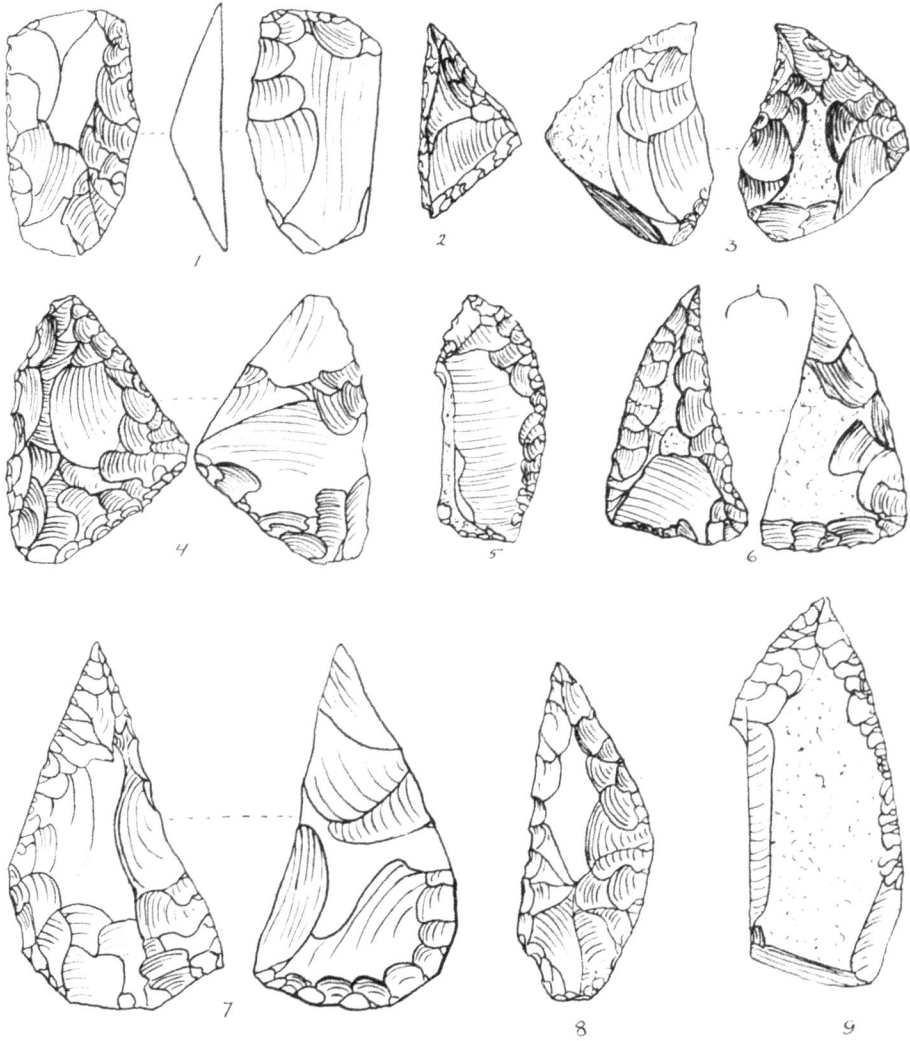

FIG. 11. Upper Layer of Kiik-Koba (IV). After H. Breuil.

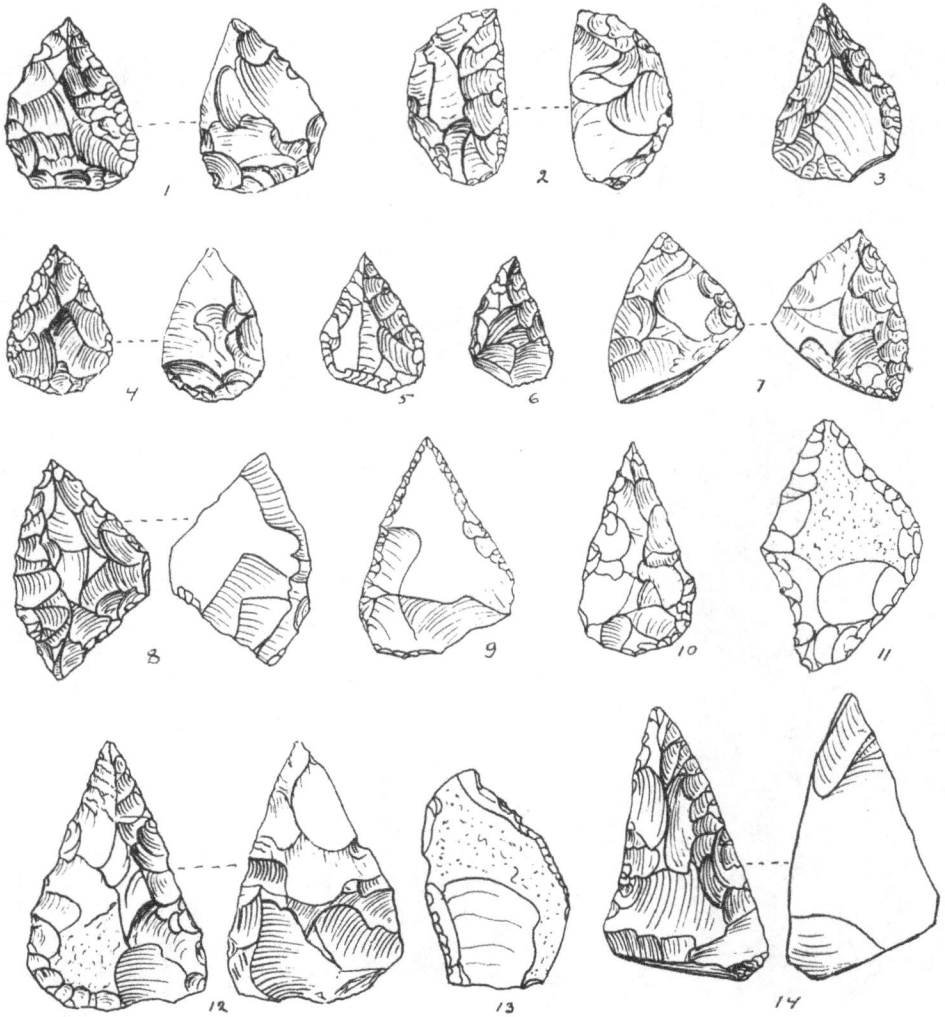

FIG. 12.   Upper Layer of Kiik-Koba (IV).   After H. Breuil.

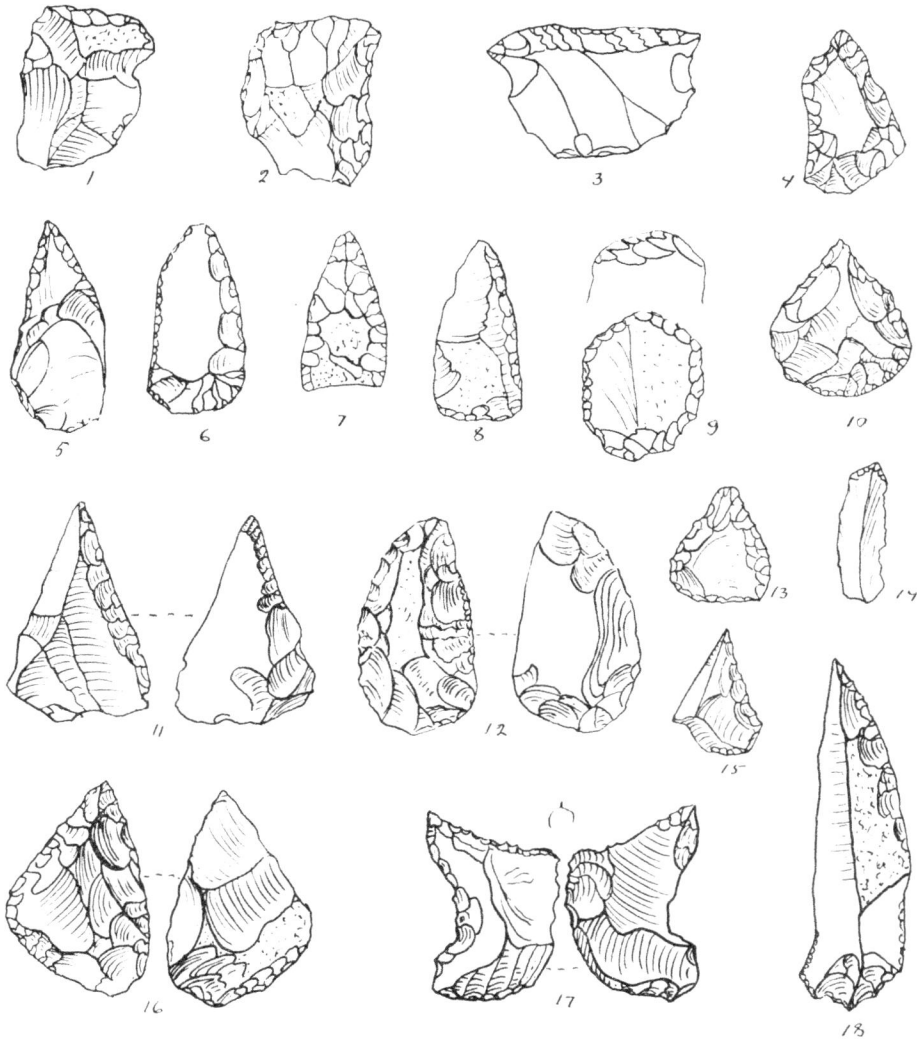

FIG. 13. Upper (IV) Layer of Kiik-Koba, Nos. 1-16. Third Layer (from the Top) of Kiik-Koba, Nos. 17-18.
After H. Breuil.

The buried body evidently had been laid on the right side with the knees slightly flexed. In the middle and the eastern part of the grave no bones were found except of a few phalanges of the hand. In this place the upper (IV) hearth layer, characterized by many animal bones, gray flint and well made artifacts, extended down to the grave. Near the bones of the feet, and among them, were found a large quantity of flint chips typical of the lower (VI) hearth layer.

The human bones were as fossilized as the animal bones. Excavating them took two days, because of their extraordinary fragility. Before they could be taken out, it was necessary to saturate them twice with a solution of shellac in alcohol. Otherwise they turned into powder at the slightest touch.

Doubtless, this burial is contemporary with the lower hearth layer; very likely it was partly destroyed by the men of the upper hearth layer, who dug a small pit here. Evidently, the rest of the skeleton was thrown down the cliff, for in 1925 several human bones were found near the base. Perhaps, a careful search may result in the recovery of other missing parts of the skeleton.

At a distance of about 1 meter north of the grave, on the rocky bed of the cave, was found the skeleton of a one-year old child. It lay on the left side in a flexed position, with the left hand under the knee. Here the IV layer was directly over the VI (the intermediate layer being absent), but, on the basis of the character of the flint chips, it could be supposed that this skeleton also is contemporary with the lower hearth layer.

The following bones of an adult were found in 1924:

1. One of the front teeth with a very worn crown and one root. Only 1 mm. of the enamel was left. This tooth is very similar to the two incisors from La Quina, but judging from the measurements, more massive; final examinations are pending.

2. Two carpal bones.

3. Several metacarpal bones.

4. Twelve phalanges of the hand, three of them having nails, the last phalanges surprisingly large.

5. The right knee cap, well preserved.

6. Right tibia and fibula, the upper epiphyses of both damaged and partly reconstructed, so that the length and degree of retroversion could be measured.

7. All bones of both feet, except the second phalange of the fifth toe of the right foot. Nine of them partially damaged, the rest well preserved. Total 55 (?) bones. All together 77 (?) bones of an adult were found; all of them very massive and rough. Even at first glance the difference from modern man is very apparent (Plate XIII, b).

In general the bones of the Kiik-Koba man are similar to the corresponding bones of the Neanderthal Man, but the measurements disclose their more primitive character. Unfortunately, the material from La Ferrassie, which would be especially helpful for comparison with Neanderthal skeletons, is still unpublished.[1] However, the following figures are given by Bonch-Osmolovsky;

[1] Written by Bonch-Osmolovsky in 1929.

PLATE XIII.

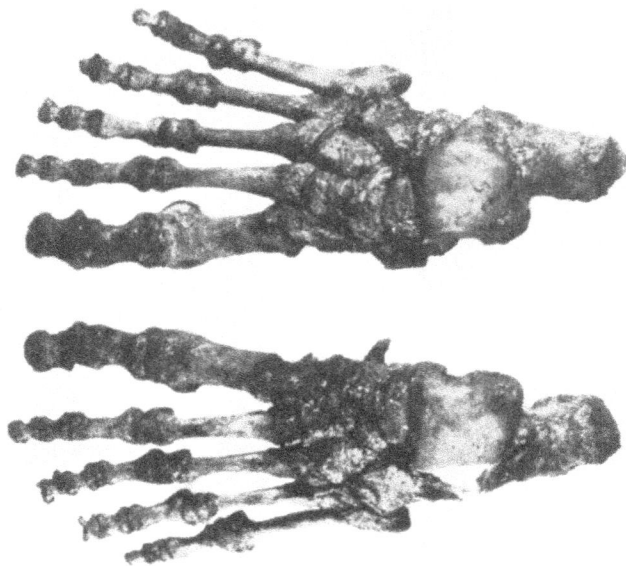

a, General View of the Excavated Portion of Kiik-Koba, Crimea, showing six layers, and human bones found in the lower layer (VI).
b, Part of the Skeletal Remains from the Lower (VI) Layer of Kiik-Koba. After the photos of G. A. Bonch-Osmolovsky.

	Kiik-Koba		Neanderthal	Maximum or minimum of Modern Man after Volkov
The Knee Cap	Left	Right	Right	
Height		52	39	
Width		52	46	
Thickness		28	21	
Tibia				
Length		34.5	34	
Strength index		24.3	24	
Astragalus				
Height-length index	62.5	64.2	61.4	
Width of inner facet	14	13.5	8–11	
Os scaphoideum				
Index of inner and outer thickness	25	27.4	32.7	
Os cuneiformia				
Index of thickness to length	149	154.9		145.6
Os cubideum				
Width-length index	28.4	28.4		27.7
Phalange of the First Toe				
Width to length	52.7	55.3		48.5

In general out of 82 indices computed after Volkov, 68 per cent. are near or over the maximum for contemporary races. From the material existing for comparison, out of twelve Neanderthal skeletons, eight are nearer to modern man than corresponding indices of the Kiik-Koba man. At present it can be said that the hand of the latter was much shorter and more massive than that of Homo Sapiens, in spite of the fact that the nail phalange of the Kiik-Koba man is much thicker. There are also reasons to believe that the opposition of the thumb was much less than it is now.

These facts, in spite of the preliminary character of the information, indicate the importance of the find. Of special interest are the almost complete bones of the feet, which, because of some of the primitive qualities of their form, may very likely add important data to the problem of the evolution of man.

The child's skeleton is nearly complete, except for the lower jaw and teeth. All bones found were in a very poor state of preservation and, consequently, do not have much scientific value.

The excavation of 1925 yielded several additional human bones not included in the above description, an enormous number of animal bones and stone chips. Several very interesting details of the mode of life were observed. A quantity of unbroken animal bones were found stored in the low passage. In one part, a pit had been dug out by the men of the upper hearth, similar to the one which disturbed the burial. Finally, on the side of the entrance, in the lower part of the upper layer of humus, microlithic implements were found.

In summary, Bonch-Osmolovsky gives the following preliminary conclusions:

1. In the cave of Kiik-Koba are two Palaeolithic cultural layers, divided by a long interval of time, but connected genetically.

2. In a culturo-chronological sense, the lower layer may be ascribed to the Lower Palaeolithic, the upper to the beginning of the Middle Palaeolithic.

3. From the point of view of geological chronology, the relative dating of cultures of Kiik-Koba with the corresponding cultures of Western Europe cannot be decided now, because of lack of knowledge of the faunal changes in the Crimea.

4. The human remains, undoubtedly buried there intentionally, are contemporary with the lower layer. Their antiquity, and the fact of the burial in the artificially excavated grave, make this find of distinct scientific interest.

In general similar to the Neanderthal Man, the skeleton shows somewhat more primitive characteristics. The possibility of explaining those primitive traits by individual deviation cannot be absolutely excluded.

These conclusions should be considered as preliminary until further detailed comparison with corresponding West European materials.

4. WOLF GROTTO

The site of the Wolf Grotto is near the village of Mazanka (Mazaiha),[1] fourteen miles east of Siempheropol on the banks of the Bieshtiretsk River. It was discovered by K. S. Merezhkovsky[2] during his palaeontological investigations in Crimea (1879–1880).

The Stratigraphy of the site is as follows:

     L. Contemporary black soil.
     II. Humus.
     III. Light greasy clay with debris of limestone.

In this last layer Merezhkovsky found cultural remains consisting of stone artifacts, charcoal ashes and bones of the following animals:

1. *Elephas primigenius,*
2. *Equus caballus,*
3. *Bos* sp.,
4. *Cervus elaphus,*
5. *Saiga tatarica,*
6. *Meles meles,*
7. *Capreolus* sp.

Bones of other animals were found but, unfortunately, Merezhkovsky does not describe them.

The *industry* found by Merezhkovsky is very small in quantity, so that only a general impression can be formed.

The tools are of considerable size, one of them chipped on one side only and the other on both.[3] This is described later by Ephimenko[4] as "a small point of fine Mousterian

---

[1] Sometimes referred to in literature as "Bieshtiretsk Cave" or "Masaiha."

[2] Merezhkovsky, K. S., "The Report of the Preliminary Investigations of the Stone Age in the Crimea," Izvestia of Imperial Russian Geographical Society, Vol. XVI, 1880, p. 110; Merezhkovsky, "Station Mousterienne en Crimée," L'Homme, 1884.

[3] Bonch-Osmolovsky, G. A., "The Prehistoric Cultures of Crimea," "Krim," No. 2, Moscow, 1926, p. 86.

[4] Ephimenko, P. P., "Traces of Mousterian Culture in South Russia," Baghalei Memorial Volume, p. 295; also, "Pre-Clan Society," Leningrad, 1936, p. 174.

PLATE XIV.

Neanderthal Grave in the Cave Kiik-Koba, after Excavations. Photo courtesy of G. A. Bonch-Osmolovsky

type," and is also mentioned by Mortillet.[1]   Gorodzov [2] and Ephimenko [3] speak of a *coup de poing* of an ancient type, though very small (Fig. 14).

Bonch-Osmolovsky [4] feels that the industry and fauna of the Wolf Grotto are very similar to those of the upper horizon of Kiik-Koba.

Ephimenko [5] points out that the lack of more definite information concerning the fauna and flint industry "aroused many doubts among Russian investigators," [6] but the discovery of a Mousterian industry in Kiik-Koba and additional excavations in Ilskaya have dispersed these doubts.   In this connection he brings to our attention the occurrence of the *coup de poing* both in the Wolf Grotto and Ilskaya (see page 266).

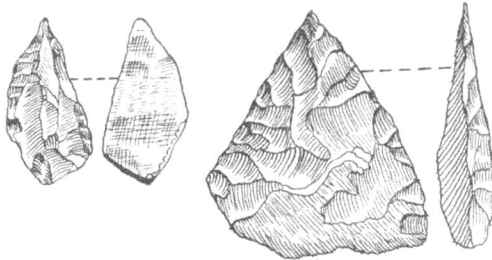

FIG. 14.   Point and Handaxe of Wolf Grotto.   After Ephimenko, "Pre-Clan Society."

5. CHOKURCHA CAVE

The Chokurcha cave is situated on the steep rocky slope of one of the Crimean chains, stretching southwest for a distance of one-half km. along the left bank of the Malii Sagir River, near the village of Chokurcha, two km. east from Siempheropol.

In front of it there is a small ledge ending in a gentle slope to the river 15 meters below. Judging by the aggregation of cultural refuse and traces of hearth on the plateau and along the slope, prehistoric man lived not only in the cave proper, but also outside of it.

This site was first discovered by Dr. V. V. Lorentz, who called it to the attention of S. I. Zabnin in 1927.   The latter made preliminary excavations there with N. L. Ernst and the geologist, P. A. Dvoichenko, in 1928–29.[7]

According to Zabnin, the excavations revealed the following stratigraphy (see ground plan and cross section on Fig. 15).

[1] Mortillet, G. A., "Le Pré-histoire," p. 506.

[2] Gorodzov, V. A., "Archaeology, The Stone Age," Vol. I, Moscow, 1925, p. 172.

[3] Ephimenko, P. P., "Some Results of the Study of Palaeolithic Period in U.S.S.R." Chelovek, No. 1, 1928, Leningrad, p. 50.

[4] Bonch-Osmolovsky, G. A., "Prehistoric Cultures of Crimea," "Krim," No. 2, Moscow, 1926.

[5] Ephimenko, P. P., "Traces of Mousterian Culture in Southern Russia," Baghalei Memorial Volume.

[6] Ephimenko, P. P., "Some Results of the Study of the Palaeolithic Period in U.S.S.R.," Chelovek, No. 1, 1928, Leningrad, p. 50.

[7] S. A. Zabnin, "The Newly Discovered Palaeolithic Site in Crimea," Izvestia of the Tavrida Society of History, Archeology and Ethnography, Siempheropol, 1928, Vol. II, pp. 146–157.

I. *The Top Layer,* consisting of:

(*a*) An upper layer of humus with a large admixture of nummulitic debris.   At a depth of 10 meters, bones of modern animals (mostly sheep) were found.

(*b*) Below it, a layer of ashes 25 cm. thick with pieces of Greek pottery, black (?)

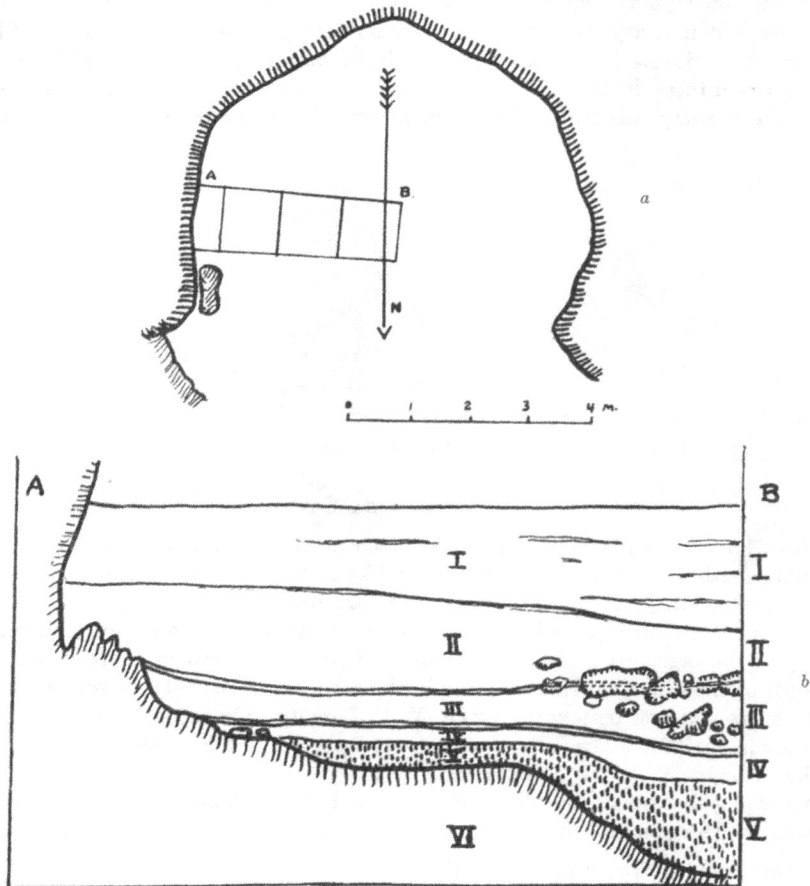

FIG. 15.   Ground Plan and Cross-section of Chokurcha Cave.   After S. A. Zabnin.

a, Ground Plan.   *b,* Cross-section Along the Line *A–B.*

pottery, two flint flakes, and bones of the sheep, ass, and dog.   Near the wall of the cave this layer is 50 cm. thick and in square no. 1–73 cm., sloping down toward the entrance.

(*c*) A thin layer of charcoal and in the square no. 1 a small "hearth" lying on the strata of Quaternary clay.   This "hearth" layer, judging from the pottery, flints and animal bones found, is similar to the Kizil-Koba culture of the Bronze Age.

II. *The Quaternary Clay of Light Yellow Color*, with an admixture of nummulites, 40 cm. thick, consisting of:

(*a*) A layer with large coprolites of the cave hyena and small bones of other animals— 10 cm.

(*b*) A layer of large bones, some broken to extract the marrow (?).

(*c*) *An upper cultural layer* (at the depth of 85 cm.) with flint industry and animal bones.

(*d*) A sterile layer 1 cm. thick.

III. *The Middle Cultural Horizon.*

This is a similar layer of Quaternary clay, with a large quantity of nummulites, and a still larger quantity of coprolites of the cave hyena, many small and large bones of the same fauna, and flint implements.

IV. *The Lower Cultural Horizon.*

This lies in the clay containing fewer coprolites, but more flint tools and animal bones.

V. *The Sterile Layer.*

VI. *The Rocky Floor of the Cave.*

Zabnin, in the above classification, distinguishes three separate cultural horizons, though the sterile layer (II-*c*), separating the middle from the upper horizon, is only 1 cm. thick, and no sterile strata between the middle and the lower.

The Lower Cultural Horizon at the IV Layer.

Fauna.—

Elephas primigenius,   *Rhinoceros tichorhinus.*

Flint Industry (13 implements and 24 flakes) (Fig. 16).

The implements are of average quality flint of chalk origin and of three colors: gray, black and yellow, the first predominating.   No patination is evident.

Nuclei.—Only one was found.   It is small, rectangular, elongated in shape, 3.5 cm. long and 1.5 cm. wide, showing the detachment of small blades in all directions.

Hammerstones.—Three river pebbles show traces of use as hammerstones, or possibly in breaking bones.

Flakes.—Flakes of three types: oval, triangular and square, were found.   They were 2-2½ cm. long and 2½ cm. wide and ¼ cm. thick.   The larger flakes have no definite form.

Points.—

1. Bifacial.

(*a*) The largest point is triangular in form with an irregular base.   It is 5.5 cm. long, 3.5 cm. thick, made of large flake.   The lower face is improved by a few strokes, the upper face retouched by rough facets. The shorter working edge is improved by the finer secondary retouch (Fig. 16, no. 1).

(b) Similar to the above, but one-half the size, and with a round base (Fig. 16, no. 3).

(c) Almond shaped point. The lower face is retouched by large flakes along the edges. The tip of the point and both edges have a secondary retouch. On the left edge of the base is a small coche (Fig. 16, no. 2).

(d) Triangular point made of thick flake, the bulb on the lower side and part of the lower face retouched. Both working edges have a secondary retouch.

2. Single faced points.

Made of small thin triangular flake, both edges improved by a fine steep retouch (Fig. 16, no. 5).

Scrapers.—Scrapers are made of both small and large flakes.

1. Made on bi-faceted flake, working edges have a flat retouch (Fig. 16, no. 6).

2. Convex scraper made on a thick oval flake, the upper face high, covered by long facets from the edges to the ridge. It has a convex working edge with a fine, steep retouch.

3. Scraper with the straight edge made of a square flake. The upper face preserves the crust. The working edge opposite the striking platform has a fine flake retouch.

Some of the small scrapers remind one of the "amorphic" culture of Kiik-Koba.

Bone Industry.

No bone objects were found.

Middle Horizon (III Layer).

Fauna.—

1. *Elephas primigenius,*
2. *Rhinoceros tichorhinus,*
3. *Equus caballus,*
4. *Saiga tatarica,*
5. *Vulpes vulpes,*
6. *Vulpes* sp.,
7. *Hyena spelaea,*
8. Large rodent—(?).

Flint Industry. (There are 20 flakes and finished tools.)

The same flint used as in the lower horizon. The flakes are smaller but of the same general form. The leading forms are points of a type somewhat different from those of the lower layer.

1. Longer triangular point, with rounded base, left working edge has a secondary retouch (Fig. 16, no. 7).

2. Similar point made of flat flint pebble preserving the original surface on the lower face (Fig. 16, no. 8).

3. Triangular point with dulled tip (Fig. 16, no. 9).

4. One sided point made from a long flat flake, both edges have a steep retouch (Fig. 16, no. 10).

No scrapers were found in this layer, but one oval pebble, with the traces of possible use as hammerstone, was found.

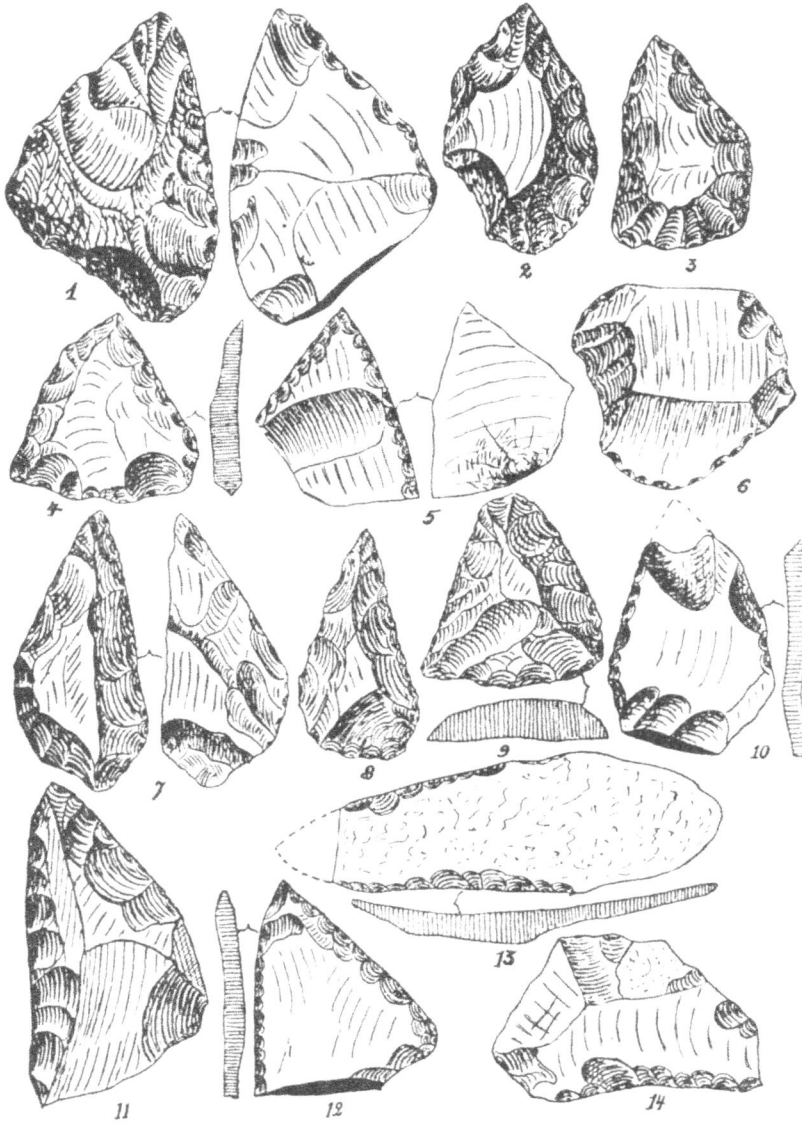

FIG. 16.   The Flint Industry of Chokurcha.   After Zabnin.

Upper Horizon (Layer IV).[1]

Fauna.—

1. *Equus caballus,*	6. *Lynx* sp.,
2. *Equus asinus* sp.,	7. *Hyaena spelaea,*
3. *Saiga tatarica,*	8. Two sp. of antelope,
4. *Vulpes vulpes* sp.,	9. Many bones of small rodents.
5. *Ursus spelaeus,*	

Flint Industry.—

In quantity the flint industry here is twice that of the middle horizon. The flint is of better quality, mostly black. The flakes are twice the size, flatter and longer.

*Hammerstones.—*Hammerstones are of the usual type, made of river pebbles.

*Points.—*Points are the characteristic tools.

1. Made of massive triangular flake with a rounded base. The straight working edge has a small, steep retouch. The other side has a flatter retouch, resulting in a sharper cutting edge (Fig. 16, no. 11).

2. Made of thin triangular flake with a sharp tip, with a similar secondary retouch along both edges (Fig. 16, no. 12).

*Scrapers.—*There is a large quantity of scrapers made of knife-like blades, both edges of which have a steep retouch (Fig. 16, nos. 13, 14).[2]

Bone Industry.—

A number of pieces of bone with the traces of work, mostly in the form of cuts and notches, were found. In one case the bone has the general form of a point.

Conclusions.—

S. A. Zabnin concludes that the flint industry and fauna of Chokurcha are analogous to the second hearth layer of Kiik-Koba with the following differences:

1. The Quaternary layer of Chokurcha has three well-pronounced horizons, while all of the 25 cm. of the cultural layer of Kiik-Koba represent really only one horizon.[3]

2. No traces of the "Amorphic culture" (Tyasian?) of Bonch-Osmolovsky was found.

As far as the Chokurcha industry is concerned, Zabnin believes that the two lower layers are late Mousterian,[4] because the points found (both single and bi-faced) are made from typical Mousterian flakes, with the characteristic retouch.

However, the long curved knife-like blades made into scrapers, found in the upper part of layer IV, are similar to those of the Aurignacian site of Suren I (see pages 289–293).

[1] While Zabnin gives a separate list of fauna for each horizon, V. I. Gromov ("Geology and Fauna of the Palaeolithic Period in U.S.S.R.," Problems of the History of Material Culture, Nos. 1–2, Leningrad, 1933) adds for the site as a whole the following:

1. *Sus scrofa,*	3. *Canis lupes,*	5. *Marmota bobac,*	7. *Cricetus cricetus,*
2. *Cervus megaceros,*	4. *Lepus timidus* sp.,	6. *Citellus rufescens,*	8. *Lagurus luteus.*

but does not mention:

1. *Vulpes* sp.,	2. *Equus asinus,*	3. *Ursus spelaeus,*	4. *Lynx* sp.

[2] Zabnin calls No. 14 a "beaked scraper," but it seems to be far-fetched. (E. G.)

[3] S. A. Zabnin is in accord with V. A. Gorodzov, who does not recognize two cultures in Kiik-Koba.

[4] P. P. Ephimenko (Pre-Clan Society, Leningrad, 1936) agrees with Zabnin in spite of the similarity to the upper layer of Kiik-Koba, mostly because it has the character of a hunting society, which according to the Soviet scheme appeared in Mousterian times.

These blades, as well as the primitive bone work, are innovations. The layer is late Mousterian with a suggestion of Aurignacian. According to Ephimenko, evidently during the subsequent excavations of 1929, more bones were found here with cuts, notches, and traces of polishing and use. Finally, and this is especially important, at that time real bone tools in the shape of filed and polished points were found.[1] The latter finds lead Ephimenko to date this site as late Mousterian. On the other hand Bonch-Osmolovsky definitely feels that the lower layers of Chokurcha "*se rapprochent de pres de l'industrie du foyer superier de Kiik-Koba.*" [2]

6. Kosh-Koba

This site is situated in the valley of the Zuya River, 30 kl. east of Siempheropol, in the same rock formation as Kiik-Koba (see pages 240–258).

Investigations show the following stratigraphy:

1. Contemporary black earth.
2. Brown clay 50–150 cm. thick, with pottery of the Kizil-Koba type of the early Iron Age.
3. Thick layer of yellow loess-like clay of Quaternary origin, containing cultural remains.
4. Rocky bottom of the cave, with depression filled with moist and compact clay.

Gromov [3] lists the following fauna, which seems to contain mixed elements of both forest and steppe, corresponding to the supposed climatic condition of the time.

1. *Elephas primigenius,*	9. *Saiga tatarica,*
2. *Rhinoceros tichorhinus,*	10. *Canis lupus,*
3. *Equus caballus,*	11. *Vulpes vulpes,*
4. *Equus asinus,*	12. *Vulpes lagopus,*
5. *Sus scrofa ferus,*	13. *Ursus spelaeus,*
6. *Bos* sp.,	14. *Hyaena spelaea,*
7. *Cérvus elaphus,*	15. *Lepus europeus timidus,*
8. *Cervus megaceros,*	16. *Marmota bobac.*

The cultural layers, according to Ephimenko,[4] contained a very large quantity of animal bones, obviously intentionally broken, the remains of two hearths with ashes and burned earth, and a few flint tools.[5]

Gorodzov [6] mentions "bone tools so primitive that, but for the presence of undoubtedly artificial holes, they are not recognizeable as such."

The site is provisionally dated as Mousterian.

[1] Compare this with Cave Castillo, Spain. Ephimenko, P. P., "Pre-Clan Society," Leningrad, 1936, p. 181.

[2] Bonch-Osmolovsky, G. A., "Le Paleolithique de Crimée," Bulletin of the Quaternary Committee, No. 1, 1929, p. 28.

[3] Gromov, V. I., "Geology and Fauna of the Palaeolithic Period in U.S.S.R.," Problems of the History of Material Culture, Nos. 1–2, Leningrad, 1933.

[4] Ephimenko, P. P., "Pre-Clan Society," Leningrad, 1936, pp. 175–176.

[5] No illustrations are available.

[6] Gorodzov, V., "The Results of the Archaeological Activities in U.S.S.R. during 1917–1930." Manuscript.

7. Gamkovo

The Gamkovo site is located near the village of the same name in the Smolensk region, on the left slope of a gully, which opens into the Ufinia River, the left tributary of the Dnieper. This locality has been known ever since 1909, when it received newspaper publicity because of the discovery of fossil bones.

Since that time, it has been visited by many amateur and professional paleontologists, who made sporadic trial excavations and found numerous remains of Palaeolithic fauna, but no flint industry.

The more careful excavations of K. Polikarpovich, G. Mirchink and A. Lavdansky [1] in 1926 and 1927 established the following geological conditions:

1. Humus,
2. Gray-yellow loess with ferreous spots,
3. Gray-yellow loess with animal bones.

The fauna is represented mostly by *Rhinoceros tichorhinus*. It is interesting that no traces of implements, flakes, chips, or charred bones were found. The large conglomeration of bones, however, may represent one of the bone piles so characteristic of many Russian Palaeolithic sites, which are usually very poor in flint finds. It is possible that the hearths may be situated nearby in some unexcavated portion.

According to Mirchink,[2] the geological age of Gamkovo is somewhere between the end of the Riss and the Würm period. Therefore it is older than the nearby sites of Suponevo, Mezine, Berdizh, Urovichi, and others,—and, perhaps, of Mousterian affinities.

8. Ilskaya

The site of Ilskaya was discovered in 1898 by the famous French archaeologist, Baron G. de Baye.[3] It is situated on the right bank of the Il River, near the village of Ilskaya, Krasnodarsk region, Caucasus, on the 17th section of the Ilskaya oil fields.

News of the discovery spread all over Europe and various authors commented on it from time to time.[4] Baron de Baye did not have the opportunity to continue the excavations, and Ilskaya remained untouched until 1902 when the geologist, N. I. Krishtaphovich,[5] visited the site for the purpose of studying its geological conditions.

[1] Lavdansky, A. N., "Archaeological Investigation of Palaeolithic Site near Gamkovo, district of Smolensk." Pratzi of Archaeological Committee of Bielorussian Academy of Science, Vol. II, Minsk, 1930, pp. 495–8.

[2] Mirchink, G. F., "Geological Correlation of River Terraces and Palaeolithic Sites in the Basin of the Desna and Sozh Rivers," Bulletin of Moscow Society of Naturalists, Vol. VII, Moscow, 1920, pp. 1–2.

[3] de Baye, G., "Au Nord de la Chaine du Caucase," Revue de Géographie, 1899, July–August, p. 13.

[4] Volkov, Th., "L'Anthropologie," Vol. IX, 1898, p. 617.

Volkov, Th., "Materials for Ukrainian Ethnology," Vol. I, 1899, p. 221.

Volkov, Th., "Annales Archeologique de la Russie du Sud," 1889, pp. 74–75.

Capitan, L., "Etudes sur les collections rapportées de Russie par M. le Baron de Baye," Bull. de la Soc. d'Anthr. de Paris, t. X (IVᵉ serie), fasc. 4, 1899, pp. 322–327.

[5] Krishtaphovich, N. I., "Geological Investigations of the Palaeolithic Site during the Summer of 1904." "Drevnosti," Trudi of Moscow Archeological Society, Vol. XXI, Part 2, p. 183.

In 1925, S. N. Zamiatnin [1] was able to identify the place of the original excavations of deBaye, carried on rather extensive excavations in 1926, and 1928. The results are published in the Revue Anthropologique, and, therefore, are available for Western European scholars. Still, because of the fact that Ilskaya has aroused quite a controversy among the Russian archaeologists (see the opinion of Gorodzov as given at the end of this chapter), it was felt that an almost literal translation instead of a broad summary should be given here.[2] According to Mr. Zamiatnin, we have the following situation:

The winding river actually passes through the alluvial region composed of clay and gravel, which forms the first terrace 5–6 m. above the river. The second terrace, about 20 m. high, is less easily visible, as it is often masked by fluviatile deposits; it is more pronounced on the right bank. The site itself is situated on the left bank of the river, on the edge of the second terrace, past the oil wells Nos. 171–172.

The fluviatile deposits differ in the nature of their composition, but consist for the most part of brown or bluish clay, sometimes mixed with chalk (?), and their thickness increases as one goes farther from the river. The cultural deposit, averaging 40–50 cm. in thickness, is almost horizontal, with a very slight inclination toward the river. It contains a large quantity of artificially broken bones, stone implements, flakes, and chips. On the edge of the terrace, this layer, where it appears near the surface, gives the impression of being subjected to water action; there one could find fewer remains of human industry.

There were no definitely determined hearths found, and the objects were quite uniformly distributed. Still it is interesting to note one peculiarity of this distribution: i.e., while the parts of the layer in which large bones predominated yielded fewer stone implements, there were usually several finished tools. On the other hand, those spots where there was a predominance of stone implements, flakes and nuclei, contained, in most cases, only a limited quantity of broken bones. The majority of the nuclei were all found in a rather small area. This peculiar fact, which can be observed more clearly in the Upper Palaeolithic stations of southern Russia, gives hope of discovering some traces of the camps of Quaternary hunters.

A surface of 170 sq. meters was excavated by Zamiatnin. According to Mme. V. J. Gromova, assistant of the Zoological Museum of the Academy of Sciences, the fauna of this station is represented by:

Elephas primigenius [3]	*Bos primigenius* [5]
Equus sp.[4]	*Hyena.*

[1] Zamiatnin, S. N., "Station Moustérienne à Ilskaia province de Kouban (Caucase du nord)," Revue anthropologique, Nos. 7–9, 1929, pp. 282–295.

[2] Mr. Zamiatnin has kindly placed at author's disposal a set of photographs of the Ilskaya flint industry. These are reproduced here (Plates XV and XVI) in addition to drawings, because of the importance of the site, in spite of the fact that there are some duplications: Specimen on Plate XV, no. 13, is reproduced in drawings on Fig. 18, no. 11; no. 14 is reproduced in drawings on Fig. 18, no. 9; no. 12 is reproduced in drawings on Fig. 17, no. 3; no. 1 is reproduced in drawings on Fig. 17, no. 8; Plate XVI, no. 1, is reproduced in drawings on Fig. 17, no. 7; no. 2 is reproduced in drawings on Fig. 18, no. 6.

[3] Gromov, V. I. ("The Geology and Fauna of the Palaeolithic Period in U. S. S. R." Problems of Gaimk, no. 1–2; 1933, p. 27), adds the following: *Equus asinus, Sus scrofa ferus, Cervus elaphus, Megaceros sp., Canis lupus, Ursus spelaeus, Hyaena spelaea.*

[4] *Equus equus*, according to Gromov, V. I.

[5] *Bison priscus*, according to Gromov, V. I.

Bos species were of a smaller size than *Bos primigenius*. The bones of *Bos* species predominate, but the almost complete absence of horns in the layer makes more precise identification of this species very difficult.

The industry of this site is represented by a collection of about 300 tools, numerous nuclei, and a very large quantity of flakes, chips, etc.

The implements are characteristically Mousterian, but with some points of differentiation. There is an abundance of implements of exceptionally small size (averaging 5–6 cm.) chipped on both sides. In spite of the small size most of these implements are rather thick. This specific characteristic of the industry is often explained by the nature of the material used. The formations of the neighbourhood of Ilskaya do not offer anything but dolomite, flint and limestone. Because of its fragility the first is not very suitable for the manufacturing of tools; it is often found in this site in the form of flakes and finely retouched blades; those of larger size are especially noticeable. Finished tools of this material are exceptional.

The chief material for the Ilskaya industry is black jasper (lydite) and "horned flint" (hornstein). Chalcedony, carnelian, quartz, quartzite, and hard sandstone were also used. But in the vicinity of Ilskaya these materials are found only in the form of small pebbles in the bed of the Il River. The size of these pebbles determines the size of the tools. Almost every implement of Ilskaya preserves part of the polished and flattened surface of a pebble. The collection contains a considerable series of nuclei (about 50) of typical discoidal form (Fig. 17, nos. 2, 3). In this series the nuclei are quite similar, but one can distinguish several groups. One includes the more massive types, of special form (Fig. 17, no. 1). Another very interesting type is those of exceedingly small size, with the flakes of so minute that they could not have had a practical use. These nuclei could be regarded as "ebauche" of small amygdaloid tools, described below (Fig. 17, no. 4).[1] Finally, it is necessary to mention a considerable group of relatively small pebbles from which only one or two large flakes have been detached (Fig. 17, no. 5).

There are many examples of small ovoid pebbles, preserving on the ends traces of numerous blows. These served as hammer stones and, perhaps, were used for making tools. Their surfaces show characteristic scratches (Fig. 18, no. 12).

The *flakes* and *blades* used for manufacturing were triangular and produced large, short and usually massive implements. The striking platform is large and the bulb of percussion is strongly pronounced; when interfering with the handling of the implements, the bulb was often chipped off by one or several blows (Fig. 18, no. 6). A large quantity of flakes and blades had facets on the striking platform. Among the large flakes several were of flint, dolomite, and quartzite, approaching the type of Levallois. On the other hand, one also finds a certain number of long straight blades, which recall Upper Palaeolithic types.

As we have already noted, part of the industry of Ilskaya was made not on flakes or blades, but directly on pebbles of appropriate size worked on both sides. Among the tools which preserve the Lower Palaeolithic traditions, it is necessary to mention a group of twenty small, carefully made amygdaloid implements, derived from the Acheulean *coup de poing* (Fig. 17, nos. 8, 9). The largest one is but 6 cm. long. The fact that one of the sides is convex and the other is more or less flat is characteristic. They are usually chipped with large facets. In making this implement the lower side was chipped first.[2]

[1] Probably disk-nuclei. (E. G.)
[2] Compare with ebauche-nucleus (Fig. 17, no. 4). (E. G.)

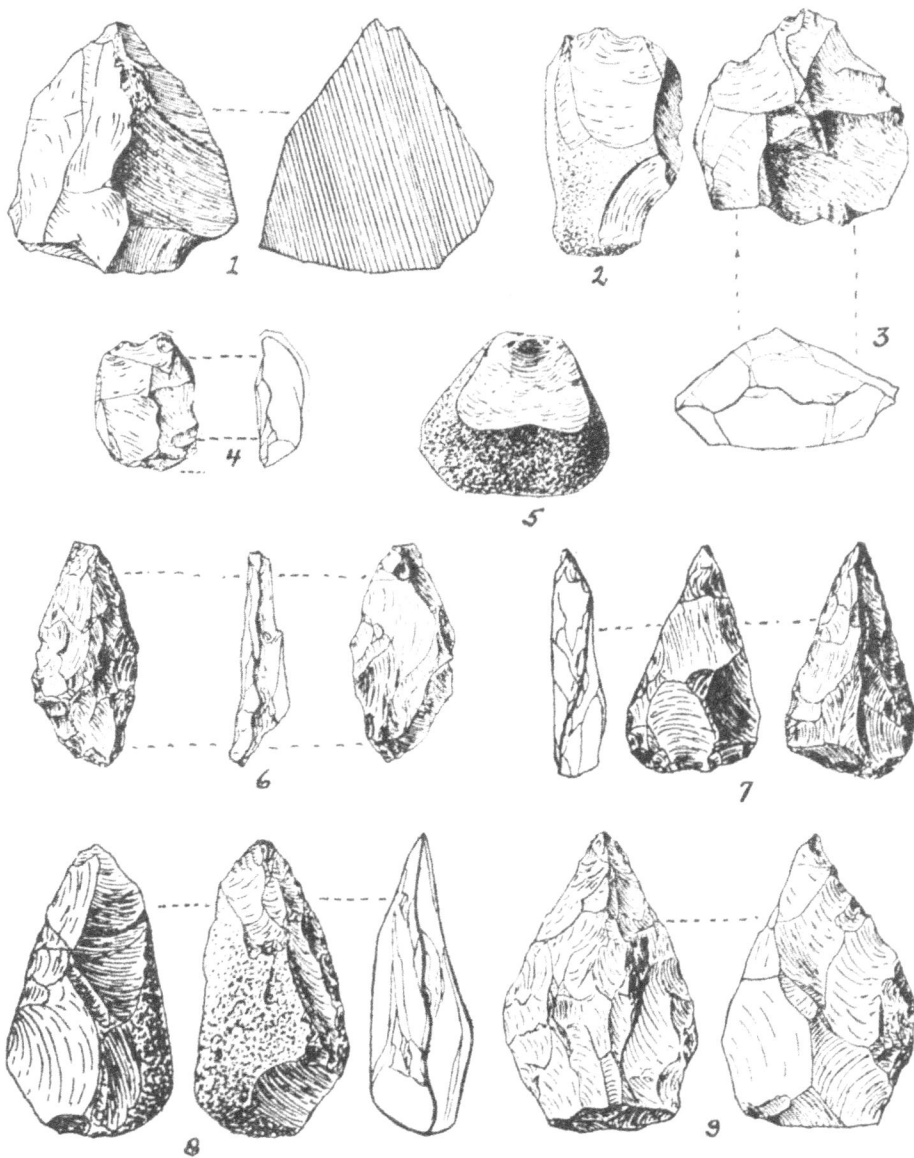

FIG. 17.   Flint Industry of Ilskaya.   After Zamiatnin.

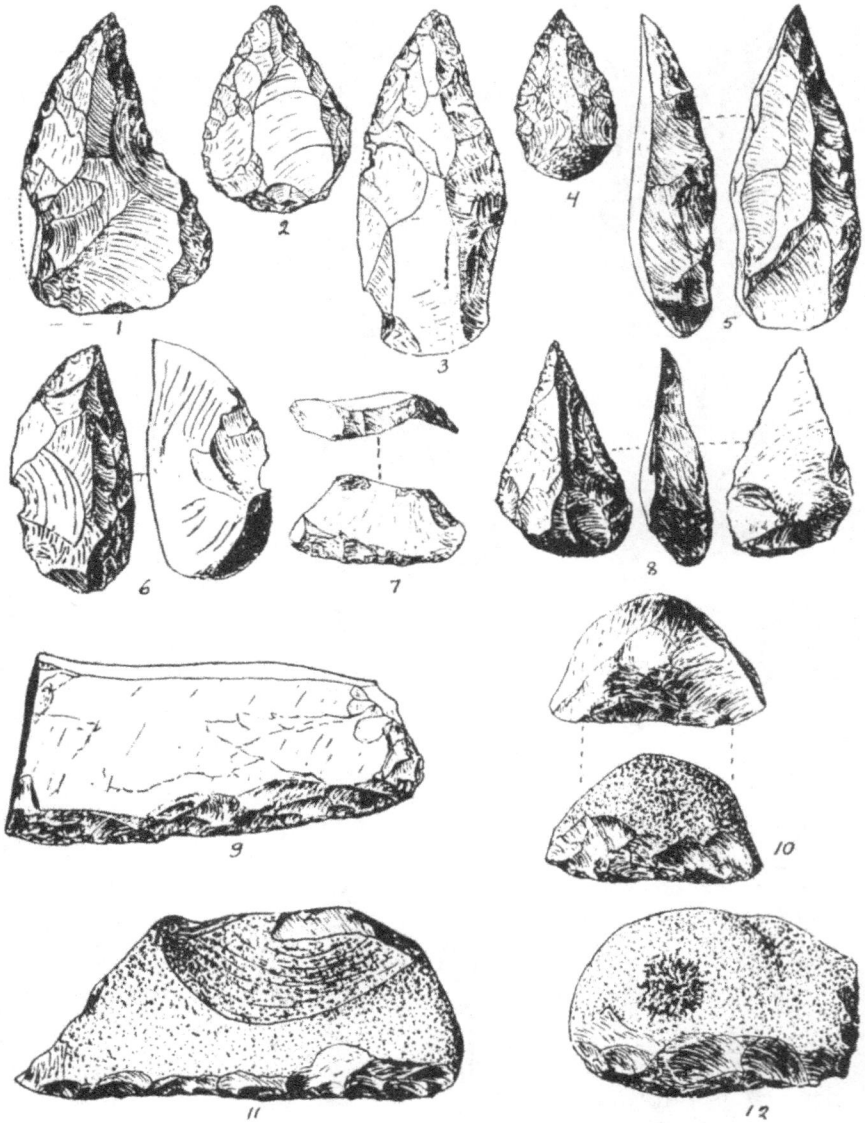

FIG. 18.   Flint Industry of Ilskaya.   After Zamiatnin.

The convex side always has a fine and careful retouch along the edges which is sometimes carried all around the implement. This peculiarity, together with the dimensions of the objects, clearly indicates that those tools, though morphologically related to the large, almond-shaped tools of the Lower Palaeolithic, nevertheless had the same purpose as Mousterian points.

Another group of implements, with a similar bifacial chipping, have asymmetrical contours. The majority of them are made of pebbles, and some of massive flakes. They are characterized by the fact that more careful retouching is found only on the point and on one end of the upper surface.

The Points (about 100) form the most numerous group of the Ilskaya industry. They are always very finely and carefully retouched at one end, and on one or both edges. Made of fine blades, or flakes with the prepared striking platform, they can be divided into the following groups:

(*a*) Implements made from triangular flakes, fairly large and thin, retouched on both edges with the point in the middle (Fig. 18, nos. 1, 2). Most of them are 4–5 cm. long and only a few range from 6–7 cm. These beautiful implements repeat in miniature the classical forms of Mousterian points.

(*b*) Long points, made usually on fine blades, are less abundant (Fig. 18, no. 3).

(*c*) Lance-like points, usually made of the first flake of the pebble with the cortex partially visible, are quite common (Fig. 18, no. 4).

(*d*) Implements (pointed side scrapers) with the point inclined toward one edge which is retouched along the whole length; the other edge is retouched only partially (Fig. 18, no. 6).

(*e*) Small tools remarkable for their thickness, with a long straight point in cross-section, like an isosceles triangle with shortened base. These tools probably were used as perforators (Fig. 18, no. 8).

Although the *Side Scrapers* of Ilskaya are almost as abundant as the points, they are less varied. They are usually slightly larger than the points (Fig. 18, nos. 9, 11). Those measuring 6–8 cm. are as numerous as the small ones. They can be divided into two groups: (*a*) those with a straight edge, (*b*) those with a curved edge. One can also distinguish some made of small flakes, the form of which is less regular (Fig. 18, no. 7). There are also many side scrapers chipped on the lower side (Fig. 18, no. 10). Some are double tools, with one of the edges retouched on the upper convex side, while the other is retouched on the lower side (Fig. 18, no. 12).

A number of flakes and blades found in Ilskaya, long and narrow in shape, and more or less regular, retouched on one or both sides, can be regarded as either a special type of side scraper, or a simple cutting tool.

Perforators are represented by a small number (10–15). They have no definite form. They are all small, made of thin flakes of accidental form and little regularity, transformed into perforators by very little retouch. Several smaller points can be included in this group.

There is also a considerable quantity of tools of indefinite and irregular form, perhaps simple flakes barely marked with retouching strokes.

Finally, some tools of specific form, which cannot be included in the above series, must be described. Among these there are, first of all, two double points which resemble the tools of this type from La Quina, Ehringsdorf and Karlstein.

The first, 10.5 cm. long, 4 cm. wide, Plate XVI, no. 8, is the largest stone tool of Ilskaya. It is made of a long flat pebble of grey-brown granular sandstone. Its outline, resembling the Solutrean laurel-leaf point, makes its appearance in Ilskaya quite unexpectedly; only the technique of its manufacturing connects it with this site. The upper part of it preserves a considerable portion of the smooth surface of the pebble and it is retouched only around the edges. The whole lower side is chipped with large facets. Its very regular shape at first glance gives the impression that this point was fixed to a handle, but closer examination shows that the one side of the upper surface is carefully retouched, forming a finely curved cutting edge. It is thus possible to assume that this tool was used as a side scraper. It may be compared with a large one found in the cave of Kulna (Moravia) and described by H. Breuil.

The second double point is small (6.5 x 1.5 cm.), relatively thick and straight. It is made of a broken pebble of grayish sandstone, its thickness reminding one of the point-perforators described above. The ends and sides are carefully retouched, the lower surface is flat.

The third implement, less common, is likewise made of a broken pebble of fine, fawn colored sandstone. The lower surface is flat and not retouched. The upper surface is very convex, preserving the cortex in one small part and is retouched with fine facets. The retouch of its sides is perfect. One of the ends is sharpened to a thick point or perforator, the other, chipped in a semi-circle, forms a regular convex scraper equally thick.

A careful examination of those three objects permits one to connect their technique very closely with that of Ilskaya; their extraordinary form is easily explained by the fact that they were made of pebbles.[1]

There are very few traces of bone work in Ilskaya. Among the large number of artificially broken bones there is only one which bears unquestionable traces of work. Typically striated, it is similar to the "anvils" or "compressors" found by Dr. H. Martin in La Quina, and known also from other sites of this period.

In addition, there were found the second phalange of an adult Bovidae, perforated along all its length, and the cervical vertebra, likewise perforated, of the same animal. The bad state of preservation of these two specimens does not permit one definitely to recognize the work of man in them.

Considering the industry of Ilskaya in its totality, represented by the massive tools, small *coups de poing* and tools worked on both sides, Zamiatnin feels that it should be included in the special group of Eastern Mousterian sites and also apparently connected with La Micoque culture.

The area of the distribution of this culture includes Germany, Austria, Moravia, Poland, Hungary, Transylvania[2] and stretches through Ukraine of South Russia to Ilskaya of the North Caucasus. But all these sites, scattered over this enormous territory are not synchronical. Except the most ancient sites of the warm fauna, they appear to belong to the different phases of Mousterian. An exact study of their chronological relationship

[1] Unfortunately Zamiatnin does not give illustrations of these interesting implements. (E. G.)

[2] A number of Transylvanian sites recently discovered resemble Ilskaya, not only by the aspect of their industry, but also by the main material employed. Unfortunately, they are represented only by a small number of objects. (Breuil, H., "Les stations paléolithiques en Transylvanie," Bull. de la Soc. des Sciences de Cluj, t. II, 1926, pp. 192–217.)

PLATE XV.

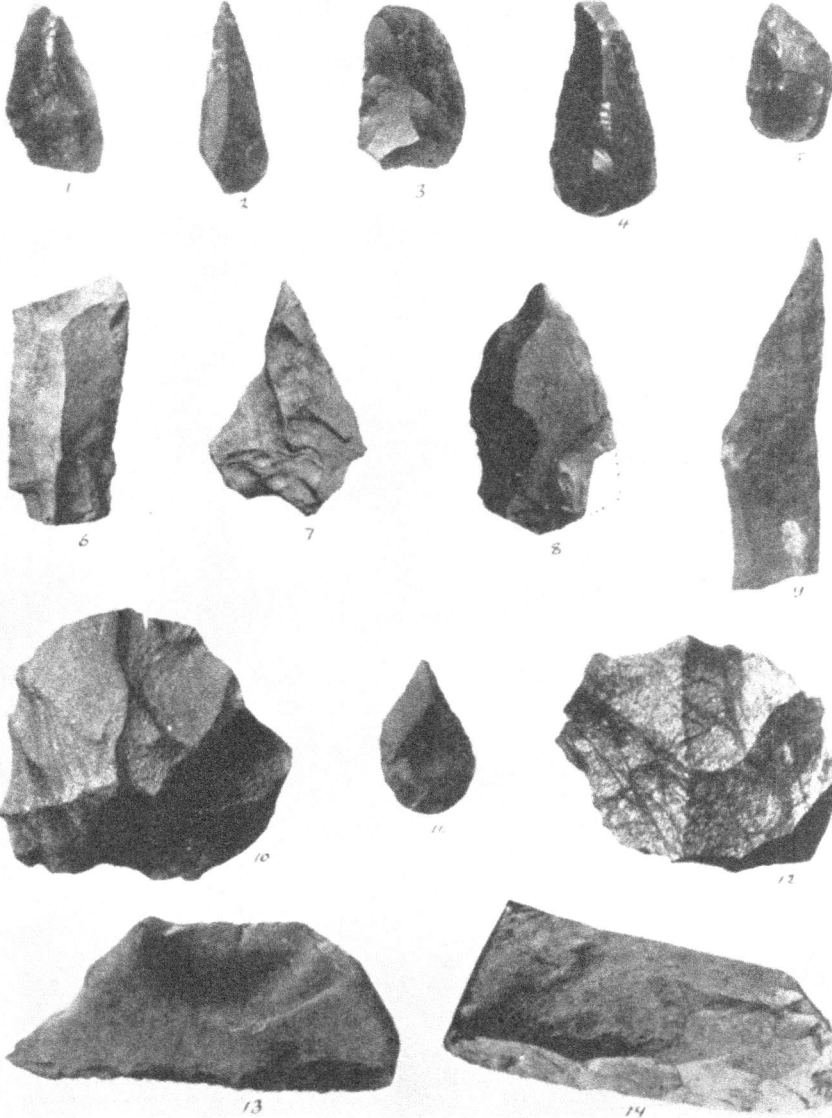

FLINT INDUSTRY OF ILSKAYA. Unpublished photographs of Zamiatnin.

presents great difficulties and, up to the present, has resulted only in contradictory conclusions. This question is also complicated by the fact that a large number of finds are represented only by isolated objects or by a very small series, which does not permit one to form an adequate idea of the cultural complex.

It is necessary to be very cautious in specifying the chronology of Ilskaya, especially because this Mousterian site, is at such a distance East, from all the other sites with which it could be compared. Still, it can be considered characteristic of the evolution of the Eastern Mousterian.

Several facts already pointed out indicate that the age of Ilskaya is not very ancient:

(a) The most archaic forms, such as the small *coup de poing* with the convex side very carefully retouched and the lower one retouched with large flakes, point to a use analogous to that of Mousterian points; they are the same implements, but made with the later technical processes.[1]

(b) The typical forms, as well as variation of Mousterian points are the most abundantly represented implements of Ilskaya.

(c) A large number of implements and flakes show a careful chipping of nuclei and the presence of facets on the striking platform.

(d) The appearance of the first traces of bone work.

The next problem is to compare Ilskaya with the geographically nearest Mousterian sites of South Russia.

I. The Wolf Grotto near Siempheropol, discovered in 1881 by K. S. Merezkovsky, yielded rich faunal remains, many worked flints, a triangular *coup de poing* and a fairly large point.[2]

II. The grotto of Kiik-Koba, discovered and investigated in 1924–25 by G. A. Bonch-Osmolovsky, consists of two archaeological layers of different ages: the lower layer with atypical industry of primitive Mousterian,[3] represented by irregular flakes very slightly retouched, does not contain any definite types outside of small *coups de poing*, crudely retouched. In the same layer were found bones of Neanderthal man. The upper layer presents an industry much richer in form, and nearer to that of Ilskaya; there is the same predominance of points, the same flat-faced *coups de poing*. On the basis of notes furnished by Mr. Bonch-Osmolovsky, Zamiatnin makes the following distinction: nuclei richly represented in Ilskaya are absent in Kiik-Koba; on the other hand, implements retouched on both sides are much more numerous. This may perhaps indicate that the upper layer of Kiik-Koba is somewhat earlier than Ilskaya.

III. A third and the most interesting find, was made in 1924 by Professor P. P. Ephimenko, in the open site on the bank of the Derkula River, not far from its junction with the Donetz in the region of the Don, almost on the border line of Ekaterinoslavl province. One very large nucleus (about 25 cm.) of greenish quartzite was found there,

---

[1] It is very interesting to note that in a much later epoch the "laurel-leaf" point of Hungary and of Predmost derived, as was well established by H. Breuil, of flint Mousterian tools, shows analogous peculiarities of manufacture. It is very probable that the technique of these "laurel-leaves" is the result of a tradition of this epoch. (Breuil, H., "Voyage paléolithique en Europe Central," L'Anthrop., t. XXXIV, p. 522, fig. 4.)

[2] Merezhkovsky, K. S., Station moustérienne en Crimée (L'Homme, 1884, pp. 300–302).

[3] Tyasian. (E. G.)

as well as many large flakes of the same material, and two large implements: a side scraper and a point of gray flint with fine, beautiful retouch.   The station of Derkula is evidently the most recent Mousterian site in South Russia.   Its industry differs from others by the large perfect implements, recalling classical Mousterian types.

All these finds permit one to follow the evolution of Mousterian cultures in South Russia.   The site of Ilskaya, as well as two caves in Crimea, apparently belong to the same group of cultures, as they are similar in character.   The most ancient stage is represented by the two layers of Kiik-Koba.[1]

Ilskaya closely resembles the upper horizon of the latter; in the opinion of Zamiatnin, it immediately follows Kiik-Koba.

The finds of the Wolf Grotto should be placed in a somewhat later, more evolved stage. However, this succession of cultures does not embrace all the Mousterian cultures of Eastern Europe.

During the fall of 1928, Mr. Zamiatnin continued his work in Ilskaya for one and a half months.   He states that his finds furnish additional interesting details and considerably increase his collection.   He believes that this additional material confirms his general conclusions.

The position taken by Zamiatnin seems to have been supported by P. P. Ephimenko,[2] who, even previous to the last excavations, classified Ilskaya as Mousterian on the basis of the material of de Baye.   In his other article,[3] knowing the results of Mr. Zamiatnins' excavations, he confirms his conclusions, and places Ilskaya somewhat later than the upper horizon of Kiik-Koba on the basis of the appearance of a scraper point made on regular prismatic flakes.

However, V. A. Gorodzov,[4] though fully aware that all previous investigators have dated Ilskaya as Mousterian, disagrees with them.   He writes: "All investigators agree in dating it as Mezolithic[5] (Mousterian).   Still, among the industrial material collected by the efforts of S. N. Zamiatnin, forms were present which definitely indicate that it belongs not to the Mousterian, but to the Upper Palaeolithic epoch.

"Such forms are:

"(a) Miniature stone nuclei with narrow facets worked by pressure, introduced not earlier than the beginning of Aurignacian.

"(b) A very well developed form of high scraper (grattoir carené) which appears for the first time in the middle horizon of Upper Palaeolithic (Upper Aurignacian).

"(c) A leaf-like blade (point de sagaie en feuille de laurier), the characteristic form for the middle horizon of the middle time (Upper Solutrean) of the Palaeolithic, found for the first time in Eastern Europe.[6]

[1] Bonch-Osmolovsky, G. A. believes the lower layer to be lower palaeolithic (Tyasian) and the upper layer La Micoque, considering it the most ancient stage of Mousterian.   (E. G.)

[2] Ephimenko, P. P., "Traces of the Mousterian Epoch in Southern Russia," Baghalei Memorial Volume, Kiev, 1927, p. 294.

[3] Ephimenko, P. P., "Some Results of the Study of the Palaeolitic Period in U.S.S.R.," "Chelovek," No. 1, Leningrad, 1928, p. 50.

[4] Gorodzov, V. A., "The Results of Archeological Activities in U.S.S.R. During 1917–1930." Manuscript, pp. 17–19.

[5] Gorodzov's classification equivalent to Mousterian.   (E. G.)

[6] Only the form, but not the technique.   (E. G.)

PLATE XVI.

FLINT INDUSTRY OF ILSKAYA.   Unpublished photographs of Zamiatnin.

"(d) On September 12, 1928, in the presence of the author, during the excavations of S. N. Zamiatnin, there was found the miniature nucleus (about 2 cm. high) of the black jasper (lydite) from one side of which had been separated several flakes of microlithic form.

"It is clear that this station belongs to the middle horizon of the middle time of the Palaeolithic (Middle Solutrean).

"The explanation of the co-existence of the archaic forms of older periods with the latter ones was given by the author [1] in the work devoted to the Yenisey Palaeolithic, especially in reference to Afontova Gora in Krasnoyarsk, where archaic forms were also found. The reasons for dating this station as Mousterian are examined and it is pointed out that this site also belongs to the middle horizon of the middle times of the Palaeolithic epoch (Magdalenian).

"It should be pointed out that in Ilskaya, as well as in Afontova Gora, the surviving archaic forms were made, not of flint, but of other rock formations which gave rougher fractures, and by its archaic appearances, led very experienced investigators into errors."

It should, perhaps, be mentioned again that Professor Gorodzov is stressing the formal typological approach, as contrasted with an evaluation of the cultural complex as a whole.

Mr. Zamiatnin reproduces in his illustrations (Fig. 17, no. 4) the miniature nucleus of black jasper with fine facets, and in his photographs, which he so kindly furnished me, he gives a "laurel-leaf" form of point, Plate XVI, no 8. Still, the rudeness of the work and the impression of accidental similarity persists. The author is inclined to accept Mr. Zamiatnin's point of view and considers Ilskaya as a peculiar Eastern development of Mousterian culture.

9. DERKULA

This site is situated on the bank of the Derkula River where it joins the Donetz River, near the border of Ekaterinoslavl province in the Don region.

The discovery was made by P. P. Ephimenko,[2] a summary of whose description is given below:

Geology.—The natural cross-section of the bank reveals a chalky bed, covered by a very uneven layer of dark marl, containing many worn chalk pebbles, and pieces of chalky flint, washed out of the lower strata.

The surface of the marl is very irregular, forming high elevations and deep depressions (Fig. 19).

The sand deposit which covers the marl is divided into two horizons. The lower part, which lies immediately over the marl, is a clean quartz sand which fills the uneven surface of the marl stratum. In it are incrustations of small worn flint pebbles.

It is evidently the bottom of an ancient river terrace (6 meters above the present river level). The Derkula cut its bed deep in the marl and deposited sands at the time when it was a large flowing stream.

At a height of 9 meters above the river there is a thin horizontal layer of small, very much worn flint pebbles, which evidently came from the lower strata; in places this layer

[1] Gorodzov, V. A., "The Determination of Age and Several Peculiarities of Yenisey Palaeolithic," Northern Asia, Vol. I, Moscow, 1929.

[2] Ephimenko, P. P., "The Traces of Mousterian Cultures in Southern Russia," The Baghalei Memorial Volume, Kiev, 1927, pp. 286–301.

is very thick, showing the size of an ancient deposit, the washed out debris of which it represents. It lies at the base of the upper horizon of sands.

Little is preserved of this upper layer of sands in this section of the shore. The surface of the sands is uneven and mixed with the ground layer. It is dark in the upper part and is cemented by iron oxides. The cultural remains were found in the pebble layer between the two horizons of sands. Consequently we have the following section (Fig. 19A):

    (a) Upper horizon of alluvial sands.
    (b) Pebble layer with the cultural remains.
    (c) Lower horizon of alluvial sands.
    (d) Washed out marls.
    (e) Chalk.[1]

The stratum C, contained large pieces of broken quartzite, quartzite flakes and implements made of flint and quartzite. That this part of the shore is continually underwashed by the action of the river, is evidenced by many pieces of quartzite found at the bottom of the steep bank. As a matter of fact, this action of the river has practically destroyed this valuable site. Consequently it appears that the river Derkula has dug a bed in the ancient (second or upper) terrace, the surface of which must have been far higher than its present level above the surface of the river.

It is very likely that the processes of deposition and subsequent washing out which took place, are closely connected with the approach and retreat of the glacier in the period of its greatest development. It is especially probable because the limit of the maximum glaciation lies not far to the north, in the Don River divide, near the source of the Derkula. Thus the upper sands are the remains of the lower (first) terrace, which here covers the upper (second) terrace, and which was formed during the last glacial epoch.

The Palaeolithic hunter left his tools on the surface of this ancient terrace, near the water, during the northward retreat of the maximum glaciation. Only one tubular bone of a large animal was found, which, because of its size, Ephimenko believes may belong to the Pachydermata (?).

Further investigations may disclose more of the faunal and cultural remains of this very interesting site, which is being rapidly destroyed by the river action. The investigations of Ephimenko in 1924–25, though preliminary, have resulted in very interesting material.

The Industry.—Most of the implements are made of clear gray, small grained quartzite, which is quite common in the Ukraine. Flint was not often used, perhaps because the flint pebbles found in abundance in the ancient river deposits are very small and there are no outcrops of good flint in the vicinity. The fact that the chips of quartzite show no evidence of water action, in contrast to the much worn flint pebbles, indicates, in the opinion of Ephimenko, that this material was brought here by man.

Besides the numerous flakes and chips, the following implements were found.

1. A large disc-like nucleus of typical Mousterian form. It is very flat, conical, 16.5 cm. in diameter and 5 cm. long, with a smooth lower surface, the upper one worked in broad

---

[1] According to Malycheff, V. ("Le Loess," Review de Géographie physique, Paris, 1929–30, Vols. II, III, p. 273), the deposits of the Donetz valley, like those of Badrak, are contemporaneous with the lower strata of the younger loess, or at least with its upper strata.

triangles, joining at the apex. It is made of quartzite (quartzite verdâtre) [1] (Fig. 19*b*).

2. Several characteristic broad massive flakes of quartzite.

3. A typical scraper (*racloir*) 50 mm. long and 45 mm. wide at the base of the flake, made of the same material (Fig. 20, *b*).

4. Part of a wide blade with side retouch.

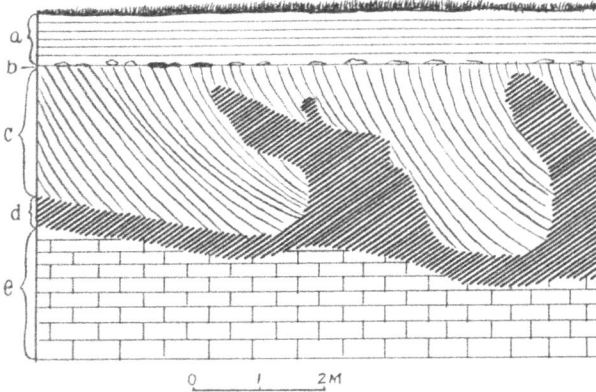

FIG. 19A.   Section of the Terrace of the River Derkula.   *a*, upper horizon of alluvial sands; *b*, pebbles with finds of Mousterian implements; *c*, lower horizon of alluvial sands; *d*, washed-out marl; *e*, chalk.   After Ephimenko.

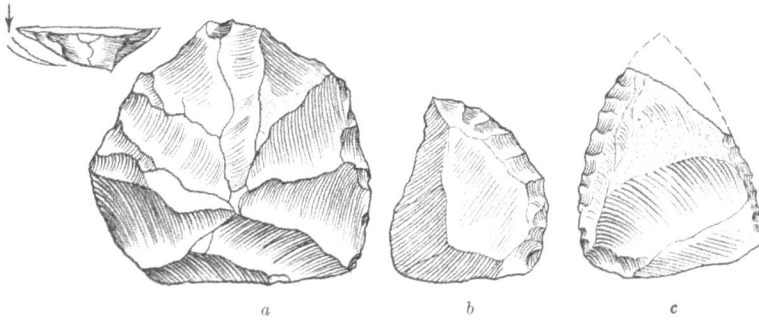

FIG. 19B.   Discoidal nucleus, scraper and a point from Derkula.   ½ Actual Size.   After Ephimenko.

5. A fine point of Mousterian type, made of flint, unfortunately with one end broken (Fig. 20, *c*).

In spite of the small number of finished tools found, Ephimenko believes them to be sufficiently characteristic to classify Derkula as a Mousterian site.   The workmanship and the size of the implements differ a great deal from those of other Mousterian sites of Russia, but are similar to the classical Mousterian of Western Europe.

[1] Levallois, unstruck core (?).   (E. G.)

10. Yashtukh Mountain

According to Field and Prostov,[1] Lower Paleolithic industry was found by Zamiatnin in 1934, in the vicinity of Yashtukh Mountain, near the town of Sukhum, of the Eastern shore of the Black Sea. (Fig. 20, no. 2.)

The site is located in the old alluvium of the third terrace of Gumnista River, on the road from Sukhum to Mikhailovka, 4 kls. north of the former, at the entrance to the gorge between the Byrts and Yashtukh Mountains.

Geology.—The cultural finds are in the stratum of the ancient alluvium of the Eastern Gumnista River. They were encountered "in the soil excavated from a ditch along the road, cutting across the ancient deposits; afterwards their place of origin was located in the gravel bed overlain by a stratum of dilluvial, argillaceous soil changing into the top soil. The gravel bed containing the stone implements has become, under the influence of the ferro-manganese combinations, a solid mass of conglomerate." [2]

Stone Industry.—The cultural finds occured sporadically, and did not form a cultural stratum. The following description of the artifacts is given by authors quoted above. "The majority are very massive, wide and short flakes of an irregularly triangular shape, with a very large striking platform occupying a considerable portion of the lower, flat side of the implement. The striking platform rarely displays preliminary flaking. In general the flints were utilised while still in this stage, without any further retouching, and have only a marginal fracture which may have been caused by usage. A number of discoidal nuclei were also collected."

Besides the nuclei, sharp-pointed implements, massive scrapers, several cleavers and discoidal implements worked on both sides are listed by the above authors. As the illustrations are not available, it is difficult to form any definite idea of this industry. It is being stressed that "comparatively few artifacts are worked by means of the Mousterian retouch" and that only "crude flaking characterizes the secondary working, which is typical of the *biface* implements of the Lower Palaeolithic Period."

Dating.—The geological observations of A. S. Soviev, collaborator of Zamiatnin, lead him to believe that these finds are "associated with the third terrace of the Eastern Gumnista River, connected in turn with the third terrace of the Sea." He feels that this site was occupied "during the Riss-Wurm interglacial period and it was denuded at the beginning of the Wurm period."

The "archaic" character of this industry, and the mixture of Clactonian and Levalloisian methods are rather puzzling. Zamiatnin is inclined to consider the finds from Yashtukh Mountain site, "the oldest Paleolithic implements yet discovered in the U. S. S. R."

11. Kiurdere

According to Field and Prostov,[3] the implements similar to the industry of Yashtokh of the Lower Paleolithic type were found by Zamiatnin at Kiurdere near Psyrtskhi (formerly Novii Afon) on the Black Sea. (Fig. 20, no 1). The stratigraphy was similar to that of the Yashtukh site. No illustrations are available.

[1] Field, H., and Prostov, E., "Recent Archeological Investigations in the Soviet Union," American Anthropologist, Vol. 39, No. 2, pp. 268–269.

[2] Ibid.,

[3] Field, H., and Prostov, E., "Recent Archeological Investigations in the Soviet Union," American Anthropologist, Vol. 38, No. 2, pp. 268–269.

12. OCHEMCHIRI

According to Field and Prostov, Zamiatnin has found on the shore of the Black Sea near Ochemchiri, eighty kls. south of Sukhum (Fig. 20, no. 4), typical Mousterian implements in the similar stratigraphical position as the finds of Yashtukh, with this exception: that the artifacts "were found not in the gravel bed, but in the lower argillaceous level."

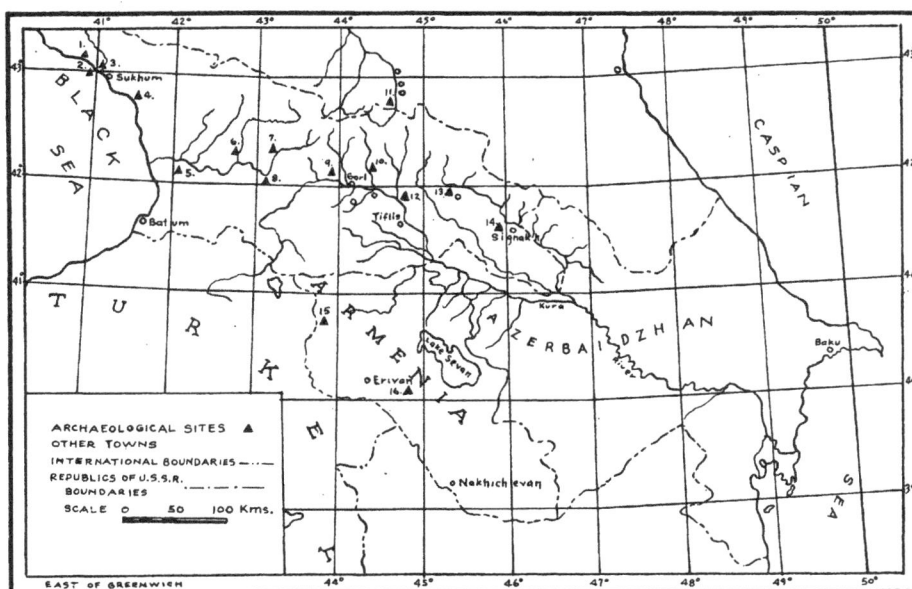

FIG. 20.   Map of Archeological sites in Caucasus. *The Paleolithic Sites:* 1. Kiurdere; 2. Yashtukh Mountain; 4. Ochemchiri; 6. Virchov Cave; 7. Khergulis Klde; Taso Klde; Gvardzhilas Klde; Mgrimevi; Tsizkhvali; Chiaturi.   After Field & Prostov.

13. SHAITAN-KOBA (BADRAK)

Location.—The rock-shelter of Shaitan-Koba was discovered in 1928 by S. N. Bibikov. It is situated in the valley of the river Badrak, 12 kl. southwest of Siempheropol, Crimea. It lies on the eastern slope of the second chain of Crimean mountains, at the base of the limestone formations, 20 mtr. above the river level and 170 mtr. above sea level.

The rock-shelter is semi-oval in form, faces southwest, and is 4 mtr. wide, 7 mtr. in its maximum depth and 2 mtr. high (Plate XVII, *a, b, c*).

The trial pit, 1 sq. mtr., has revealed a flint industry of "typical Mousterian type," made up largely of points and side scrapers.   About twenty specimens were found, made from wide flakes, a few long, massive flakes and a large nucleus.

In 1929 G. A. Bonch-Osmolovsky,[1] leading the Crimean Expedition of the State

[1] Bonch-Osmolovsky, G. A., "The Shaitan-Koba, Crimean Site of the Abri-Audi Type," Bulletin of the Quaternary Committee, No. 2, 1930, p. 65.

Zoological Museum of the Russian Academy of Sciences, made systematic excavations of the rock-shelter of Shaitan-Koba proper (25 sq. mtr.) and of the slope leading to the bottom of the valley, where only a large trench was made (Fig. 21, I, II, III).

According to the author the excavations revealed the following conditions:

Geology.—The cultural level of Shaitan-Koba must have been much thicker at one time because cultural remains in the lime deposits near the walls of the cave reach a considerable height.[1] However, sometime during the Middle Ages, when the nearby settlement of Bakla was founded, the half filled rock-shelter of Shaitan-Koba was partly cleared of the cultural debris. The cave was apparently used by the inhabitants for domestic purposes, as it is clearly seen from the hitching ring worked in the rock below the ancient cultural layer. Only a comparatively thin layer of undisturbed cultural deposit, about 30 cm. in thickness, was left. The cultural remains in the trench along the slope testify to the clearing (Fig. 21).

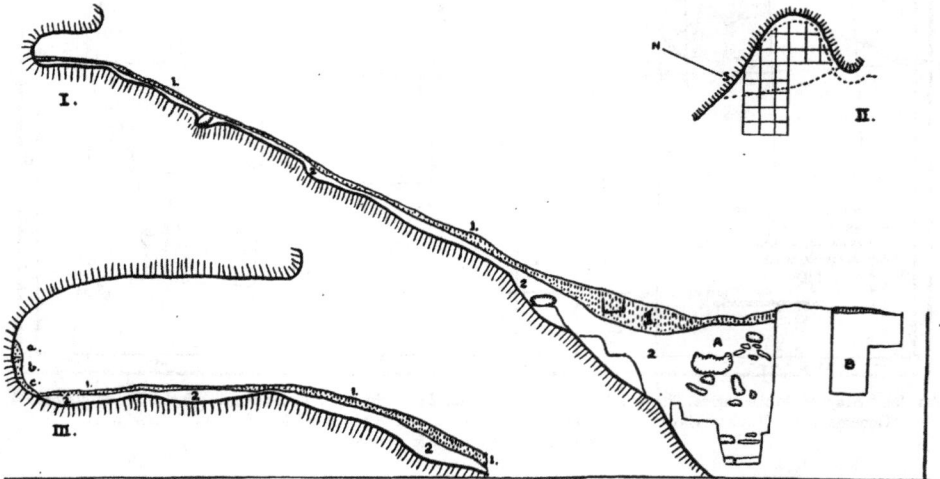

Fig. 21. Rock Shelter, Shaitan-Koba. I, General Profile of the Rock Shelter and the Slope; II, Ground Plan of the Rock Shelter; III, Detailed Profile of the Rock Shelter. A, Scythian Burial; B, Trial Pit. A, upper cultural layer; b, intermediate layer; c, lower cultural layer. After Bonch-Osmolovsky.

From the few remains found in the lime deposit along the walls, one gets the impression that we are dealing with two separate cultural horizons, but, according to the author, the scarcity of material does not permit any such classification, and he deals with all finds of Shaitan-Koba as belonging to one level.[2]

[1] According to Vera Malycheff (Le Loess, "Review de Géographie physique," Paris, 1929–30, Vols. II, III). "La formation des couches de l'abri de Badrak pourrait correspondre a la mise en place du niveau inferieur du complexe loessique récent."

[2] This fact should be borne in mind at all times in determining the place of Shaitan-Koba in the scale of stone industries, as the material from the trench may very well belong to the second horizon, if the latter existed. (E. G.)

PLATE XVII.

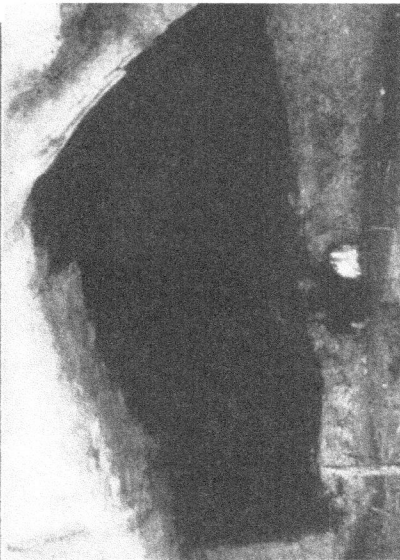

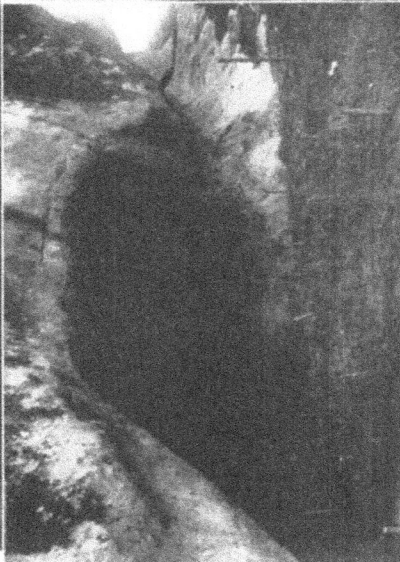

A. GENERAL VIEW OF ROCK-SHELTER, SHAITAN-KOBA. B, C. ENTRANCE INTO THE CAVES. Photos after Bonch-Osmolovsky.

Fauna.—According to Gromov: [1]

Elephas primigenius,	*Felis leo,*
Equus caballus,	*Hyaena spelaea,*
Equus asinus,	*Cricetus (Cricetulus) migratorius,*
Sus scrofa ferus,	*Mus sylvaticus,*
Saiga tatarica,	*Arvicola amphibius,*
Canis lupus,	*Lagurus luteus,*
Vulpes vulpes,	*Alactaga jaculus.*
Vulpes lagopus,	

In the cave proper there was a very thin modern level of ashes, with occasional flint chips at the top. Between it and the rocky bottom of the cave was an undisturbed Quaternary layer, yellow, and, in spots, grey-green in color. It is in the upper part of this level that most of the finds of flints and charred bones were made. About 10 cm. below the surface was a thin layer of dark ashes and charred bone (hearth). The author is certain that the cultural deposit in the cave was undisturbed, as he found many very fragile unbroken hyena coprolites. He also states that neither in the material found in the lime deposits adhering to the walls, nor in the trench below, was anything found which was not present in the main "undisturbed deposit" of the cave proper. Consequently, the following observations are related to the finds of the cave deposit, but apply to the site as a whole.

Industry.—Most of the tools found were made out of dark, almost black flint of a very good quality, quite similar to that found now only 2 kl. from the cave, where pieces weighing up to 20 kilogrammes are encountered on the surface.

Nuclei (Fig. 22, nos. 1–3).—There is evidence of the abundance of good material, as, out of a total of 650 objects, about 150 nuclei varying from 5–13 cm. in length were found. The nuclei are for the most part discoidal in shape, worked on one side, the other preserving the original surface. According to the investigator, "this surface is preserved usually only in the middle of the lower side. Around the edges there are always the small facets resulting from the improving of the striking platform, thus assuring a more precise blow." [2] Only a few nuclei were worked on both sides. The other types of nuclei are either prismoidal, elongated, or cube-shaped, with the parallel ridges formed by the longer flakes. The striking platform is usually on the acute angle. The cube-shaped nuclei form the exception with the striking platform on the right angle, bearing the traces of faceting.

Flakes.—Flakes and blades are about equal in proportion. The latter are unusually fine and regular. The blades are hardly distinguishable from the Upper Paleolithic ones, except for the faceting of the striking platform and their width and massiveness. They form about 15 per cent. of the industry (Fig. 22, no. 4).[3] About 10 per cent. of the tools are prepared on the long flakes which preserve the original surface of the nucleus on one side. The presence of *eclat primitif* [4] and the long flakes (Fig. 22, no. 4), in the opinion of Bonch-Osmolovsky, takes Shaitan-Koba out of the pure Mousterian class, and gives its

[1] Gromov, V. I., "Geology and Fauna of the Palaeolithic Period in U.S.S.R.," Problems of the History of Material Culture, Nos. 1–2, 1933, pp. 23–33.

[2] Bonch-Osmolovsky, G. A., "The Shaitan-Koba, Crimean Site of the Abri-Audi Type," Bulletin of the Quaternary Committee, No. 2, 1930, p. 65.

[3] Really a very wide and thin flake. (E. G.)

[4] Which in itself is not a reason. (A. Kelley.)

industry a transitional relation to the upper Paleolithic, corresponding to that of the Abri Audi in Europe.

Bifaces, according to Bonch-Osmolovsky, are the most archaic forms and are as rare as in Abri Audi. Out of 500 tools, only 14 had traces of bifacial work, and, of this number, only 6 can be considered as finished tools. Such are 2 examples of hand-axes. The first (Fig. 22, no. 5) is 14 cm. long, 8.5 cm. wide and 4 cm. thick, made out of a large flat flint nodule, very roughly chipped on both sides, but preserving the original surface of the block, which "as result of the long use is much rubbed and striated." [1] It has very little in the way of the secondary retouch. In its archaic form and technique of making, it is nearer to the typical Chellean (Acheulean?) *coup de poing*.

The second biface (Fig. 23, no. 6) is a regular almond shaped tool (9.5 cm. long, 5.7 cm. wide and 1.5 cm. thick.) of unusually fine workmanship, being flaked by skillful blows from the periphery. The straight working edge is finely retouched, the opposite edge backed for the hand. There are five other tools with bifacial techniques, reminding one of scraper-like hand axes, of small size. The rest of the nuclei, with traces of use or improvement, are unfinished and appear accidental in character.

Scrapers.—About fifty per cent. of all the tools are "typical Mousterian side-scrapers," half of them made on flakes, half on very thin blades. The majority of them are single scrapers with little retouch on the working edge (Fig. 23, no. 4). Some have retouch on both ends (Fig. 23, nos. 1, 5); a few have a convex edge and coche (Fig. 23, no. 3).

Points.—Next in importance are the points, mostly made on blades (Fig. 25). There are a few asymmetrical points with one straight and another curved edge, the latter usually being steeply retouched (Fig. 25, nos. 1, 3, 4).

About 25 per cent. of all the tools are made on long flakes and primitive blades (Fig. 25, nos. 6, 7). Bonch-Osmolovsky also mentions the points and asymmetrical points of the type of Abri-Audi, eight in number, and, according to him, quite typical (Fig. 24, no. 2; Fig. 25, no. 3).

The investigator stresses the presence of tools of the Upper Palaeolithic type, such as the prototype of a scraper of high form (Fig. 24, no. 7), the scraper-like tools made on the flakes and blades (Fig. 24, nos. 5, 6), the tool which he considers either an end scraper or a nuclei-form graver (Fig. 24, no. 3), the scrapers on the end of a blade with a straight working edge (Fig. 24, no. 6), and perforators made of flakes with a fine point formed by an alternating retouch (Fig. 23, no. 2a, b).

Gravers.—Among the number of gravers which are not so well made, and do not have a typical appearance, there are found, however, some ten specimens of very definite form. Such are a middle graver on a nucleus (Fig. 26, no. 1), the same on an original flake (Fig. 26, no. 2), an angle graver (Fig. 26, no. 3) and graver flakes (Fig. 26, no. 4).

There are cases of vivifaction [2] of points with *burin* strokes. Such are the points represented on Fig. 26, nos. 5 and 6. [3]

Bone Industry.—No bone implements were found in Shaitan-Koba. Many fragments of bone with traces of use as anvils or retouchers were found, however, some with traces of

---

[1] Bonch-Osmolovsky, G. A., "The Shaitan-Koba, Crimean Site of the Abri-Audi Type," Bulletin of the Quaternary Committee, No. 2, 1930, p. 65.

[2] The term used to indicate additional work done on a dulled or broken point to produce a new working edge. (E. G.)

[3] In the case of No. 6, the blow may be for the reduction of the bulb. (E. G.)

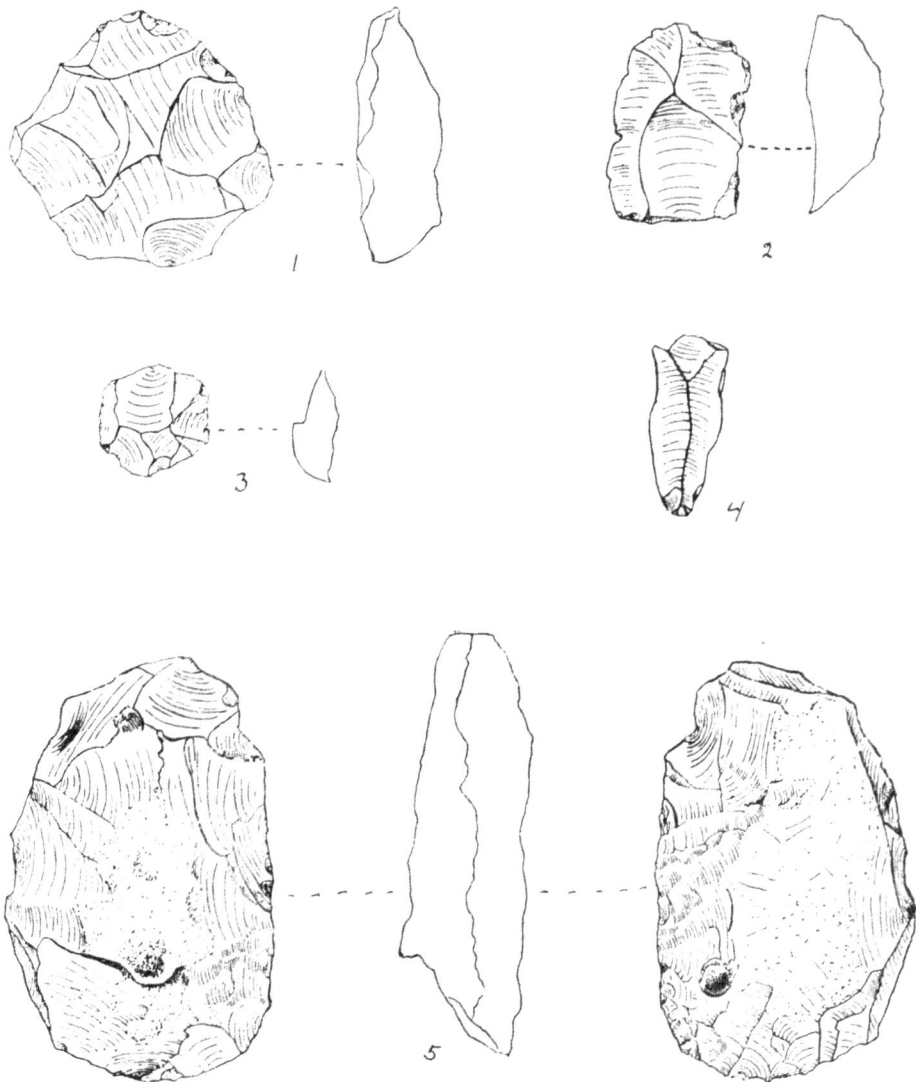

FIG. 22.   Flint Industry of Shaitan-Koba.   After Bonch-Osmolovsky.

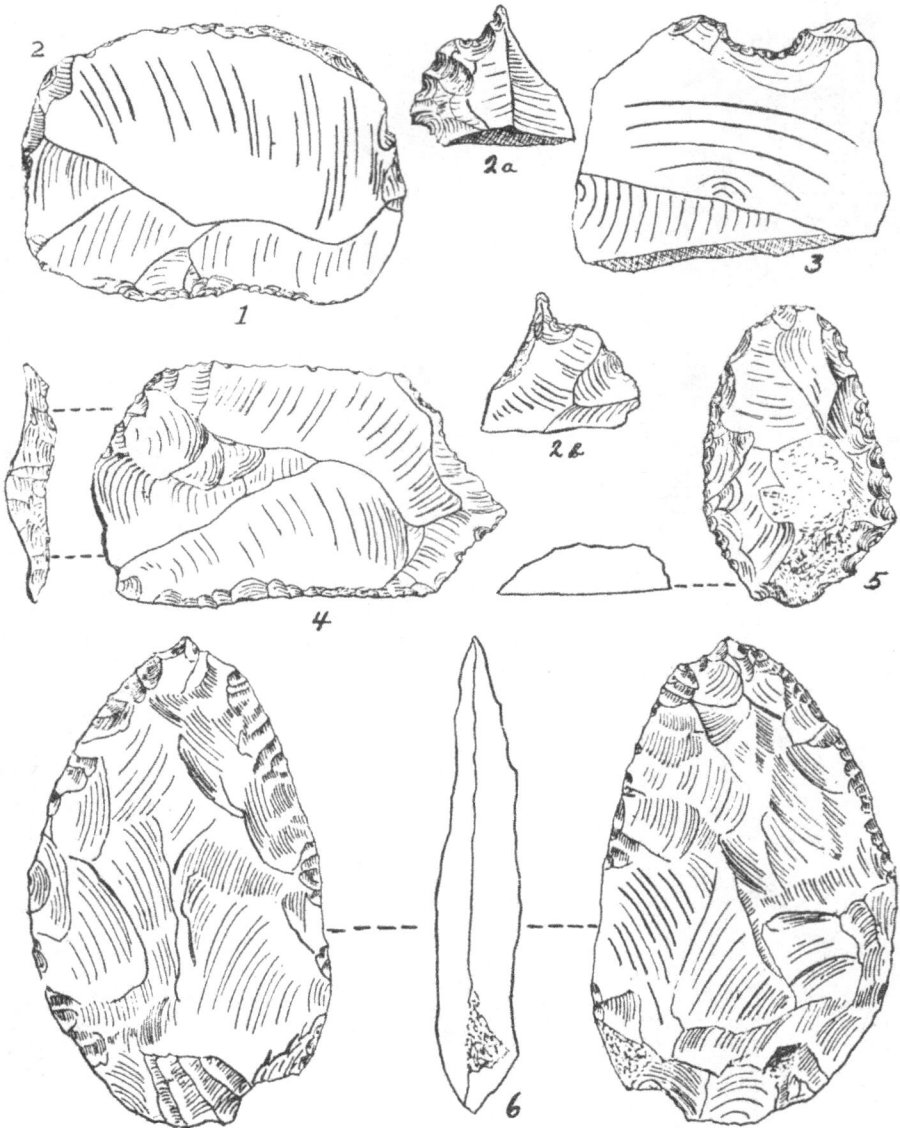

FIG. 23.   Flint Industry of Shaitan-Koba.   After Bonch-Osmolovsky.

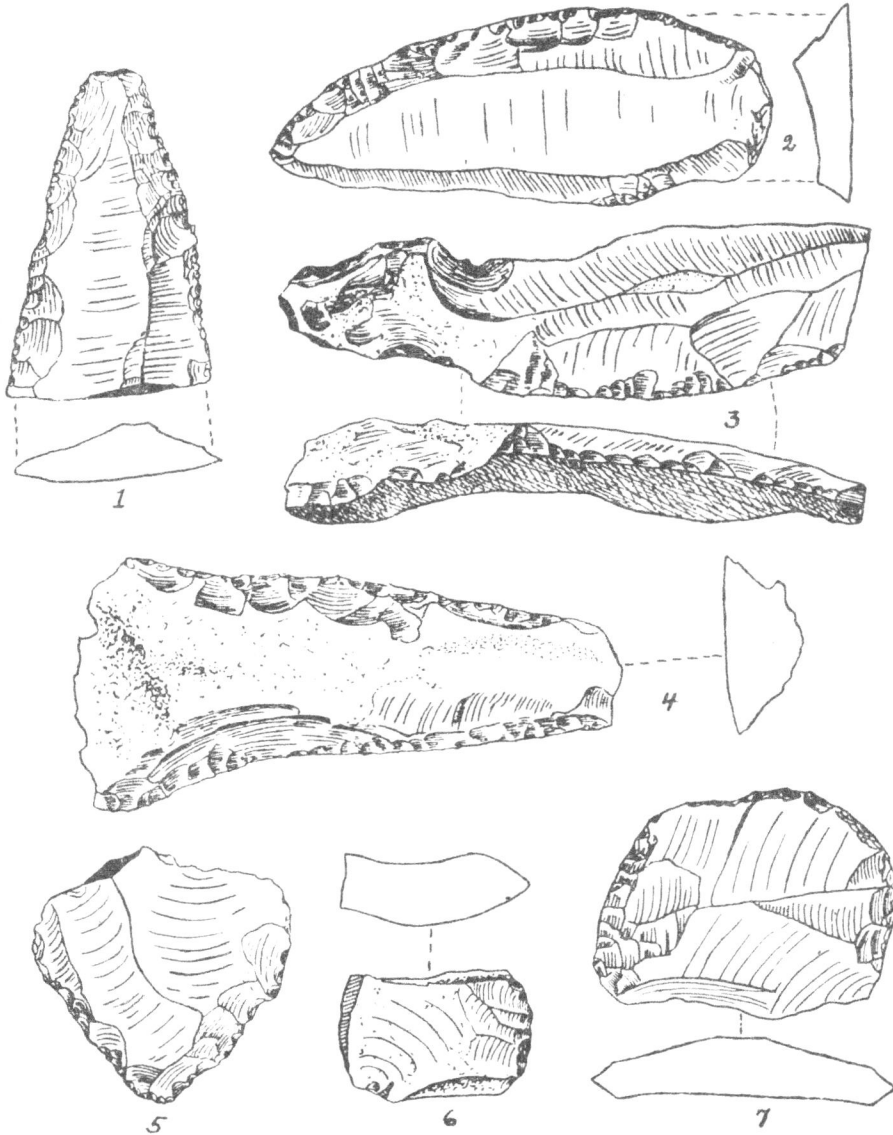

FIG. 24.   Flint Industry of Shaitan-Koba.   After Bonch-Osmolovsky.

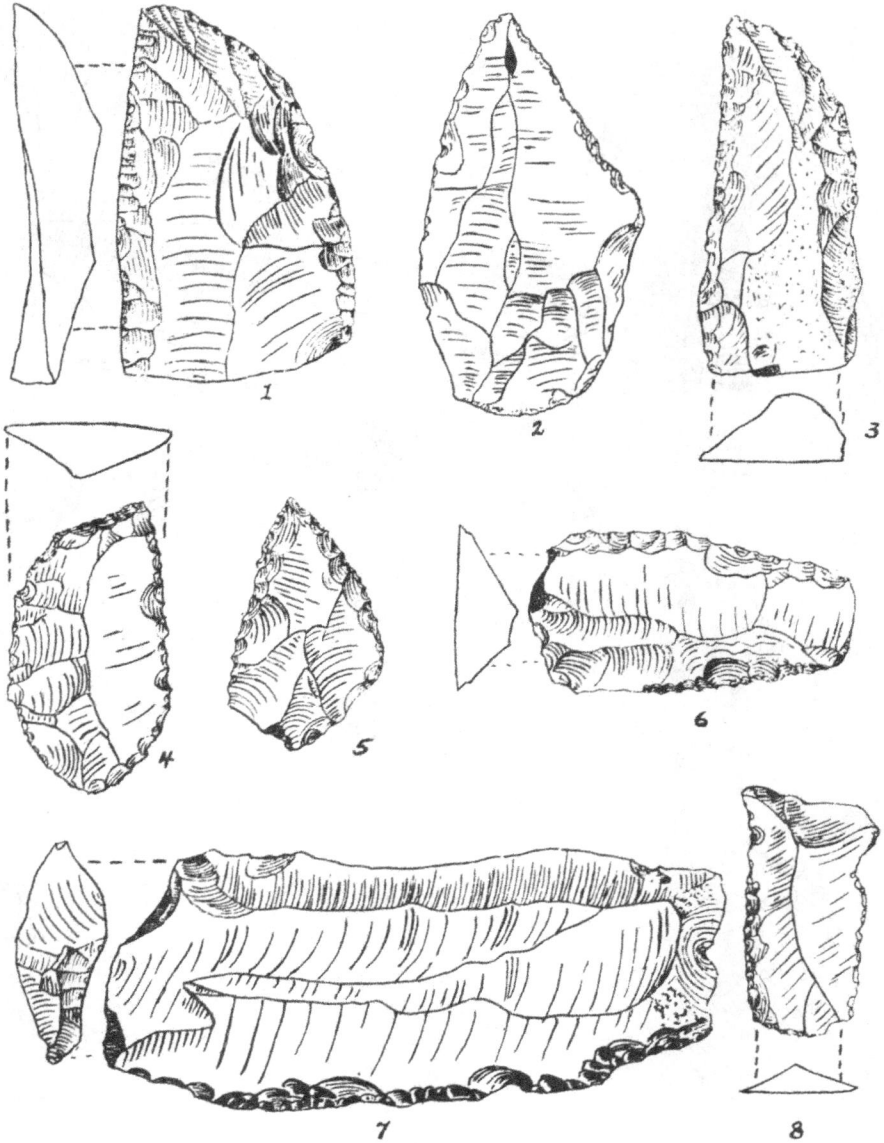

FIG. 25. Flint Industry of Shaitan-Koba. After Bonch-Osmolovsky.

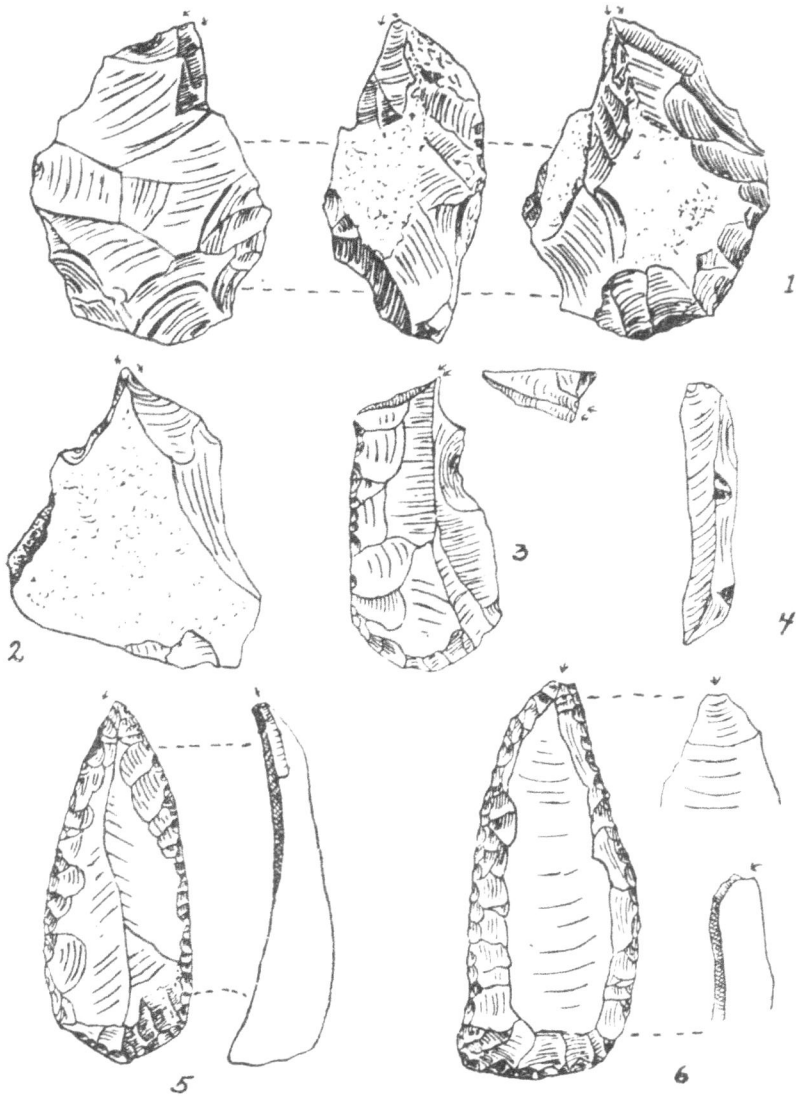

FIG. 26.   Flint Industry of Shaitan-Koba.   After Bonch-Osmolovsky.

smoothing, some with typical ligament cuts, etc. In general the bone material is quite similar to that found in Chokurcha cave (see pages 259–265) and Kiik-Koba (see pages 240–258).

Dating.—The analogy to Abri-Audi, which Bonch-Osmolovsky finds both in the finished implements, and in the unworked flakes and blades (the secondary material) of Shaitan Koba, places it in between the late Mousterian and early Aurignacian. The excavator is inclined to consider Shaitan-Koba slightly older on the basis of the larger predominance of typical Mousterian flakes and methods. He feels that the presence of *burins*, as well as the large use of long flakes and blades, makes it a link between the Mousterian and Upper Palaeolithic industries, and, as such, it is of very great importance.

The geology of the site does not give any clue, although, in the future, it may be possible to tie it in with the river deposits of Badrak. The fauna itself, according to Bonch-Osmolovsky, is slightly older than that of Suren (the Aurignacian site).

Up to now, Shaitan-Koba is the only site in Eastern Europe which presents the characteristics of transition between the Lower and Upper Palaeolithic industries, and in this lies its importance.[1]

It was suggested by Mrs. Alice Kelley that Shaitan-Koba, except for the presence of the *burins* is very much like La Quina, though somewhat older.

[1] One should not forget in following the argument of Bonch-Osmolovsky that we may still be dealing with two horizons and that his layer in the cave was disturbed or mixed. To quote the author in reference to the find in the lime deposit on the walls: "They are all either thin Mousterian blades or primitive blades and as a whole produce a somewhat more perfect impression than the finds in the basic (cave layer) culture." This is especially important because, according to his table, there is an intermediate sterile layer between the two layers of concretions. (See Fig. 21). Can it be that we are really dealing here with two horizons, one Mousterian and the other Aurignacian? (E. G.)

D. UPPER PALÆOLITHIC SITES OF EUROPEAN RUSSIA

SUREN I

- The cave of Suren I is located near the village of Biuk-Suren of the Bahchisarai region, off the highway between Bahchisarai and Yalta in Crimea. It is in the valley of the Belbek River, 12 km. southwest of Bahchisarai, and was discovered in 1897 by K. S. Merezhkovsky.[1] Later excavations were made by G. A. Bonch-Osmolovsky and N. L. Ernst in 1926–29.

Suren I is a very large, typical rock-shelter, situated 110 m. above the sea level and 25 m. above the floor of the valley, in the second chain of the Crimean mountains, on the eastern slope of the Belbek canyon. Its formation is the result of weathering (Plate XVIII).

The excavations of Merezhkovsky resulted in the discovery of the following stratigraphy:

1. A layer of decayed excrement.
2. A white layer 75 cm. thick of limestone debris fallen from the ceiling.
3. A gray layer of the same with admixture of clay. In the lower part of this layer were two thin "hearth" layers, where flint tools, small fragments of bone and charcoal were found.

The list of fauna remains given by Merezhkovsky include: *Saiga tatarica*, *Bos* sp., *Sus* sp., *Cervus elaphus*, *Canis* sp. (?).

The flint industry found by Merezhkovsky in the two horizons [2] belongs, according to G. A. Bonch-Osmolovsky, to Aurignacian and Magdalenian, respectively. It includes a group of nuclei-like forms, very characteristic of Early Aurignacian, such as nuclei-like and *rabot* type scrapers and a series of high nuclei-like gravers. There are also a number of usual Upper Palaeolithic forms made of long flakes, scrapers, gravers (usually multifaceted), blades with dulled back, and large Aurignacian blades with a round retouch.

According to P. P. Ephimenko,[3] there were also found simple bone tools, appearing for the first time in the Aurignacian. He feels that both horizons contained a more or less similar industry.[4] This is perhaps due to the incompleteness of the account.

The excavations of Bonch-Osmolovsky, in 1926 and subsequent years, have resulted in a fuller, though somewhat different picture. He finds the following stratigraphy.[5]

1. A black sterile layer of decomposed excrement and ashes 5.20 cm.
2. Next a layer 6 cm. in thickness formed of slabs of limestone fallen from above, sometimes very large, with three Palaeolithic layers situated at different depths. In some spots, owing perhaps to the water action, the space between the stones was filled with dark clay or with a mixture of lime and sand, whitish grey in color.
3. Below this, and extending 9 cm., almost down to the floor of the cave, is a sterile layer of detritus of fine limestone, rolled with humid brown clay.

[1] Merezhkovsky, K. S., "The Report of the Preliminary Investigations of the Stone Age in Crimea," Izvestia of Imperial Russian Geographical Society, Vol. XVI, pp. 106–142, 1880; Vol. XVII, pp. 104–115, 1881.

[2] Bonch-Osmolovsky, G. A., "The Prehistoric Cultures of Crimea," " Krim," No. 2.

[3] P. P. Ephimenko, "Some Results of the Study of Palaeolithic Period in U.S.S.R.," " Chelovek," No. 1, 1928, Leningrad, p. 51.

[4] *Ibid.*

[5] Bonch-Osmolovsky, G. A., "Le Paléolithique de Crimée," Bulletin de la Commission pour l'Étude du Quaternaire, No. 1, 1929, pp. 27–34.

The formation of this layer is best explained by the action of water flowing on the bottom of the cave and destroying the lower layers of stone.[1]

All three layers yielded flint tools, bone objects, animal bones, Quaternary fishes, and small pieces of charcoal.

The lower cultural horizon extended over almost the entire surface and yielded most of the finds. The middle horizon covered only part of the area, while the upper horizon is represented by small deposits vaguely indicated and with a depth of 0.5–1.5 meters.

Fauna and Flora.—Gromov [2] gives the following list of mammals:

	Lower Horizon (Layer IV)	Middle Horizon (Layer III)	Upper Horizon (Layer II)
Elephas primigenius	×	—	—
Equus caballus	×	×	×
Sus scrofa	×	—	×
Bos sp.	×	×	×
Cervus elaphus	×	×	×
Cervus megaceros	×	×	×
Rangifer tarandus	×	—	—
Saiga tatarica	×	×	×
Canis lupus	×	×	×
Vulpes vulpes	×	×	×
Vulpes lagopus	×	×	×
Vulpes corsac	× (sp.)	× (sp.)	×
Putorius evesmanni	×	—	—
Putorius ermineus	×	×	—
Putorius nivalis	×	—	×
Ursus arctos	—	—	×
Ursus spelaeus	×	—	—
Hyaena spelaea	×	—	—
Lepus timidus	×	×	×
Ochotona pusilla	×	×	×
Marmota bobac	×	×	×
Citellus rufescens	×	×	×
Castor fiber	×	—	—
Cricetus cricetus	×	×	×
Cricetus (Mesocricetus eversmanni)	×	×	×
Cricetus (Cricetulus)	×	×	×
Mus sylvaticus	×	×	×
Arvicola amphibius	×	×	×
Microtus arvalis	×	×	×
Evotomys glareolus	×	×	×
Lagurus luteus	×	×	×
Ellobius talpinus	×	×	×
Alactaga jaculus	×	×	×
Alactaga elater	×	×	×
Scirtopoda telum	×	×	×

[1] According to Malycheff, Vera ("Le Loess," Revue de Géographie physique et de Géologie dynamique, 1929–30, Vols. II and III, pp. 272), the formation of the layers of Suren I "est contemporaine en partie tout au moins de la mise en place du niveau loessique supérieur qui, en effet, a fourni de l'Aurignacien, tout comme les couches de l'abri."

[2] Gromov, V. I., "Geology and Fauna of the Palaeolithic Period in U.S.S.R.," Problems of the History of Material Culture, Nos. 1–2, Leningrad, 1933.

PLATE XVIII.

GENERAL VIEW OF CAVE SUREN I, CRIMEA. After photo of Bonch-Osmolovsky.

M. Tichii [1] studied the fish bones, and identified the following:

	Lower Horizon (Layer IV)	Middle Horizon (Layer III)	Upper Horizon (Layer II)
Salmo sp.	×	—	×
Salmo trutta laborax	×	—	×
Rutilus frisii	—	—	×
Leuciscus cephalus	—	—	×

A. Hammermann [2] identifies the flora by the study of charcoal and ashes of Suren I: (Fig. 27.)

	Lower Horizon (Layer IV)	Middle Horizon (Layer III)	Upper Horizon (Layer II)
Sorbus aucuparia	·×	—	—
Populus tremula (?)	×	×	×
Rhamnus cathartica	×	—	—
Betula sp.	×	×	×
Salix sp.	×	—	—
Juniperus sp.	×	—	×
Taxus baccata (?)	×	—	—

Industry.[3]—The industries of all three horizons are of the Aurignacian Age. The fundamental type is rough, consisting of prismatic flakes characteristic of this stage. The differences permit us further to date the horizons.

In the lower horizon one finds:

1. Mousterian forms—not numerous.
2. Large flakes with lateral retouch.
3. Nuclei-like scrapers with round edge.
4. Scrapers on the end of blade.
5. Gravers (*burin droits*), angle and multifaceted (few).
6. Small blades with alternative or side retouch.

Those forms definitely indicate lower Aurignacian industry.

In the middle horizon the Mousterian forms almost entirely disappear, the nuclei-like scrapers acquire the contour of "*grattoirs caréné*," while gravers, which increase in quantity, approach the form of the "*burin busqué special*," and finally the small blades almost disappear.

In the upper horizon the industry has a more definite character. Besides a large number of multi-faceted gravers, among which the most massive can be classified as nuclei-like,

[1] Tichii, M., "Fische aus dem Paläolithikum der Krim," Bulletin of the Quaternary Committee, Vol. 1, 1929, pp. 43–46.

[2] Hammermann, A., "Kohlenreste aus dem Paläolithikum der Krim, Höhlen Ssjuren I und II," Bulletin de la Commission pour l'Étude du Quaternaire, No. 1, pp. 39–42.

[3] Unfortunately Bonch-Osmolovsky does not give any illustration or detailed description of the industry.

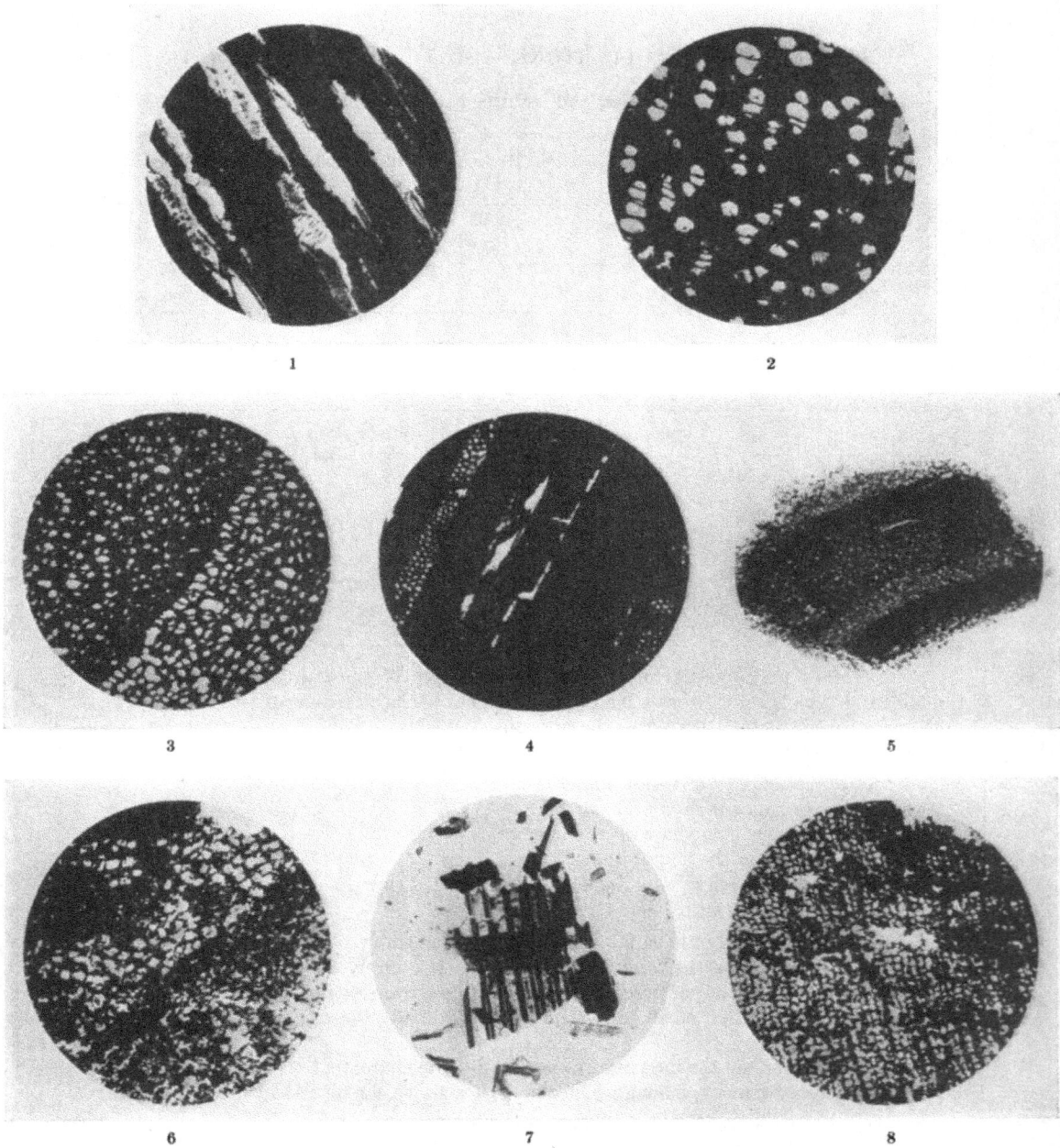

FIG. 27.   Flora from Suren I.   After Hammermann.

1. *Betula* sp. Längsschnitt.   Kohle aus Ssjuren I.   2. *Betula* sp. Querschnitt.   Kohle aus Ssjuren I.   3. *Populus* sp. Querschnitt.   Kohle aus Ssjuren I.   4. *Populus* sp. Längsschnitt.   Kohle aus Ssjuren I.   5. *Sorbus aucuparia*.   Querschnitt.   Kohle aus Ssjuren I.   6. *Rhamnus cathartica*. Querschnitt.   Kohle aus Ssjuren I.   7. *Juniperus* sp. Abschabepräparat.   Kohle aus Ssjuren I.   8. *Juniperus* sp. Querschnitt.   Kohle aus Ssjuren I.

the presence of several points similar to La Gravette permits us to assign it, without a doubt, to Upper Aurignacian.

An interesting fact is the appearance here of small blades in large quantities, but these are exceedingly blunt on the edges.

The bone industry of all three horizons is less typical. Most of the objects have the appearance of points (*poinçons*) and one from the lower horizon can, perhaps, be regarded as an Aurignacian point with a slight notch feebly marked on the end. No objects of art except an ornamented fragment of horn were discovered.

Bonch-Osmolovsky considers the industry analogous to the corresponding industries of Western Europe, not only in regard to similarity of various types of implements, but also in the general complexity of the layers.

2. Devis–Hvreli

This cave is situated in the district of Sharopan, Kutais region, Caucasus, half way between the railroad stations of Charaguli and Dzirùla, 4 km. from the quarries of Chandebi, on the slope of Chandebi mountains.

It is of crystalline limestone formation, lies 340 m. above the sea level, and some 80 m. above the level of the river Chkherimèlla. The cave was discovered in 1926 and investigated by Prof. George Nioradze,[1] assisted by students, during the years 1926–32. The material from his excavations was displayed in 1932 at the International Archaeological Congress in Leningrad and is now preserved in the State Georgian Museum in Tiflis. The geology of the site was studied by A. Sorokin, S. Semenov and P. Gamkrelidze.

According to Nioradze, the cave is 25 m. long, with a width at the entrance of 4.5 m. to 6.5 m., opening to the sourth-west (Fig. 28).

Geology

The excavations revealed the following stratigraphy:

    I. Humus and animal bones, dark in color.
    II. Brown layer, likewise with animal bones.
    III. Cultural layer.
    IV. Rocky Bottom.

Fauna.—Remains of the following fauna were found in the cultural level in association with the stone artifacts:

1. *Sus scrofa ferus*,	7. *Rupicapra tragus*,
2. *Bos primigenius*,	8. *Capreolus* sp.,
3. *Bison priscus*,	9. *Ursus spelaeus*,
4. *Cervus elapus* aff. maral,	10. *Ursus arctos*,
5. *Rangifer tarandus*,	11. *Mesocricetis kownigi*.
6. *Capra* sp. (*Capra cylindricornis?*),	

In general, according to Nioradze, the fauna is like that of more recent times, (?) although somewhat richer in species.

Human Remains.

Nioradze indicates that in the cultural horizon there was found the right side of a human mandible with second and third molars missing, and a *ramus mandibularis*, broken

[1] Nioradze, G., "The Palaeolithic Industry of the Cave Devis-Hvreli," Travaux du Musée de Géorgie, Vol. VI, Tiflis, 1933, pp. 1–109.

in the upper part. The *corpus mandibularis* is in good condition. Its height, in the region of the second molar is, including the alveolar parts, 2.8 cm. The distance between the *spina mentalis* and the *foramenum mandibularis* is 6.5 cm.

Stone Industry.

The majority of the stone artifacts were made of local flint, only a few artifacts were made of obsidian. The numerous rejects and flint chips show that the artifacts were made on the spot.

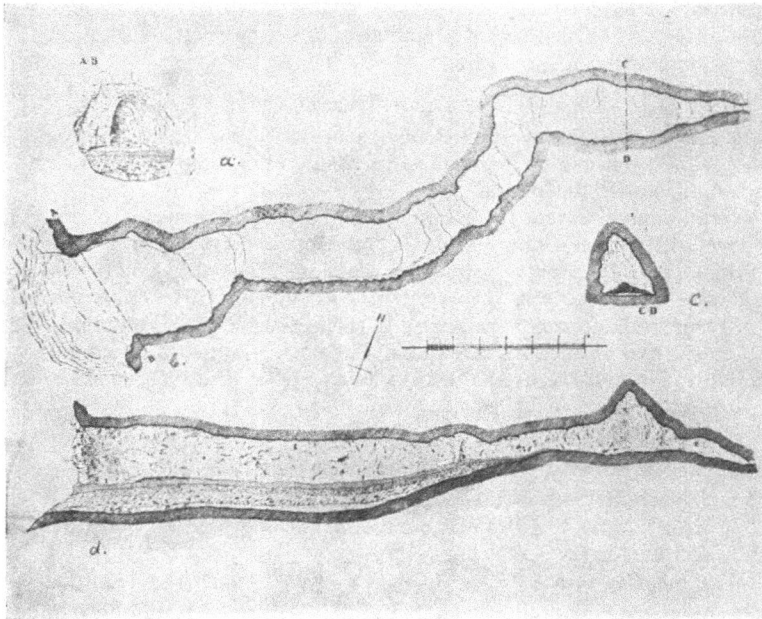

Fig. 28. The Cave Devis-Hvreli, Cutais Region, Caucasus. *a*, entrance, section along *A–B*; *b*, ground plan; *c*, section along *C–D*; *d*, vertical section, showing cultural layers. After Nioradze.

Nuclei.

These were very numerous, about two hundred, almost all of irregular shape, with the tendency to become conical. The majority of them are four and a half centimeters in height, while some are six centimeters. Nioradze distinguishes three groups:

1. With two striking platforms. . . . . 92 specimens
2. With one striking platform. . . . . 56 specimens
3. Others. . . . . 52 specimens

Scrapers.—Scrapers constitute a very large percentage of the industry. The majority of them are of Upper-Palaeolithic type, the end-scrapers being definitely predominant. While Nioradze does not give the number of each type, the fact that there are forty-three nucleiform examples and some eight *grattoirs carénées*, may be significant. In general there is a great variety of forms, ranging from the rough side scrapers to the elegant *grattoirs museaus carénées.*

Burins.—Although bone work is very scarce, and no engraving was reported, *burins* are quite numerous.   They fall into several types: *d'angle, bec de flûte,* and nucleiform. Their technique is somewhat rough though not uncertain.   Double forms were encountered, but these are not numerous.

Points.—What Nioradze calls *Châtelperron* (Fig. 33, no. 1) is at the best a very poor example.   The majority fall into the group of *La Gravette,* diminishing at times to microlithic dimensions, and some developing the side notch like the *pointe a cran atypique.*   The points do not form the larger group of tools, in Devis-Hvreli as is the case of some other sites.

Perforators.—Several good examples of perforators with well-made points were found (Fig. 33, nos. 21–24).   Others (Fig. 33, nos. 19–20) are somewhat atypical.   The following implements are reproduced from the illustrations of Nioradze, which unfortunately are far from being perfect.

Fig. 29.

    1. Conical nucleus with one striking platform.
    2. Prismoidal nucleus with two striking platforms.
    3. *Grattoir* on the end of a broken blade.
    4. Notched side scraper.
    5. *Grattoir* on the end of a broken blade.   Nioradze calls it *grattoir bec de perroquette atypique* (?).
    6. Nuclei-form scraper.
    7. *Grattoir caréné.*
    8. *Grattoir caréné.*
    9. Massive scraper of rectangular shape.
    10. Scraper of rectangular form with rounded corners.

Fig. 30.

    1. End scraper and *burin bec de flûte.*
    2. Double *grattoir.*
    3. Round scraper.
    4. Triangular scraper with round working edge.
    5. Triangular scraper with round working edge.
    6. Semi-oval side scraper of obsidian (Nioradze calls it "Bogen-formige").
    7. Notched scraper.
    8. *Grattoir* on the end of long blade.
    9. Small end scraper.
    10. *Grattoir* on the end of a long blade.
    11. *Grattoir* on the end of a long blade.
    12. Oval scraper with fine peripheral retouch.
    13. Fragment of the long oval scraper.

Fig. 31.

    1. *Grattoir museau.*
    2. *Grattoir museau.*
    3. *Grattoir* on the end of a long blade.
    4. *Grattoir* on the end of a long blade.
    5. *Burin d'angle.*
    6. Double *burin bec de flûte.*
    7. *Burin d'angle* (?).
    8. *Burin bec de flûte.*
    9. *Burin bec de flûte.*
    10. Multi-facetted *bec de flûte.*

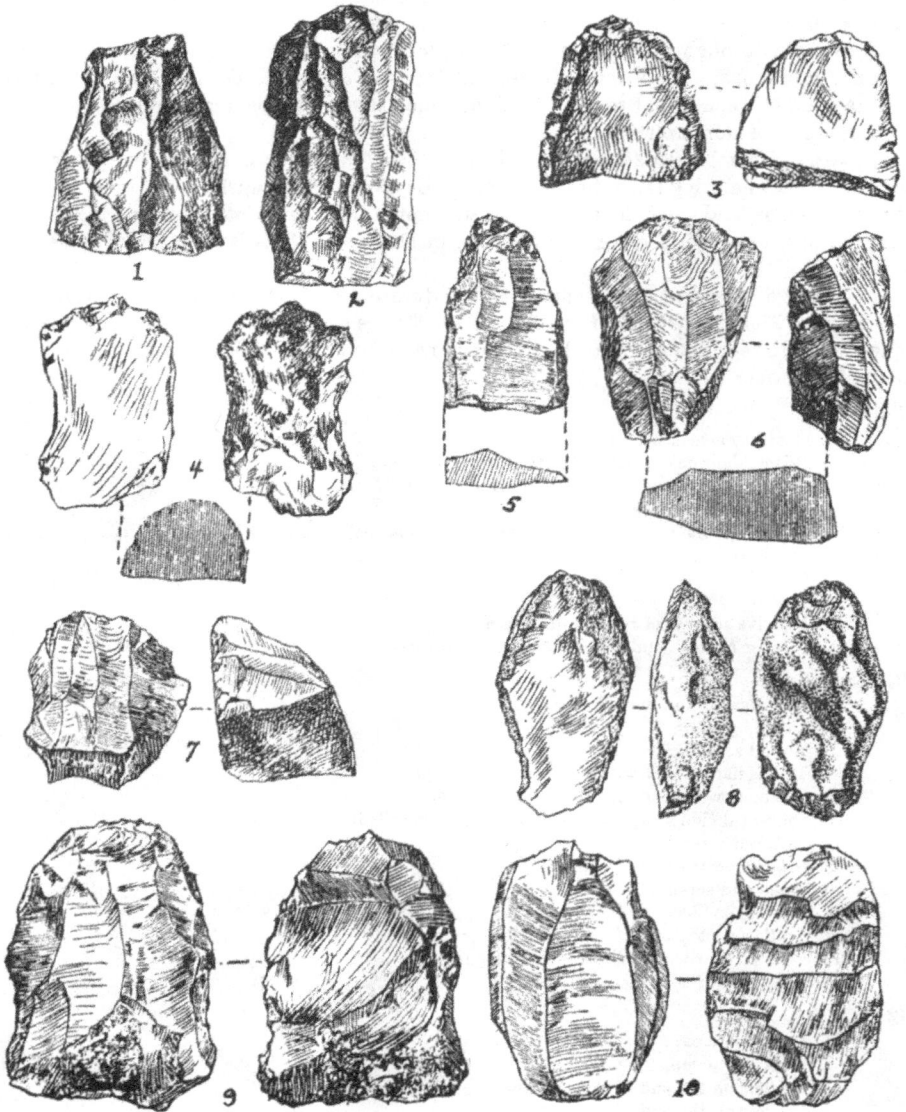

Fig. 29. Flint Industry of Devis-Hvreli. After Nioradze.

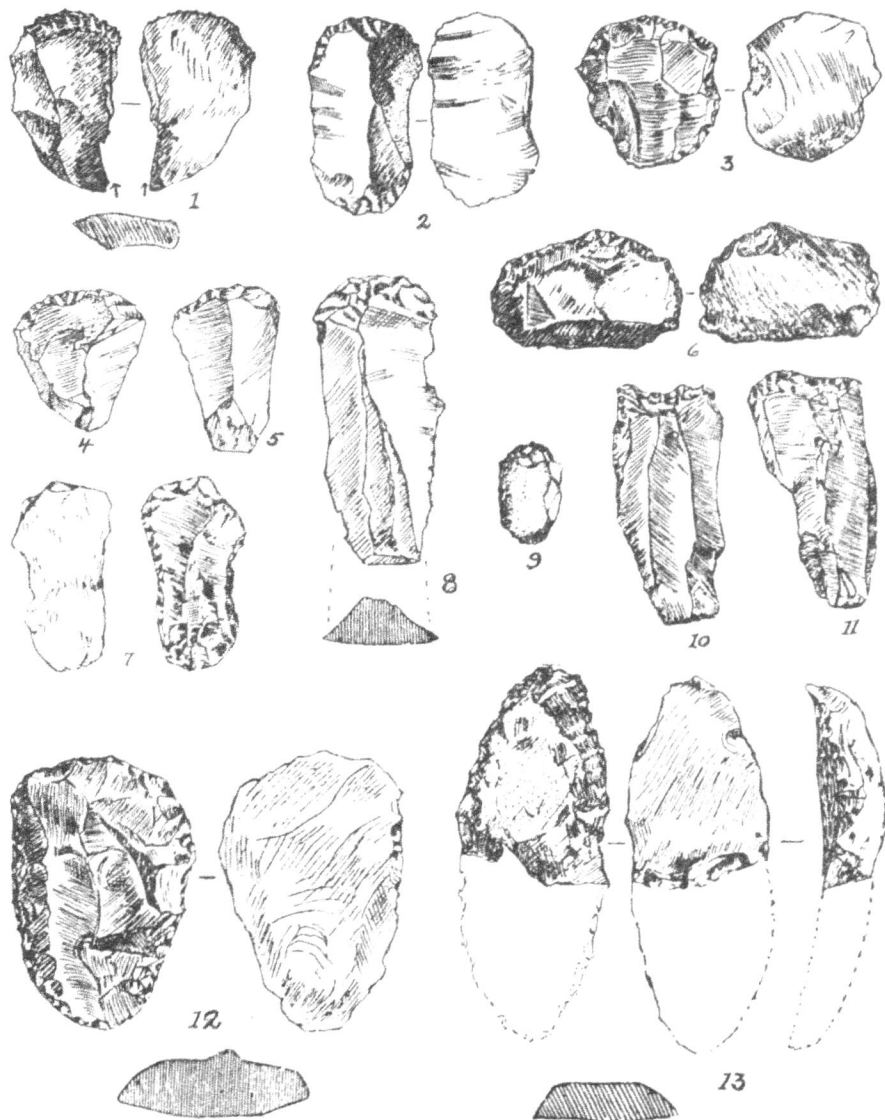

FIG. 30.   Flint Industry of Devis-Hvreli.   After Nioradze.

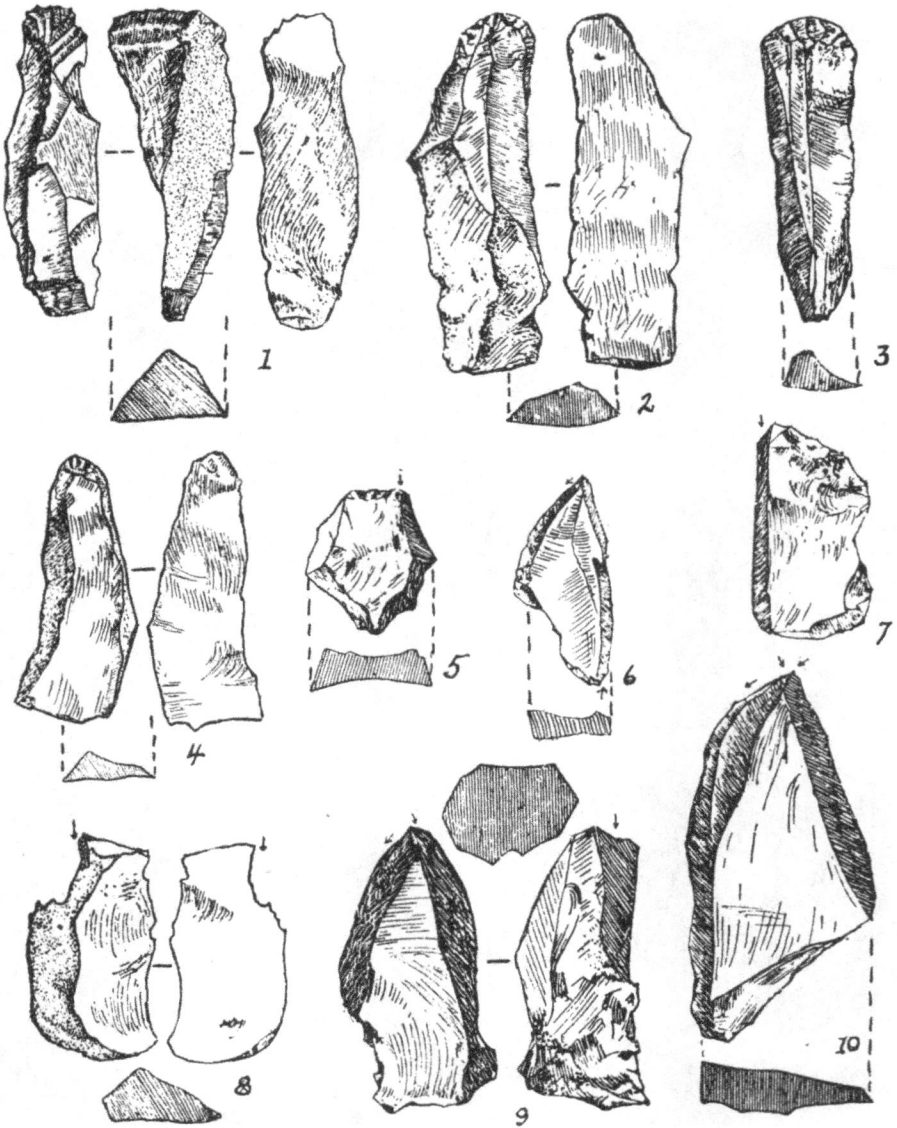

FIG. 31. Flint Industry of Devis-Hvreli. After Nioradze.

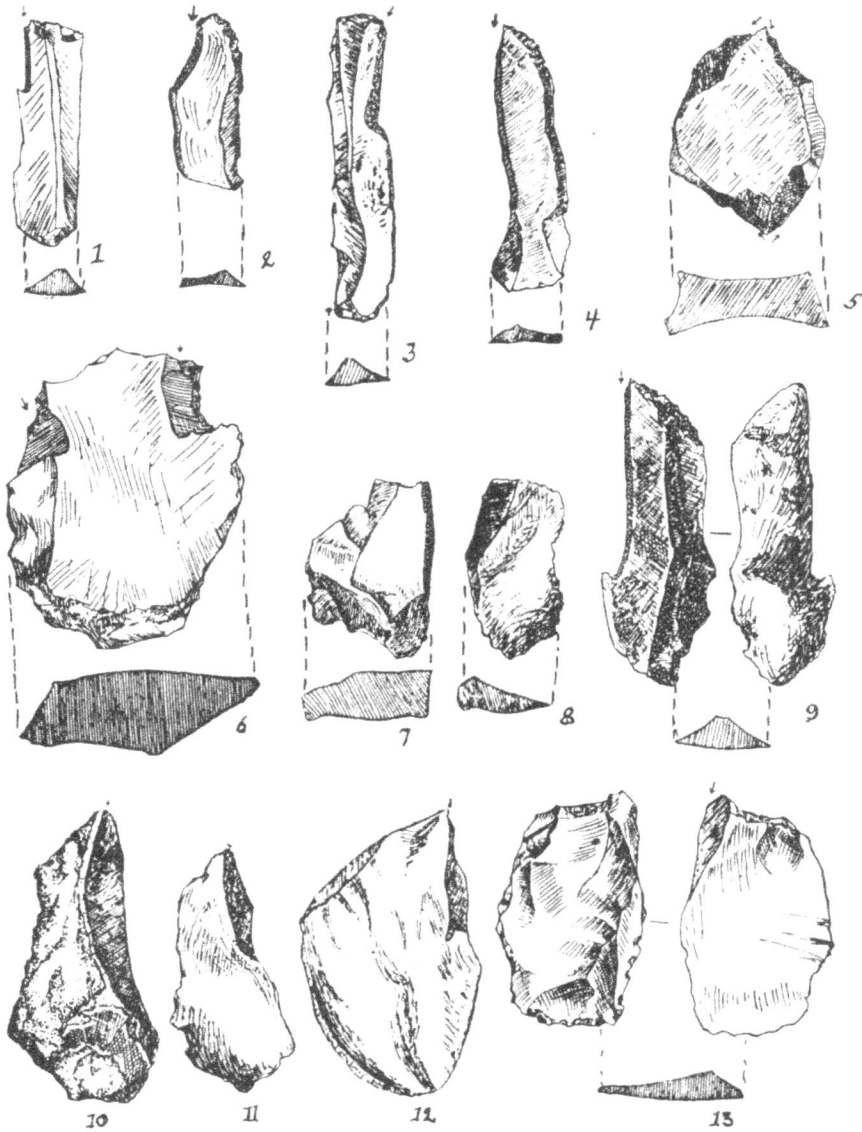

FIG. 32.   Flint Industry of Devis-Hvreli.   After Nioradze.

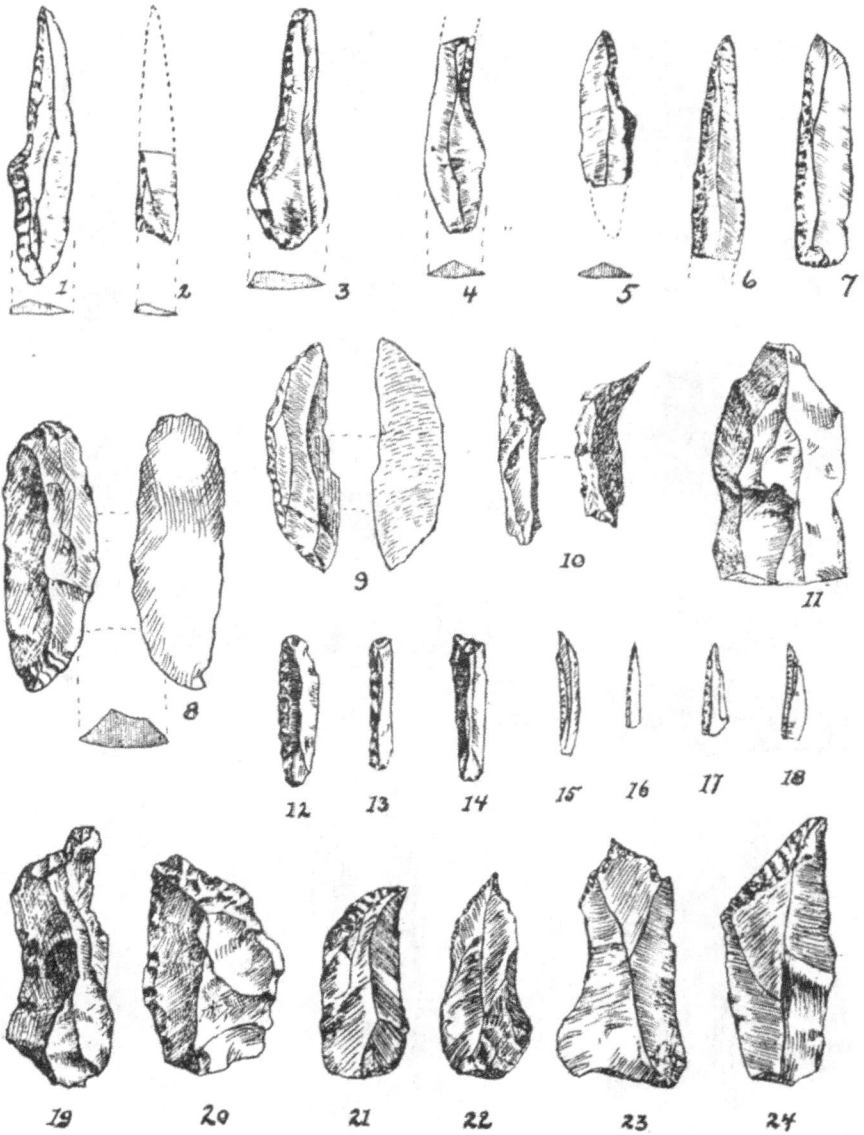

FIG. 33.   Flint Industry of Devis-Hvreli.   After Nioradze.

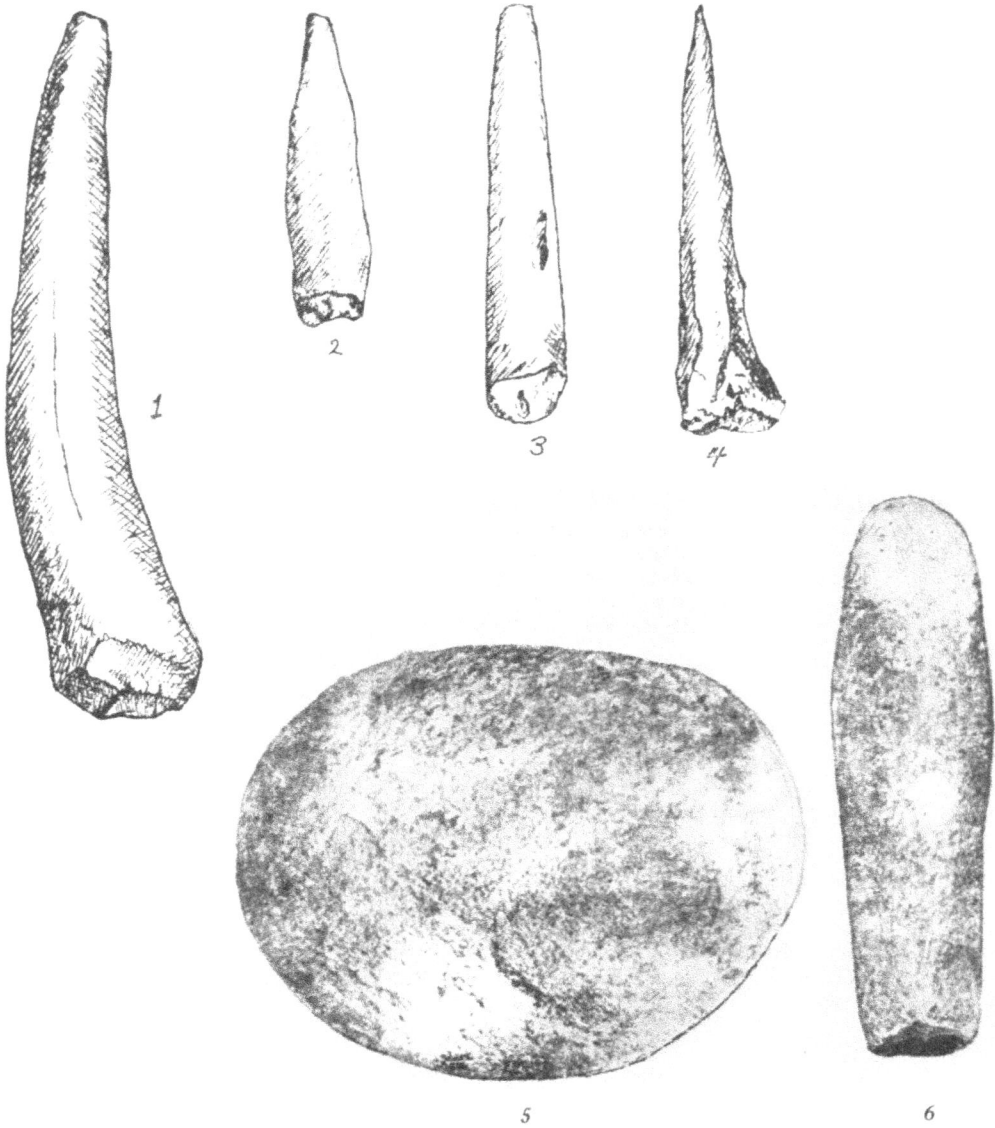

Fig. 34.   Bone and Stone Implements of Devis-Hvreli.   After Nioradze.

Fig. 32.

    1. *Burin d'angle.*
    2. *Burin d'angle.*
    3. *Burin d'angle.*
    4. *Burin d'angle* forming a point.
    5. Double *bec de flûte.*
    6. Nuclei-formed *burin.*
    7. *Bec de flûte.*
    8. *Bec de flûte.*
    9. *Burin d'angle.*
 10–13. *Burin bec de flûte.*

Fig. 33.

    1. *Pointe a cran atypique* (?).
    2. Fragment of *La Gravette* point (Nioradze calls it *pointe a cran atypique*).
    3. *Pointe a cran atypique* (?).
    4. *Pointe a cran atypique* (?).
    5. Fragment of *pointe a cran atypique.*
    6. *La Gravette* point.
    7. *La Gravette* point.
    8. Blade with retouch on both ends forming a point.
    9. *Chatêlperon* point (?).
   10. Flake (Nioradze calls them *lames a dos rabattu*).
   11. An implement of peculiar form described by Nioradze in Georgian as "*shubis-tzverl*" (shubis meaning hitting implement; tzverl, end).
 12–18. *Lamelle* of *la Gravette* type.
   19. Retouched flake (Nioradze calls it "perforator").
   20. Perforator.
   21. Perforator on the end of a blade.
 22–24. Perforator on the end of a blade.

Fig. 34.

    1. Compressor of reindeer horn.
    2. Bone awl.
    3. Bone awl.
    4. Bone awl.
    5. Stone anvil.
    6. Hammer stone (?).

Dating.—Nioradze hesitates to place any definite dating for Devis-Hvreli, feeling that it requires further research. He, however, feels that it may represent a special regional development of the Upper-Palaeolithic, corresponding and related to late Aurignacian and Capsian cultures.

3. BORSHEVO I

This site is situated near the village of Borshevo, Voronezh province, on the right bank of the Don River, 35 km. south of Voronezh, a few miles from the site of Kostenki I (see page 308). Here the Don River is separated from the high bank by the narrow terrace on which is situated the village of Borshevo. A number of small but wide gullies cut the river bank. On the left slope of so-called "Kuznetzov's Gully," near its mouth at the river, is the site Borshevo I.

PLATE XIX

EXCAVATIONS OF THE CULTURAL LAYER, BORSHEVO I.   After photograph of Zamiatnin.

It was discovered in 1905 by A. A. Spitzin,[1] who found the bones of a mammoth in association with flint chips at a depth of 85 cm. Peasants digging for a cellar came across these bones, and the trial excavation pit made by Spitzin resulted in similar finds. He, however, felt that "the layers were very much mixed and that it is evident that all cultural remains fell down from above."[2]

Small excavations by S. N. Zamiatnin in 1922 resulted in the discovery of "hearths" with a large quantity of mammoth bones and a rich paleolithic industry. In 1923 and 1925 P. P. Ephimenko, continuing the excavations of Zamiatnin, found that the "hearths" without implements and bone heaps occupy a considerable area (Plate XIX).

The first "hearth" was in the backyard of one of the local peasants; the second, just beyond it. The cultural layer here had the same general appearance as that of Kostenki I, but the tools found were somewhat more crude.[3]

In 1925, larger excavations were conducted by P. P. Ephimenko, but as yet no account of them has appeared in print. The best illustrations of the finds, given by Sawicki,[4] are reproduced here, the short article of the excavator [5] forming the basis for the summary given below.

According to Ephimenko, Borshevo I is similar both in its general plan and stratigraphy to a series of other sites in the Kostenki-Borshevo region. The cultural horizon is in a layer of yellowish loam of diluvial origin, covered by a stratum of humus. At the implacement of the "hearths" the cultural finds begin even in the lower part of the humus, but usually are situated 120–135 cm. below the surface. Higher up the slope, though, where the humus is more heavily eroded, the cultural finds are much nearer the surface, and in places even exposed on the surface. The cultural layer increases in thickness as it proceeds upward.

The cultural complexes of this site consist of:

1. A number of "hearths" filled with broken and charred bones, masses of flint chips, and some finished tools.

2. Enormous accumulations of mammoth bones (long bones, broken to extract the marrow, shoulder blades, jaws, vertebrae, tusks, etc.). Usually flints, ochre, and charcoal are absent in these bone heaps.

The "hearths," as in the case of Kostenki I, have a more or less definite red color, due to the presence of ochre.

[1] Spitzin, A. A., "Russian Paleolithic Period," Zapiski of the Imperial Russian Archaeological Society, Slavic Div., Vol. XI, Petrograd, 1915, p. 164.

[2] Spitzin, A. A., Otchet of Imperial Archaeological Committee, 1905, p. 184.

[3] Vishnevsky, B. N., "Pre-Historic Man in Russia," Appendix to Translation of Men of the Stone Age by Osborn, H. F., 1924, Leningrad, p. 460.

[4] Sawicki, L., "Materials for the Study of Russian Archaeology," Przeglad Archeologicznii, Vol. III, pp. 2, 3, 1926–1928.

[5] Ephimenko, P. P., "Some Results of the Study of Paleolithic Period in U.S.S.R.," Chelovek, No. 1, 1928, Leningrad, pp. 50–56. "Pre-Clan Society," Leningrad, 1936, pp. 357–359.

Fauna.[1]—

1. *Elephas primigenius*...............................numerous.
2. *Rhinoceros tichorhinus*...........................very scarce.
3. *Equus* (sp.)....................................numerous.
4. *Rangifer tarandus*
5. *Canis* (sp.)

As is seen from the list the fauna is not very varied, and consists mostly of the mammoth and a species of horse. Ephimenko mentions also several teeth of the marten and lynx.

Flint Industry (Figs. 35, 36, 37).—

The flint industry of Borshevo I consists mostly of points of *La Gravette* type, points *à cran atypique*, *burins* and scrapers, made of dark flint of chalk origin.

Points of la Gravette type (Nos. 1–5).

According to Sawicki, no. 2 is less typical and may be *lame en bias.*[2]

Lamelles à dos rabbatu (Nos. 6–26).

These are very numerous. Some of them approach the points, being retouched on both edges (nos. 6, 12, 26). The second edge forming the point is only partially retouched, and the end is often truncated (nos. 14, 16–18, 20–22).

Pointe à cran atypique.

These, with the exception of one (no. 28), are made on narrow blades; and the retouching sometimes is limited only to the notch (nos. 30–32, 35) or partly covers the other edge (nos. 29, 33, 34). In one case, there is a trace of retouching on the lower face (no. 35a). No. 36 may be a fragment of *Font Robert* type.

Burins.

Burins form an important group. There is a large proportion of nucleiform, multifacetted *burins* and double forms.

Burin d'angle double.................................	No. 39
Burin d'angle on the retouched blade.....................	Nos. 40–44
Burin d'angle opposed to lateral *burin*....................	No. 45
Burin d'angle on a wide retouched blade	Nos. 46–48
Burin d'angle on a thick flake.........................	No. 49
Burin d'angle with notch.............................	No. 50
Burin lateral, nucleiform, multifacetted....................	Nos. 51–53
Burin d'angle opposed to multifacetted *d'angle*............	No. 54
Burin bec de flûte double	No. 55
Burin bec de flûte single	No. 56
Burin bec de flûte with revivification strokes	No. 57

[1] Gromov, V. I., "Geology and Fauna of the Palaeolithic Period in U.S.S.R.," Problems of the History of Material Culture, Nos. 1–2, Leningrad, 1933.

[2] Ephimenko, P. P. ("Pre-Clan Society," Leningrad, 1936, p. 260), mentions as a special type flint points with dulled retouch on one side, which he considers to be of *Chatelperron* type, and decisive for the dating.

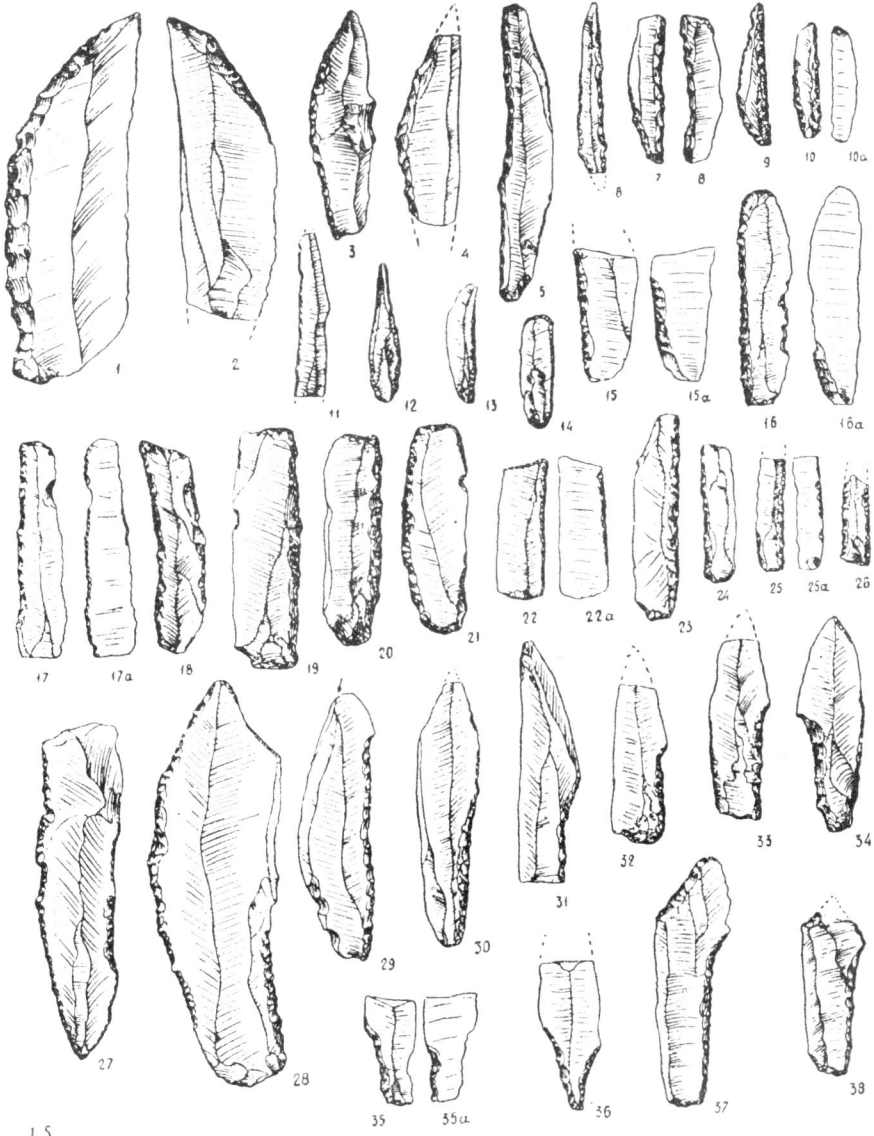

FIG. 35.   Flint Industry of Borshevo I.   After L. Sawicki.

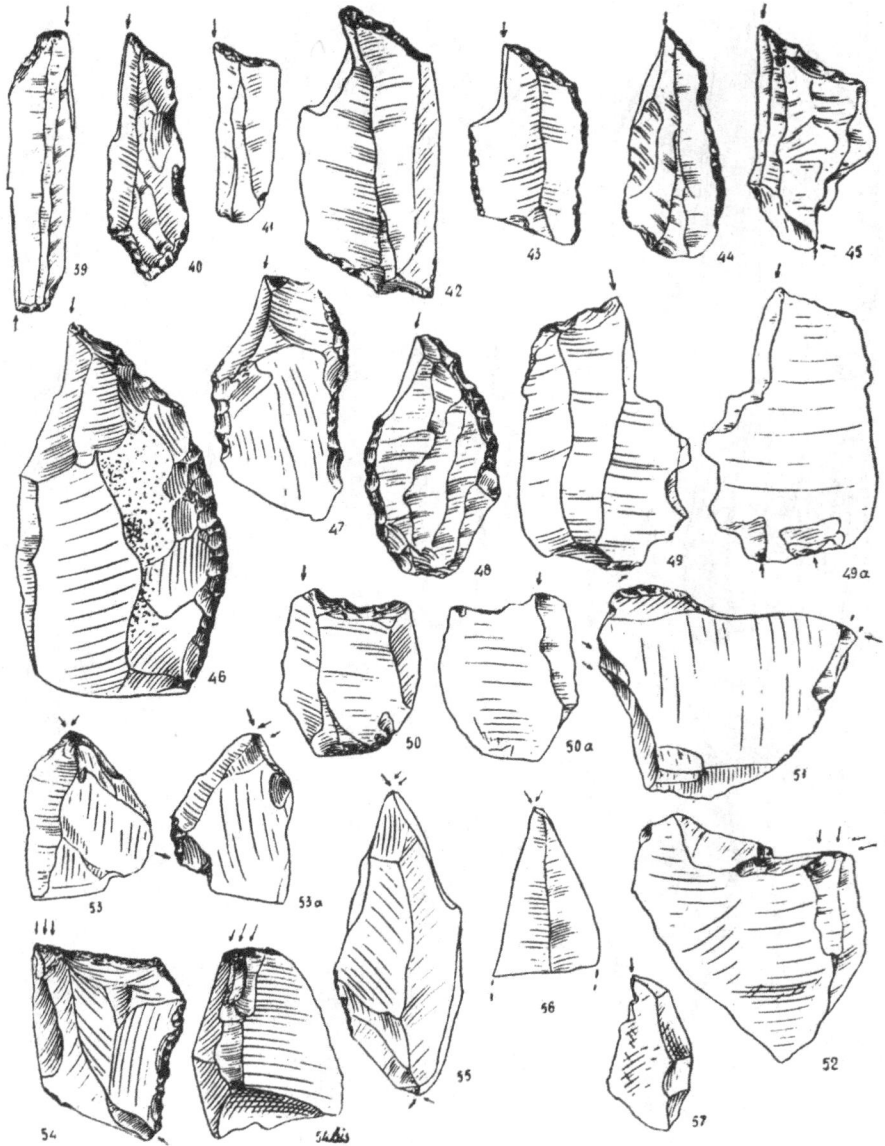

FIG. 36.   Flint Industry of Borshevo I.   After L. Sawicki.

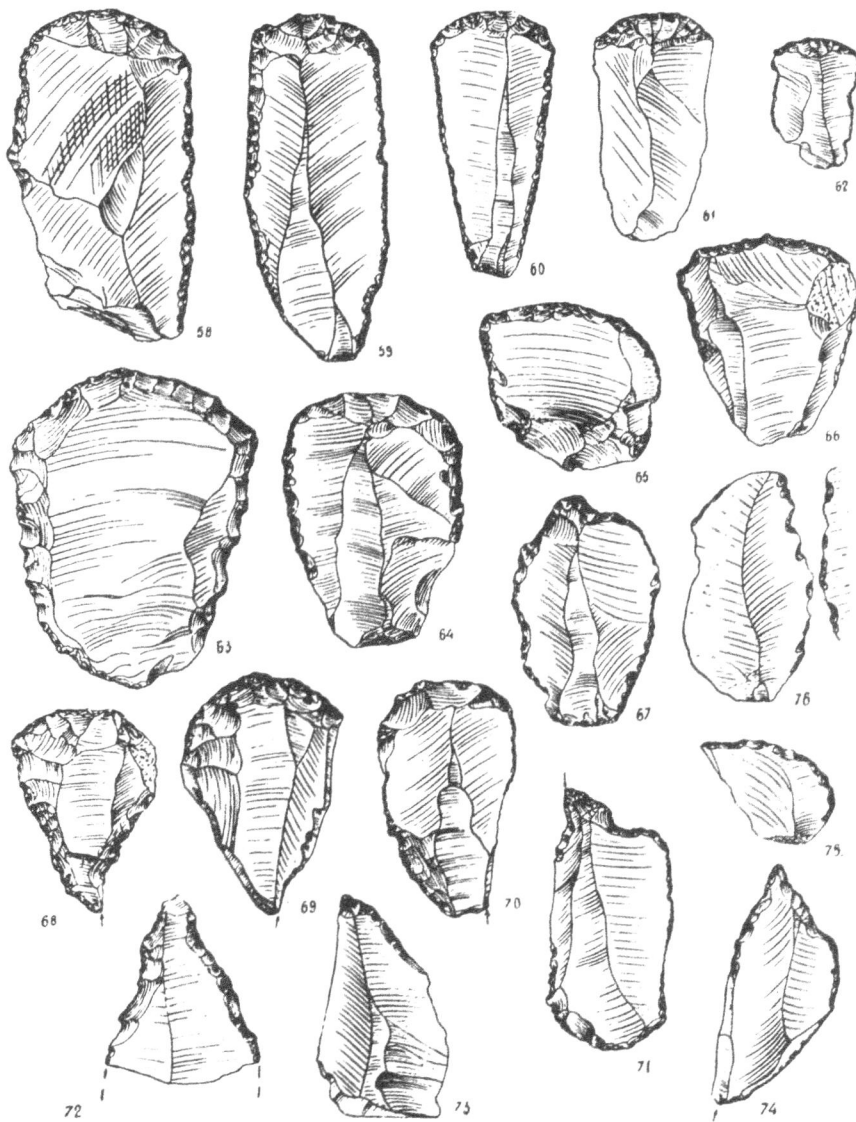

FIG. 37. Flint Industry of Borshevo I. After L. Sawicki.

Scrapers.

Scrapers are of various sizes, made on the end of retouched blades or wide flakes, often double, or in combination with *burins.*

Grattoirs.

Grattoirs made on retouched blades .... Nos. 58–60, 62
Grattoirs made on blade without retouch.............. No. 61
Grattoirs made on retouched flakes.................. Nos. 63–67
Grattoir and *burin d'angle*........................ Nos. 68, 70, 71
Grattoirs and *burin bec de flûte* ....No. 69

Percoirs.

Percoirs are not typical (nos. 72–74). No. 74 has *burin d'angle* on the opposite end.

In general, according to Ephimenko,[1] the flint industry of Borshevo I is quite similar to that of Kostenki I, consisting of the same general types of tools, but of a more archaic form. Thus in Borshevo I, the end scrapers have more often side retouch than those in Kostenki I. Likewise, Borshevo I abounds in massive, nucleiform *burins; point à cran* is smaller, less typical, nearer the type found in the early loess sites such as Willendorf and does not have Solutrean retouch.[2]

The Bone Industry.

Ephimenko [3] mentions the find of a fragment of a large pointed tool, made of mammoth ivory, but gives no illustration. S. N. Zamiatnin has found also three tiny mother of pearl discs,[4] perforated in the middle, which were apparently used for decoration. He also found a pendant made from the perforated incisor of a horse, and a tertiary shell filled with light red iron pigment which was previously well powdered and mixed with some other substance.[5]

Dating.

Savitsky considers Borshevo I as belonging to the Upper Aurignacian culture, especially on the basis of similarity with Du Ruth, Noaille, La Ferrassie and Font Robert. He does not differ essentially from Ephimenko, who thinks that Borshevo I belongs to the same age as the group of "Aurignacian-Solutrean" sites such as Willendorf, Predmost and others. He points out however that Borshevo I is somewhat earlier than any of the sites mentioned, because of the presence of the *Chatelperron* type of points which he considers characteristic for the Aurignacian.

4. KOSTENKI I

This very important station is situated in the village of Kostenki ("kost"—meaning "bone" in Russian; Kostenki-bone village), 35 km. south of Voronezh, on the right bank of the river Don. The history of this site is very romantic. The finding of many mammoth-bones is responsible for the name of the place. Local legends attribute the bones to the mythical giant animal "Inder," who lived under the ground, and whose bones, after his death, appear on the surface. Another legend describes, in picturesque detail, the

[1] Ephimenko, P. P., "Pre-Clan Society," Leningrad, 1936.
[2] Ephimenko sees in these points the prototypes of the Solutrean type of *point à cran* as found in Kostenki I.
[3] Ephimenko, P. P., "Pre-Clan Society," Leningrad, 1936, p. 360.
[4] Ephimenko, P. P., "Pre-Clan Society," Leningrad, 1936, p. 360.
[5] Ephimenko, P. P., "Pre-Clan Society," Leningrad, 1936, p. 360.

death of this monster.   The giant animal was passing the territory of Kostenki with his young and came to the river Don.   The river was too deep for him to wade across and he decided to drink it dry, which he did.   When he turned around to call his young, his body could not withstand the strain of so much water, and he burst, the bones flying in different places where they are now found.[1]

Another tradition held that the bones were the remains of the exhausted elephants of the army of Alexander the Great.

A report by De Bruin [2] attracted the attention of Peter the Great, and later Catherine II, who directed Gmelin, a member of the Academy of Sciences, to make an investigation. Gmelin visited Kostenki in 1768–69, and reported the presence of mammoth bones in the village and its vicinity.[3]

In the summer of 1879, I. S. Poliakov found many bones both in the secondary deposition, as well as *in situ*, where they were accompanied by cultural remains.   Two years later, the excavations of Poliakov [4] were continued by A. E. Kelsiev,[5] whose more extensive excavations revealed similar material.

Next followed a number of visits by various investigators, Stukenberg, Krishtophovich, etc., who, however, made only small trial pits.

In 1915, S. Krukovsky made extensive excavations, resulting in very rich material. S. N. Zamiatnin, in 1922, working on the adjoining site of Borshevo I, made a few trial pits in Kostenki I, and continued his work in 1923 with P. P. Ephimenko, who later devoted himself more systematically to the study of this site, returning there in 1926, 1931, 1933.

As a result of the various investigations the quantity of the material gathered by the investigators is large and complex in nature, but descriptions are not always available.   I will endeavor to describe the most important results of each investigator, as far as material at hand allows, before giving the general summary of the site.

Geology.—The vicinity of Kostenki is part of a large plateau, which forms the steep right bank of the Don in the province of Voronezh.   This bank is a formation of chalk covered by typical glacial clay with small boulders of diorite, granite, and quartzite.   The valley of the Don has two well pronounced terraces: the ancient high terrace, which is not flooded in the spring by the river, and the recent, lower terrace, which is annually flooded. The base of the terrace is sand, which is covered with thick deposit of reddish or grey clay, schistous in structure, sometimes with pebbles of white chalk.   All this is covered by the layer of humus, fully five feet thick.   The present village of Kostenki is situated on this ancient terrace.   In the upper layers of the terrace clay were found the palaeolithic remains.

Location.—A number of deep gullies divide the village into several sections:

    (*a*) Popov Log.

    (*b*) Pokrovsk Gully, the center of the village, the site of Kostenki I.

    (*c*) Anosov Log—Kostenki II.

    (*d*) Alexandrovsk Gully—Kostenki IV.

[1] Zamiatnin, S. N., "Notes on Prehistory of Voronezh Region," Voronezh, 1922, p. 6.

[2] De Bruin, "The Travel in Moscowie," 1701, p. 130.

[3] Gmelin, S. G., "Travel Through Russia for the Investigation of the Three Kingdoms of Nature," 1769.

[4] Poliakov, I. S., "Anthropological Trip to Central and Eastern Russia," Zapiski of the Imperial Academy of Sciences, Vol. XXXVII, pt. 1, St. Petersburg, 1880.

[5] Kielsiev, A. E., "The Palaeolithic Kitchen Refuses in Village Kostenki, Voronezh Province," Trudi of the Moscow Archaeological Society, "Drevnosti," Vol. IX, pp. 154–179.

Excavations of Poliakov and Kielsiev.—Although Poliakov did not find the spot indicated by Gmelin, he, nevertheless, discovered the palaeolithic remains in the center of the village at the mouth of "Pokrovsk Gully." Thus, he writes: "In the clay immediately below the humus I began to encounter the first typical flint flakes; then I noticed clay mixed with ashes and flint implements and, finally, with these, the fragments of the teeth and shoulder blade of a mammoth."[1] A considerable number of flint implements and bones of fossil animals were collected. This material is now preserved in the Museum of Anthropology and Ethnography of the Academy of Sciences in Leningrad.

Fauna.—Most of the bones found by Poliakov were those of a mammoth, chiefly ribs, shoulder blades, teeth and tusks. They were scattered, absent in some places and numerous in others. Sometimes they appeared to be purposely sorted, shoulder blades in certain places and tusks in others. The long bones had been broken to extract the marrow. The bones of other animals were rare, only teeth of bear and small *carnivorae*, marten, and skunk being identified. All these were concentrated near the remains of the open "hearth" and represent typical kitchen refuse. In spite of the small size of the excavations, bones of at least sixteen mammoths were found.

Flint Industry.—The tools were made of semi-translucent, smoke-colored or almost black flint which is found along the right bank of the Don River in the form of pebbles, covered with ochreous patination. Traces of similar crust could be seen on many implements. They are large and were obtained by a skillful stroke on the flat surface of a flint pebble so that the lower surface is usually flat, producing a flat, long, and wide implement with two cutting edges.

The flint implements showed a considerably advanced technique. The flakes were clean, the lower side smooth, the upper side worked, with several well-made facets. The edges were sharp, sometimes with fine secondary retouching. The best finds were near the "hearth."

Hearths.—At least three "hearths" are found, which are characterized by a mass of ashes, charred bones, and stones with traces of fire action. According to Ephimenko,[2] these "hearths" are the cultural accumulations on the sites of former dwellings. They are easily distinguished in the light ground by their dark coloring, and by the presence of bones. They have a characteristic reddish tint, due to the presence of powdered ochre, large supplies of which were found in Kostenki I.

No mention of worked bone is made by these investigators though the large number of gravers indicated a bone industry. This supposition was confirmed later.

In 1912 part of this collection was studied by R. R. Schmidt, who intended to publish the results. On the basis of his manuscript, Max Ebert wrote in 1921 concerning Kostenki I in "Sud Russland in Altertum."

Flint Industry.

The flint material from the Poliakov and Kielsiev collections was described by P. P. Ephimenko,[3] whose summary is given below.

[1] Poliakov, I. S., "Anthropological Trip to Central and Eastern Russia," Zapiski of the Imperial Academy of Sciences, Vol. XXXVII, St. Petersburg, 1880.

[2] Ephimenko, P. P., "Palaeolithic Site on Kostenki," Annual Report of the Russian Anthropological Society, Vol. V, St. Petersburg, 1915.

[3] Ephimenko, P. P., "Palaeolithic Site on Kostenki," Annual Report of Russian Anthropological Society, Vol. V, St. Petersburg, 1915.

Nuclei are absent in Poliakov's and rare in Kielsiev's collection, but Ephimenko considers this accidental.   One of Poliakov's nuclei is really a hammer stone.

Blades (*lames*) have either three or four facets, are irregular in form, and are rarely over 10 cm. long, the majority being smaller.   Many have retouched sides (saws) or are similar to "*lames à dos rabattus*" of Kielsiev's collection.   Really, they are nearer to "*pointes à cran atypique*" with certain massive, wide and short flakes reminding one of Levallois.

Scrapers.—Two types can be distinguished: (*a*) Convex scrapers (over twenty specimens) are made on the ends of wide flakes by means of curved retouch which at times is continued on the sides.   Often, the other end is made into a graver (Fig. 38, no. 12).   Straight end-scrapers and double scrapers were also found.   In general, they are different from Magdalenian forms and much nearer those of the Solutrean.[1]

(*b*) Concave scrapers were found, usually in the form of irregular flakes with small segment-like retouch on one side, more rarely on the end.

Perforators.—Perforators (*percoirs*) are absent in Poliakov's collection, and only two were found in Kielsiev's material in spite of their common occurrence in all Upper Palaeolithic stations.   This may be accidental, for in the case of Mezine (see pp. 335–346) most of the perforators were found in one spot of the site.

Gravers (*burin*) are represented by a numerous group, including such types as simple middle gravers (*burins ordinaires*), the side gravers (*burins lateraux*), the angle gravers (*burins d'angle*), and often *burins nuclei* on the end of broken blades (Fig. 38, nos. 6–9).   There is a certain negligence in their manufacture, distinguishing them from the fine forms of the Magdalenian type.   According to Ephimenko no "*burins busqué*" were found in Kostenki I, contrary to the assertion of R. R. Schmidt.

Points form the most interesting group, especially for the purpose of dating.   While the rest of the implements are retouched on one face only, the points have bifacial retouch (the lower side being worked near the tip) forming wide flakes, perhaps made by pressure.   In typical cases, these points are on large, massive blades, one or both ends of which are sharpened from the lower face by secondary retouch.   Usually, one end is flatter and sharper than the other.   Often, on one edge there is a notch (*cran*) quite analogous to the "*pointes à cran atypique.*"

Ephimenko further states that the absence of the *burin busqué*, the thumb scraper, and the *Châtellperron* points does not permit one to accept the Aurignacian dating of R. R. Schmidt.   The bilateral chipping of points and the types of notched points may be either proto-Solutrean or the degeneration of Solutrean technique.   The arguments pro and con are about equal.   Thus, the *pointes à cran atypique* are quite similar to those of La Ferrassie, La Font Robert, Grotto de St. Font, Noailles, Combe-Capelle, Mentone, Trou Magrite, Munzingen, Sirgenstein, Willendorf, and Hundssteig.   They could be considered the result of the evolution of the blades with retouched side of Aurignacian type.   On the other hand in La Madeleine and other stations, Peyrony found many similar points, suggesting a degeneration of the typical Solutrean in pre-Magdalenian (IV–?) times.

Max Ebert considers Kostenki I as belonging to the transition period between the Aurignacian and Solutrean, of the so-called Font Robert culture, this agreeing with the proto-Solutrean theory of Ephimenko.

---

[1] Ephimenko tries to indicate post-Aurignacian age at these scrapers, using the term Solutrean in a temporal sense.   (E. G.)

Very good illustrations of flint industry from Poliakov and Kielsiev excavations are given by L. Sawicki, who had the opportunity in 1924 to study and make drawings of the typical implements from the material of the Moscow State Historical Museum. He reproduces after Ephimenko from Poliakov's collection the following types:

Poliakov Excavations (Fig. 38, nos. 1–13)

Small backed blades . . . . no. 1
Small *pointe à cran* . . . . : . . . . no. 2
Large *pointe à cran* . . . . nos. 3–7
Burin d'angle . . . . nos. 8–10
Burin bec de flûte . . . . no. 10
Burin plane . . . . no. 11
Grattoir—burin bec de flûte on massive flake . . . . no. 12
Grattoir on massive wide blade . . . . no. 13

Kielsiev Excavations (Figs. 39–44, nos. 1–57)

Nos. 1–9.   Backed blades of la Gravette type.
Nos. 10–11.   *Pointe à cran atypique.*
No. 12.   Long wide blade pointed by retouch on the right edge. The lower face presents the Solutrean type of retouch, from both edges. The lower half of the right edge is roughly retouched, suggesting *pointe à cran.* The bulb is removed by retouch of the lower face.
No. 13.   *Pointe à cran atypique.*
No. 14.   Blade with partial bifacial retouch. (*Pointe à cran atypique ?*).
No. 15.   Fragment of retouched blade with slight inverse retouch.[1]
No. 16.   *Burin d'angle-grattoir* with left edge retouched forming a notch.
No. 17.   Asymmetrical blade with partial retouch on both faces. Fine steep retouch on the right edge forms a notch. *Burin* stroke on the end (?).
No. 18.   Fragment of a long blade retouched on one side. The bulb is removed by flat retouch on the lower face.
No. 19.   Same as No. 18, with truncated end retouch. (*Burin* stroke ?).
No. 20.   Leaf-shaped blade, with partial bifacial retouch, mostly on the lower face, on each end. Recalls proto-Solutrean technique.[2]
No. 21.   Asymmetrical blade with retouch principally on the lower face.
Nos. 22–23.   Pointed blades, bifacial retouch toward the point.[3]
No. 24.   *Burin plan* (?) on a long blade.
No. 25.   Fragment of blade with alternate bifacial retouch.[4]
No. 26.   Leaf-shaped blade with broken point (?) retouched on both edges.
No. 27.   Blade with partial bifacial retouch on both edges.
No. 28.   Wide flake retouched on left edge, both ends scaled by use.

[1] Compare Sawicki, L., "La Grotte Nietoperzowa à Jerzmanowice près Ojcow (District Olkusz)," Revue Archeologique Polonaise, Vol. III, Pt. 1, pp. 1–8, Plate II, no. 11.
[2] Compare Sawicki, L., "La Grotte Nietoperzowa à Jerzmanowice près Ojcow (District Olkusz)," Revue Archeologique Polonaise, Vol. III, Pt. 1, pp. 1–8, Plate III, no. 4, 4a.
[3] Compare Sawicki, L., "La Grotte Nietoperzowa à Jerzmanowice près Ojcow (District Olkusz)," Revue Archeologique Polonaise, Vol. III, Pt. 1, pp. 1–8, Plate IV, nos. 13, 14.
[4] Ibid., Plate II, no. 14.

No. 29.   Long blade beautifully retouched on both edges.[1]

No. 30.   Fragment of asymmetrical blade with alternative retouch and a notch formed by bifacial retouch on one edge.

No. 31.   Long blade retouched along the left edge, one end forms the end scraper, the other shows scaling or inversed retouch on the lower face.

Nos. 32–33.   Tools scaled by use (*utiles écaillées*).

Nos. 34–37.   End scrapers on a broken blade.

No. 38.   Almost rectangular, double end scraper with fine retouch along both edges, made on a wide, flat blade.   The lower face shows the reduction of the bulb.

No. 39.   Tool made of wide flake, completely retouched on one face only, with secondary peripheral retouch.

No. 40.   The same, one end showing traces of retouch or use on the lower edge.

No. 41.   Large flake.

No. 42.   *Burin d'angle.*

No. 43.   *Burin d'angle* with the *burin* stroke on the opposite end (?).

No. 44.   Multifacetted *burin* on broken retouched blade.

No. 45.   *Burin d'angle* and multifacetted *burin* on the other end.

No. 46.   *Burin d'angle* and *burin plan* on retouched blade.

No. 47.   Same as 45.

No. 48.   Nucleiform multifacetted *burin.*

No. 49.   *Burin d'angle.*

No. 50.   Multifacetted nucleiform *burin.*

No. 51.   *Burin* on the end of a retouched blade.

Nos. 52–53.   Double *burin.*

No. 54.   Multifacetted (*plan, d'angle,* etc.) nucleiform *burin.*

No. 55.   Multifacetted nucleiform *burin.*

Nos. 56–57.   Nuclei, probably re-utilized as *grattoirs.*

No. 58.[2]   Stone figurine representing mammoth.

No. 59.   Small ivory female figurine of Kozlovsky's excavations.

No. 60.   Ivory tool.

No. 61.   Bone implement of unknown use.

Sawicki[3] calls nos. 10–21 *pointes à cran*, distinguishing several types, characteristic for Kostenki.   He uses the term of *Font-Robert* as the equivalent of Proto-Solutrean in describing nos. 4, 7, 8, 22, 25.   It is possible that some of his *pointes à cran* are very distant eastern relatives of the typical implements.   He considers Kostenki as the eastern phase of the middle European Solutrean with Hungary as the center.

Krukovsky's Excavations.—This Polish archeologist made in 1915 large excavations in Kostenki I, finding the same general conditions as did the previous investigators.   He has also found traces of "a hearth" with a number of flint implements and bones of fossil animals.   It is very unfortunate that the results of his excavations are as yet unpublished as his material is very rich, consisting of several thousand flakes and several hundred finished tools.   It was preserved in the Voronezh State Museum.

[1] Ibid., Plate II, no. 15.

[2] Nos. 58–61 are from P. P. Ephimenko collection, according to L. Sawicki.

[3] Sawicki, "Le Grotte Nietoperzowa à Jerzmanowice près Ojcow (District Olkusz)," Revue Archeologique Polonaise, Vol. III, Pt. 1, pp. 1–8.

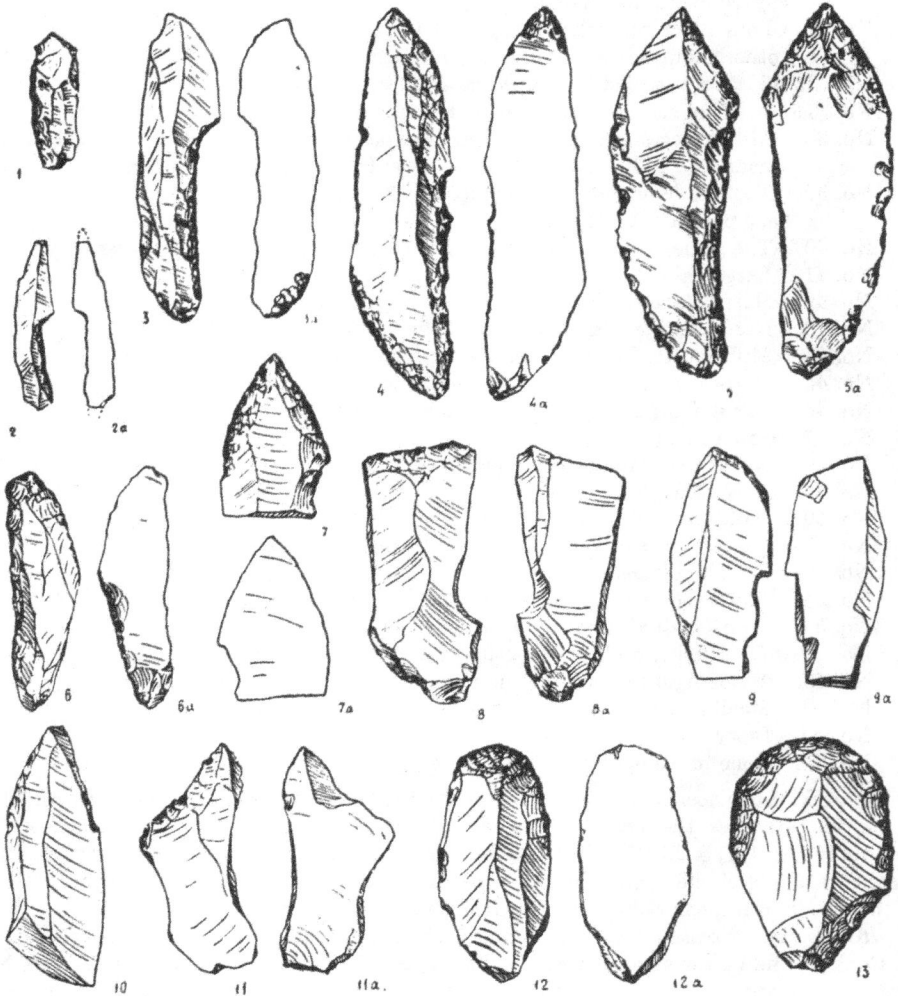

FIG. 38. Stone Industry of Kostenki I. Poliakov's Excavations. After L. Sawicki.

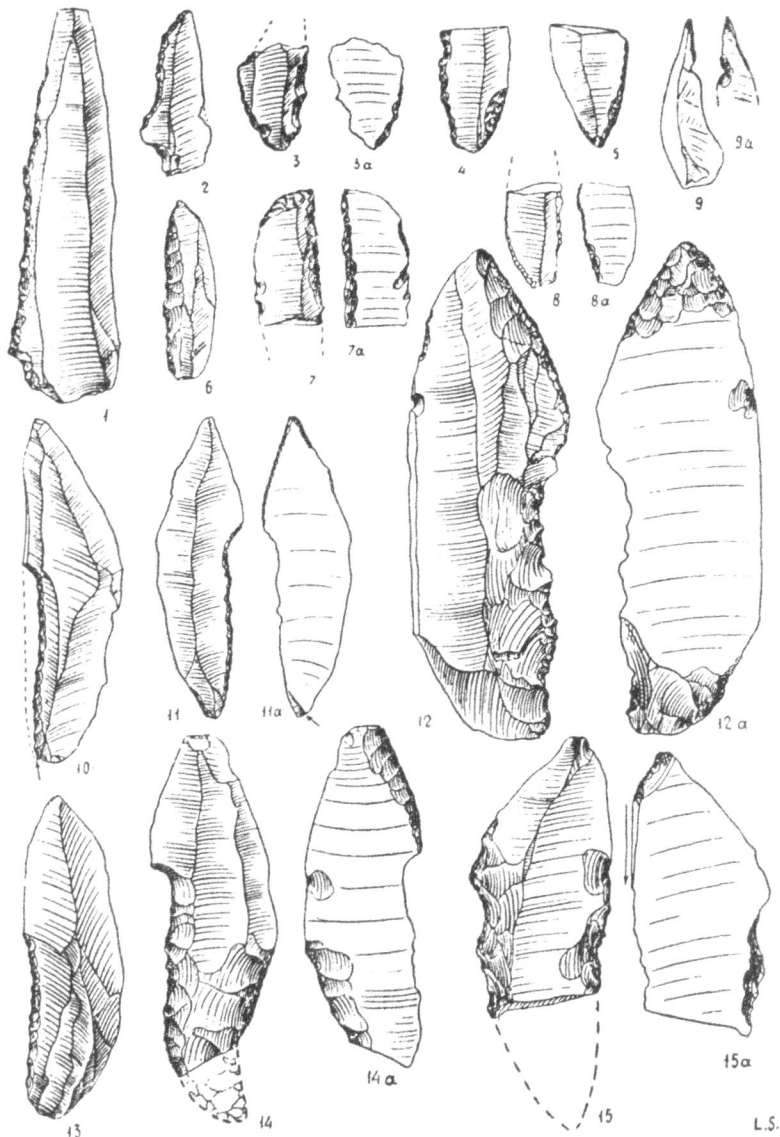

FIG. 39. Stone Industry of Kostenki I. Kielsiev's Excavations. After L. Sawicki.

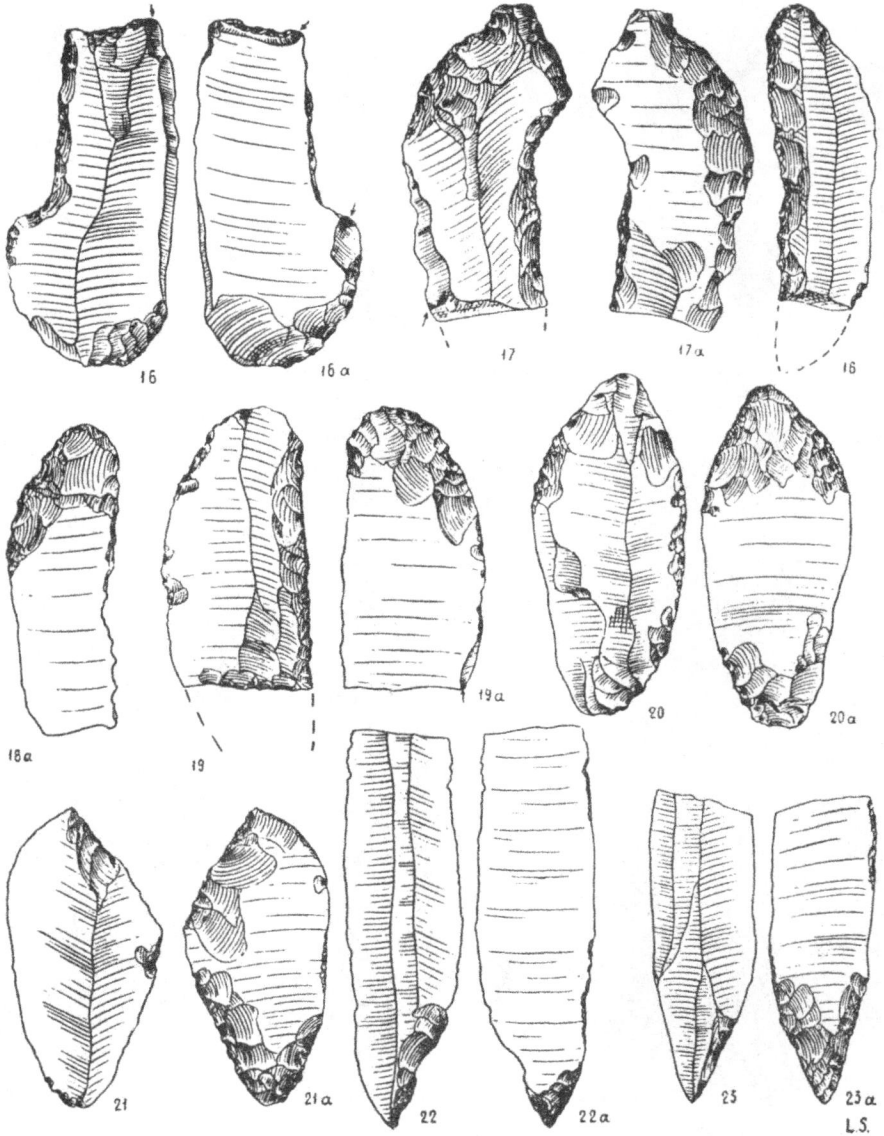

FIG. 40. Stone Industry of Kostenki I. Kielsiev's Excavations. After L. Sawicki.

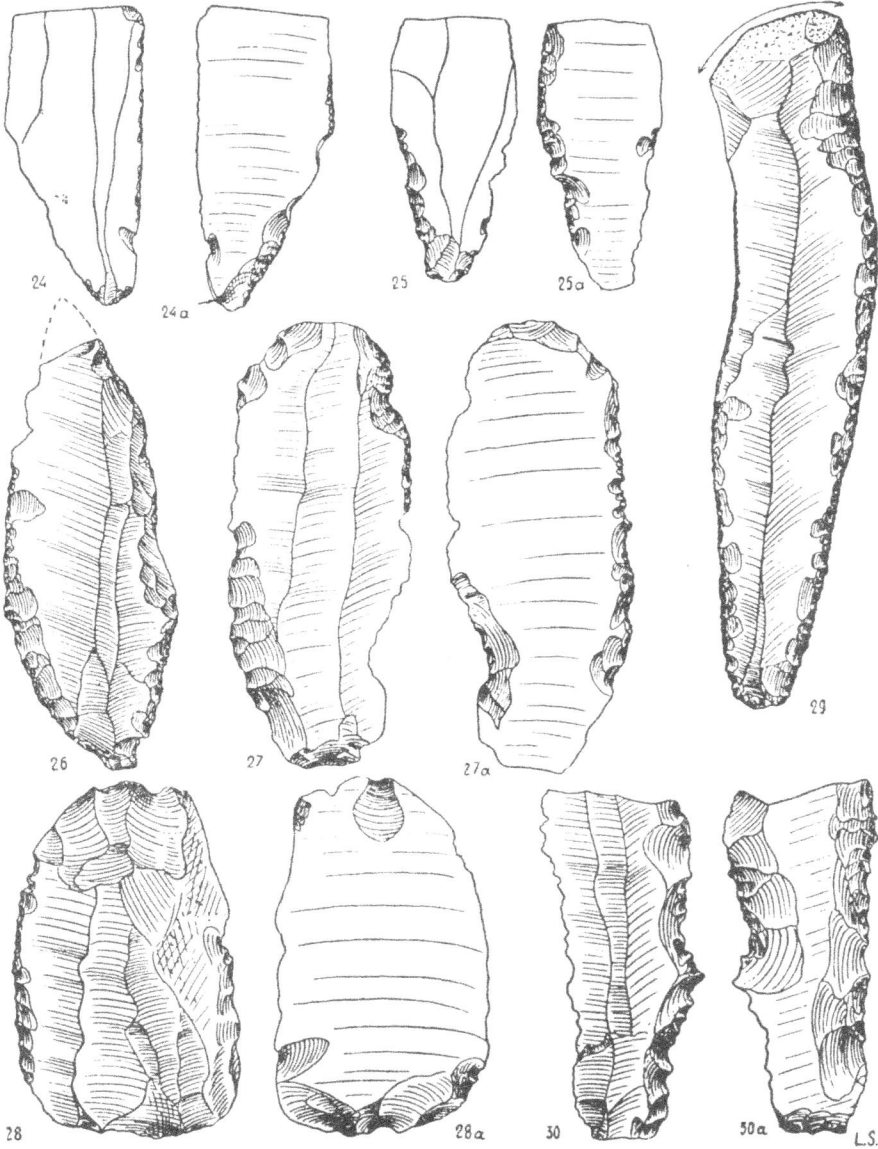

FIG. 41. Stone Industry of Kostenki I. Kielsiev's Excavations. After L. Sawicki.

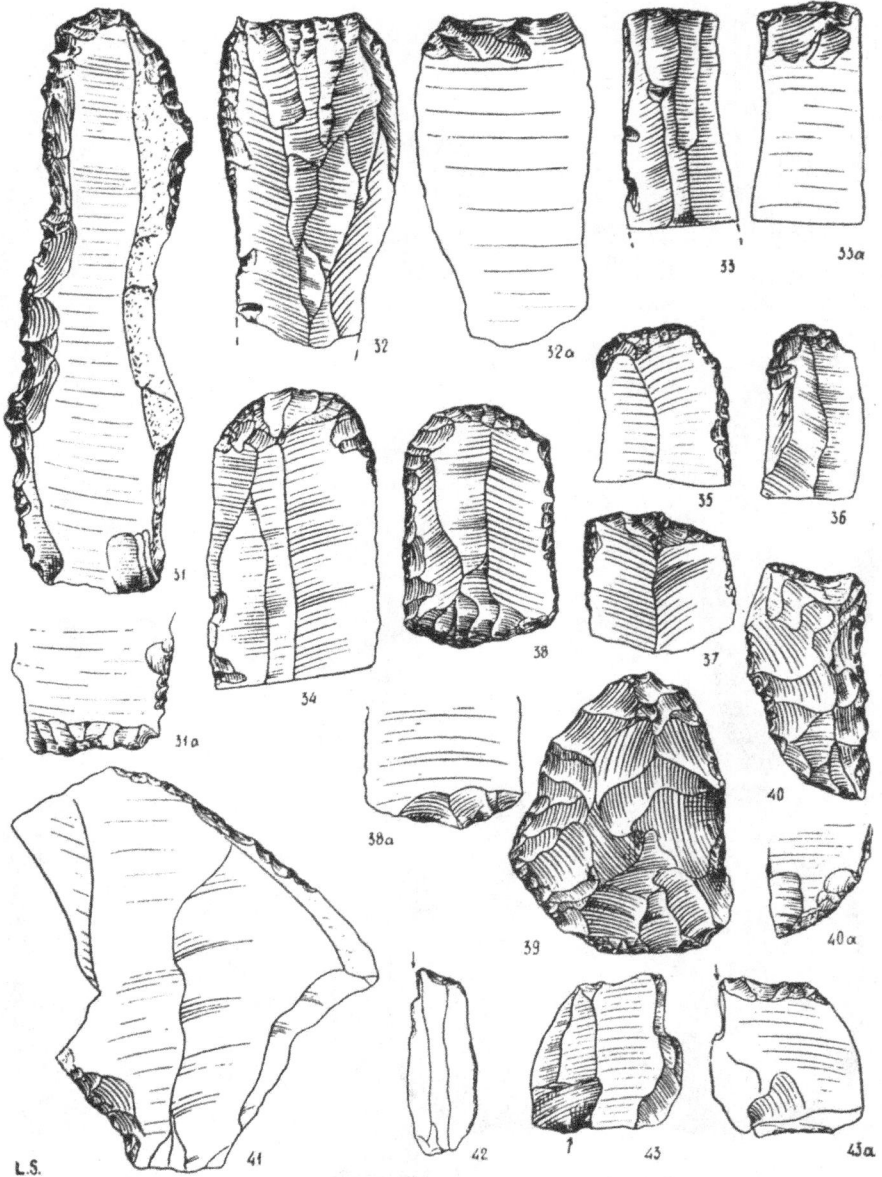

Fig. 42.   Stone Industry of Kostenki I.   Kielsiev's Excavations.   After L. Sawicki.

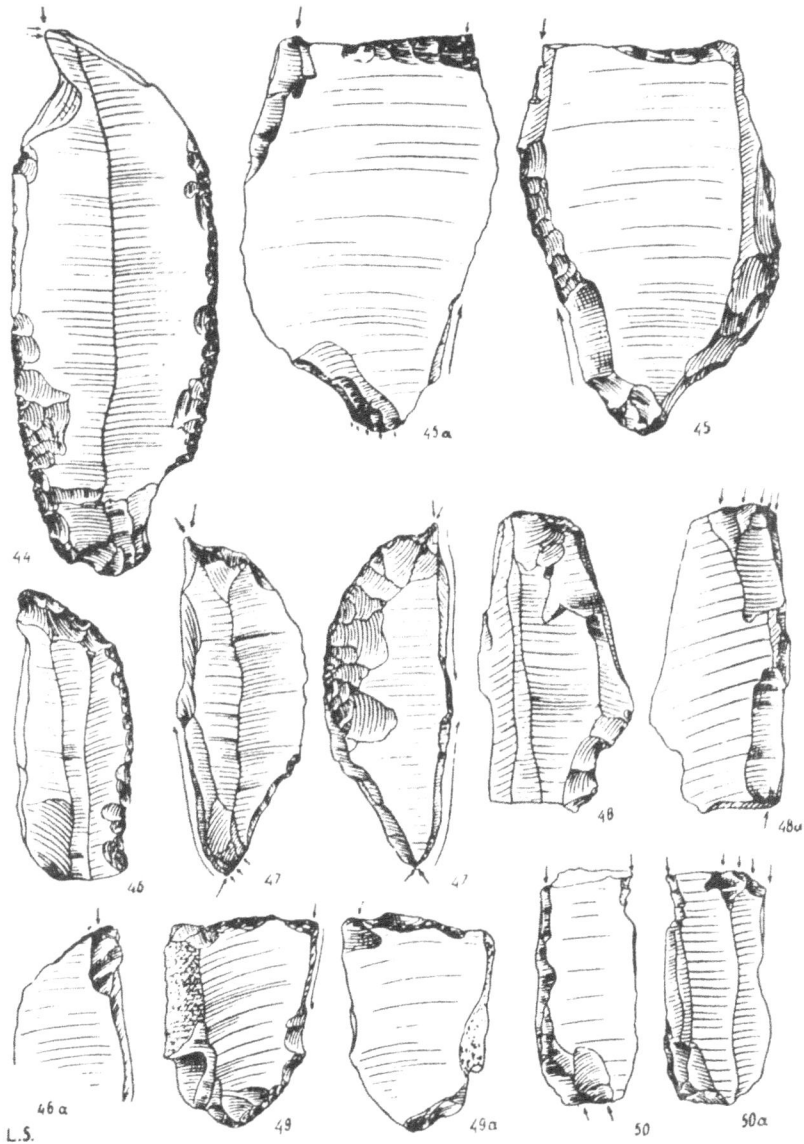

FIG. 43. Stone Industry of Kostenki I. Kielsiev's Excavations. After L. Sawicki.

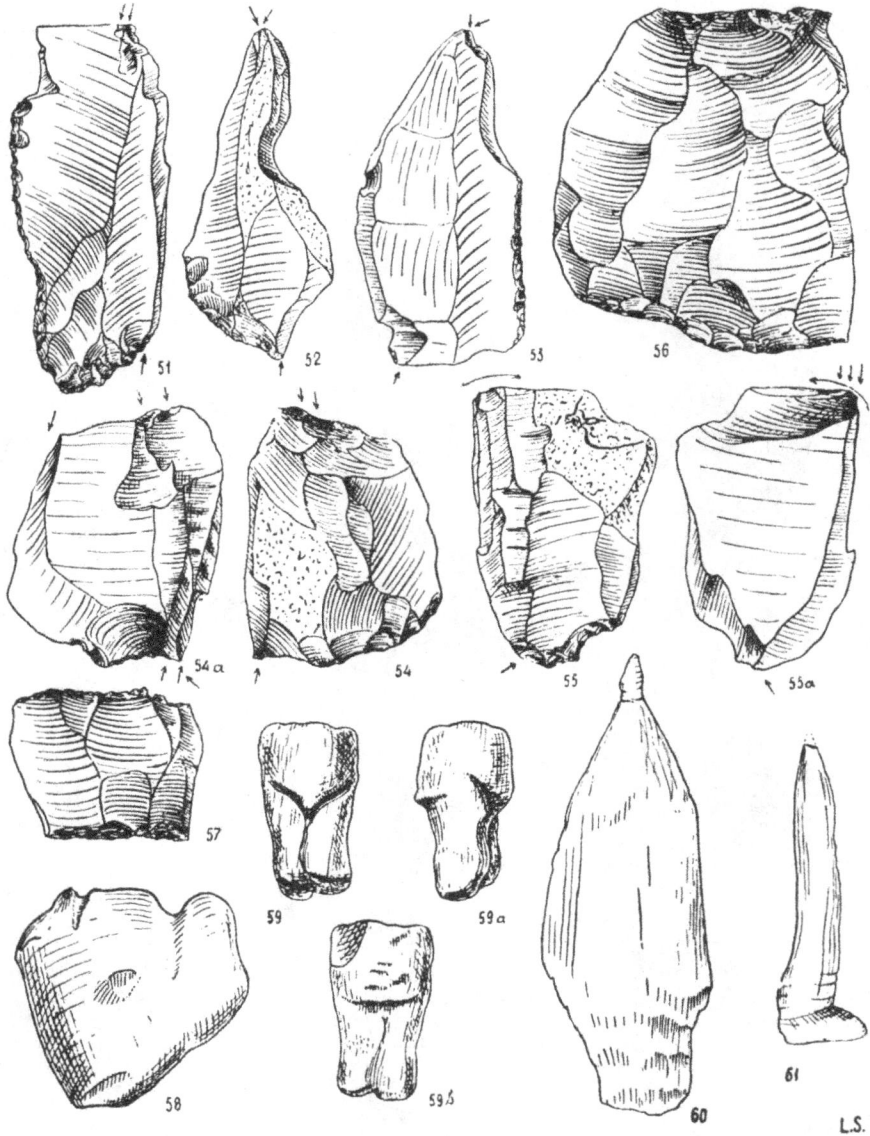

FIG. 44. Stone and Bone Industry of Kostenki I. Nos. 51–57 of Kielsiev's Excavations; Nos. 58–61 of P. P. Ephimenko. After L. Sawicki.

PLATE XX

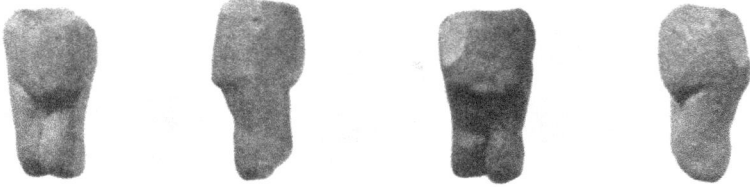

a. KOSTENKI I. FRAGMENT OF STONE (MERGEL) FEMALE FIGURINE FOUND BY KRUKOVSKY. Photo from the cast by the author. 1/1.

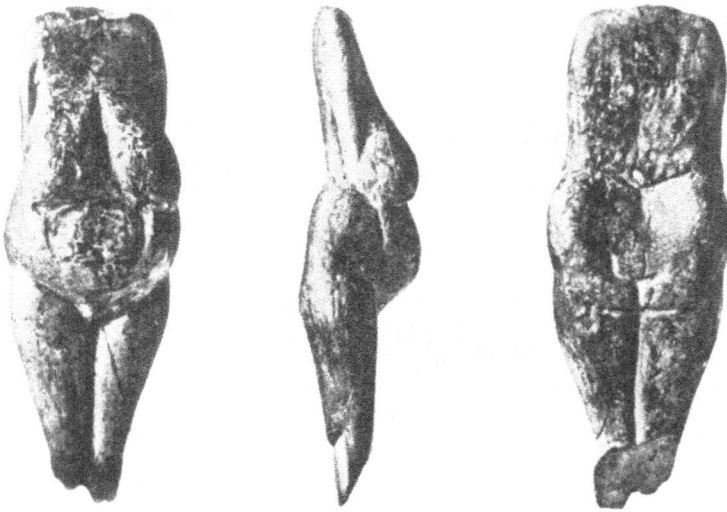

b. KOSTENKI I. IVORY FEMALE FIGURINE FOUND BY P. P. EPHIMENKO IN 1923. 1/1. After Ephimenko.

According to Zamiatnin,[1] the material for the stone industry of Krukovsky's excavations is dark, almost black, flint, found at present in the nearby chalk outcrops. It is covered by light blue patination. Chips of brown ferrous flint are rare. The tools were manufactured on the spot, as evidenced by the number of chips. Zamiatnin points out among other forms large knives of Magdalenian type, scrapers, small points, very fine points of Solutrean type with a notch in the lower part, and "even some typical Aurignacian tools such as high scrapers."

There are also small prismatic flakes with three or four facets retouched on one side. Contrary to the assertion of R. Schmidt who calls them "*petites pointes,*" "*Borsteinspitzen,*" or "*lames à dos rabattu*" they are not finished tools. Even though some of them may have been utilized, they were simply the results of manufacturing gravers. In order to obtain a massive point, a long splinter was detached from the side of a blade by a longitudinal, slightly oblique, blow (*coup de burin*). What appears to be retouch on these splinters is a part of the retouch of the mother blade, made not only on the end but also on the side in order to obtain a more regular facet. Similar flakes were observed in Mezine.

Bone Industry.—Among the charred and purposely split bones of a mammoth, there was found a jaw with scratches and cuts, made by stone knives or scrapers, in the removal of the flesh.

Female Figurines.—Of special interest is the find of a small statuette made of "*mergel*" [2] (Plate XX, *a*), representing a nude female.[3] The upper part including head and thorax is not preserved. The legs are present only as far as the knees which are slightly bent, and the protruding abdomen sags down to the well-pronounced pubic triangle.

Excavations of Ephimenko and Zamiatnin, 1923, 1926, 1931.—For a period of seven years no further work was done in Kostenki. Then, in 1922, S. N. Zamiatnin, who at that time was Curator of the Archaeological Section of the Voronezh Museum, made a few trial excavations while working in the Borshevo site. Finally, in the summer of 1923, very large excavations were made by P. P. Ephimenko with S. N. Zamiatnin at Kostenki I and Borshevo I. Ephimenko continued these in 1926 and 1931.

The trench made by the investigators on the spot of Kostenki I (Pokrovsky Gully) gave very interesting results. Near the former excavations of Krukovsky, there was found the border line of the cultural layer, which consisted of loam and contained many flint implements. Nearer to what was perhaps the center of a "hearth," the excavations revealed more and more interesting finds. On the floor of the square pit, 0.5 m. deep, which widened toward the bottom and was undoubtedly purposely dug out by Palaeolithic man, were found mammoth tusks and the ribs of smaller animals.

Ephimenko [4] states that these bones were stored as material for manufacturing bone tools, and include the tips of mammoth tusks, several horse ribs, some showing traces of breakage at the cut. Beside these were found elongated tools made of longitudinally split horse ribs with spatulated ends, so-called "smoothers." There was also a small point or awl of mammoth ivory, round in cross-section.

---

[1] Zamiatnin, S. N., "Notes on the Prehistory of Voronezh Region," Voronezh, 1922, p. 6.

[2] Marly limestone.

[3] This is really the first sculptured representation of woman found in Russia as Mezine statuettes are variously interpreted. (E. G.)

[4] Ephimenko, P. P., "Pre-Clan Society," Leningrad, p. 357, 1936.

The most interesting object was an ivory female figurine, of very fine workmanship (Plate XX, *C*). "This statuette," writes Ephimenko, "executed very artistically, permits us for the first time to link Russian finds with the series of pre-Magdalenian statuettes of Western Europe. It is 9.05 cm. long; the head is lost, having evidently been broken before its discovery. The legs end just below the knees and taper down to a point. This is often the case in similar figurines, perhaps for the purpose of fixing them on a stand or simply on the ground. The figure is represented in full length and treated very formally. As is seen from the photograph, the physical type is represented realistically. The large, pear-shaped breasts and enlarged abdomen suggest the over-development of the fatty tissues. However, the statuette has no steatopygy, thus differing from Mentone, Brassempouy, Willendorf and others. At the same time there is a certain slimness, with the thin arms delineated along the sides of the body and folded on the abdomen; the narrow thighs with a wide pelvis, the definitely indicated umbilicus and a line down the back, give the characteristic lines of a corpulent body." [1]

Slightly above the breasts there is a thin engraved triple line with oblique notches, which according to Ephimenko,[2] represent a ribbon or decoration. He considers that this statuette represents a different physical type from Willendorf, being less corpulent.

The later investigations of P. P. Ephimenko in 1923, 1926, and 1931 [3] yielded new and very important facts indicating the existence here of a very developed bone industry, objects of art, and traces of habitations in the shape of hearth pits, etc.

No complete description of the excavations of 1923, 1926, and 1931 has appeared as yet. However, the results of the 1931 season were especially interesting. According to Ephimenko, Kostenki I was a very large hunting camp. In 1931, an area of 145 sq. meters was excavated, and almost all of it was occupied by a large "dwelling" not less than 15 meters long. The extent of it could easily be determined by a more or less thick layer of kitchen refuse, consisting of broken bones of animals, charcoal, worked flint, etc. A considerable portion of this layer had a pronounced red color due to the presence of ochre. In spots where the pits were located, this layer reached a thickness of 40–50 cm.

Ephimenko distinguished two cultural complexes here, though not very sharply delineated, consisting of large depressions in the loess, oval in form with the long diameter up to six meters, the bottom being 40 cm. below the surface of the floor of the "dwelling." In both groups were found several hearths of regular round form, the walls of which were burned red, and in one case a trace of "shoulder" was preserved. These hearths were filled with a thick layer of charred bones, thus confirming according to the investigator, the theory that the accumulations of mammoth bones, so usual in Russian Palaeolithic sites constituted the supplies of fuel.

A number of various depressions and pits dug out in the floor, were used as storage pits and for other domestic purposes. There were several holes for posts supporting the roof, similar to those found by Bayer in Lang-Mannersdorf.

[1] Ephimenko, P. P., "The Statuette of Solutrean Epoch," Materials for the Ethnography of Russia, The State Russian Museum, Vol. 3, Pt. I, 1926, pp. 139–142.

[2] Ephimenko, P. P., "The Statuette of Solutrean Epoch," Materials for the Ethnography of Russia, The State Russian Museum, Vol. 3, Pt. I, 1926; also in "Pre-Clan Society," 1936, p. 315.

[3] Ephimenko, P. P., "Pre-Clan Society," Leningrad, 1936, p. 352. Also: "Kostenki I," Soobschenia of the State Academy for the History of Material Culture, 1931, no. 11–12, pp. 59–60.

PLATE XXI

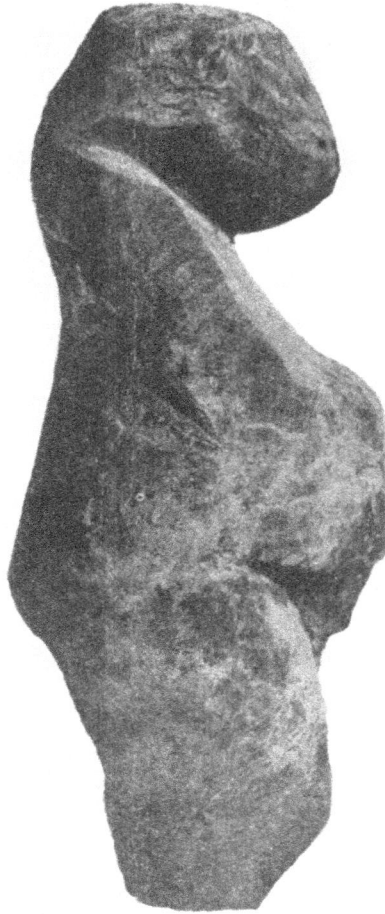

KOSTENKI I. LARGE FEMALE FIGURINE OF SOFT STONE, FOUND BY P. P. EPHIMENKO IN 1931. After P. P. Ephimenko.

Some pits were larger towards the bottom. Ephimenko interprets them as primitive ovens for the roasting of enormous pieces of meat; others may have been used as storage pits.

Next to the fireplace, in the northwestern corner of the dwelling, he has discovered a niche-like, round pit, 0.80 meters wide, 0.50 meters high, and 1 meter deep.[1] In the process of clearing, it became evident that it did not contain any refuse, thus differing from the storage pits. On the bottom of it were found several interesting objects, figurines, etc.

Female Figurines.—At the entrance to this niche Ephimenko found a large ivory statuette representing a nude female (Fig. 45). It was in a poorer state of preserva-

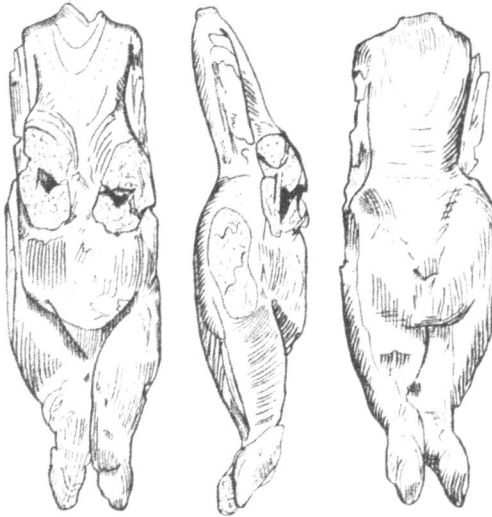

Fig. 45. Large Female Figurine of Mammoth Ivory Found in Kostenki I. 3/4 natural size. Excavation of P. P. Ephimenko after Ephimenko.

tion than the one found in 1923, yet it presents some very interesting details. There is an engraving, representing a necklace, which hangs down the shoulders and breasts. The head is missing, but on the upper part of the back where it joins the neck, traces of series of small elevations may be observed. These show that originally there was a cap or hairdress like those of Willendorf, Gagarino and others.

Next to this statuette, there was found another one, broken in four pieces. It was made of stone, much larger, some 15 cm. in height, very massive, and carved in rather schematic fashion (Plate XXI). On the whole it gives the impression of a very primitive looking creature. When viewed in profile, the thick neck and protruding supraorbital ridges give the impression of a reconstruction of Neanderthal Man. Ephimenko [2] compares it with the stone statues [3] of South Russia.

[1] According to Ephimenko it is 1.60 meters. "Kostenki I," Soobschenia of the State Academy for the History of Material Culture, 1931, nos. 11–12, pp. 59–60.

[2] Ephimenko, P. P., "Pre-Clan Society," Leningrad, 1936, p. 315.

[3] So called "Babas." (E. G.)

By the side of the last figurine was found an object carved out of mammoth tusk, in the shape of a handle, terminating in a round spherical head. This is, perhaps, a rough and schematic representation of a human being. Ephimenko speaks of a number of fragments, heads, parts of torsos, and several small female figurines made of soft stone, one of which is decorated on the back by a series of vertical lines like those of the Lespugue statuette. A sandstone slab with an engraved representation of a nude female was found in another depression dug out in the floor. During the same season a number of pendants and different figurines were discovered, one of which is made of stone and represents a mammoth, showing a well defined head and legs. In this respect, it is similar to the one found at Predmost. Among other examples of bone work, Ephimenko [1] speaks of a well preserved *baton de commandement* of reindeer horn found in 1931 and of a large tool with a pointed end made of mammoth tusk which he calls a digging tool (Fig. 44). According to a Russian newspaper report, some forty-two bone and stone carvings were found, but as yet no complete list has been published.[2]

Dating.—The Aurignacean-Solutrean of Ephimenko (proto-Solutrean of some authors) or Eastern phase of Solutrean seems to be the acceptable dating of Kostenki I.

5. GAGARINO

The site of Gagarino [3] was discovered in 1928 by S. N. Zamiatnin in the village of Gagarino, formerly of Lipetsk district Tambov Region, on the river terrace of the Don.

According to Ephimenko, the cultural deposit discovered here was concentrated on a small area, under the humus, in a layer of loess-like loam and has the appearance of a flat lense, 4.50–5.50 meters in diameter, oval in form.

The area of cultural remains has sharply outlined borders, beyond which the finds disappear, except for a small quantity of pieces of flint and bone, probably brought there by small rodents. On the periphery of this accumulation were rows of slabs of local Devonian limestone, some partly remaining in their original vertical position. The thickness of the cultural layer averaged 0.40–0.50 meters, and, as is usually the case with the Russian sites of the Upper Palaeolithic period, the cultural layer has a rose shade, due to a considerable admixture of ochreous pigment, mostly marked in the lowest part (0.08–0.12 meters) around the lower edge of the stratum.

This "hearth," which in every detail is similar to one which is described by Ducrost

---

[1] Ephimenko, P. P., "Pre-Clan Society," Leningrad, 1936, p. 248.

[2] According to Field, H., and Prostov, E. ("Recent Archaeological Investigations in the Soviet Union," A. A., Vol. 38, No. 2, pp. 285), "Two ancient dugouts were disclosed in Kostenki I. Besides flint implements and bones of *domestic* animals, stores of minerals used possibly for dyeing were found. Fragmentary female figures of marl, also those of animals, probably connected with totemic beliefs, came to light. A mammoth, the head of a cave lion, a bear, and a *camel* could be identified. This is first discovery of a camel effigy reported from a Palaeolithic site." The domestic animals and camel's effigy are most puzzling finds. (E. G.)

[3] Ephimenko, P. P., "Pre-Clan Society," Leningrad, 1936, pp. 291–293.

Gorodzov, V., "The Results of Archaeological Activities in U.S.S.R. during 1917–1930." Manuscript.

Golomshtok, E. A., "Trois Gisements Paléolithique Supérieur, en Russe et Sibérien," L'Anthropologie, T. XLIII, 1933.

Zamiatnin, S. N., "Gagarino," GAIMK, 1935, not available to me, review in Antiquity, June, 1936, pp. 242–245.

PLATE XXII

a. REMAINS OF DWELLING. BEFORE CLEANING.

b. REMAINS OF DWELLING. AFTER CLEANING. Photograph of S. Zamiatnin.

in Solutré, represents, in the opinion of P. P. Ephimenko, the remains of a dwelling.[1]   He believes that "in the majority of the open stations of the Aurignacian-Solutrean period there are indications of the existence of massive permanent huts, of the type of semi-subterranean dwellings of tribes of the polar regions.   They are found in Timonovka (Central Russia), in Predmost (Moravia), where one was incorrectly called a 'common grave' in Lang-Mannersdorf (Lower Austria), in Solutré, in Fourneau du Diable, etc."   He further states that "besides the sites, where the walls of the foundations dug out in the ground were in most cases strengthened by stone slabs or mammoth bones and, therefore, could be easily detected, there are in Eastern Europe many stations, where the walls of dwellings were also dug out of the ground but strengthened with wood.   These often pass as "hearths," and usually have the appearance of large spots, generally oval in shape, sharply outlined, and abundant in refuse." [2]

Fauna.—According to Gromov [3] the remains of the following animals were found:

1. *Elephas primigenius*....many.
2. *Rhinoceros tichorhinus*..few.
3. *Bos* (sp.)............very few.
4. *Vulpes vulpes*...........
5. *Alopex lagopus*.........many.

Dwellings.—We have a similar condition in Gagarino, where S. N. Zamiatnin discovered the lower part of a habitation in the shape of a shallow pit, the walls being strengthened with a row of stones (Fig. 46).   A considerable quantity of large animal bones, the long bones of rhinoceros and mammoth and the tusks and lower jaws of the latter, were encountered on the periphery of the cultural layer and were probably used for the same purpose as in Predmost (Plate XXII).   The wooden part of the structure, of course, was not preserved.   Zamiatnin does not mention the hole for the central post supporting the roof because this pole was situated in the center of the hut; when some domestic repairs were made on the farm where this station is situated, this part of the deposit was destroyed previous to the excavation.

Thus, this construction of stones (Plate XXIII), evidently the foundation and walls of the hut, gives us the possibility of forming an idea of the type of dwellings used by Palaeolithic man when he could not use caves and natural shelters.

The space inside the pit was filled with all sorts of remains of habitation, small fragments of bone, charred bones, pieces of red ochre, flint flakes, and a considerable quantity of flint and bone implements.   Altogether Zamiatnin found six hundred flint tools there, more than one thousand flakes, and a large quantity of remains of flint manufacturing.

Among the bones bearing the marks of industry, Zamiatnin found pieces of mammoth tusk in preparatory stages of manufacture, bones with traces of work, dart points, awls, needles, needle cases, pendants carved out of mammoth tusk, perforated teeth of arctic fox, and, finally, seven figurines made of mammoth tusk (Fig. 46, Plates XXIV, XXV).

Two figurines were found intact, a third well preserved, except for the loss of its legs, three others only in fragments and a seventh in unfinished form.   They are small, and all

[1] In the shape of a shallow pit with walls strengthened by stones.   (E. G.)

[2] P. P. Ephimenko, "Female Statuettes of Aurignacian and Solutrean epochs," Soobschenia of the State Academy for the History of Material Culture, pp. 4–5, July, 1931.

[3] Gromov, V. I., "Geology and Fauna of the Palaeolithic Period in U. S. S. R.," Problems of the History of Material Culture, No. 1–2, Leningrad, 1933.

evidently represent females. Two of them (one whole and one with broken legs) have much in common with the famous Willendorf statuette of Upper Austria, except for the treatment of the arms. The thin arms of Willendorf are placed on the breasts, while in one of the Gagarino figurines the arms are at the sides, and in the second they are stretched forward, the palms opened and joined in such a way as to suggest a praying attitude, which gives it an unusual aspect.[1] It is interesting to note that the second whole statuette represents a

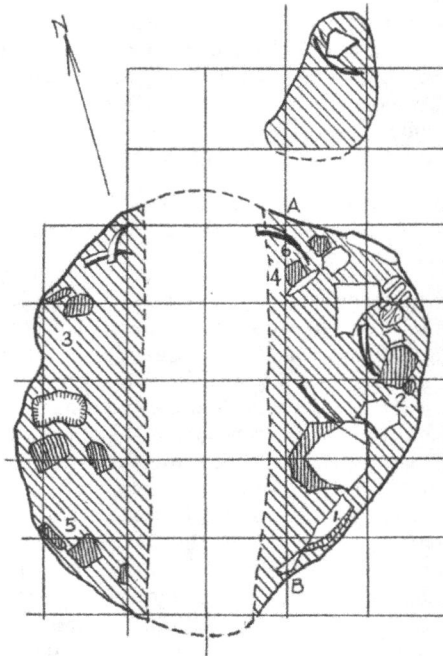

FIG. 46.   The Ground Plan of the "House Complex" in Gagarino.   1–6, the position of the statuettes.   A–B, sectionline.

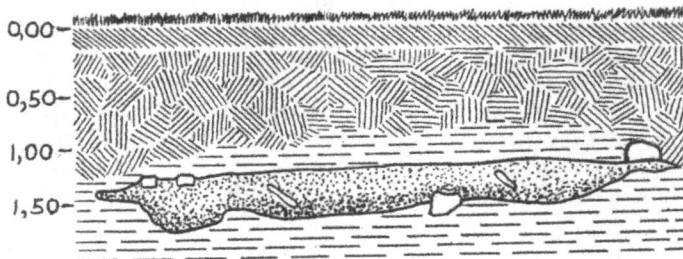

FIG. 46a.   The Section Along the A–B Line of the "House Complex" in Gagarino.   After P. P. Ephimenko, "Pre-Clan Society," Fig. 89a, 89b.

[1] Compare with the bas-relief of Laussel, France.   (E. G.)

PLATE XXIII.

THE STONE FOUNDATION AT DWELLING.   Photograph of S. Zamiatnin.

PLATE XXIV.

GAGARINO. MAMMOTH TUSK FEMALE FIGURINES. Photograph of S. Zamiatnin.

PLATE XXV.

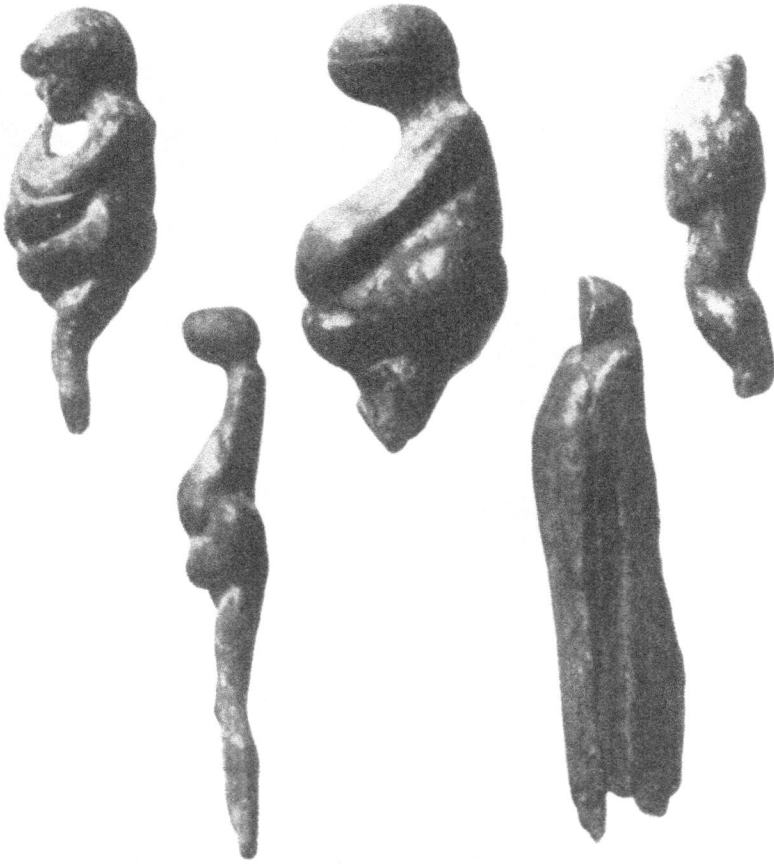

GAGARINO. MAMMOTH TUSK FEMALE FIGURINES. Photograph of S. Zamiatnin.

thinner woman and thus differs from the corpulent one of Willendorf, but has its arms placed similarly on its breasts.

These sculptures, representing figures of nude women, find their nearest analogies in the statuettes of Willendorf, Mentone, Brassempouy, and the bas-relief of Laussel (Dordogne) and represent the most ancient sculptured remains that have reached us. The first find of this sort in the U. S. S. R. was made in 1923, in Kostenki, by P. P. Ephimenko, greatly widening area of distribution of Aurignacian sculptures. Still a number of characteristics connect the Gagarino statuettes with the Venus of Willendorf, rather than with that of Kostenki, thus making them somewhat older than the former though belonging to the same general chronological period. This is also confirmed by the character of the rich flint industry of Gagarino, which is not so stabilized and has not so settled an aspect as the industry of Kostenki I.

It is interesting to note that all the statuettes of Gagarino were found along the walls of the hut, whereas the other cultural remains were found more or less scattered evenly on the floor. In one case only, was there found a cache of flint and bone objects in a little depression in the floor.

The time of the Gagarino site is determined fairly accurately by the whole complex of finds, especially by the characteristic flint points with a lateral notch, and the beginning of the Solutrean retouch, encountered also in Kostenki I and the upper layer of Willendorf, where female statuettes were also found. The remains of this remarkable palaeolithic station are preserved in the Museum of Anthropology and Ethnology of the Academy of Sciences of U. S. S. R.[1]

According to Gorodzov, the leading forms from the point of view of dating are high scrapers (*grattoir caréné*) and crooked gravers (*burins busqué*), which in his opinion, place this station in the Aurignacian period.

Gagarino, by its importance, will undoubtedly occupy one of the first places among the Upper Palaeolithic stations of Eastern Europe.

6. Berdizh

This site was discovered in 1926 by K. M. Polikarpovich[2] near the village of Berdizh, near Chechersk district, Gomel region, and represents the first find of a Palaeolithic culture in the territory of Bielorussia. (See map, Fig. 47.) This discovery was made during a systematic survey of Neolithic sites in the valley of the Sozh River. Preliminary excavations by Polikarpovitch yielded a large quantity of fossil bones most of which were those of mammoths. Over seventy fragments were found in an area of eleven square meters, together with flints of undoubtedly human workmanship and fragments of charred bone.

In 1927, the State Academy of the History of Material Culture sent Mr. S. N. Zamiatnin, who, under the auspices of the Institute of Bielorussian Culture, conducted excavations in which K. M. Polikarpovich, S. A. Dubinsky, and others participated. The geological conditions of the site were studied by a Professor of the Moscow State University, G. F.

---

[1] Ephimenko, P. P. ("Pre-Clan Society," Leningrad, 1936), on the basis of similarity of sculpture, the presence of *pointe à cran atypique*, and the beginning of Solutrean bifacial retouch, places it in the final Aurignacian or early Solutrean times (Proto-Solutrean?).

[2] Polikarpovich, K. M., "Prehistoric Sites of the Middle and Lower Part of the Sozh River," Pratzi of the Archaeological Committee of the Institute of Bielorussian Culture, Vol. I, Minsk, 1928, pp. 156–158.

Mirchink,[1] who has given a definite geological dating of the site, on the basis of the stratigraphy.

During the three weeks of excavation a total area of sixty square meters was opened. An account of the excavation was published by S. N. Zamiatnin,[2] a more or less complete summary and translation of which is given below.

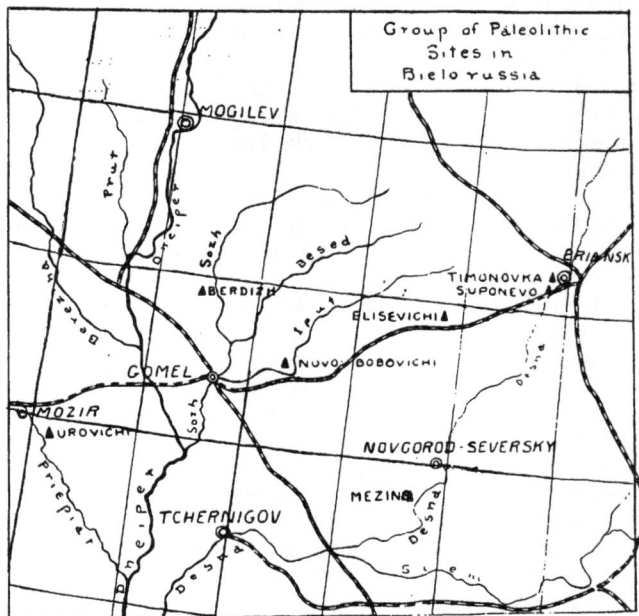

FIG. 47.

Location.—The Palaeolithic site of Berdizh is situated on the bank of the Sozh River about 1.5 km., S–SW., of the village of Berdizh, in the Kalodezhki Valley. (See map, Fig. 47.)   The cultural remains were discovered by Polikarpovich on the hill which rises 3.5 meters above the Sozh (Plate XXVI, *a*).

Later investigation showed that the cultural remains are found at the mouth of an old gully, which is in the southern end of the Sozh valley.   Zamiatnin's work began with the trial pit in 1926 (no. 1) where it was determined that the bone layer spreads inward from the shore.   Another pit (no. 2) 5 x 6 meters was dug due west from the first one.

The unusual looseness of the upper sandy layers and the rainy weather made it necessary to dig the walls of the pits in an oblique manner to avoid their collapse.   For this reason, the exposed cultural area of pit No. 2 was only 5.5 x 3.4 meters, or less than twenty square meters.   (See plan, Fig. 48.)

[1] Mirchink, G. F., "The Geological Condition of the Palaeolithic Site of Berdizh on the River Sozh," The Pratzi of the Archaeological Committee of the Bielorussian Academy of Sciences, Vol. II, Minsk, 1930.

[2] Zamiatnin, S. N., "The Excavation of the Palaeolithic Site of Berdizh in 1927," Pratzi of the Archaeological Committee of the Bielorussian Academy of Sciences, Vol. II, Minsk, 1930, pp. 479–490.

PLATE XXVI.

a. BERDIZH. GENERAL VIEW AT EXCAVATIONS.

b. BERDIZH. CROSS SECTION, SHOWING FOLIATION OF SANDS. Photograph of S. Zamiatnin.

Geology.—The stratigraphy of the site appears to be as follows:

A. A turf layer.

B. Dark-gray, fine grained sand, colored by the humus.

C. Light gray, uniform, fine grained sand with streaks of the ortsand.

D. Foliated white fine grained sand in spirals and convolutions.

E. Greenish-gray loam with streaks of sand and gravel, containing the cultural remains, 35–60 cm. thick.

F. Coarse-grained, horizontally foliated sand with gravel (Plate XXVI, *b*).

The cultural layer E has a sharp incline southward toward the shore of the Sozh, and westward toward the gully, so that one end of the excavated area was 2.9 m. from the surface and the other 4.4 m. In pit no. 2 the cultural layer was 3–4 m. above the water level.

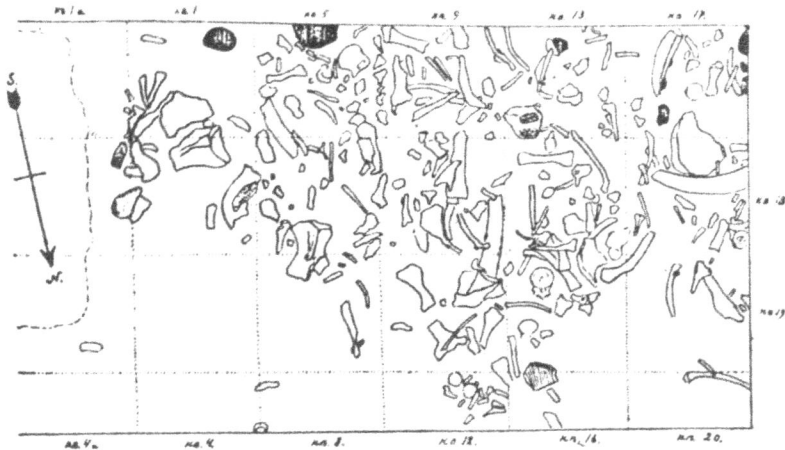

FIG. 48. Berdizh. Plan of the Excavated Area of Pit No. 2 in Berdizh, Showing Square Numbers. After S. Zamiatnin.

The investigations showed that the cultural remains in the excavated area were not evenly distributed. They increase in the southwest part of the pit, almost disappearing in the southeast part. The trial pit, no. 3, which was sunk two meters south of pit no. 2, proved to be sterile, thus confirming this observation. Consequently all material obtained came from pit no. 2.

Fauna.—The cultural layer was found to be literally filled with animal bones which were piled up, without any apparent order, in a small area (Plate XXVII). The exact boundary of this area, which extended in a westerly direction, could not be determined. Zamiatnin definitely feels that this group of bones was found *in situ*, as they were concentrated in a small area of eleven square meters, often lying one on top of the other despite the considerable slope of the shore.

This concentration and the absence of the bones beyond the border of this area indicate, in the opinion of the excavator, that these bones were piled here by prehistoric man for some definite purpose. He states that this condition could not exist if the bones had been moved about to any extent, for in this latter case both their quantity and size would decrease from the center of the pile down the slope.

Most of the bones found were those of mammoths. V. I. Gromov[1] indicates the remains of sixteen individual animals and established the presence of the following fauna:[2]

1. *Elephas primigenius,*	*7. *Rangifer tarandus,*
*2. *Elephas primigenius minor,*	*8. *Canis lupus,*
*3. *Sus scrofa ferus,*	9. *Canis canis* sp.,
4. *Equus equus,*	10. *Alopex lagopus,*
5. *Bos* sp.,	11. *Ursus* aff. *spelaeus,*
*6. *Cervus elaphus,*	12. *Citellus* sp.

A small quantity of charred bone was found after clearing the bone pile.

Flint Industry.—A large quantity of flints, consisting mostly of whole nuclei, chips, and flakes, was found in the bone layer. Only twenty-five could be unmistakably recognized as being produced by man. These were either finished implements or flakes resulting from their manufacture.

All of these were found in the southern and western part of the pit. (Squares 1, 5, 9, 13, 17, 18, and 20. See plan, Fig. 48). Large pieces of flint and pebbles were also found in these sections of the pit.

Material for the stone industry was furnished by the local flint of chalk origin, often found in the neighborhood in secondary deposition on the surface or as an outcrop in the chalk near by.[3] The flint is dark gray in color, of slight transparency, and has many incrustations, which differ from the basic color.[4] The nodules have a white crust on the surface. The patina is absent on most of the flints.

The finished tools of Berdizh equal in number the chips and flakes found. Zamiatnin stresses the special care taken in excavation, in which he endeavored to collect every bit of stone. On the face of it he believes that the actual manufacturing of the tools was not done in the excavated portion of the site but elsewhere.

As the total number of flints found in Berdizh is very small, it was thought advisable to cite almost verbatim Zamiatnin's description, given by squares. He describes the following artifacts:

PIT No. 2.

Square 1.—1. A small point (Fig. 49, no. 3*a*, *b*) with side notch (*pointe à cran atypique*), of dark-gray flint without patination, the upper end of which had been broken. The length

[1] Gromov, V. I., "Fauna of The Paleolithic Site of Berdizh," Pratzi of the Archeological Committee of the Bielorussian Academy of Sciences, Minsk, 1930, Vol. II, pp. 7–30.

[2] Those with asterisks are added from Gromov, V. I., "Geology and Fauna of the Palaeolithic Period in U.S.S.R.," Problems of the History of Material Culture, No. 1–2, Leningrad, 1933.

[3] Zamiatnin points out one such location, 4 klm. southeast of Berdizh near the village of Gorodovichi on the left bank of the Sozh.

Polikarpovich, K. M. ("Prehistoric Sites of the Middle and Lower Part of the Sozh River," Pratzi of the Archaeological Committee of the Institute of Bielorussian Culture, Vol. I, Minsk, 1928, p. 129), cites several locations near Berdizh.

[4] Zamiatnin points out that the poor quality of this flint in comparison with the fine, uniform material of Kostenki-Borshevo region may account for the somewhat rougher appearance of the Berdizh forms.

PLATE XXVII.

B<small>ERDIZH</small>. A<small>GGLOMERATION OF</small> A<small>NIMAL</small> B<small>ONES IN</small> B<small>ERDIZH</small>. Photograph of S. Zamiatnin.

is 58 mm. (originally about 70 mm.). The notch (35 mm. long) on the right side is well pronounced and carefully made with a steep, dulling retouch. The presence of the graver facet on the right side indicates that after being broken, the tool was reworked and used as a graver.

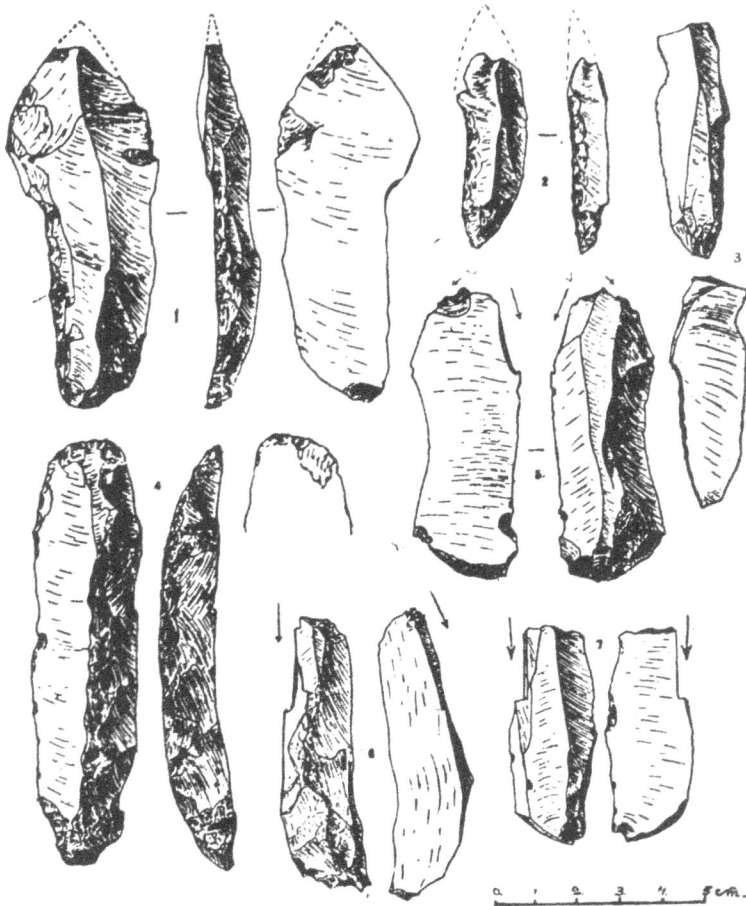

FIG. 49. Stone Industry of Berdizh. After S. Zamiatnin.

Square 5.—2. A double lateral graver (*burin d'angle double*) (Fig. 49, nos. 6, *a, b*) made on the opposite end of a long, massive flake detached from the side of a nucleus. The flint is dark-gray without patination, the length 68 mm.

3. The lower part of a regular knife-like blade with three sides, made of yellow-gray flint with light patination. The surface is very shiny, due to rolling. The chipping along

the edge is accidental, as it was done after the rest of the surface had become patinated. Length 30 mm. (Fig. 50, no. 12).

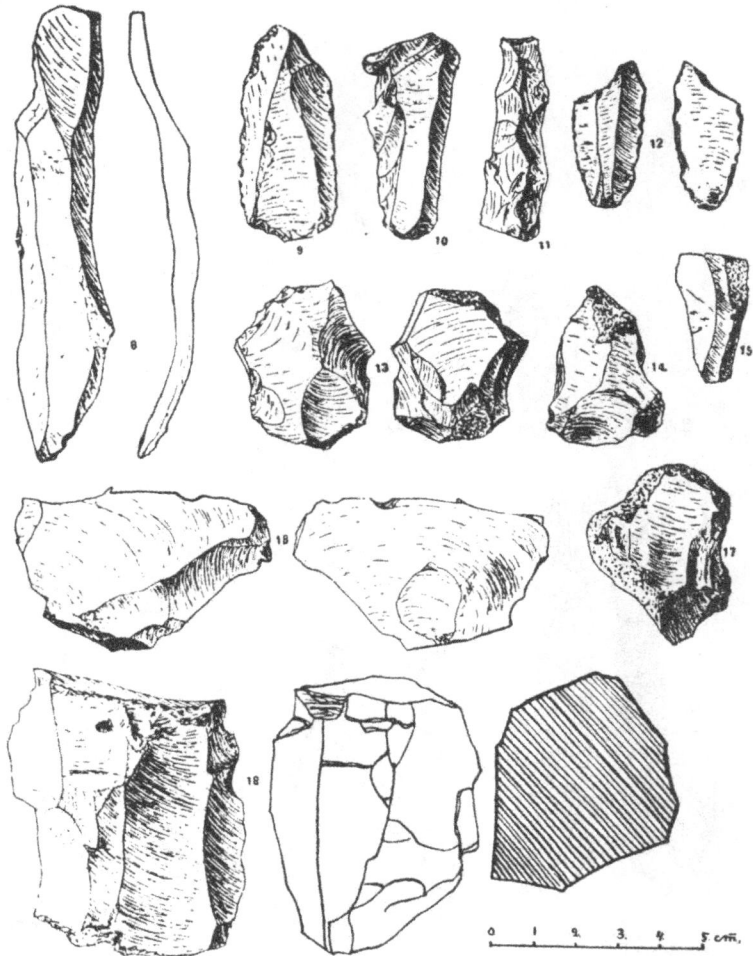

FIG. 50. Stone Industry of Berdizh.   After S. Zamiatnin.

4. A small triangular flake of coffee-colored flint, 32 x 20 mm.

5. A hammer stone, a massive nodule of tabular flint with the crust partly preserved. On the upper working end are traces of numerous blows 42 x 32 mm. (Fig. 50, no. 17).

Square 9.—6. A long slender flake of coffee-colored flint, 30 mm. long.

Square 13.—7. A rather large atypical point (90 mm. long) with lateral notch, with the tip broken off.[1] The flint is dark-gray without patination. The notch (54 mm. long) on the left side of the tool is well pronounced. The dulling retouch of the notch is large and rough. The lower end is carefully rounded by retouch, and the lower surface of the upper end as well as the back is chipped with fine flat facets. This is a very nice example, rivaling the best of this type from Kostenki and Willendorf (Fig. 49, no. 1, *a*, *b*, *c*).

8. A scraper on the end of a long massive blade, detached from the side of a nucleus. The flint is dark with many incrustations. The upper part is slightly retouched on the left side. The very steep, dulling retouch of the working edge shows that the tool served for cutting rather than for scraping. Length 105 mm. This is the largest tool from Berdizh (Fig. 49, no. 4).

Square 17.—9. A triangular flake of dark transparent flint, 41 mm. long (no. 14).

10. The lower part of a small regular three-faceted blade. The crust of the nodule is partly preserved. Length 37 mm. (Fig. 50, no. 15).

11. A small fragment of a three-faceted blade of dark flint. Length 25 mm.

Square 18.—12. A massive, knife-like, three-sided blade of dark flint without patination. On the upper end are traces of retouch and an unskilled graver's facet. This implement represents an unfinished side graver. Length 70 mm. (Fig. 49, no. 5).

13. A massive flake (45 x 30 mm.) of grayish-cinnamon color, an irregular oval in shape, roughly chipped on both sides. One of the surfaces is chipped by large facets from the sides to the middle. The opposite, very convex side, is worked by rough retouch on the edges.[2] 45 x 30 mm. (Fig. 50, no. 13).

14. A nucleus of irregular prismatic form, slightly worked, with one striking platform. It is made of dark-gray flint with light-bluish patination in spots. Length 80 mm. (Fig. 50, no. 18).

15 and 16. Two wide flakes of irregular form. On one are traces of the crust. Sizes 40 x 38, 30 x 42 mm.

Square 20.—17. A small flake of gray flint without patination, 50 mm. long. The lower part with the bulb of percussion is broken off (Fig. 50, no. 11).

18. A very wide flake of dark-gray flint from the surface of the nodule. On the upper side and on the striking platform the lime crust is preserved. 60 x 80 mm.

SPACE BETWEEN PITS No. 2 AND No. 3.

19. A small "atypical point with lateral notch"[3] 48 mm. long, of dark-gray flint with bluish patination. The upper part of the implement is broken. The notch on the left side is 3 mm. long and is retouched by large facets. The lower end of the tool is likewise retouched (Fig. 49, no. 2).

20. A long, narrow, knife-like blade, very much curved, of dark-gray flint without patination. On the back, a small part of the crust remains. Length 112 mm. (Fig. 50, no. 8).

21-22. Two small, three-sided flakes, both of dark flint with spots of light-gray and bluish patination. The shiny surface and dulled edges indicate much rolling. The first one shows traces of use along the sides. Lengths: 53 and 51 mm. (Fig. 50, nos. 9 and 10).

---

[1] *Pointe à cran.?* (E. G.)

[2] This specimen, like the preceding one, is unfinished. Zamiatnin considers this tool (*ebauche*) very important for the purpose of dating.

[3] *Pointe à cran.* (E. G.)

23. A flake, irregular in form, made of dark flint. (55 x 30 mm.)

24. A large, wide, irregular flake of cinnamon-colored flint. 88 x 66 mm. (Fig. 50, no. 16).

PIT No. 3, SOUTHERN END.

25. A graver with simple lateral facet (*burin d'angle*) on the end. The flint is covered by light, gray patination. The upper surface is polished, and the edges are smooth. Length 58 mm. (Fig. 49, no. 7).

These constitute the finds. Even this small quantity of material permits us to form a sufficiently definite conception of the flint industry of Berdizh.

Conclusions.—Summarizing, Zamiatnin gives the following picture: The site was located on the slope of the gully near its mouth. This condition is characteristic and is repeated in a number of sites. It is interesting that in the Don region a similar situation occurs in a group of older sites: Kostenki I, Borshevo I, Gagarino; while the later sites: Kostenki II, Kostenki III, Borshevo II are situated in the actual valley of the river. Thus, the very location of the site gives us some data for chronological calculations.

The cultural remains consist of a large quantity of animal bones. It is important to note, that according to V. I. Gromov, among sixteen individual mammoths, the remains of which are found in Berdizh, fourteen were young ones and only two adults. This preponderance of young animals already noted for other sites confirms the fact that the animal remains are the results of hunting activity on the part of man.

The small number of tools found seems to be incompatible with the large number of animal remains. This inconsistency may be understood if one considers the actual distribution of flint implements in other East-European sites with large accumulations of animal bones. Usually the remains of hearths do not contain large bones but are composed of a large quantity of small bits of bone, purposely broken tubular bones, charred bone, numerous flint tools, flakes and quantities of small flint chips. On the periphery of each of these "hearths" are one or several piles consisting almost entirely of large bones and containing no other cultural remains. Such accumulations were very likely refuse from feasts or perhaps material for fires. A similar distribution, is well expressed in Kostenki II, Mezine, etc. An especially good illustration are the large heaps of bones found by K. Absolon in Lower Wistirnitza in Moravia.

Evidently we are dealing with the same situation in Berdizh. The excavation uncovered only the bone pile, which here, as well as in other sites, is very poor in flint finds. The hearth, which probably lies nearer the bottom of the gully, as yet is not uncovered; it should contain much more flint material and thus will enrich our knowledge of the Berdizh industry.

The flint industry of Berdizh man is characteristic of the Upper Palaeolithic Age. Thus we have (1) the irregular prismatic nucleus, (2) the long blade with not very careful retouch of the back, (3) the graver, the most characteristic tool of the Upper Palaeolithic, is present in its basic forms: (*a*) the side graver (*burin d'angle*) made of a blade retouched on the end, (*b*) made on the end of a broken blade, (*c*) double graver, (4) the scraper on the end of a long flake.

But the most interesting form of the Berdizh site is the point with the lateral notch. This tool, represented in the Berdizh material by three characteristic examples, immediately places this industry in a definite archaeological division, connecting it on one hand with the group of early Upper Palaeolithic sites of the Don region (Kostenki I, Borshevo I, Gagarino), and on the other hand with the late Aurignacian sites of Moravia, and Lower Austria (Predmost, Wisternitz, Pekarna, Willendorf, etc.)

The Berdizh point, is especially similar to the tools of the upper horizon of Willendorf and Kostenki I, in its large size and its definitely worked form, which distinguish it from "atypical" points of Western-European sites of this period. It is interesting that in Berdizh (as far as may be said considering the limited material) this point with the lateral notch constituted a considerable percentage of the finds, as is the case in Kostenki I.

Besides the point with the lateral notch, another form also gives interesting indications of the cultural connection of Berdizh. This is the tool with bifacial technique (described as no. 18).[1] The forms nearest to it, which are connected with the industry of the Hungarian Solutrean, were found recently in some East-European sites. Unfortunately, the scarcity of material does not permit us at present to speak of this more fully.

In 1928 excavations in Berdizh were made by K. M. Polikarpovich. In his material, among other forms, is found an interesting tool, which has the shape of a characteristic Mousterian *coup de poing*, with the upper side convex and the lower one flat, worked by large flakes. Considering the above-mentioned connections with the Hungarian Solutrean, it is not necessary to seek an explanation of this from in the accidental mixture of the cultural layers of the site.

It is interesting to note that the excavations of 1928 produced a large quantity of the points with the lateral notch, which are represented by a fine series.[2]

Dating.—In conclusion it is necessary to underline the important facts in regards to the geological and cultural age of this site. Professor G. F. Mirchink,[3] on the basis of his study of geological conditions indicates that it was formed "right after the maximum retreat of the Würm glaciation and before the formation of the end moraines of the Bühl stage." This dating corresponds quite accurately with the placing of the Berdizh culture in the late Aurignacian, which for Western Europe corresponds to the Achen stage.

The investigation of Berdizh, of course, cannot be considered final. Further work there could bring out much that is valuable (hearths, a larger series of the stone and bone industries, perhaps, objects of art) to substantiate and confirm the present dating. But even as it is, the site of Berdizh can enter prehistory as one of the classical sites of Eastern Europe, which forms an important link between the early group of the Don and the rich sites of Moravia and Lower Austria. It clearly confirms the correct chronological corellations of the geological and cultural stages of Quaternary Man.

7. MEZINE

The site of Mezine is situated on the right bank of the Desna River in the village of Mezine, Kroletz district, Chernigov region. (No. 19 on the map, Fig. 51.) It was

---

[1] Zamiatnin does not give any details or picture of this tool.

[2] No actual material or illustrations were seen by the author. (E. G.)

[3] Mirchink, G. F., "Geological Condition of the Palaeolithic Site of Berdizh on the River Sozh," The Pratzi of the Archaeological Committee of the Bielorussian Academy of Science, Vol. II, p. 5, Minsk, 1930.

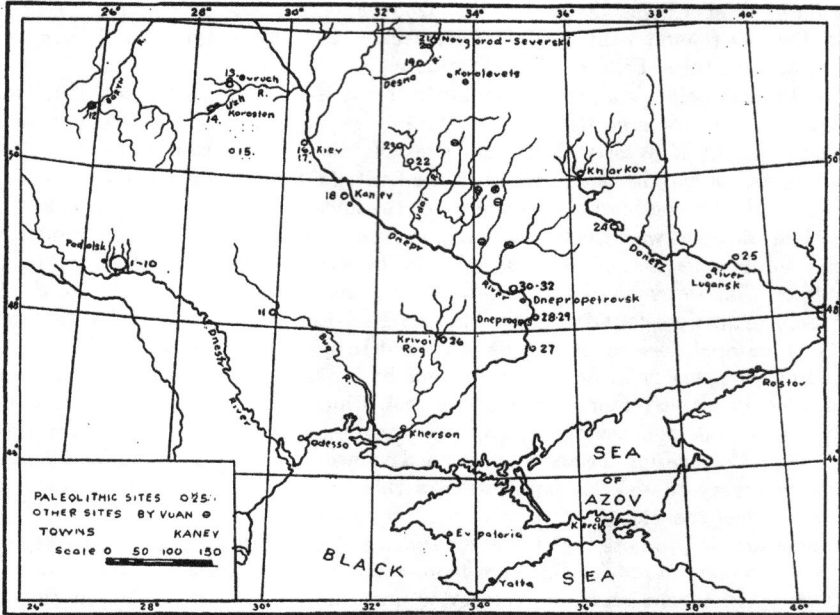

Fig. 51. Map of the Palaeolithic Sites of South Russia.

1. Sokol.	9. Kosheleva.	17. Kiev, Protasov Yar.	25. Lugansk.
2. Vrublintzi.	10. Kalus.	18. Selis'che.	26. Krivoy Rog.
3. Kitaigorod I and II.	11. Semenki.	19. Mezine.	27. Kodak.
4. Demshyn.	12. Gorodok.	20. Pushkari.	28. Dubova Gully.
5. Kolachkovichi.	13. Dolginichi.	21. Novgorod Seversky.	29. Kastrova Gully.
6. Studenitza.	14. Korosten.	22. Gontzi.	30. Osokorovka.
7. Bakota.	15. Kolodiazhonoye.	23. Zhuravka.	31. Miorka.
8. Staraya Ushitza.	16. Kiev, Cyrill Street.	24. S'churii Rog.	32. Yamburg.

discovered in 1907, excavated by Th. K. Volkov [1] and later in 1909 by P. P. Ephimenko.[2]

Location and Geology.—This site is located at the base of the river bank, overlooking the valley of the creek, a tributary of the Desna. In a complete geological cross-section of the shore, according to Armashevsky,[3] we have the following situation:

[1] Volkov, Th., " The Palaeolithic Site in the Village of Mezine, Chernigov Province," Trudi of XIV Archaeological Congress, Moscow, Vol. 3, pp. 262–270, 1911.

Also, "Nouvelles découvertes dans la station paléolithique de Mezine," Congres International d'Anthropologie et d'Archeologie Préhistorique. Compte-rendu de la XIV Session, Geneve, 1912, Vol. I, pp. 415–428, and the article in Bulletin et Memoires de la Société d'Anthropologie de Paris, 1909, Nos. 4–5, p. 501.

[2] Ephimenko, P. P., "The Stone Implements of the Palaeolithic Site in the Village of Mezine, Chernigov Province," The Annual Report of the Russian Anthropological Society of the Imperial St. Petersburg University, Vol. IV, pp. 62–102, St. Petersburg, 1913.

[3] Armashevsky, P., "The Geological Survey of Chernigov Province," Zapiski of the Society of Naturalists, Kiev University, Vol. VII, Pt. I, p. 30.

1. Humus.
2. Loess.
3. Boulder clay.
4. Tertiary green-gray glauconite sands.
5. Chalk-foundation.

The layer of chalk forms the common base, the rest of the layers of the site are not in primary, but in secondary diluvial positions, in consequence of which their thickness and contents, in comparison with the basic deposits, were changed. Thus Th. K. Volkov gives the following cross-section of the spot of the find:

1. Plowed earth, sandy, sometimes of clayish consistency, with remains of
   vegetation.. 0.80 meters
2. Loess with small chalky formations............................... 0.80 meters
3. The cultural layer of reddish clay with fragments of bones and flints.... 2.70 meters
4. Yellow clay, perhaps loess or loess-like..........................0.10–0.15 meters
5. Green-gray, fine sand with grains of glauconite...................... 1.05 meters
6. Orange-yellow sand.. 0.25 meters
7. Chalk... 0.07 meters

Especially interesting is the description of the cross-section of the Mezine site given by the geologist G. F. Mirchink: [1] "I have recorded during my visit with A. P. Pavlov and A. N. Soboliev in 1914, during excavations, conducted by the collaborators of Mr. T. Volkov, the following layers:

1. Slightly porous loess, diminishing southward toward the valley, increasing
   toward the mountains.
2. Cultural layer with large bones.
3. Sandy, loess-like, pale yellow clay................................ 0.56 meters
4. Layer of sand...0.3–0.05 meters
5. Pale yellow, slightly porous, loess-like sandy clay.................... 0.7 meters
6. Green-gray glauconite sand...................................... 0.2 meters
7. Yellow sand... 0.2 meters
8. White chalk, the base of which is 10.3 meters over the level of the Desna
   River.

Mr. Mirchink considers the first layer of loess of post-glacial diluvial origin; the third, of loess-like clay, belonging to the alluvial or diluvial deposits of the fourth glaciation.

P. P. Ephimenko [2] in 1909 made additional excavations. Enlarging the trial excavations of Volkov from 9 square meters to 18 square meters, he found the following stratigraphy:

1. Humus.
2. Loess-like, yellowish loam (3–4 m. thick) of the same character as the valley loess,
   but differing from the latter in being rougher and of schistous formation.
3. A cultural layer of dark loam, colored by organic remains, particles of charred bone
   and wood. It had, besides, a considerable admixture of red ferrous matter.

---

[1] Gorodzov, V. A., "Archaeology." The Stone Age, Vol. I, Moscow, 1925, p. 279, quoting from Mirchink's manuscript.

[2] Ephimenko, P. P., "The Stone Implements of the Palaeolithic Site in the Village of Mezine, Chernigov Province." Annual Report of the Russian Anthropological Society of the Imperial St. Petersburg University, Vol. IV, St. Petersburg, 1913, pp. 9–102.

This layer had an incline of 15–20°, corresponding to the direction of the layers of loam. It increases in thickness toward the southeast and contains flints and numerous bones of the mammoth, reindeer, musk-ox, and horse.

4. The same as layer 2.

5. A layer of green glauconite and ochreous sands.

6. A layer of chalk.

According to Ephimenko, the geological history of the deposit was as follows: With the retreat of the glacier there followed a period of loess formation which corresponds either to the interglacial, or to the period of the final retreat of the ice. The loess was deposited in horizontal layers on the plateaux, but in the valleys in sloping layers, and there it is somewhat coarser in structure. Long before the end of this process, palaeolithic man selected this bank for his settlement, the cultural remains left by him being covered by the subsequent deposition of the loess. The presence of such animals of the cold climate, as *Ovibos moschatus, Alopex lagopus, Rangifer tarandus*, suggests that the ice was not very far away.

Ephimenko also asserts that the cultural remains were undoubtedly discovered *in situ* and not as a secondary deposition. This is evident from the absence of rolling action on the implements, and the fact that alongside the bulky and large bones of animals were found small flints and particles of charcoal and ashes.

The cultural layer is not uniform in contents. In places among the mass of chips there is a definite predominance of finished implements of certain types, in other spots there is an agglomeration (not accidental) of animal bones of large size (compare Gontzi, Kostenki, etc.). Thus, in the northwest corner of the excavated area was found a large pile of mammoth tusks, reindeer antlers, and other bones, as though it were a store of raw material for manufacturing of bone implements.[1]

Fauna.—

1. *Elephas primigenius.*
2. *Rhinoceros tichorhinus* . . . . very few.
3. *Equus caballus* . . . . very numerous.
4. *Ovibos moschatus* . . . . numerous.
5. *Bison (priscus) (?).*
6. *Rangifer tarandus* . . . . numerous.
7. *Canis lupus.*
8. *Alopex lagopus.*
9. *Ursus arctos.*
10. *Lepus (variabilis?).*
11. *Gulo borealis (?).*

About sixty shells of Tertiary age were found, the majority of which were perforated for stringing.

The faunal remains were identified by Mme. M. V. Pavlova and the Swiss zoologist, Dr. Schtuder. The bulk of the bones belong to the *Elephas primigenius, Ovibos*, and *Cervus tarandus*. Among the bones of *Canis lupus*, according to Schtuder, there was a fragment of lower jaw having certain characteristics similar to *Cyon europeus*, a diluvial wild dog.

[1] Ephimenko, P. P., "The Stone Implements of the Palaeolithic Site in the Village of Mezine, Chernigov Province," The Annual Report of the Russian Anthropological Society of the Imperial St. Petersburg University, Vol. IV, St. Petersburg, 1913, pp. 9–102.

The Stone Industry.

Material.—The chief material for the stone work was flint, obtained in the vicinity, where even now many flint nodules are found in the gullies. This flint is of chalk origin, dark gray, almost black in color. The patination is either altogether absent or very thin, only dulling the surface or covering it with white or lilac spots.

Out of many thousands of flint chips, filling the cultural layer in the form of rejects, remains of manufacture, or blades prepared for the final work, over one thousand finished tools were found.

A detailed description of the flint industry is given by P. P. Ephimenko in the article cited above, from which are reproduced the most important forms and a description of the types.

I. *The Nuclei.*—These are encountered often; several hundred were obtained. They are usually not large, about 5 cm. in length (Fig. 52, nos. 13, 14).

II. *Hammerstones.*—Hammerstones are not very typical (Fig. 52, no. 15).

Flakes.—Flakes are very numerous, but regular forms with parallel sides are absent. They were usually small, 5–6 cm. long (Fig. 52, no. 3) and in most cases were utilized as cutting tools, a few being retouched on the side (Fig. 52, nos. 1, 6).

Scrapers.—*Grattoirs*, mostly made on small blades up to 6 cm. in length, terminating with a convex retouched edge, are typically Magdalenian. The retouch is very accurate (Fig. 53, nos. 1–3, 5–7). In two cases they are combined with *burins* (Fig. 53, nos. 6, 2), or form double scrapers (Fig. 53, no. 7). Besides the described forms there were found oblique side scrapers, twenty-three of which were made on irregular flakes with one side roughly retouched, and some very small ones ranging from 1–3 cm. in length and 1–4.8 cm. in width, with fine retouch.

Concave scrapers (Fig. 53, nos. 4, 12, 13, 14, 15), which according to Ephimenko were used mostly for work on bone, are of two types:

1. *Lames* and flakes with a carefully retouched segment-like notch on one side. Sometimes two or three of these notches (*coches*) are found.
2. The usual concave scraper with the notch on one end.

Perforators.—There were found perforators (*percoirs*) in a considerable quantity (260) besides triangular flakes, obtained accidentally and slightly retouched for the same use (*outils de fortune*). The following types can be distinguished:

1. Perforators with working point in the middle (Fig. 54, nos. 1–4, 7–10, 15–20, 23–25)
2. Side Perforators (*percoirs lateraux*) (Fig. 54, nos. 5, 6, 12, 13, 14)
3. Angle Perforators (*percoirs d'angle*) (Fig. 54, nos. 21, 22, 26, 27).

Lames and Lamelles (Fig. 52, nos. 7–9).—Some 150 small blades with a retouched side (*petites lames a tranchant lateral abattu*) were found. They are usually 2.3 cm. long, and 0.3 cm. wide, finely retouched, some of the geometrical forms reminding one of Tardenosian types.[1] One side is backed, the other (*tranchant vife*) left intact. Both ends are carefully retouched to a point and could be used either for drilling the eye in a bone needle, engraving, or tattooing. They often have a definite *coche* in conjunction with a perforator which may have been used as a hollow scraper.

[1] The forms illustrated are not typical geometrical forms of the Tardenosian period. (E. G.)

Gravers.

Gravers form the largest group of tools. *Bec de flûte* type is very rare. Most of them are made on massive irregular flakes with the facet formed by two longitudinal blows at an angle (Fig. 56, nos. 1, 2).

Side gravers (*burin d'angle, burin lateral*), are very common (250) (Fig. 56, nos. 11–14). Usually they are made from a long blade, the end of which is carefully retouched; then by a longitudinal blow a flake is detached, forming a cutting edge. Often the sides are dulled by a hand-protecting retouch. Double forms are numerous (Fig. 56, nos. 15–16).

Middle gravers (*burins medianes*) are much less numerous (25), usually made on a wide thick flake (Fig. 54, nos. 28, 31).

Parrot-beak gravers (Fig. 55, nos. 13, 14) (*bec de perroquet*) are not the typical French forms, but very similar to the Magdalenian VI. In most cases they are small graver points with a beak-like, curved end. Many have a dull, flat back suggesting perhaps a good place for the index finger.[1]

The *percoir-burin* is usually a short massive flake, the end of which is obliquely chipped to form a massive three-facetted point. (Fig. 55, nos. 1, 5).

Ephimenko concludes his analysis by pointing out the similarity with the French Upper Palaeolithic types. "The general character of the Mezine stone industry, the technique of the preparation, the small size of the implements as well as the types and forms, connect this station with the Upper Palaeolithic, and above all with the Magdalenian sites of Western Europe."[2]

The following implements are reproduced here from illustrations given by P. P. Ephimenko:

Fig. 52:

    1. Flake retouched on both edges.
    2. End scraper (?).
    3. Long triangular blade.
    4. *Burin d'angle* (?).
    5. Side scraper on thin flake.
    6. Same as 1.
  7–9. *Lames a dos rabbatu.*
   10. *Burin* flake.
11–12. Nucleiform scraper.
13–14. Nuclei.
   15. Hammer stone.

[1] While both Ephimenko, P. P. ("The Stone Implements of the Palaeolithic Site in the Village of Mezine, Chernigov Province," Annual Report of the Russian Anthropological Society of the Imperial St. Petersburg University, Vol. IV, pp. 62–102), and Volkov, T., ("The Palaeolithic Site in the Village of Mezine, Chernigov Government." Trudi of XIV Archaeological Congress, 1909, Vol. 3, pp. 262–270), definitely speak of *bec de perroquet* type, neither of them reproduce any tools with clearly pronounced *burin* stroke. (E. G.)

[2] Ephimenko, P. P., "The Stone Implements of the Palaeolithic Site in the Village of Mezine, Chernigov Province," Annual Report of the Russian Anthropological Society of the Imperial St. Petersburg University, Vol. IV, pp. 62–102, St. Petersburg, 1913.

Fig. 53:

    1. *Grattoir museau.*
    2. End scraper.
    3. End scraper.
    4. Convex end scraper.
    5. Concave end scraper.
    6. Concave end scraper and *burin bec de flûte.*
    7. Double *grattoir.*
    8. Scraper (?).
    9. *Percoir-burin.*
  10. Notched scraper.
  11. Triangular scraper (?).
  12. Notched scraper.
  13. Notched scraper.
  14. *Burin d'angle.*
  15. Notched scraper.
  16. *Burin d'angle.*

Fig. 54:

   1–4. Single perforator.
    5. Double perforator.
   6–7. Single perforator.
  8–10. Double perforator.
11–13. Single perforator.
  14. Double perforator.
15–24. Single perforator.
  25. *Burin d'angle* (?).
26–27. Notched scrapers.
28–29. *Burin d'angle.*
  30. Perforator.
  31. *Burin d'angle.*

Fig. 55:

   1–5. *Burin d'angle.*
    6. Double perforator.
    7. *Burin d'angle.*
    8. *Burin d'angle* and *bec de flûte.*
    9. *Burin bec de flûte.*
10–11. Perforator.
  12. *Burin d'angle.*
13–16. *Burin bec de perroquet.*
17–18. Perforator.

Fig. 56:

    1. *Burin d'angle.*
    2. *Burin d'angle.*
    3. *Burin bec de flûte.*
    4. Nucleiform *burin.*
    5. *Burin bec de flûte.*
  6–16. *Burin d'angle.*[1]

[1] The description is given on the basis of illustration, and often differs in nomenclature, from one given by Ephimenko.

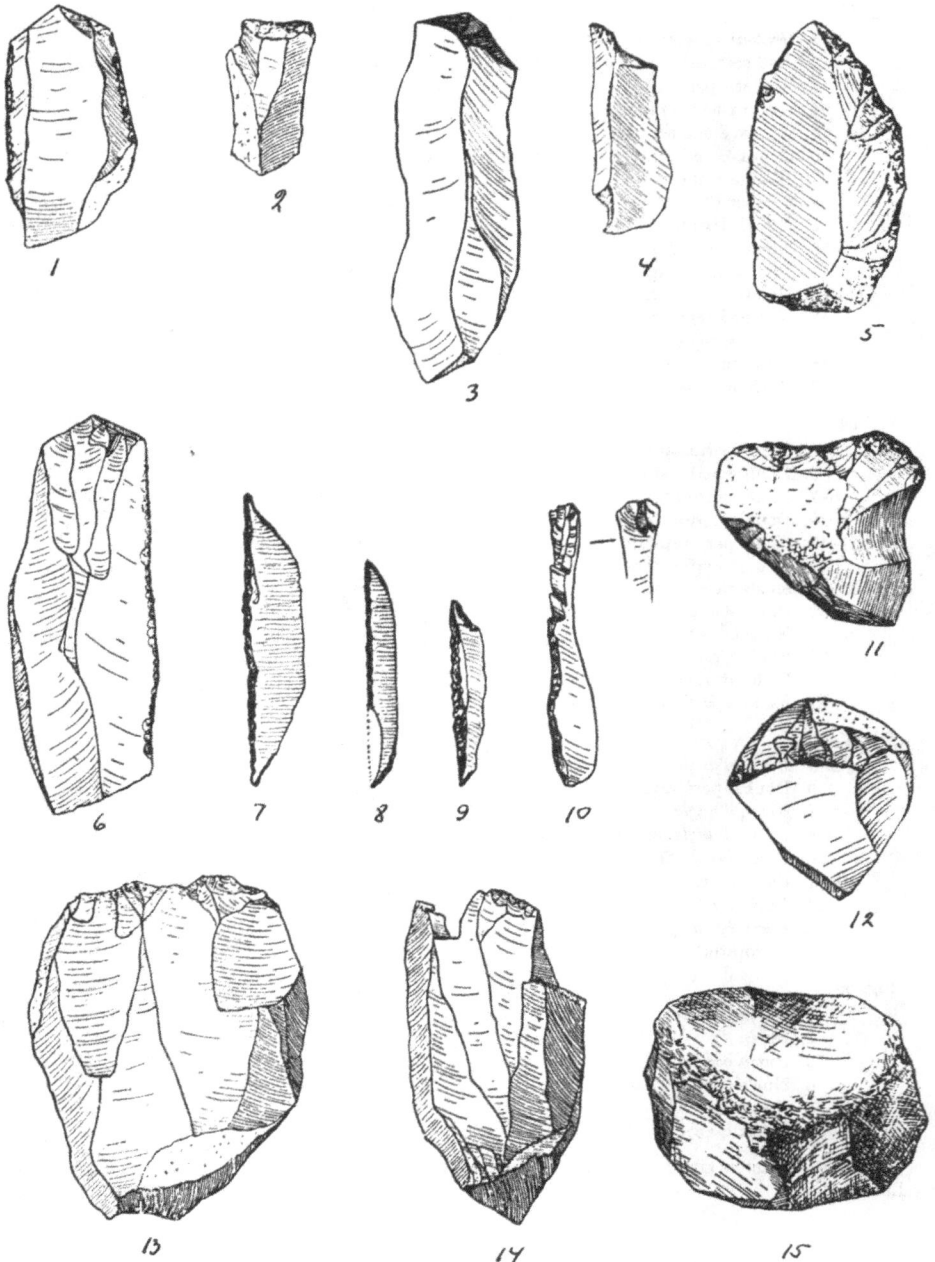

Fig. 52. Stone Industry of Mezine. After Ephimenko.

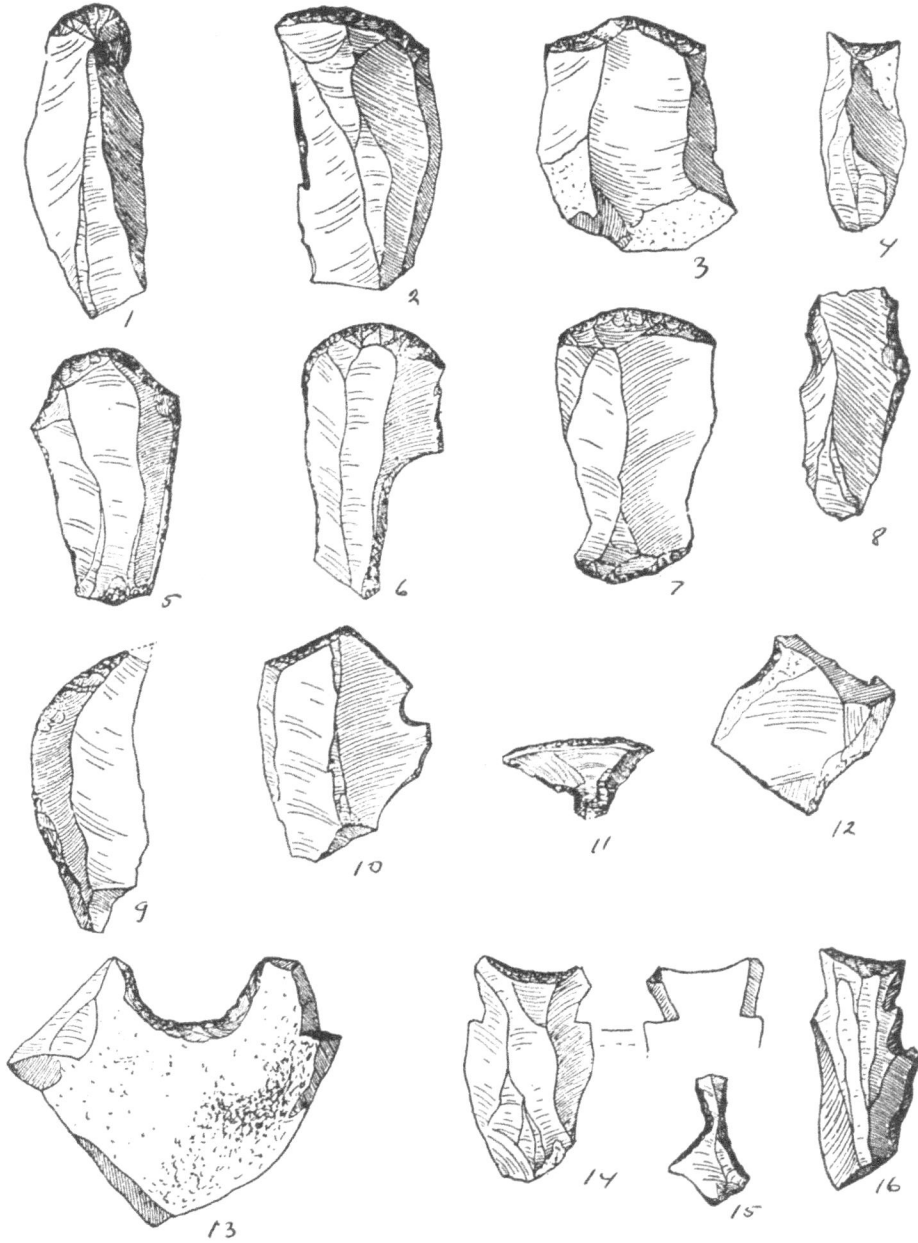

FIG. 53. Stone Industry of Mezine. After Ephimenko.

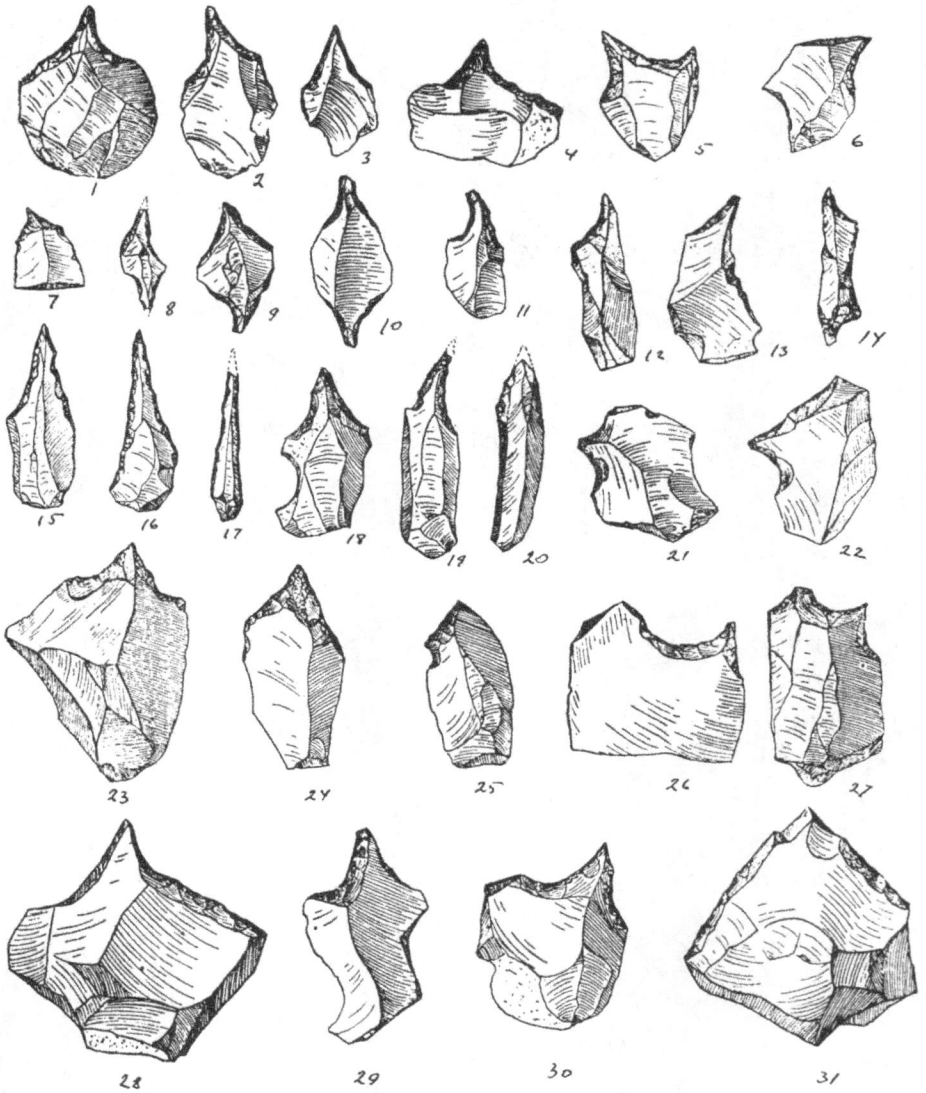

Fɪɢ. 54. Stone Industry of Mezine. After Ephimenko.

FIG. 55. Stone Industry of Mezine. After Ephimenko.

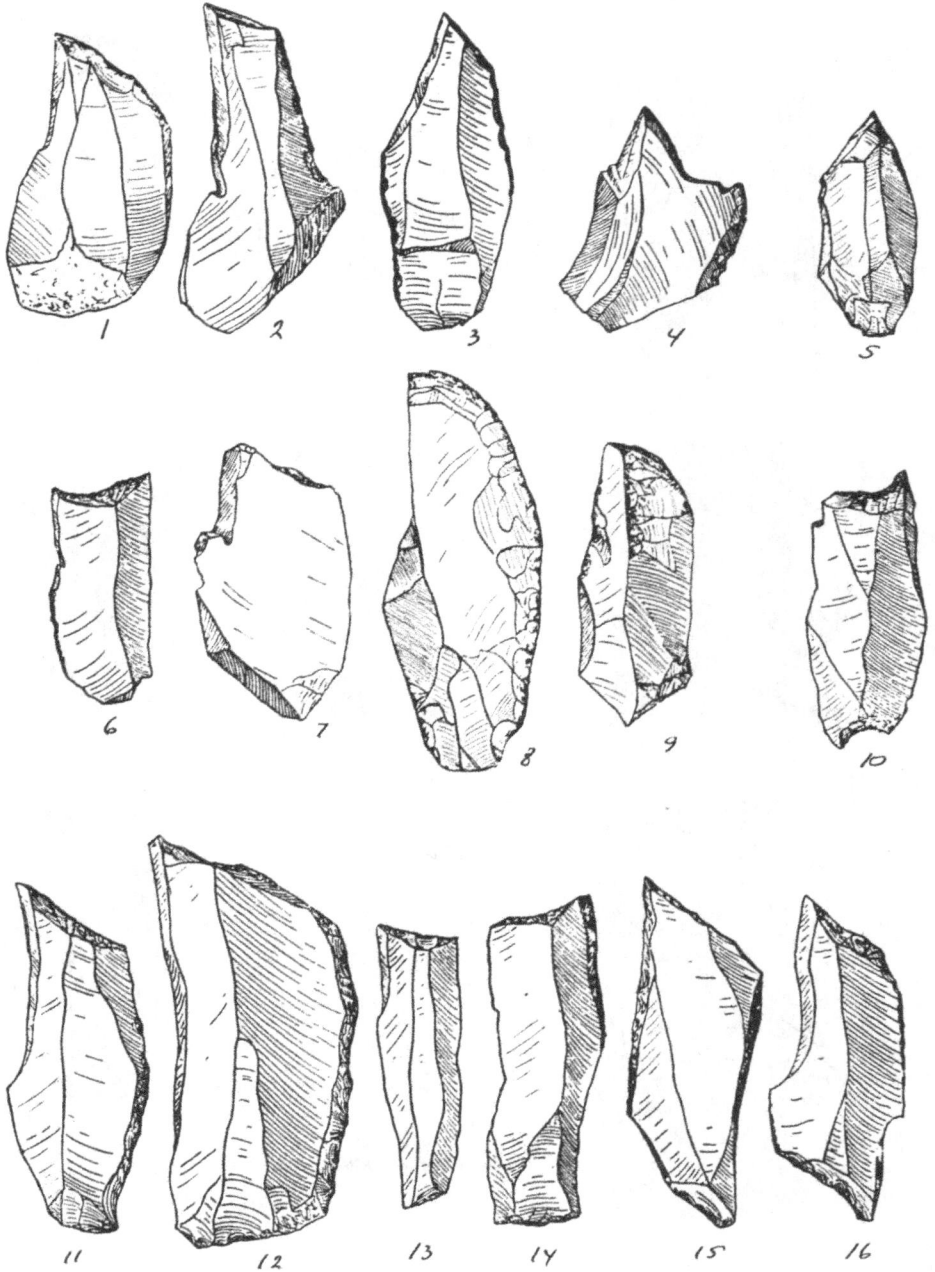

Fig. 56. Stone Industry of Mezine. After Ephimenko.

Bone Industry.

The bone industry of Mezine by its richness and variety corresponds to the development of its stone work.   Unfortunately, the most complete and recent publication [1] gives rather unsatisfactory reproductions.   The following description is based on the above work with additional information furnished by other authors.[2]

Awls.—A large number of awls were made out of long splinters of the tubular bones of animals, with sharp polished points (Fig. 57, nos. 1–6, 14–18).   In some cases the articulation of the bone was preserved (Fig. 57, nos. 1, 6, 16, 17).

Needles.—According to investigators, needles were usually perforated.   One, found by Ephimenko in 1909, was of the size of a small steel needle (Fig. 57, no. 15), but the reproduction of the fragment does not show the eye.

Dart Points.—Dart points were made of mammoth ivory and reindeer horn, round in cross-section, tapering down on both ends, with two longitudinal grooves (Fig. 57, nos. 8, 9).

Mallets.—The mallets were made of reindeer horn (Fig. 57, nos. 10, 11).   The complete specimen shows adaptation for hafting.

"Dart-shaft Straightener."—An object made of mammoth ivory, more or less pointed and round on one end, and flat and round on the other.   The flat end has a slightly oval hole.   Volkov sees the analogy with the tools employed by the Chukchi of Siberia to soften the strips of hide, while P. Ephimenko calls it a dart-shaft straightener or *baton de commandement* (Fig. 57, no. 13).[3]

Saw (?)—There is an implement of unknown use made out of rib with a serrated, saw-like edge (Fig. 57, no. 7).

Polisher (lissoir).—A polisher made of the mammoth tusk has one end (the handle) hacked, the other polished.   In the geometrical design on it, Volkov sees the schematic representation of an animal, the legs of which are represented with double lines.   Over it, the series of chevrons reminds him of the North American Indian manner of representing a summer dwelling (Fig. 58, no. 11).[4]

Perforated Point.—There is a point made of mammoth tusk, perforated on one end (the other is broken), with a series of double oblique lines on both faces, called a needle by Volkov (Fig. 58, no. 3).

Plaques.—A group of round and square plaques, richly decorated with engraved lines, chevrons, and Greek key designs are made of mammoth tusk.   Most of them have traces of perforation (Fig. 58, nos. 1–15).

Pendants.—There is also a series of oval pendants with a perforation on the obtuse end, made out of mammoth tusk.   Ephimenko remarks that they imitate the shape of a reindeer tooth.[5]   The other pendants are of mammoth tusks very schematically representing animals sitting on their haunches.[6]

[1] Rudinsky, M., "Bone Industry of the Palaeolithic Site of Mezine in the Interpretation of T. Volkov," Ukrainian Academy of Sciences, Anthropological Laboratory of T. Volkov, Kiev, 1931, pp. 1–65, 32 plates.

[2] P. P. Ephimenko, "Pre-Clan Society," Leningrad, 1936, p. 393.

[3] Compare with *"baton percé* from Predmost, Breuil, H., "Voyage Paléolitique en Europe Central," L'Anthropologie, t. XXXIV, fig. 13, nos. 1 and 7.

[4] Compare Ibid., fig. 16, no. 3.

[5] See Rudinsky, M., "The Bone Industry of the Palaeolithic Site of Mezine in Interpretation of T. Volkov," Ukrainian Academy of Sciences; Anthropological Laboratory of T. Volkov, Kiev, 1931, Fig. XXI; p. 150; also: Ephimenko, P. P., "Pre-Clan Society," Leningrad, 1936, Fig. 129.

[6] See Rudinsky, M. (opus cit.), Table XIV.

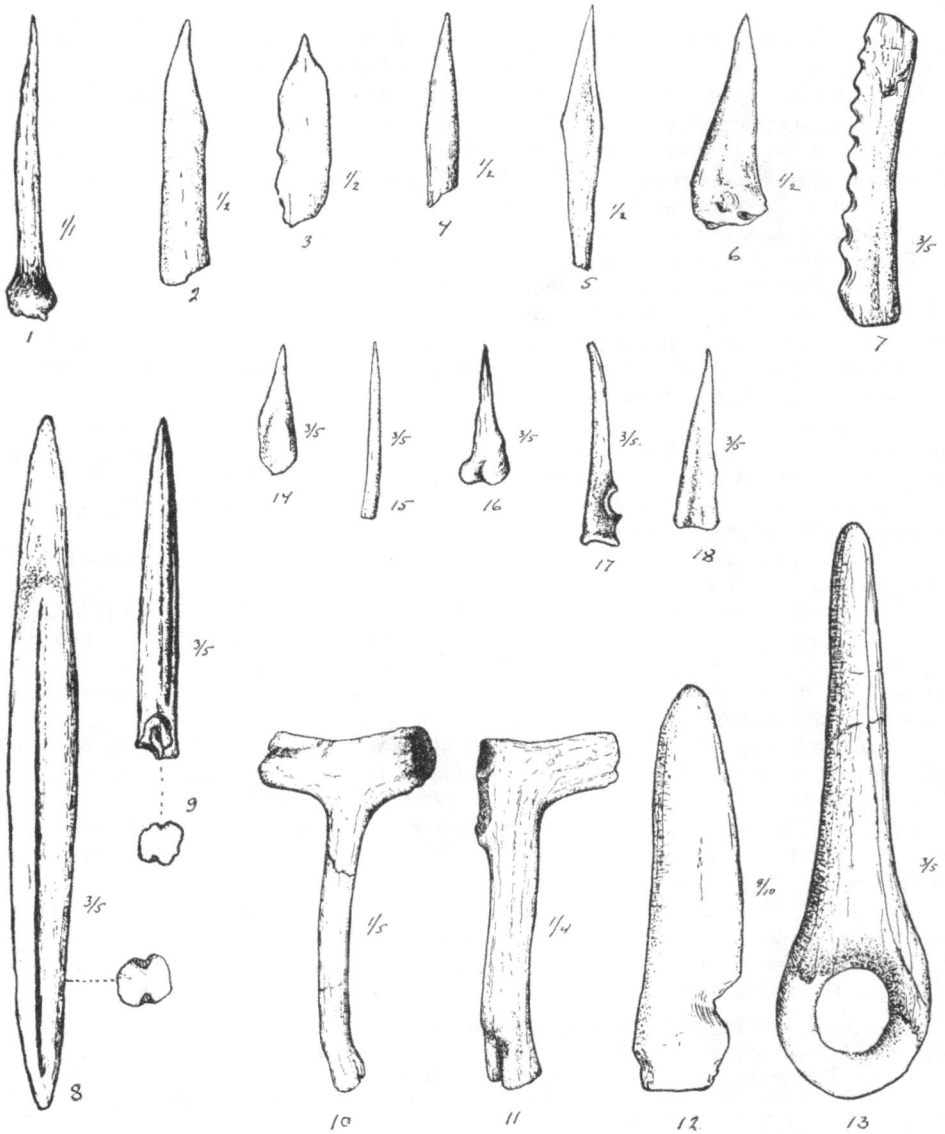

FIG. 57. Bone Industry of Mezine. After M. Rudinsky.

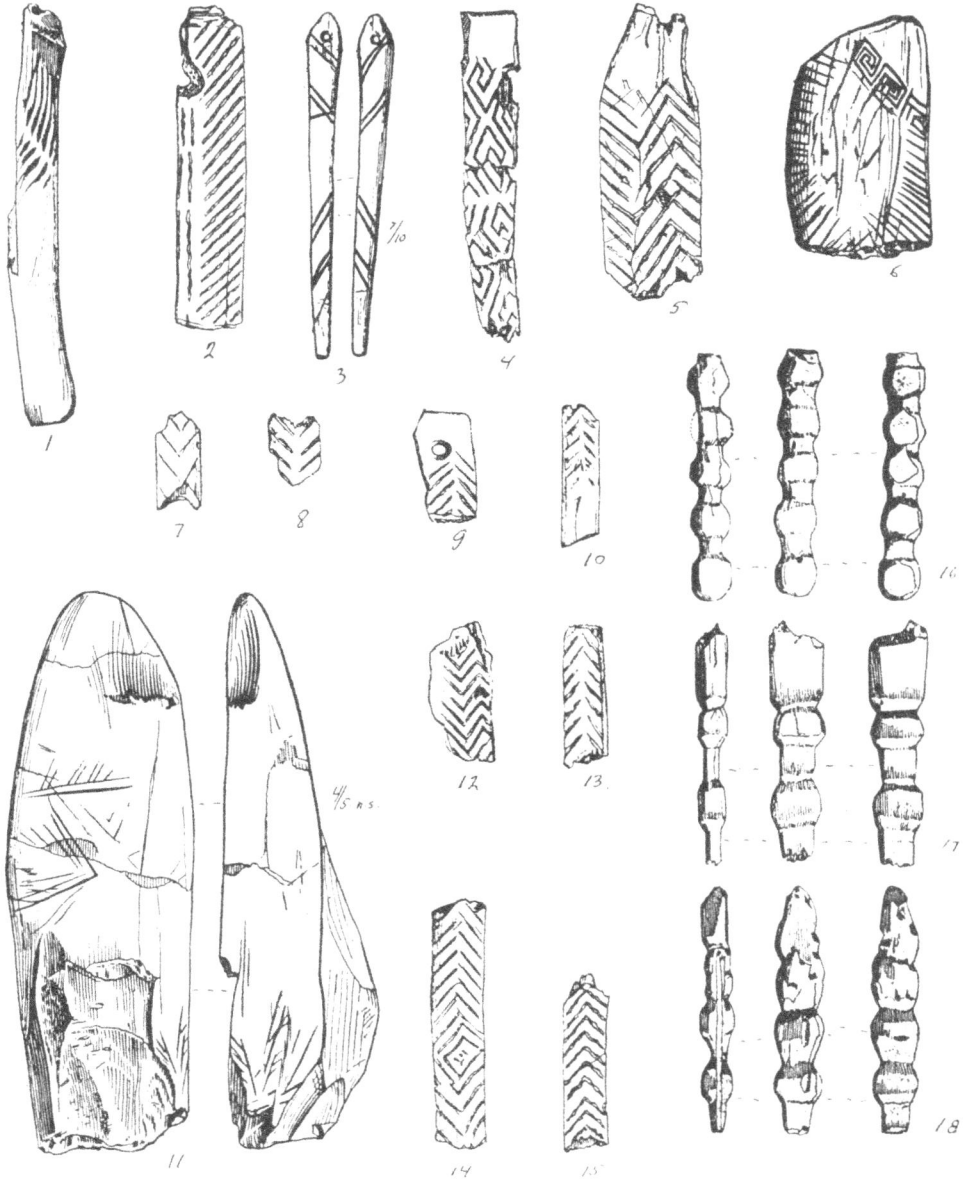

Fig. 58. Bone Industry of Mezine. After M. Rudinsky.

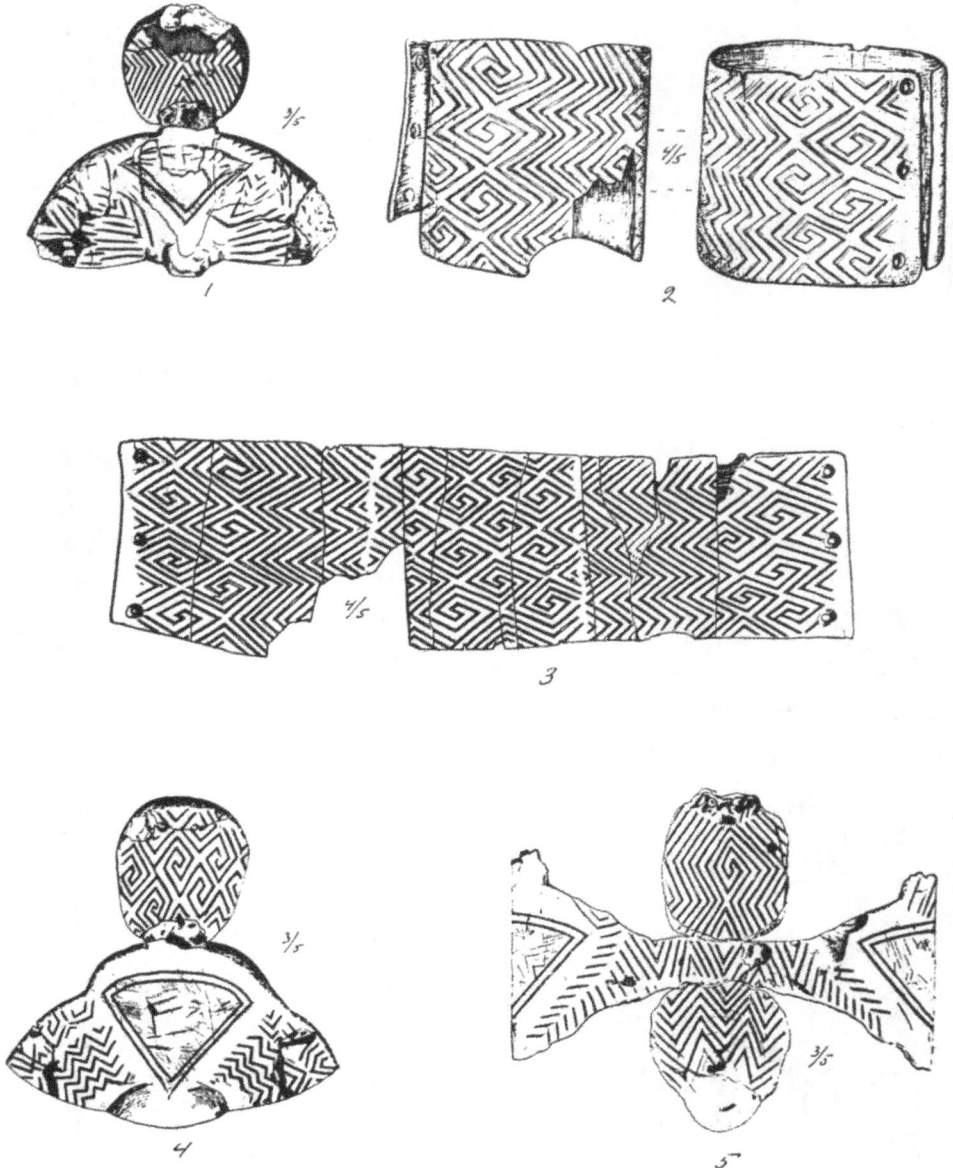

FIG. 59.   Mezine Ivory.   After M. Rudinsky.

PLATE XXVIII.

IVORY SCULPTURAL REPRESENTATIONS FROM MEZINE. Photo from the casts in the University Museum, by author.

Rods.—There are four ivory rods carved in a series of roughly shaped spheres, spaced at more or less regular intervals (Fig. 58, nos. 16–18).[1]

Bracelet (Fig. 59, nos. 2, 3).—There is a bracelet made out of mammoth tusk which reveals an unusual degree of skill in workmanship. This superb example of the art of Mezine man is unparalleled in Western Europe. It has the form of a wide curved plaque, the ends of which almost meet, with three perforations on each end, possibly for lacing it on the arm. The entire surface is covered with very finely executed meander, romboid, and zigzag ornaments of unusual accuracy.

"Phallic Representation" (Figs. 60, 61).—Nine objects, made out of mammoth ivory, deserve special attention. As may be seen from the illustrations, they resemble bird beaks in form, especially when viewed in profile.[2] One can distinguish two parts, an elongated, slightly curved, tapering point and a shorter butt which starts with a rather abrupt swelling and gradually subsides to a blunt end. In at least one case (Fig. 60, no. 6) the longer part is nicely polished, the thicker end being roughly chopped, the very end showing the tracer of breaking from the original block.

There are usually two types of engraving: heavy deeper incisions which usually coves the side with the protuberance, and lighter, thinner lines which cover the flat side.

Rudinsky[3] distinguishes a combination of two principal motives in ornamentation which repeat themselves in all examples, differing in form:

1. An elongated figure which he called a "feather," above which are more or less pronounced traces of a second long figure, partly covered by the first motive, forming at the base a sort of sector.

2. The flat side is usually covered by geometrical ornaments, consisting of a series of double chevrons with their bases toward the pointed end (Fig. 61). These are sometimes replaced by rows of smaller chevrons, occasionally combined with straight lines and parallelograms.

It is hard to decide definitely what these objects represent. Volkov, while calling them provisionally "phallic representations," indicates the possibility of their being merely ornamented handles. Rudinsky does not commit himself as to their purpose but is inclined to see a purely ornamental motive of decoration, suggesting that its uniformity may represent a tribal, family, or artist's mark.

It was suggested by Breuil, Obermaier, and others that we have here very crude, or perhaps very much stylized, representations of human beings.[4] In this case the shorter, thicker figures would represent steatopygious women with the pronounced swelling representing the buttocks and the fine "geometrical ornament" on the flat side indicating the nose, eyes, breasts, thin arms and pubic triangle. On the other hand, the second group of thinner and longer objects would represent males with the same facial features, but with the swelling obviously less pronounced, and the "feather" of Rudinsky, depicting a very much exaggerated phallus. In both cases, the decorations of the other side would represent tattooing.

---

[1] Compare ivory rods from Predmost. Breuil, H., "Voyage Paléolithique en Europe Central," L'Anthropologie, t. XXXIV, 1924, fig. 16, nos. 1 and 2.

[2] They are of two types: one long and thin, the other shorter and thicker. (E. G.)

[3] Rudinsky, M., "Bone Industry of the Palaeolithic Site of Mezine in Interpretation of T. Volkov," Ukrainian Academy of Sciences, Anthropological Laboratory of T. Volkov, Kiev, 1931.

[4] This description is more acceptable. (E. G.)

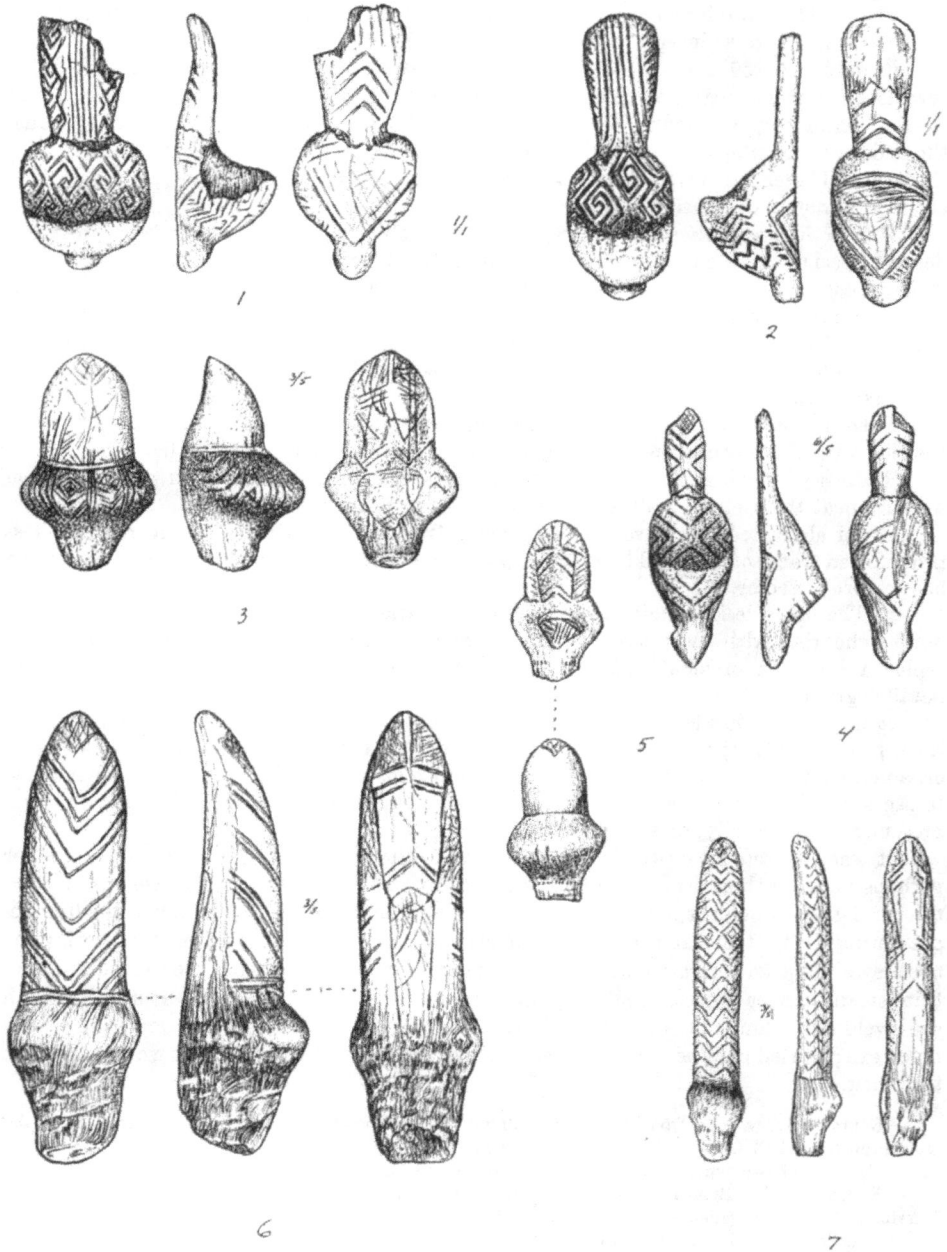

Fig. 60. Mezine Ivory. Drawings from photos of M. Breuil and reproductions of Rudinsky.

FIG. 61.   Mezine Ivory.   After Rudinsky.

Ephimenko does not accept this explanation as altogether proved, but, remarks, that either phallic, or anthropomorphic interpretation would fit into the general tendency of the art of the time (female representations).

Birds (Plate XXVIII, Figs. 59, 60, 61).[1]—Another series of very unusual objects were described as birds by most authors. They are made likewise of mammoth tusk and have very elaborate geometrical ornamentations.

Gorodzov states that several of them were found, but four are of special interest. Two of them are well preserved (Figs. 60, nos. 1, 2), and two have broken tails (Plate XXVIII). These "birds" are executed in a decadent conventional style, perhaps indicating a very wide distribution and a very long period of existence for this art.

The heads are barely indicated by adze-like protuberances without any details. The bodies are round, with softly outlined "breast" and "back" surfaces. The upper part of the "tail" is decorated with chevrons; below this in one case are perpendicular lines, in another a criss-cross geometrical design.

The breast and the end of the abdominal part is variously ornamented. The breast of one bird has a design of single lines drawn from the neck down, evidently representing the position of the feathers, while the lower part of the abdomen is decorated with four groups of chevrons with the apices pointing to a common center. Another one has on the breast and on the lower part of the abdomen a well-engraved Greek-key design.

However ingenious and suggestive the bird theory may be, it does not explain a number of features encountered. The animal motive is rare enough in the sculptural art of Palaeolithic man of European Russia, while the bird representation would be quite unique. The well pronounced triangle on the "breast" serves as a clue for our interpretation.

It seems more plausible to see in these "birds" much stylized female figurines, in which case the objects should be viewed upside down, as they are reproduced in (Plate XXVIII, Figs. 59, 60). The "swelling breast" then becomes the familiarly exaggerated buttocks; the tail—the upper part of the body with the usual tattooing on the back, (the "feather" of the first interpretation). The lack of details in the "heads" is then natural, as they represent only extremely schematicized legs tapering down to a point, a feature characteristic of the female figurines of Palaeolithic times. The well pronounced and otherwise unexplainable pubic triangle, as the only remaining, and therefore very much emphasized sexual attribute, completes the picture.

Dating.—The question of dating the Mezine site has received much consideration among Russian and Western European authorities. Thus, V. A. Gorodzov considers that the geological conditions of Mezine and Cyrill Street, Kiev, are very similar.

In the flint industry, the oblique gravers *burins busqué* and small blades sharply retouched on one side, similar in type to *La Gravette* are especially interesting for the purpose of dating. Both types appear for the first time in the upper horizon of the Aurignacian. Evidently, on the basis of these forms, Obermaier[2] considers that Mezine belongs to the end of the Aurignacian and the beginning of the Solutrean period. Still, continues Gorodzov,[3] these Mezine forms are not typically expressed and may be only survivals of ancient forms in later times.

[1] In accordance with the anthropomorphic interpretation of the author, these figurines are reproduced upside down, contrary to the usual way. (E. G.)

[2] Obermaier, H., "Fossil Man in Spain," Yale University Press, 1925, p. 222.

[3] Gorodzov, V. A., "Archaeology, The Stone Age," Vol. I, pp. 232–239, Moscow, 1925.

Burkitt,[1] however, considers Mezine a specialized Upper Aurignacian culture which may have existed in Eastern Europe longer than in the West.   While *La Gravette* points, and some gravers link it with the Upper Aurignacian of Western Europe, he feels that the bone work and the development of art "has no contemporaneous Western analogy."   He is rather inclined to see a special Russian school of primitive art, with great emphasis on geometrical designs as found in Cyrill Street and Mezine, and notes the Russian influence in similar Predmost objects.

Both Volkov and Chikalenko date Mezine as Magdalenian.

Ephimenko, on the basis of the art development in comparison with Predmost, Kostenki I and Gagarino, assigns Mezine to the Solutrean-Magdalenian transition,[2] which fits into the accepted Russian scheme of unilinear development of culture.

8. NOVO-BOBOVICHI

This site is located near the village of Novo-Bobovichi, on the right bank of the Iput River, 15 km. northwest of the town of Novozibkor, Klinzov district, Briansk region. According to Polikarpovich,[3] the finds of large animal bones by peasants, digging for a well attracted the attention of G. E. Giterman and D. P. Diatlov who visited the place and made a trial pit 6 x 9 meters.   This resulted in the discovery of many mammoth bones showing the characteristic accumulations which are so typical of the Russian Upper Palaeolithic sites.

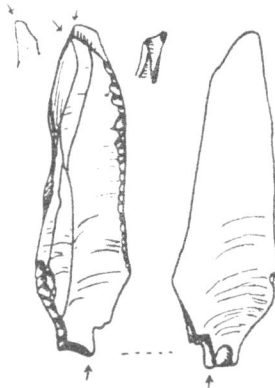

FIG. 62.

The pit is situated in the village, 70 meters above the level of the Iput River, on the bottom of a gully mouth, which opens into the river.   It is 22 meters long and 5 meters wide.

[1] Burkitt, M. C., "Prehistory," pp. 130.

[2] Menghin, O. ("Die Weltgeschichte des Steinzeit," p. 163), dates Mezine as Late Solutrean or early Magdalenian.

[3] Polikarpovich, K. M.; "The Palaeolithic Site of Novo-Bobovichi," Pratzi of Archaeological Committee of Bielorussian Academy of Science, Vol. II, pp. 491–494, Minsk, 1930.

The following stratigraphy was revealed:

1. Wash out material—0.15 to 1.0 meters.
2. Gray sandy loam and coarse grained sand with gravel—1.0 to 2.0 meters.
3. Gray-yellow sandy ferrous loam—2.0 to 3.0 meters.
4. Black, greasy cultural layer.
5. Green-gray, loess-like loam.

The cultural layer abounds in mammoth bones, about half of which belong to young animals. This is typical of the other Palaeolithic sites.

Industry.—The only implement found is a graver, made of dark-gray flint of chalk origin similar to that often found in the vicinity of the site (Fig. 62). It is made on a three faceted blade, 6.78 cm. long, 2.05 cm: wide and 0.72 cm. thick. The upper pointed end has two facets of *burin "bec de flûte"* type; the lower one is a *burin d'angle*. On the lower face, traces of either a *burin* stroke or of reduction of the bulb are found. Along the right edge is fine steep secondary retouch. The left edge has a retouched notch suggesting *point à cran atypique*, which could be utilized later as a *burin*.

Dating.—On the basis of the above, Polikarpovich indicates Novo-Bobovichi to be an Upper Palaeolithic site but feels that further dating is impossible with such insufficient information.

9. UROVICHI

This site is located on the left bank of the Priepiat River, a tributary of the Dnieper, 20 km. from the town of Mazir. In 1929, during the course of archaeological work in the region, the attention of K. M. Polikarpovich [1] was attracted by the find of fossil animal bones. The trial excavation (5 square meters) established the presence of the cultural horizon with broken bones of very young mammoths and points of *La Gravette* types. These were found in the dark yellow sand which lay on a strata of greenish sands and was covered in turn by a series of light yellow sands.

In 1929, the Bielorussian Academy of Science delegated G. E. Giterman, S. A. Dubinsky, A. N. Lavdansky and K. M. Polikarpovich to further investigate this site. The expedition excavated an area of 30 square meters (5 x 6) and disclosed the following stratigraphy:

1. Dark gray layer—0.10 meters to 0.15 meters thick.
2. Light yellow loess with slight foliation—6.40 meters thick.
3. Diagonally foliated sands—4.80 meters thick.
4. Dark-yellow, fine grain sand which becomes bright green in the upper part.

The animal bones were encountered in a layer 0.5 to 0.6 meters thick in the middle of the fourth layer which is bordered by greenish sands.

Fauna.—The majority of the bones were those of *Elephas primigenius*, and a few *Equus* (sp.). As the material is now being studied, the complete list of fauna is unavailable. It has been established, nevertheless, that the majority of mammoth bones are those of young animals. The bones here, unlike those of Berdizh, are in a poor state of preservation.

[1] Polikarpovich, K. M., "Excavation of Palaeolithic Site of Urovichi in 1929," Pratzi of Archaeological Committee of Bielorussian Academy of Science, Vol. II, pp. 499–501, Minsk, 1929.

PLATE XXIX.

EXCAVATIONS IN KOSTENKI II. Photograph of S. Zamiatnin.

Industry.—Besides the above mentioned two points of *La Gravette* type, four fragments of similar types, as well as one small flat flake, were found during the excavation.

Investigators believe that they have touched the very periphery of the site, the center of it being further south.

It was planned to continue the investigation in 1930, but no further information is available.

Dating.—Polikarpovich does not feel justified in assigning any more definite time than the upper Palaeolithic Age for this site.

10. KOSTENKI II

Kostenki II is situated on the northern side of the village Kostenki, Voronezh region, on the ancient terrace underwashed by the Don River. During construction work, animal bones and flint implements were found in the so-called "Anosov Log," next to the "Pokrovsky Gully," where the site of Kostenki I is situated.

The trial excavations of P. P. Ephimenko in 1923 [1] resulted in the discovery of a compact cultural layer, situated at the depth of two meters, between the layers of loam and sand. It contained a large quantity of animal bones, mostly those of mammoth, and occasional stone implements. The small quantity of tools found precluded the possibility of definite dating at that time.

Subsequent excavations by S. N. Zamiatnin in 1927 [2] confirmed the preliminary investigations. The excavators assumed, on the basis of their previous experiences, that they were dealing with one of the conglomerations of bones, which in this region accompany the hearths. Consequently, several trial pits were made, resulting in the discovery of the remains of a hearth almost immediately next to the pit made in 1923. It had an aspect typical of the sites of this region and a very rich cultural content. About two hundred flint tools, a number of bone objects, flint chips, split and charred bones, and pieces of ochre were found there.

While this preliminary report by Zamiatnin does not give any details, Gromov [3] lists the following fauna:

Elephas primigenius, numerous	*Felis* sp. (*leo?*).
Sus scrofa, few.	*Marmota* sp.
Equus (*equus*).	*Lepus* sp.
Ovibos moschatus.	*Ellobius talpinus*.
Alopex lagopus (?).	

The conglomeration of bones was, as usual, situated around the edge of the "hearth." According to Vishnevsky,[4] the forms of stone implements were different from those of Kostenki I, and were made of red, rather than black, flint.[5]

[1] Ephimenko, P. P., "Pre-Clan Society," Leningrad, 1936, p. 431.

[2] Zamiatnin, S. N., "The Expedition for the Study of Palaeolithic Cultures in 1927," Soobschenia of GAIMK, 1929, Vol. II, pp. 210–211.

[3] Gromov, V. I., "Geology and Fauna of the Palaeolithic Period in U. S. S. R.," Problems of the History of Material Culture, No. 1–2; Leningrad, 1933.

[4] Vishnevsky, B. N., "Pre-Historic Man in Russia" (Appendix to the translation of Osborn's "Men of the Stone Age," Leningrad, 1924, p. 460).

[5] Unfortunately, no further details or illustrative material pertaining to the stone industry of Kostenki II are available to the author (E. G.).

Ephimenko [1] dates this site, as well as the Kostenki III, as early Magdalenian on the basis of the general archaic, decadent character of its industry which, in his opinion, is a degeneration of Aurignacian-like technique in Magdelenian times.

11. KOSTENKI III

This site was discovered by P. P. Ephimenko in 1923 on the northern border of the village of Kostenki, Voronezh region, in the so-called "Glinistche" on the high bank of the Don River. Here the river approaches the high shore and washes out the ancient terrace, thus considerably damaging the site. At present it occupies a small plateau bordered on one side by the shore slope which, gradually elevating, reaches the heights of the right bank of the Don.

The trial excavations produced very little material. Further excavations in 1926 by Ephimenko, and in 1927 by Zamiatnin, were made next to the pit opened in 1923. A total of forty square meters was uncovered.

According to Ephimenko,[2] the natural cut of the shore near the spot where the cultural remains were found, shows the following profile:

    I. Thick layer of humus.
    II. Light-yellow loess-like loam.
    III. Compact, yellow-gray loam with lime concretions. In this layer, at the depth of 180–190 cm. from its surface, were found the cultural remains.
    IV. Gray layer with sand lenses—of earlier glacial times.

The cultural layer was colored red, due to the presence of ochre. Flint implements and a quantity of small broken animal bones were very unevenly spread over the opened area.

Fauna.[3]

Elephas primigenius, numerous.	*Alopex lagopus*, numerous.
Sus scrofa ferus.	*Ursus spelaeus.*
Equus (equus).	*Marmota* sp.
Ovibos moschatus.	*Lepus* sp.
Canis lupus, numerous.	*Ellobius talpinus.*

The fauna, in which the mammoth predominates, is similar to that of Kostenki I, with the addition of the Arctic fox.

Stone Industry.

Unlike Kostenki I, where only dark flint of chalk origin and of very good quality was used, 25–30 per cent. of the implements of Kostenki III were made of red and yellow pebble flint, much inferior in quality.[4] The flint industry (one hundred finished tools and a large number of blades and chips) is rather poor as to variety of types. The blades and

---

[1] Ephimenko, P. P., "Some Results of the Study of the Palaeolithic Period in U. S. S. R.," "Chelovek," No. 1, Leningrad, 1928, pp. 54–55; also "Pre-Clan Society," Leningrad, 1936, p. 346.

[2] Ephimenko, P. P., "Pre-Clan Society," Leningrad, 1936, pp. 429–431.

[3] According to Gromov, V. I., "Geology and Fauna of the Palaeolithic Period in U. S. S. R.," Problems of the History of Material Culture, No. 1–2, Leningrad, 1933.

[4] Menghin, O. ("Die Weltgeschichte des Steinzeit," p. 173), thinks that the use of glacial pebble flint indicates "the culture which has arrived from the north."

PLATE XXX.

EXCAVATIONS AT KOSTENKI III. Photograph of S. Zamiatnin.

finished tools are very small, the largest rarely over 5 cm.   Blades of regular form are absent.   The tools are usually irregular, made on rough, wide, and massive blades.   The majority of the implements are middle and side gravers of rather crude technique.   There are many polyhedric forms.   Ephimenko definitely stresses the unusual paucity of types; with the exception of gravers, which constitute 90 per cent. of all tools, only small end-scrapers and nucleus-like tools of the *rabo* form were found.

Bone Implements.

Many bone implements were found, but they are not characteristic for dating.   There are awls made from tubular bones, and one point of mammoth tusk, round in cross-section.   Further details are not available.

Comparing this site with Kostenki I, Ephimenko adds:[1] "In general, the industry is much more crude, primitive, monotonous, and tools are smaller, clearly showing the decline of skill in flint work, and poor choice of material, which, in Western Europe, is characteristic of early Magdalenian.   It impresses the observer with its crudity and archaism.   It seems to be representative of the culture of a backward people arriving probably from the north (as suggested by the presence of pebbles), and replacing the Solutrean inhabitants of Kostenki I.   This industry seems to be a survival in Magdalenian times of the exceedingly poor and degenerated industry, similar in its general character to that of the Aurignacian."

Dating.—On the basis of the "degeneration of technique" and the monotonous character of the industry, Ephimenko[2] dates Kostenki III as Magdalenian.

Menghin[3] considers Kostenki II and III a degeneration of Aurignacian technique, admits the possibility of influence of "bone cultures," and is inclined to derive this culture from the north, due to the use of "poor glacial flint."   He considers the industry of Kostenki II and III similar to that of Karacharovo (see page 360) but somewhat younger.

12. MEKAMET CAVE (VIRCHOW'S CAVE)

This site is situated near Motzamet (Mekamet) monastery, Kutais region.   It is known in literature as Mekamet or Virchow's Cave.

According to Vishnevsky,[4] it was discovered, in 1914, independently, by S. Kozlovsky and R. R. Schmidt.   The latter asserts that the flint industry is similar to the late Aurignacian of Western Europe.   According to Field and Prostov,[5] the types of flint industry were published by Nioradze,[6] who states that, "the excavations at Rudolf Virchow's Höhle have yielded Upper Palaeolithic cultures, mainly Aurignacian" (Fig. 62a).

Ephimenko[7] states that here, as well as in other Palaeolithic stations of the Transcaucasus, obsidian, which was the basic material for the neolithic industry, was not used

[1] Ephimenko, P. P., "Some Results of the Study of Palaeolithic Period in U. S. S. R.," "Chelovek," No. 1, Leningrad 1928, pp. 45–59.

[2] Ephimenko, P. P., "Pre-Clan Society," Leningrad, 1936, p. 430.

[3] Menghin, O., "Die Weltgeschichte des Steinzeit," p. 173.

[4] Vishnevsky, B. N., "Prehistoric Man in Russia" (Appendix to the translation of Osborn's "Men of the Old Stone Age"), Leningrad, 1924, p. 470.

[5] Field, Henry, and Prostov, Eugene, "Recent Archaeological Investigations in the Soviet Union," American Anthropologist, Vol. 38, No. 2, 1936, pp. 262–264.

[6] Nioradze, G. K., "Das Paläolithicum Georgiens," Transactions, Second International Conference of the Association on the Study of the Quaternary Period in Europe, Facsimile 5, Leningrad, 1935, pp. 225–236.

[7] Ephimenko, P. P., "Pre-Clan Society," Leningrad, 1936, pp. 492–493.

at all, its place being taken by colored flint. The types of implements vary but little: there are flakes with retouched sides, scrapers, and gravers, which by their archaic form remind one of Kostenki II and III. It is interesting that with these are met the nuclei-like forms of the *rabo* type, found also in the Voronezh sites. Characteristic of Mekamet cave, in comparison with the former group, are the flakes with retouched sides which here vary greatly in shape. Their forms remind one of the very early geometrical microliths in the stations of the Southern Mediterranean province of the Upper Palaeolithic of Western Europe.

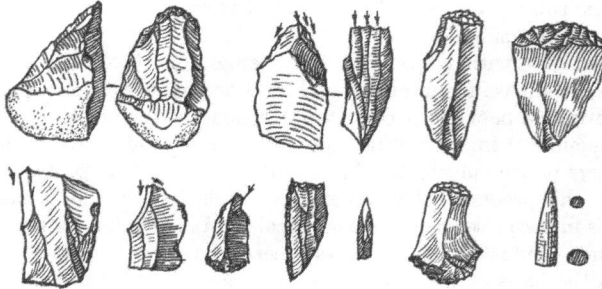

FIG. 62a. Stone Implements at Virchow's Cave. After Field and Prostov.

Dating.—Both Ephimenko and Zamiatnin, who has made an extensive study of Caucasian Palaeolithic cultures, classify Mekamet Cave with the middle period of Upper Palaeolithic cultures, which corresponds to Solutrean and Early Magdalenian. They both feel that no definite Western European parallels can be found, and stress the lack of characteristic types and the gradual changes, which took place throughout the Upper Palaeolithic period in the Caucasus.

Uvarov's Cave, Bartashvilli Cave, and Devis Hvreli (see page 293) belong to the same cultural division.

13. KARACHAROVO

Location.—The site of Karacharovo is situated near the estate of Count A. S. Uvarov, north of the village of Karacharovo, two kl. above the town of Murom, Vladimir province, on the Oka River.

The high bank of the Oka is cut here by a number of gullies leading to the valley of the river. Among the largest are the Melemkov gully, south of the village; the Karacharovo gully, between the village and the estate; and the Buhchinsk gully, at the southern end of Murom.

Discovery and History.—The site was discovered in 1877 by A. S. Uvarov, whose attention was attracted by the finds, in Karacharovo gully, of mammoth bones which were washed out by spring waters. Two other gullies proved to be also ossiferous. Uvarov's excavations, in which the geologists V. B. Antonovich and I. S. Poliakov participated, resulted in the discovery of a cultural layer with more than 500 stone artifacts.[1]

The literature on Karacharovo is very large, but due to the fact that the technique of excavating was far from perfect, and that the untimely death of Uvarov prevented

[1] Uvarov, A. S., "The Archaeology of Russia. The Stone Age," Moscow, 1881, Vol. I, pp. 112–120.

him from publishing a complete account of his find, this site was somewhat neglected and the collection scattered.[1]

The question of the geological age of the Karacharovo deposit was more or less settled by prominent scientists who examined the site at the invitation of Count Uvarov.

Geology.—V. V. Dokuchaev [2] gives the following picture of the nearby vicinity: the basic formation is (I) the Permian limestone in which here and there Jurassic deposits have been preserved; it is followed by (II) a layer of cross-bedded fluvio-glacial sands, at times containing pebbles. Next follows (III) a brown-red moraine clay with pebbles. Above this is (IV) yellow loess-like clay covered at times by (V) a thin layer of sand.

At the site itself, situated on the left slope of the Karacharovo gully under the layer of humus, was a yellow, sandy, loess-like clay, two to three meters thick, which contained the cultural remains. Beneath it was red pebble clay. The observations of Poliakov [3] showed that in the natural cut of the gully the layer of loess-like clay gradually increases toward the river, decreasing toward the head of the gully, where it disappears. On the other hand, the red pebble clay increases in the direction of the gully. The cultural remains were found in the layer of diluvial (?) loess-like clay. The pebble clay, which is cut by diluvial action is a moraine of the maximum (Riss) glaciation.[4]

Fauna.—The fauna is represented by the remains of the following animals:

(a) *Elephas primigenius.*   (c) *Bos* sp.
(b) *Rhinoceros tichorhinus.*   (d) *Cervus* sp.

Cultural Remains (Figs. 63, 64, 65).—The most complete description of Karacharovo is given by S. N. Zamiatnin,[5] who has made a thorough study of the available literature and collections.[6] The following is a more or less full extract of the above work:

The size of the excavated area is small. The cultural remains lay 1.5 meters from the surface in the lower part of the yellow loess-like clay (50–60 cm. above the moraine clay) and present a picture usual for Russian sites: a conglomeration of various animal bones mostly broken to extract the marrow, with pieces of flint, implements, and small pieces of charcoal.

According to Poliakov,[7] most of the bones and flint implements were found in a small

[1] Zamiatnin, S. N. ("The Palaeolithic Site of Karacharovo"), indicates three specimens in the State Historical Museum in Moscow, twelve in the Anthropological Department of the First Moscow University, sixty-five in the Museum of Anthropology and Ethnography in Leningrad, fifty-six in the Geological Department of Leningrad University, fourteen in the Museum of Kiev University, a few in Kazan Museum, and perhaps a few in Poland and Czechoslovakia.

[2] Dokuchaev, V. V., "The Archaeology of Russia by A. S. Uvarov," Trudi of the St. Petersburg Society of Naturalists, St. Petersburg, 1882. Vol. XIII, pt. I, p. 54.

[3] Poliakov, I. S., "Investigations of the Stone Age in Olonetzk Province, in the Valley of the Oka River, and the Upper Part of the River Volga," Zapiski of the Russian Geographical Society, Division of Ethnography, St. Petersburg, 1882, Vol. XI, p. 121.

[4] Malycheff, Vera, "Le Loess" (Review de Géographie Physique, Paris, 1929–30, t. II–III, p. 271), writes "Leurs conditions de gizement ne sont pas précisés, mais il est fort probable que les industries se trouvent ici dans des dépôts contemporains de la partie supérieure du conflexe récent." (Young loess. E. G.)

[5] Zamiatnin, S. N., "Palaeolithic Site of Karacharovo."

[6] Figs. 63 and 64 are reproduced from the above work of Zamiatnin; Fig. 65 from Sawicki, who gives some additional types. (E. G.)

[7] Poliakov, I. S., "Investigation of the Stone Age in Olonetzk Province, in the Valley of the Oka River and the Upper Part of the River Volga," Zapiski of the Russian Geographical Society, Division of Ethnography, Vol. IX, St. Petersburg, 1882, p. 121.

area of 1.5 square meters. Outside of this area separate pieces of bone and flint tools were encountered, widely scattered. Thus, evidently, most of the finds came from one hearth.

The implements of Karacharovo were made of local Permian flint, usually yellow-brown in color with red or gray spots, though some are dark-rose, light gray or yellow. The flint is homogeneous in structure without incrustations. It is encountered in the form of pebbles in the underlying moraine layer. A few tools were made of a hard, white sandstone and diorite.

Nuclei.—Nuclei are quite numerous, according to Uvarov and Poliakov, but only four are found in collections. They are of irregular prismatic form with one surface of percussion, small in size (4–6 cm.), and very much worked. No nuclei corresponding to the large flakes of the collection were found.

Flakes.—Flakes are usually large, short, and wide, often preserving the original surface of the core. The smaller ones were called by Uvarov "scrapers," and the large ones "axes." No small flakes and products of secondary retouching are found in collections. Though found in abundance, due to the methods of excavation they were not preserved.

Blades.—Blades are usually 6–7 cm. long, quite wide and often slightly curved. A few specimens are larger (10–11 cm.). Most of the best large examples are improved along one or both edges by fine retouch (sometimes also on the lower, flat side) or show traces of utilization [1] (Fig. 63, nos. 1–3). Several blades are chipped on one end by detaching large flat flakes, so that the working edge looks like a planing tool (gouge) (Fig. 64, nos. 34–36).[2]

The dented blade (*lame dentelée*), used for the preparation of bone tools (Fig. 64, no. 38), is made on a wide flake, and has three semi-circular notches on the right edge.

Gravers.—Gravers (Fig. 63, nos. 4–8, 11–16) are the most commonly used tools, and include middle gravers, side gravers, angle gravers (Fig. 63, nos. 4, 5, 7, 8), and double gravers (Fig. 63, no. 6). The number of graver flakes found (Fig. 63, nos. 9, 10) indicates the extensive use of these implements.

Scrapers.—Scrapers of two types were found: (1) Large end scrapers made on the end of a long blade, the form usual for the Upper Palaeolithic in general (Fig. 64, nos. 22–25, 31), (2) small scrapers of peculiar form and of several types: (*a*) Small, made on a short wide flake (Fig. 64, nos. 21, 26, 30); (*b*) made on a broken blade (Fig. 64, nos. 20, 22, 32); (*c*) small round scrapers (Fig. 64, nos. 27, 28).

According to Zamiatnin, all types of this second group are found in large quantity in Gontzi and also in Borshevo II, where they constitute leading forms.

Some scrapers of the first, as well as the second of the above groups, do not have a regular arc-like edge, but are somewhat elongated or sharpened. The working edge of these tools (Fig. 64, no. 31, lower and 37) has a steep, dulling retouch often made at an

---

[1] Zamiatnin refers to the chipping of what appears to be the base. (E. G.)

[2] Zamiatnin compares them with similar blades, described by P. P. Ephimenko in Kostenki I ("The Palaeolithic Site of Kostenki," Annual report of Russian Anthr. Soc., Vol. V, pp. 22–23) where they are well represented. He also indicates that H. Breuil ("Voyage paléolithique en Europe Central," L'Anthropologie, t. XXXIV, pp. 529–30) calls attention to the large number of these tools encountered in the flint industry at Predmost (Moravia) and points out that their abundance is characteristic of East-European sites.

obtuse angle to the lower surface, which makes the use of them as scrapers impossible. Consequently it is rather a preparatory stage of the scraper than the tool itself.[1]

Perforators.—Perforators are represented by two examples (Fig. 64, nos. 29, 37). One is made on the end of a long blade with a finely retouched edge.   The other is made on a rough, massive, flake, and the point may, or may not, have been accidental.[2]   To the same class, perhaps, belongs the group of tools with natural points, slightly improved by the retouch.

Bifaces.—The tools with bifacial retouch are very interesting.   The two most typical are shown on Fig. 64, nos. 18 and 19.   The first (Fig. 64, no. 18) is made of flat, wide, leaf-shaped flake with a pronounced point; the upper surface is retouched with small facets; the lower surface has several larger facets, and secondary retouch along one edge.   The second example (Fig. 64, no. 19) is oval and very massive.   It has the same characteristic features—the lower, flat side is retouched with large facets, the upper side is very convex, is worked over with smaller and longer facets, and has secondary retouch.   Two other examples of bifacial technique are the rough "hand-axes" and are less characteristic. They may be placed in the same class with the rough laurel-leaf points of Predmost.[3]

The following implements are reproduced on Fig. 65 from Sawicki's illustrations of artifacts.

  1. Same as Zamiatnin (Fig. 63, no. 2).
  2. Same as Zamiatnin (Fig. 64, no. 38).
  3. Thick flake with partial retouch on one end.
  4. Fragment of a wide blade.
  5. Same as Zamiatnin (Fig. 64, no. 30).
  6. Same as Zamiatnin (Fig. 64, no. 24).
  7. Long blade with some evidence of retouch on one edge.
  8. Same as Zamiatnin (Fig. 63, no. 1).
  9. Long blade with blunt retouch along one edge.
10. Same as Zamiatnin (Fig. 63, no. 8).
11. Same as Zamiatnin (Fig. 63, no. 13).
12. Same as Zamiatnin (Fig. 63, no. 13).
13. Same as Zamiatnin (Fig. 64, no. 27).
14. Same as Zamiatnin (Fig. 63, no. 12).

[1] It was suggested by Mrs. Kelley, that what Zamiatnin calls the working edge may be a blunting of a base, in which case Fig. 34, No. 7, would be basal fragments of other tools.

[2] Neither of the two tools mentioned could be accidental, and they appear to be points rather than perforators, at least as much as one can judge from the illustrations, which do not show the characteristic wear.   (E. G.)

[3] Zamiatnin again refers to H. Breuil (*op. cit.*, pp. 523–525), who points out their massiveness, the convexity of the upper side, and the flatness of the lower, which is retouched by large facets and has no secondary retouch on the edges, as the peculiarity characteristic for these tools.   H. Breuil thinks that the rough leaf-like point of the Hungarian Solutrean has developed from the small *coup de poing* of the Eastern Mousterian. In this connection Zamiatnin points out the find in Berdizh, in 1928, of a typical Mousterian *coup de poing* in the characteristic industry of the final Aurignacian.

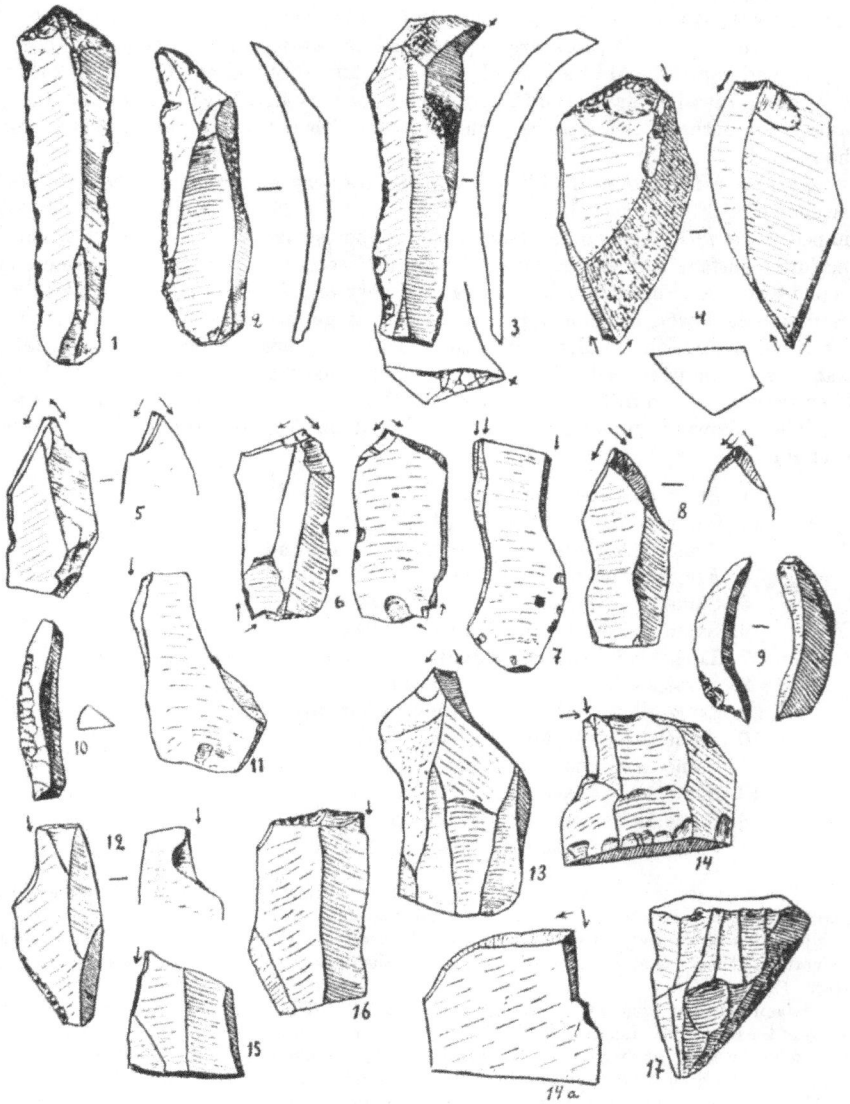

FIG. 63.   Stone Industry of Karacharovo.   After S. Zamiatnin.

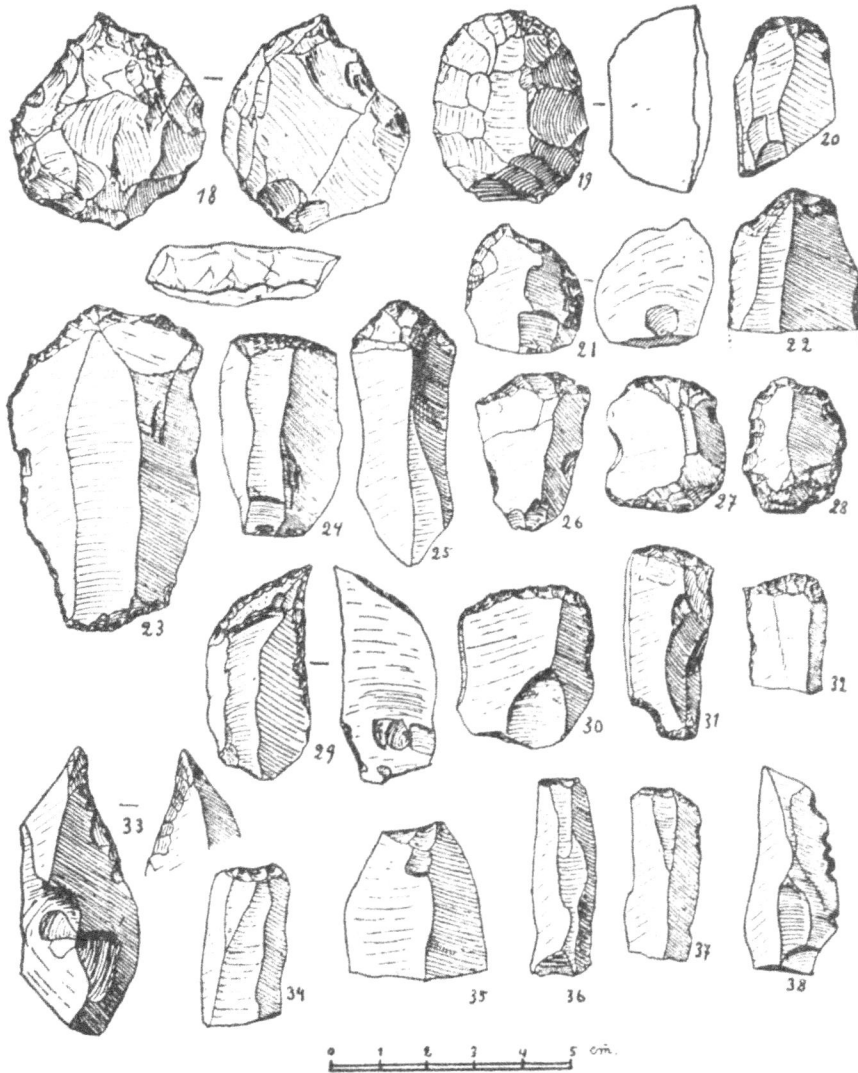

FIG. 64. Stone Industry of Karacharovo. After S. Zamiatnin.

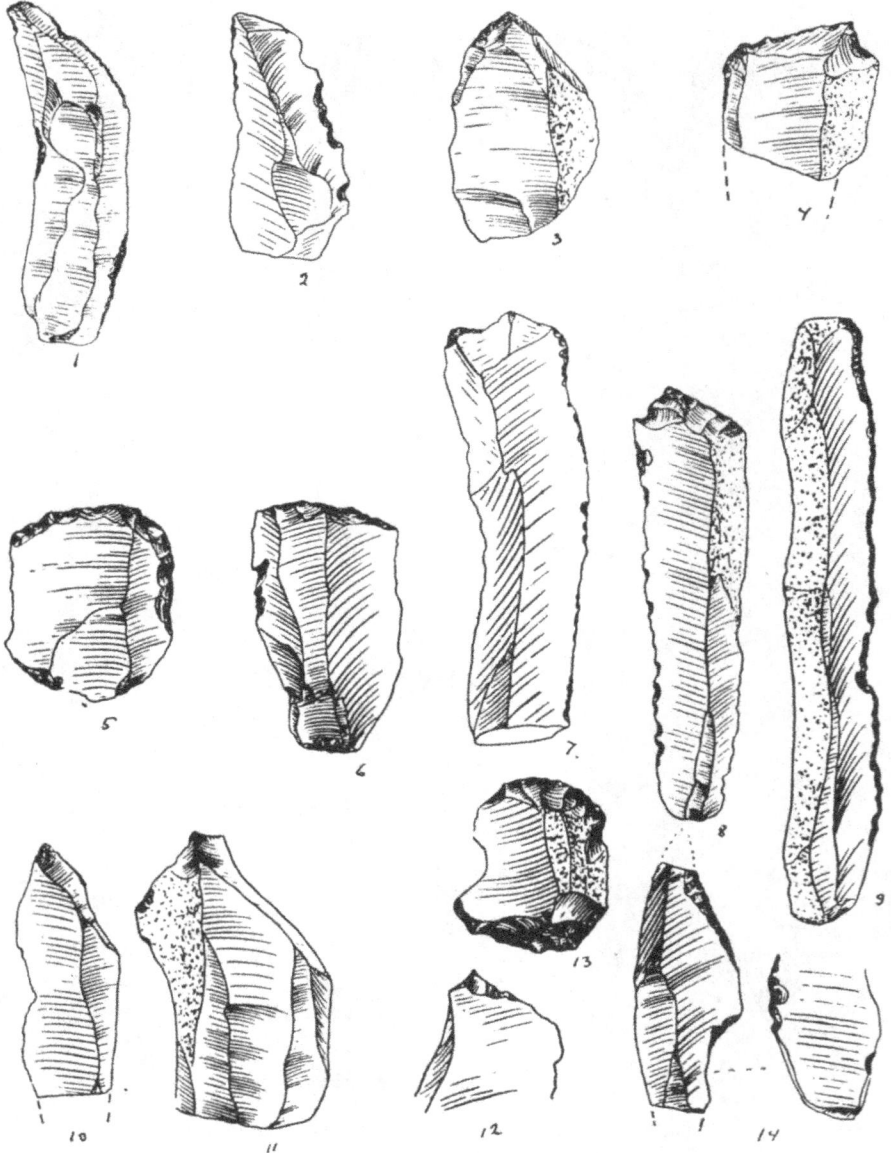

FIG. 65. Stone Industry of Karacharovo. After L. Sawicki.

Bone Work.—The complete absence of examples of bone work, as well as of certain types of flint tools, should be attributed, in the opinion of Zamiatnin, to the defective technique of excavating, as well as to the fact that the material has passed through several hands during which time some of it may have been lost.

Dating.—There is no possibility of dating Karacharovo on the basis of geological data, as it does not furnish a close enough criterion. Likewise the faunal material is undoubtedly incomplete. A possible hint for the dating may be seen in the presence of remains of the Siberian rhinoceros. This animal died out before the mammoth and is not found in a number of other sites which abound in Probiscidean remains. Beyond this fact the only remaining criterion for dating Karacharovo is its flint industry. L. Sawicki [1] considers the general character of the flint work of Karacharovo to be similar to that of Borshevo II, Gontzi, and Cyrill St. and, therefore, he dates Karacharovo as Aurignacian.

According to Zamiatnin, however, this can hardly be accepted. He points out that the fine, beautifully made tools of the latter sites differ markedly from Karacharovo, where the tools are more massive and larger; the blades as a rule do not have parallel sides and are much wider, all of which gives them a more archaic character. The only form which seems to connect Karacharovo with Borshevo II and Gontzi, is the type of scraper (Fig. 64, nos. 27–28), which is round and short. This similarity is not decisive, as in Karacharovo these scrapers do not predominate as they do in Borshevo II and Gontzi. It should also be pointed out that this form in smaller quantities is found in earlier sites such as Kostenki I, Predmost, in the Upper Aurignacian horizon of Lacoste, La Ferrassie, etc.

The gravers are of little use for dating, though the absence of the polyhedric form should be noted. Some help can be derived from the blades with the transverse flaking of the edge (Fig. 64, nos. 34–36). The evolution of this type can be followed from the characteristic specimens of Kostenki I, made on a long blade with the carefully chipped end forming the concave edge, through a number of intermediate stages to *pièce esquille* of Kostenki II and III. The forms of Karacharovo occupy an intermediate position.

It is the presence of the bifacial technique which, in the opinion of Zamiatnin, should be considered decisive for dating. Unfortunately the series represented is small, and as far as can be seen from the material at hand it represents a later stage, the degeneration of the culture.

On the basis of the above considerations, Zamiatnin places Karacharovo later than the well-expressed forms of Final Aurignacian and Early Solutrean of Eastern Europe, and of Kostenki I, Borshevo I, Gagarino, Berdizh and Predmost, but somewhat earlier than the Magdalenian sites of the Voronezh group. Ephimenko [2] considers it early Magdalenian.

14. GONTZI

Location and History.—This site is located at the base of the high right bank of the Udai River, a tributary of the Sula, between the villages of Gontzi and Duhovaya, Lubni district, Poltava region (Fig. 68). It was discovered in 1871 by F. I. Kaminsky, whose attention was attracted by the finds of mammoth bones made by the owner of the land.

[1] Sawicki, L., "Materials for the Archaeology of Russia," Przeglad Archaeologicznii, Vol. III, Poznan, 1928, p. 110.

[2] Ephimenko, P. P., "Pre-Clan Society," Leningrad, 1936, pp. 346 and 429.

Kaminsky's Excavations.—In 1873, Kaminsky undertook the first investigations, making a trench 2.5 feet deep. During the preliminary investigation he definitely established the presence of a Palaeolithic site characterized by an agglomeration of broken, split, and sometimes charred bones of mammoth, reindeer, and other animals.

In a small area of 36 square feet the remains of six mammoths were found; the bones were in layers and the skeletons incomplete. The long marrow bones were all broken, the upper jaws without tusks or frontal bones. In the same layer were found the shells of *Pupa muscorum, Succinea oblonga* and *Helix hispida.*

On one side, slightly higher than this layer of bones, was a deposit 1.5 feet thick, with a large quantity of charred bones of all sizes, small bits of tusks, dental enamel, pieces of flint, and bones of small animals. Most of the stone implements were found in the lower part of this deposit. According to the investigator,[1] 47 flint tools, numerous flakes and hammer-stones were found.[2] He states that all tools were made on flakes and were well preserved, the edges of some being so sharp that a pencil could easily be sharpened by them. Pebbles derived from the nearby boulder clay furnished the raw material. Kaminsky mentions scrapers, knives and other "cutting" tools, one bone awl, and one bone perforator.[3]

Further investigations of Gontzi were made by the geologist K. M. Pheophilaktov,[4] who gives the following cross-section of the site:

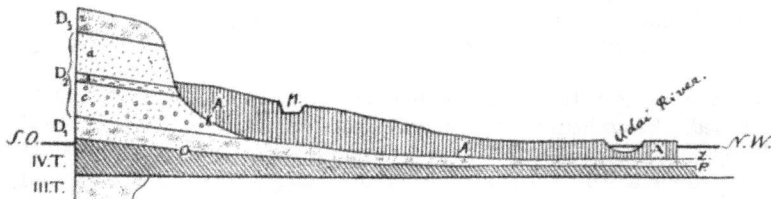

Fig. 66.   Cross-section of Udai Valley After Pheophilaktov.

M, cultural finds; *D₃*, upper pebble layer; D_2, *a, b, c*, loess; *D₁*, lower pebble layer; *IV. T.*, multicolored Tertiary clays; *III. T.*, white Tertiary clays; *K–L*, profile of the valley in Quaternary times; *O–P*, profile of the valley in Tertiary times; *A*, alluvium of the river.

He dates the deposit of Gontzi on the basis of his understanding of the geological history of the Ukraine. According to him, there are remains of two principal Ice Age deposits in South Russia. The lower, more ancient, consists of sands, gray loams, with a large admixture of pebbles, and other stone material derived from Silurian, Devonian, and Cre-

[1] Kaminsky, F., "Traces of the Most Ancient Period of the Stone Age along the Sula River and Its Tributaries," Trudi of the III Archeological Congress in Kiev; Kiev, 1878, Vol. I, pp. 147–152. It should be borne in mind that Gontzi was the first Palaeolithic site discovered in Russia, and at the time Kaminsky considered it Mousterian.

[2] See also Uvarov, A. S., "Archeology of Russia, The Stone Age," Moscow, 1881, Vol. II, no. 2591–2595.

[3] Kaminsky, F., "Traces of the Most Ancient Period of the Stone Age along the Sula River and Its Tributaries," Trudi of the III Archeological Congress in Kiev; Kiev, 1878, Vol. I, pp. 147–152. See table VII, Fig. 1–47.

[4] Pheophilaktov, K. M., "The Location of the Finds of the Stone Flint Tools with the Bones of Mammoth in the Village Gontzi, on the Udai River," Trudi of the III Archeological Congress in Kiev; Kiev, 1878, Vol. I, pp. 153–159.

taceous formations, all brought from the north into this region by the glacial waters. This layer contains remains of mammoth.[1]

The upper layer (the second for Poltava, and the last one for the Kiev region, which is situated further south) consists of loess, which attains in places a thickness of 21 meters. Both layers (the sands or loams with pebbles, and the loess) are found in the same position in the Poltava region as in Kiev, with this difference, however: that along the Udai and Sula Rivers there is also a third layer composed partly of loose quartzitic sands, and partly of yellow loam and sandy clay very similar to the loess in color and composition.

Thus, according to Pheophilaktov, in the Kiev region, the formation of the loess concluded the Ice Age, while in Poltava, which lies further north, this layer of loess was covered by the new advance of the ice (Würm?). The terminal moraine, consisting of gravel, boulders of crystalline, and other formations, formed the third layer. This layer is somewhat mixed with the underlying layer of loess, much in the same way, as the lower pebble (Riss moraine) layer was mixed with the underlying Tertiary formations.

In conclusion Pheophilaktov states that layer A (Fig. 66), in which the cultural finds were made, is a river-deposited formation, yellow in color, lithographically identical with loess, which in the Poltava region, lies above the Würmian glacial deposits, and consequently is of postglacial times.

Excavations by Scherbakovsky and Gorodzov.—After the investigations of Kaminsky and Pheophilaktov, Gontzi was forgotten until V. Scherbakovsky, the curator of the Poltava State Museum, made the first really scientific investigation of this site in 1914–1915. In 1915 Prof. Vernadsky, geologist, and V. A. Gorodzov, archaeologist, participated in larger excavations, the results of which, unfortunately, are as yet very inadequately published [2] (Fig. 69). The following summary is made mainly on the basis of the work of Scherbakovsky and Gorodzov, with the differences between the two pointed out when necessary.

Geology.—The investigators excavated an area of 60 square meters and made a number of trenches for the geological profiles, which reveal the following stratigraphy (Fig. 67):

[1] Perhaps a Riss moraine. (E. G.)

[2] (a) Scherbakovsky, V., "The Excavations of the Paleolithic Settlement in the Village Gontzi, Luben County, Poltava Region in 1914–1915," Zapiski of the Ukrainian Society for the Study and Preservation of Antiquities in Poltava region, Poltava, 1919, pp. 62–75. (Poor illustrations.)

(b) Burkitt, M. C., "Archaeological Work in Ukraine by Prof. Scherbakovsky," The Antiquaries Journal, July, 1925, Vol. V, no. 3, pp. 273–277.

(c) Gorodzov, V. A., "Archaeology, The Stone Age," Moscow, 1925, Vol. 1, pp. 285–290.

(d) Gorodzov, V. A., "The Investigations of the Palaeolithic Site of Gontzi in 1915"; RANION, Moscow, 1926, Vol. I, pp. 5–35. (Geology, industry, and very inadequate illustrations.)

(e) Breuil, H., Review of "Les fouilles de la station paléolitique (Magdalenien) dans le village de Hontzi, district de Loubne, gouv. Poltava (Ukraine) en 1914 et 1915;" "L'Anthropologie," Vol. XXXIV, p. 427.

(f) Rudinsky, M., "The Archaeological Collection of the State Poltava Museum," Poltava, 1928, pp. 7–8, reproduces twelve specimens of flint industry.

(g) Sawicki, L., "Materials for the Study of Archaeology of Russia," Przeglad Archeologicznii, Poznan, 1926–1928, Vol. III, pp. 2–3. (Gives the best illustrations.)

(h) Ephimenko, P. P., "Pre-Clan Society," Leningrad, 1936, pp. 440–443. (Reproduces several stone and bone implements.)

At the base of the slope of the Udai valley is a layer (*IV*) of multi-colored clays (*IV. T.* of Pheophilaktov and *D* of Gorodzov), over which lies a stratum (*III*) of gray sands (*C* of Gorodzov, absent in Pheophilaktov). This is covered in turn by a layer (*II*) of boulder

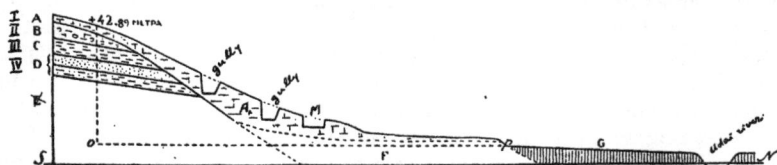

FIG. 67.   Cross-section of Gontzi Site.   After V. Gorodzov.

G, alluvium of Udai River; *A₁*, diluvial deposits; *F*, diluvial terrace; *A*, humus; *B*, pebble clay; *C*, Tertiary quartzite sands; *D*, multicolored clays; *E*, marl; roman figures, those of the author.

clay, moraine of the Würm glaciation (*B* of Gorodzov, *D* of Pheophilaktov). Over this moraine clay there lies a layer (*I*) of yellow loess, or a sort of loess-like clay or loam (*A* of Gorodzov, *A* of Pheophilaktov), which, sloping downwards, covers the ancient terrace of the Udai River.

Pheophilaktov indicated (Fig. 66) the presence of still another layer of the boulder clay (*D2*) which, according to both later investigators, was nowhere observed.

The structure of the terrace itself, according to Scherbakovsky, is as follows:

I. Loess-like loam with a large admixture of humus, 60 cm. thick.

II. Loess, 2.20 meters thick, in the lower part of which was the cultural deposit, 60 cm. thick.

III. Sandy-clayish, schistous river deposit, the ancient alluvial of Udai, reaching a thickness of ten meters.

IV. Marl, which forms the floor of the valley.

Gorodzov thinks that both II and III are of diluvial origin, and Pheophilaktov attributes the formation of both to alluvial processes.[1] Barring these differences, it is evident that the cultural deposits were found in a layer which was formed during the last retreat of the Würm glaciation, which fact seems to be substantiated also by the fauna.

Fauna.—According to Gorodzov, supplemented by the list given by Gromov,[2] bones of the following animals were found:

1. *Elephas primigenius*, very numerous.
2. *Elephas primigenius minor*.
3. *Bos* sp.
4. *Cervus* sp.
5. *Alces alces*.
6. *Rangifer tarandus*.
7. *Canis lupus*.
8. *Ursus arctos*.[3]
9. *Lepus timidus*.
10. *Lepus variabilis*.
11. *Sus scrofa ferus*.

[1] According to Gorodzov, this site existed during the strong diluvial activities of the last glaciation. At that time the high bank covered by the loess of the Third glaciation was washed by strong precipitations, and, becoming the object of the action of alluvial processes, moved down on the gentle slope of the Udai river valley. (Gorodzov, V. A., "The Investigations of the Palaeolithic Site of Gontzi in 1915," RANION, Moscow, 1926, Vol. I, pp. 5–35.)

[2] Gromov, V. I., "Geology and Fauna of the Palaeolithic Period in U.S.S.R.," Problems of the History of Material Culture, No. 1–2, Leningrad, 1933, pp. 28–29.

[3] According to Ephimenko, P. P. ("Pre-Clan Society," Leningrad, 1936, p. 441), *Ursus spelea.*

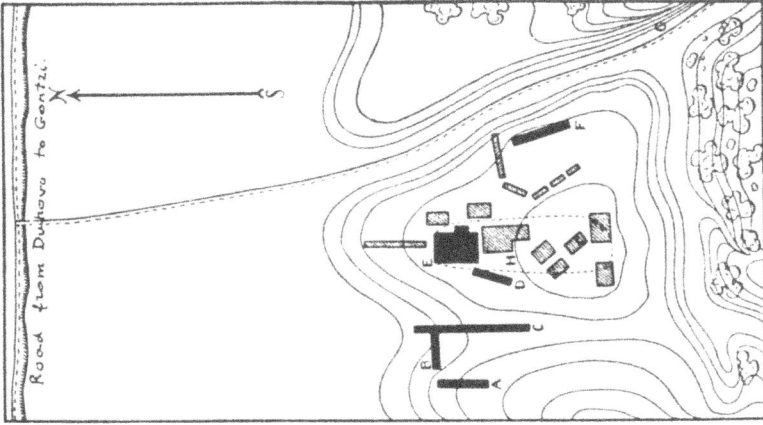

FIG. 69. The Site of Gontzi. Solid portion indicates excavations of 1915 of Gorodzov; shaded portion indicates excavations of previous years. After V. A. Gorodzov.

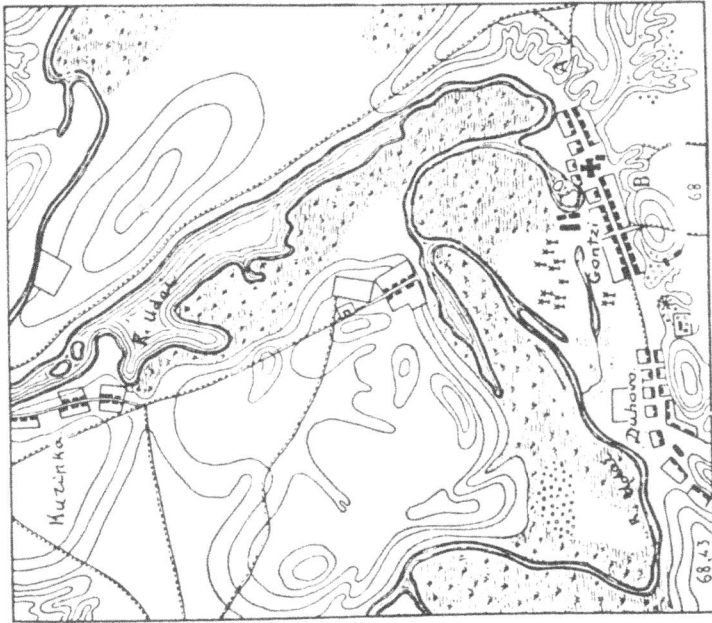

FIG. 68. General Map of Gontzi Site and the Valley of the Udai River. After V. A. Gorodzov.

Cultural Remains.—The excavation of Scherbakovsky resulted in the discovery of five groups of animal bones, varying in size from 25 sq. m. for the largest, to 7 sq. m. for the second group. The rest were about 1 sq. m. each (Fig. 70).

As is usually the case in several other Upper Palaeolithic sites of European Russia, the cultural finds were almost totally absent in the areas between these bone accumulations.

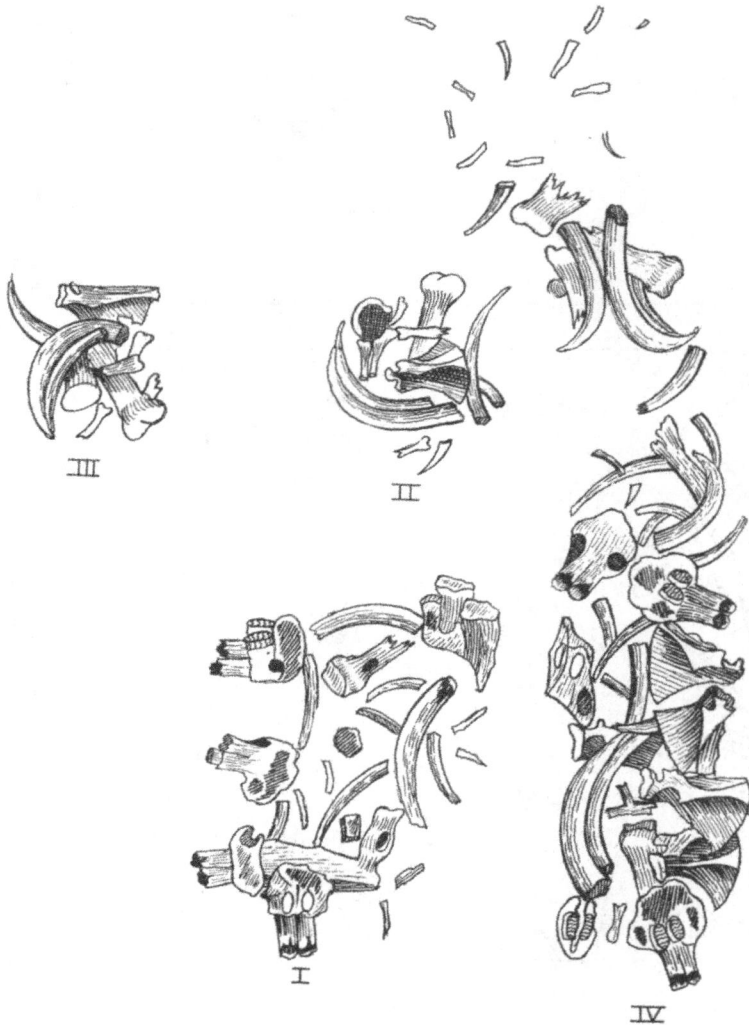

FIG. 70. The Bone Piles of Gontzi. After V. A. Gorodzov.

The middle portion of each pile consisted of small bone splinters, charred bone, and what the excavator calls "sweepings." This material was covered around the periphery by larger bones such as skulls, hip-bones, shoulder blades, and tusks of mammoth.

The actual significance of these bone-heaps is not clear. Gorodzov sees in them sweepings from dwellings, regular kitchen middens.[1] Scherbakovsky suggests that the large bones, covering the piles of small bone splinters, could have served to protect the contents of the pile from the wind(?).[2] Ephimenko interprets these piles quite ingeniously in accordance with his firm belief in the existence of Upper Palaeolithic dwellings. In his opinion,[3] the kitchen-middens in the center of the piles represent the floor of the dwelling where all sorts of rubbish easily accumulate. The larger bones were originally placed on the roof of the dwelling, possibly as a ready supply of raw material for the making of bone tools;[4] later, with the sagging of the abandoned or burned hut, these large bones slid down from the roof, forming the peripheral zone. This hypothesis does not, however, explain the similar conditions encountered in the piles of only one square meter in area (too small for a hut) and the too regular character of the "protecting" position of the larger bones. In this connection the observation of Gorodzov is quite interesting. According to him,[5] in pile no. 1 (Fig. 70) several tusks were found, lying alongside the slope of the pile, with the end of their curves turned down. He rightly remarks, that this testifies to intentional placing of tusks in this position on the soft pile of sweepings. Had they been just thrown down, or had they fallen or slid down, as suggested by Ephimenko, the law of gravity would have caused the ends to turn upwards.

Scherbakovsky describes further the remains of a hearth, characterized by the layer of very small charred bones, 2 cm. thick. He definitely states that nowhere on this site has he found any charcoal, though according to Gorodzov,[6] the bone piles were usually accompanied by a few pieces of charred wood. Both investigators agree however that Gontzi Man may have used bone instead of wood as the fire material.

The investigations seemed to show that the animals were killed elsewhere and brought to the site, as there was found only one lower jaw of mammoth to 27 upper ones, many shoulder blades, but few long bones, only a couple of vertebrae, and very few ribs. Mammoth seems to have been the main game for the inhabitants of Gontzi. It was estimated that there were bones of at least forty mammoth, the second largest number after the Cyrill Street site. The majority of bones belong to young individuals. After mammoth, reindeer and hare bones are the most numerous.

[1] Gorodzov, V. A., "The Investigations of the Palaeolithic Site of Gontzi in 1915," RANION, Moscow, Vol. I, 1926, pp. 5–35.

[2] Scherbakovsky, V., "The Excavations of the Palaeolithic Settlement in the Village Gontzi, Luben County, Poltava Region in 1914–1915," Zapiski of the Ukrainian Society for the Study and Preservation of Antiquities in Poltava region, Poltava, 1919, p. 68.

[3] Ephimenko, P. P., "Pre-Clan Society," Leningrad, 1936, p. 441.

[4] Or as the fire supply, which would also explain the large number of shoulder blades, not as good for bone work. It should be pointed out, that in spite of the abundance of tusks, very few objects were made out of ivory. (E. G.)

[5] Gorodzov, V. A., "The Investigations of the Palaeolithic Site of Gontzi in 1915," RANION, Moscow, 1926, Vol. I, p. 21.

[6] Gorodzov, V. A., "The Investigations of the Palaeolithic Site of Gontzi in 1915," RANION, Moscow, 1926, Vol. I, p. 20.

Stone Industry.

Dark gray or yellow flint of chalk origin furnished the principal material for the stone industry. Perhaps it is due to the limitation of material (small pebbles from the boulder clay) that the size of the finished tools, blades, and nuclei, is comparatively small. The majority of blades is only 4–5 cm. long.

The illustrative material of the stone industry is very limited, and because of the importance of the site, it was thought advisable to reproduce all that is available to the writer, in spite of occasional duplication and inadequacy of representation.

The following types of tools were found in Gontzi:

Nuclei.—Gorodzov reproduces two nuclei (Fig. 71, nos. 1, 2) of conical form.

Blades.—Blades are not numerous in the material illustrated. Usually they have a well retouched or blunted back (Fig. 71, nos. 3–4; Fig. 73, nos. 12–14; Fig. 74, nos. 1–3). Some are of the type of *lame de canif* (Fig. 71, nos. 5–7), and some are very small (Fig. 71, nos. 9–15).

Scrapers.—Scrapers form the largest group. In the excavations of Scherbakovsky, out of some 200 tools, there are 95 scrapers. They are usually very small and are often made from a broken blade. . There are small round *grattoirs*, very high, with semicircular retouch (Fig. 71, no. 17; Fig. 73, nos. 17, 18). One example has almost a straight edge. The single and double *grattoirs* are made on both ends of a broken blade (Fig. 73, no. 15; Fig. 74, nos. 18–20) often in combination with burins (Fig. 71, nos. 20, 21).

Burins.—Burins are also numerous (75). There are many examples of *burin d'angle* (Fig. 74, nos. 10–12) and *burin bec de flûte* (Fig. 71, nos. 22–28; Fig. 74, nos. 13, 17). There are also double forms such as *bec de perroquet-grattoir* (Fig. 74, nos. 7, 8), and examples of *bec de peroquet atypique* (Fig. 71, nos. 29, 30; Fig. 74, no. 11).

Perforators.—Perforators are rare (four in Scherbakovsky's excavations) made on narrow blades, finely retouched on both sides to form a point (Fig. 73, no. 16; Fig. 74, nos. 21–22). Gorodzov also mentions flat stones with traces of use, which he thinks were used for sharpening the bone needles and awls. Scherbakovsky speaks of a dozen round pebbles found in one place.[1]

Bone Industry.

Awls (Fig. 71, nos. 32–41; Fig. 72, nos. 1, 2).—Awls are made usually out of the long bones of a hare, split and sharply pointed on one end. To the same category may belong the fragments of larger ivory points, though Ephimenko calls them dart points.

Needles.—Needles are much less numerous, there being only two whole ones, very beautifully made, and perforated at one end (Fig. 72, no. 1).

Hammer.—A peculiar object, with a round hole in the middle, is called a hammer by Gorodzov. Scherbakovsky sees in it a sort of hair ornament, while Ephimenko considers it a miniature dart-shaft straightener or an implement for the softening of strips of hide (Fig. 72, no. 3). According to Scherbakovsky, a piece of reindeer horn, an unfinished *baton de commandement*, was found. (No illustration is given.)

Decorated Objects.

Contrasted with the splendid examples from Mezine, Gontzi yielded very little in the way of art, with the exception of one awl of mammoth ivory with traces of lineal ornament (Fig. 72, no. 4) and a tip of mammoth tusk with several engraved lines crossed

[1] Sling-stones ? . E. G.

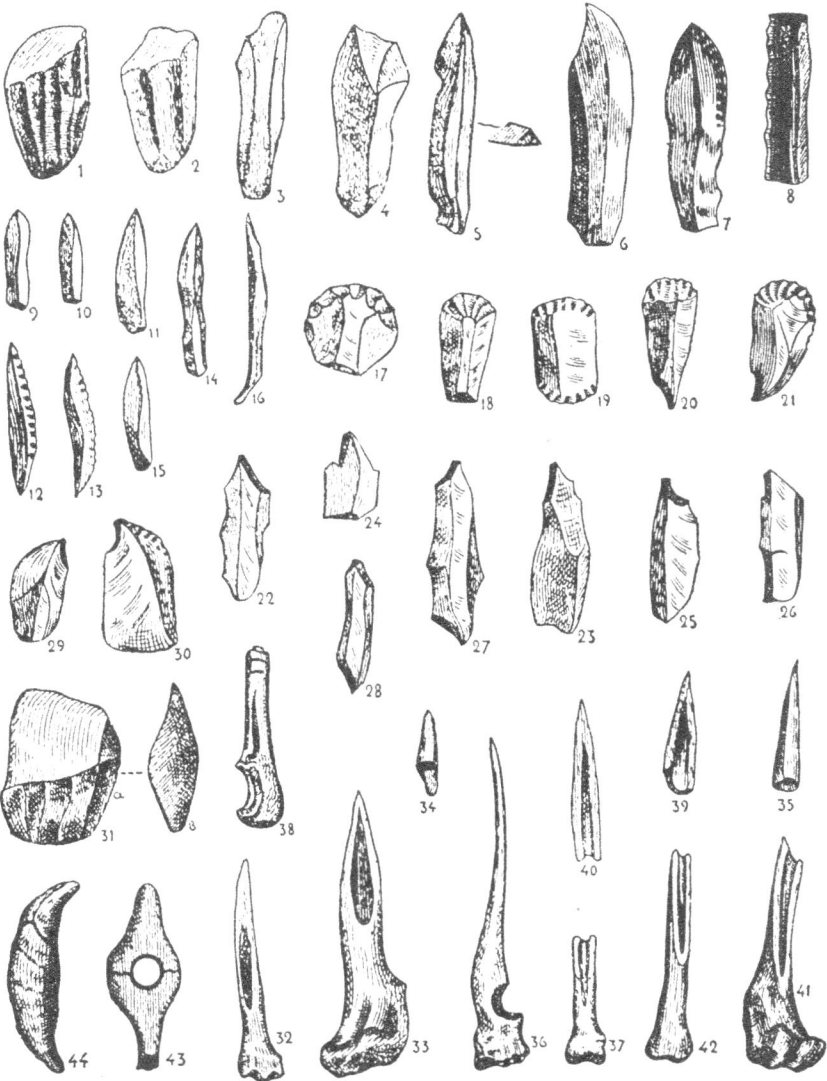

FIG. 71.   Stone and Bone Industry of Gontzi.   After V. A. Gorodzov.

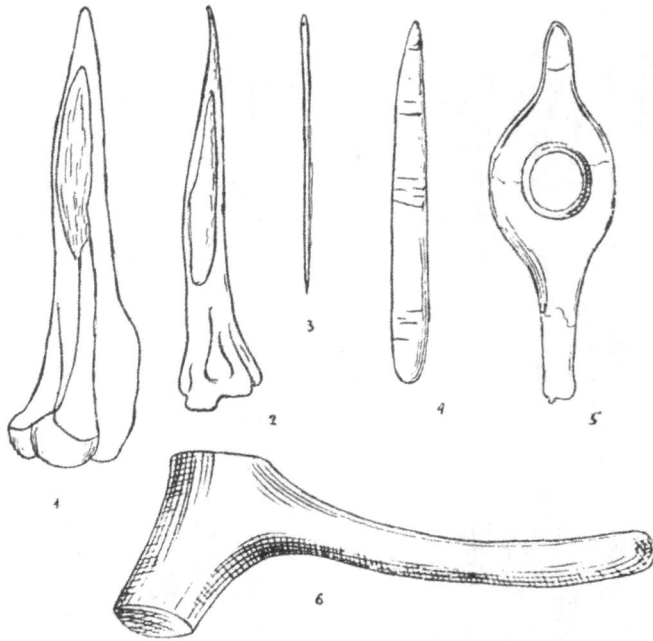

Fig. 72. Bone Industry of Gontzi. After V. Scherbakovsky.

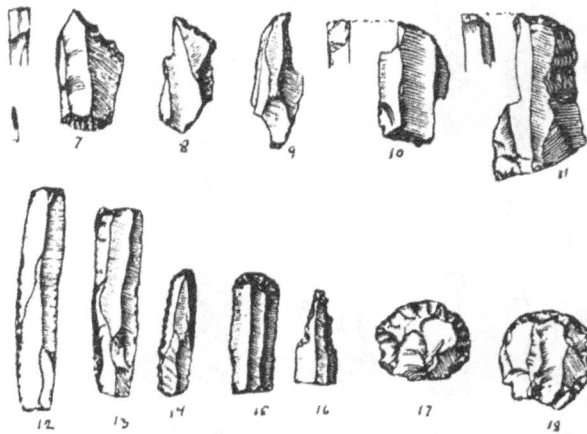

Fig. 73. Stone Industry of Gontzi. After M. Rudinsky.

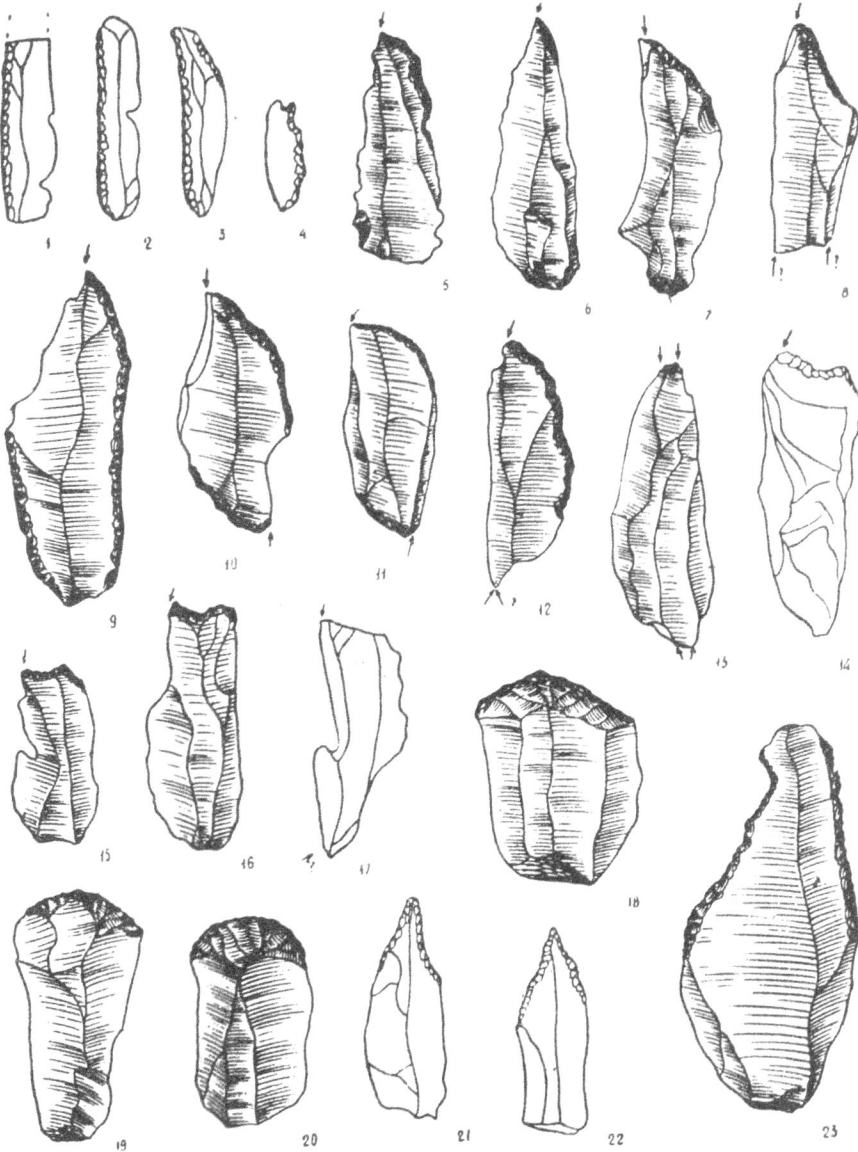

FIG. 74. . Stone Industry of Gontzi. After L. Sawicki.

at right angles and at regular intervals by smaller lines of various length (Fig. 75). The latter may be an attempt at ornamentation or, what is more likely, some sort of hunting record.[1]

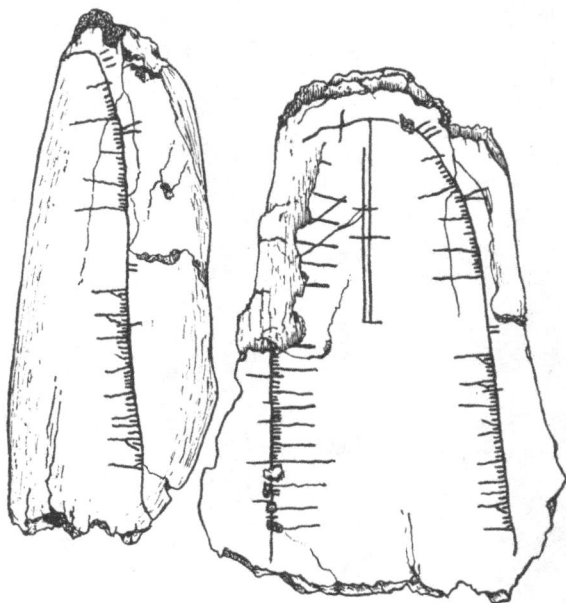

Fig. 75.   Tip of Mammoth Tusk with Lineal Engravings from Gontzi..   After V. Scherbakovsky.

Dating.—On the basis of the general character of the stone and bone industries of Gontzi, most of the investigators date this site as Magdalenian.

According to Gorodzov, Gontzi is somewhat later than Mezine, and considerably later than the site of Cyrill Street.

Scherbakovsky likewise dates it as Magdalenian, and points out the general similarity to Mezine, with one notable exception: horse remains, so abundant in Mezine, are absent in the kitchen refuse of Gontzi.

On the basis of the presence of miniature forms, such as small round scrapers, *lame de canif*, and *bec de perroquet*, Ephimenko places Gontzi in the very last phase of Magdalenian; almost the transition to the Azilian.

M. C. Burkitt [2] quite wisely suggests the term of "Eastern European Upper Palaeo-lithic" which, according to him, occurs in an area "where the Magdalenians, a French folk, never penetrated, although their influence was no doubt felt." Menghin [3] thinks

[1] Burkitt, M. C., "Archaeological Work in Ukraine by Prof. Scherbakovsky," The Antiquaries Journal, July, 1925, Vol. V, No. 3, Fig. 2.

[2] Burkitt, M. C., "Archaeological Work in Ukraine by Prof. Scherbakovsky," The Antiquaries Journal, July, 1925, Vol. V, No. 3, p. 273.

[3] Menghin, Otto, "Die Weltgeschichte des Steinzeit," p. 174.

that the industry of Gontzi is related to Kostenki I, but is younger.   The geometrical forms show, according to him, the Tardenosian influence.

15. BORSHEVO II

Location.—Borshevo II is located between the villages of Kostenki and Borshevo, at the point where the shore of the Don River approaches the mountains of the right side of the valley, leaving just enough space for the narrow terrace of the steep bank.   The chalky bank is furrowed here by numerous gullies.   At the wide mouth of one of them lies the site of Borshevo I.   A short distance further down the river the shore is even wider, forming a more or less even plateau, only 4–5 meters over the river level.   The site of Borshevo II, located here, is periodically covered by spring waters.

History.—Borshevo II was discovered, in 1923, by P. P. Ephimenko, who found a cultural layer running for 100 meters along the bank of the Don between two gullies.

Geology.—According to the excavator,[1] the geological structure of the terrace is as follows:

    I. Humus.

    II. Yellow loam with many incrustations of chalk, about 1 meter thick, composed apparently of diluvial water deposit and the material washed out of the high shore.

    III. A dark layer of ancient humus 5–10 cm. thick, containing the upper cultural horizon.

    IV. A compact clayish loam containing a very large quantity of small fresh-water mollusks; it is an alluvial deposit, reaching the river level.   This layer contains the middle and lower cultural horizons.

    V. Gray tertiary clay.

Fauna.—After Gromov:[2]

	Lower and Middle Horizons.	Upper Horizon.
Elephas primigenius	X	absent
Sus scrofa ferus	X	X
Equus (equus)	numerous	very numerous
Bos sp.	traces	absent
Ovibos moschatus	X	X
Alces alces	X	absent
Rangifer tarandus	X	X
Canis (canis) (sp.)	X	X
Vulpes sp.	X	absent
Gulo gulo	X	X
Felis leo	X	X
Marmota	X	X
Lepus sp.	very numerous	X

[1] Ephimenko, P. P., "Pre-Clan Society," Leningrad, 1936, pp. 443–446.   A full account of this site is as yet unpublished by Ephimenko.

    Sawicki, L. ("Materials for the Study of Russian Archaeology," Przeglad Archeologicznii, Poznan, 1926–28, Vol. III, pp. 2, 3), however, gives quite a bit of information and the series of good plates which are reproduced here.

[2] Gromov, V. I., "Geology and Fauna of the Palaeolithic Period in U.S.S.R.," Problems of the History of Material Culture, Leningrad, 1933, pp. 1–2.

The fauna in these three separate horizons has a more or less similar arctic character; *Gulo gulo,*[1] *Rangifer tarandus* and *Equus* sp. are the most common and are found in large quantities. In the lower horizon, situated in the ancient river deposits, mammoth are quite numerous, diminishing in quantity in the middle horizon, and completely disappearing in the upper.

The lower horizon abounds in remains of hare; among which was found a leg of this animal which, judging by the position of the bones, had evidently been cut off intact. This custom is quite prevalent among modern Siberian hunters.

The upper cultural horizon was found in the third layer, and consists of the remains of a hearth, characterized by nests of charcoal, charred bones, numerous flint chips, and lumps of red and yellow ochre. No cultural finds higher than the limits of the "ancient humus" were made.

Slightly further up the river the cultural finds in level III cease, but in layer IV was encountered the middle horizon situated about 25–30 cm. below layer III. No remains of the hearths were found here; bones and flints were scattered. Below the middle horizon lies the lower horizon which consists of a thin intermittent streak of charcoal. Ephimenko feels that both the lower and middle horizons are the remnants of a large site destroyed here by fluvial action.

Flint Industry.[2]

The majority of stone implements of Borshevo II were made of dark flint derived from chalk. Only 25 per cent. were made of pebble flint of different colors.

During the excavations, the following types of implements were found:

> *Lower Horizon.* .... Figs. 79–80, nos. 1–14.
>> *Nuclei* .... No. 14
>> *Grattoir* on broken blade .... Nos. 6, 7
>> Double *grattoir* on broken blade .... No. 8
>> *Grattoirs* on flake .... Nos. 9, 10, 11, 12
>> *Burin d'angle* .... Nos. 1, 2
>> *Burin bec de flûte* and *grattoir* .... Nos. 3, 4, 5
>> *Burin medial* .... No. 13

Ephimenko points out the presence of nucleiform tools (*rabot*) and the general massiveness of *burins*.

> *Middle Horizon.* .... Figs. 78–79, nos. 1–33.
>> Blades with partial retouch .... Nos. 1–4
>> Backed blades .... Nos. 5, 6, 7
>> Points (?) .... Nos. 8, 9
>> *Burin d'angle* .... Nos. 10–19, 25
>> Double multifacetted *burins* .... Nos. 20–24, 28
>> *Burin bec de flûte* .... No. 26
>> *Burin bec de flûte* and *burin d'angle* ...... No. 27
>> *Burin bec de flûte* and polyhedric .... No. 29
>> *Burin d'angle* and *grattoir* .... No. 30
>> *Grattoir* on retouched blade .... Nos. 31, 32
>> *Grattoir* .... No. 33

---

[1] Ephimenko, P. P. ("Some Results of the Study of Palaeolithic Period in U.S.S.R.," " Chelovek," No. 1, 1928, Leningrad, p. 56), mentions instead *Gulo borealis.*

[2] The stone industry of Borshevo II is well published by L. Sawicki, whose plates are reproduced here (E. G.)

This industry in general is quite similar to that of the lower horizon, with the exception of the predominance of double tools.   The flint in this level is vari-colored.   Ephimenko is inclined to believe that the color of the material meant more than its quality to the inhabitants of Borshevo II.[1]

Upper Horizon.	Figs. 76–77, nos. 1–37.
Nuclei	Nos. 36, 37
Nuclei of discoidal type	No. 34
Rabot or tool of nucleiform type	No. 35
Points of *La Gravette* type	Nos. 1–3
Points of *La Gravette* type-broken	Nos. 4, 7
Lame de canif atypique with partial side retouch.	Nos. 5, 6
Burin d'angle	Nos. 8–12
Nucleiform *burin*	No. 13
Burin bec de flûte	Nos. 14–15
Burin lateral-d'angle (?)	No. 16
Burin bec de flûte-grattoir	Nos. 17, 19
Burin d'angle on broken blade and *grattoir*	No. 18
Grattoir	Nos. 20–24
Grattoir sur bout de lame courte	Nos. 25, 26
Grattoir on thin wide flake and blade	Nos. 27–32
Double high *grattoir*	No. 33

The tools of this horizon are made exclusively from dark flint of chalk origin.   The cultural finds are concentrated in areas, 3–5 meters in diameter, consisting of ashes, charcoal, charred and split bones, flint chips, and pieces of ochre.   In one of these "hearths" of this horizon was found a half-shell of a fresh-water mollusk filled with bright red ochre, which, according to Ephimenko,[2] was previously powdered and mixed with some other matter.

On the excavated area of 250 square meters Ephimenko found ten such "hearths" which he considers the remains of huts.   It is interesting to note, that here, as in Gontzi (see page 367), no cultural remains were found in the area between these "hearths," which, themselves, were not very rich in finds, averaging several dozen implements in each.

The characteristic feature of the stone industry of this horizon, according to Ephimenko is the larger size of the tools (7–9 cm.) and their regularity of form.

Bone Industry.

Lower Horizon.—According to Ephimenko [3] the following implements were found: [4]

1. Awls made of split, tubular hare bones, similar to those of Gontzi.
2. Bone points.
3. A bone needle perforated at the end.
4. A plaque of mammoth tusk showing the process of needle making.

[1] The utmost utilization of good blades and a large quantity of double tools point out to the general scarcity of good raw material, necessitating the use of pebble flint which just happened to be vari-colored. (E. G.)

[2] Ephimenko, P. P., "Pre-Clan Society," Leningrad, 1936, p. 251.

[3] Ephimenko, P. P., "Pre-Clan Society," Leningrad, 1936.

[4] There are no illustrations available.   (E. G.)

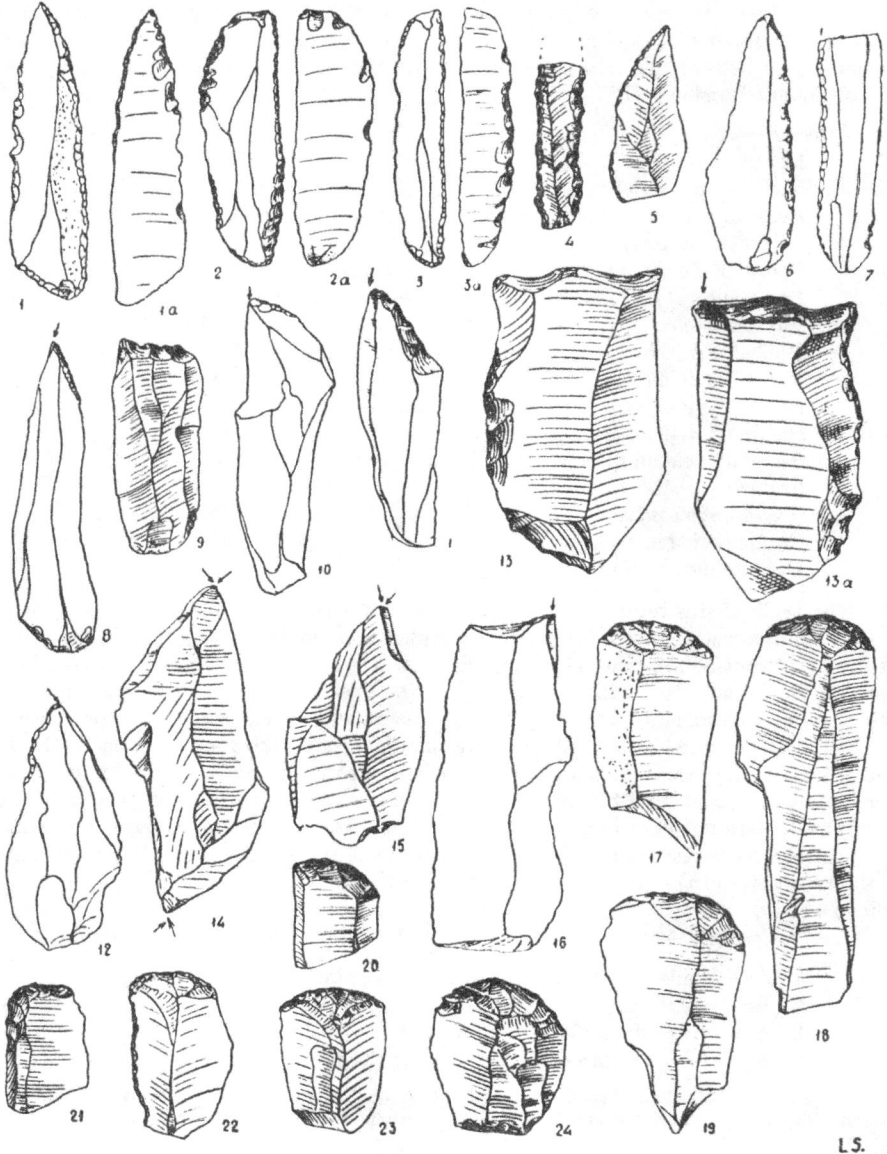

FIG. 76. Stone Industry of Borshevo II, Upper Layer. After L. Sawicki.

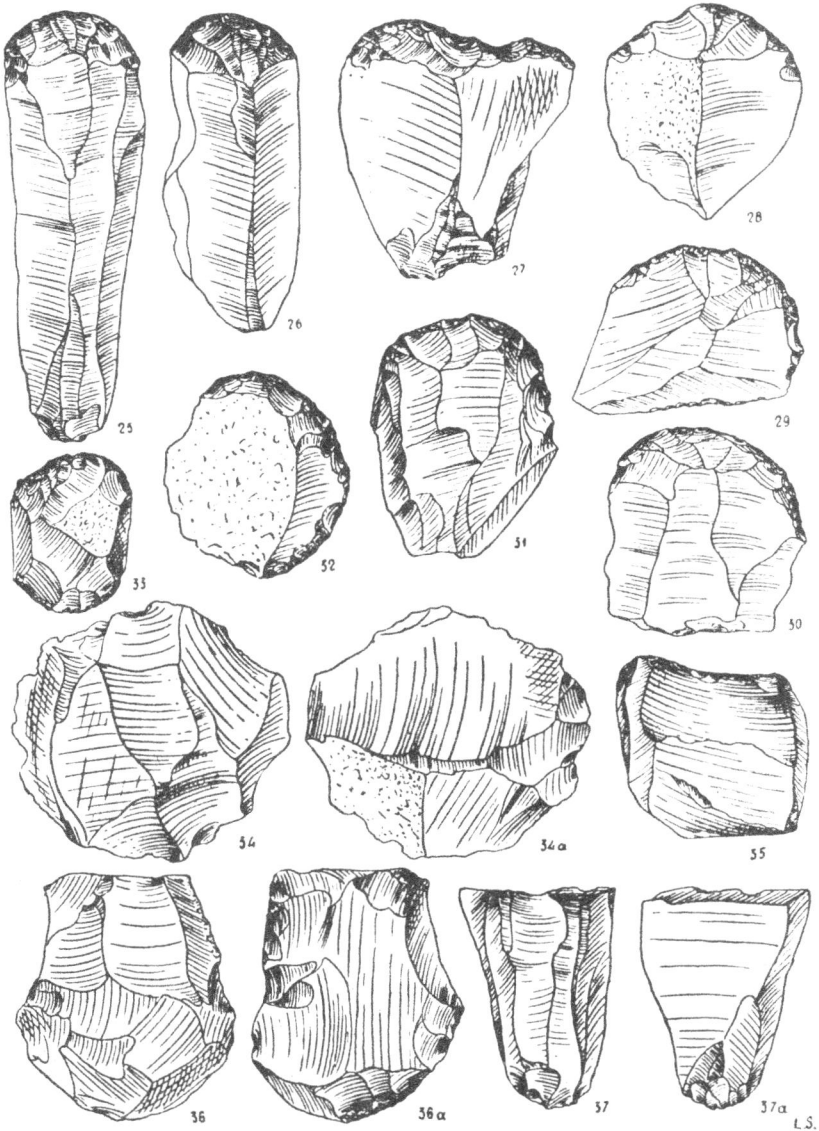

FIG. 77.   Stone Industry of Borshevo II, Upper Layer.   After L. Sawicki.

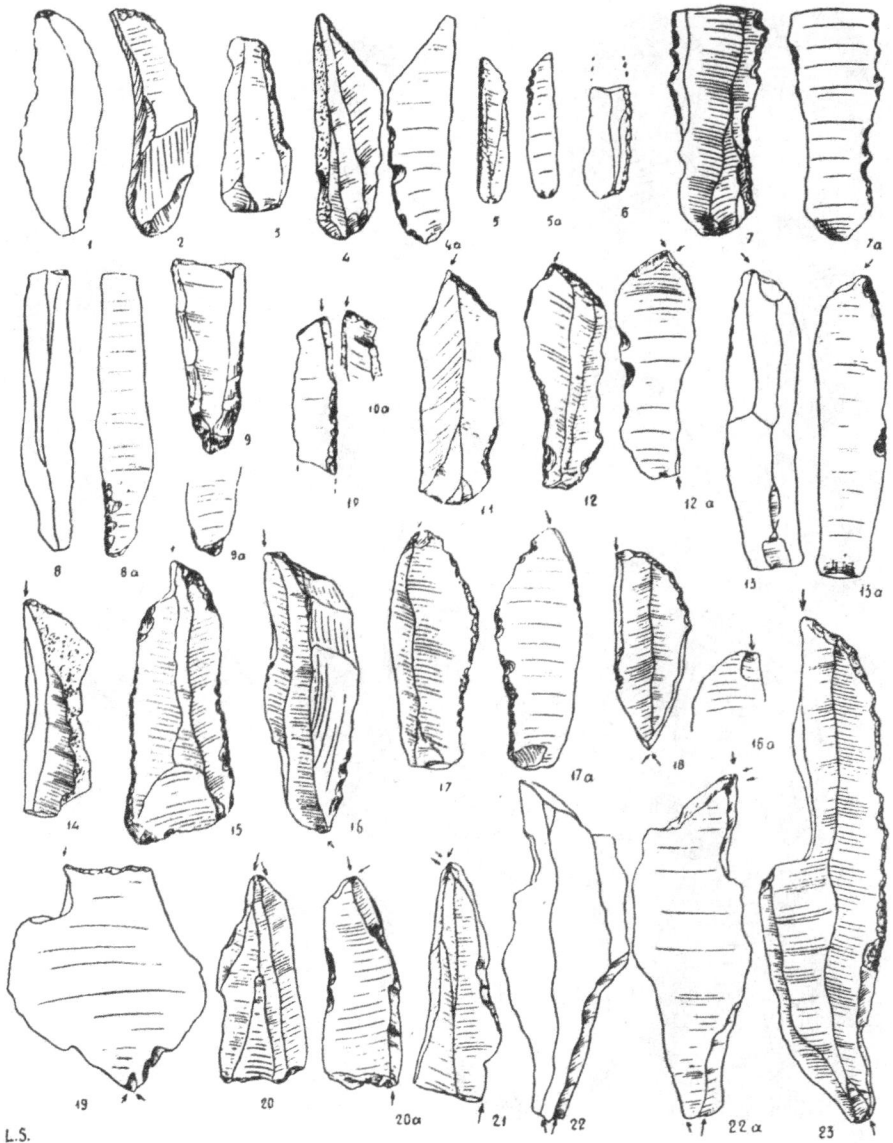

FIG. 78. Stone Industry of Borshevo II, Middle Layer. After L. Sawicki.

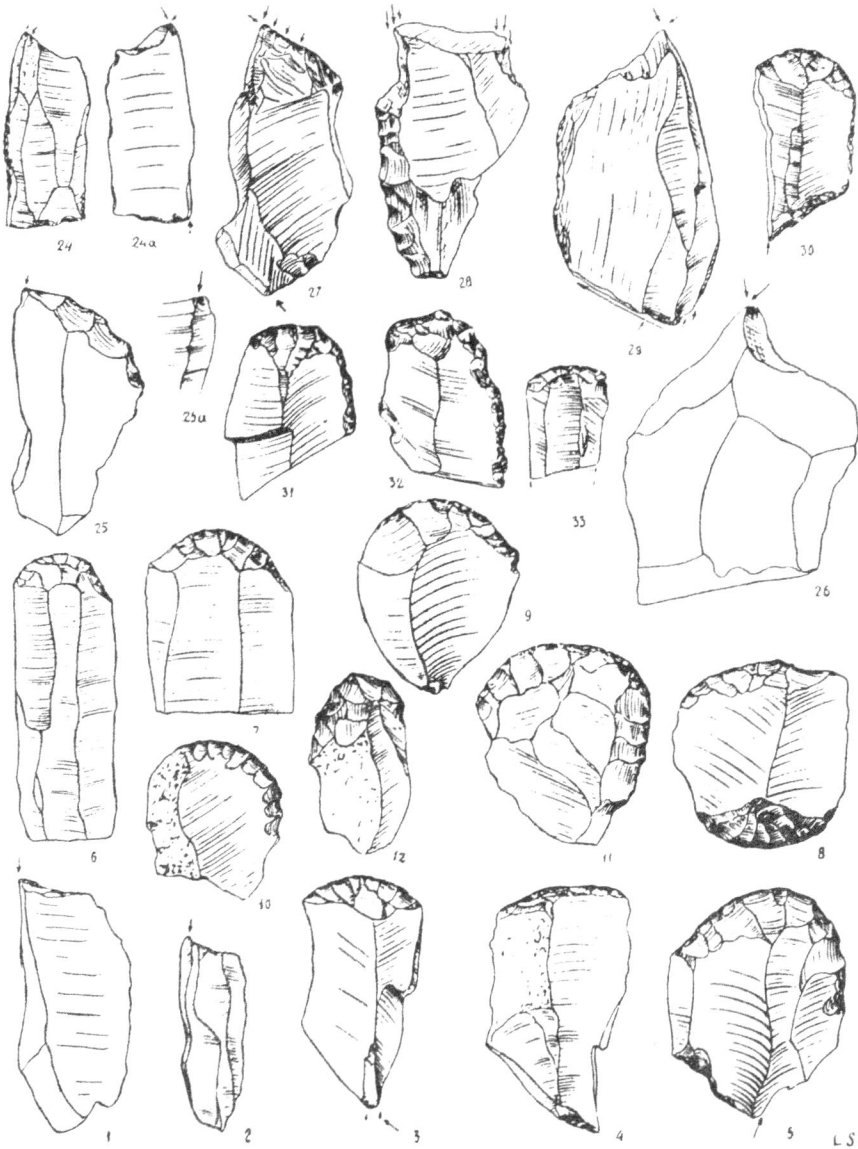

FIG. 79. Stone Industry of Borshevo II, Middle (24–33) and Lower (1–12) Layers. After L. Sawicki.

Middle Horizon.—There is no bone in the middle horizon.

Upper Horizon.—

    1. Flat points made of tubular bone (hare?).

    2. A fragment of a bone plaque decorated with a geometrical design.

According to Ephimenko shells of *Dentalium* made into a small tube and cut on both ends (necklace?) were also found in this horizon.

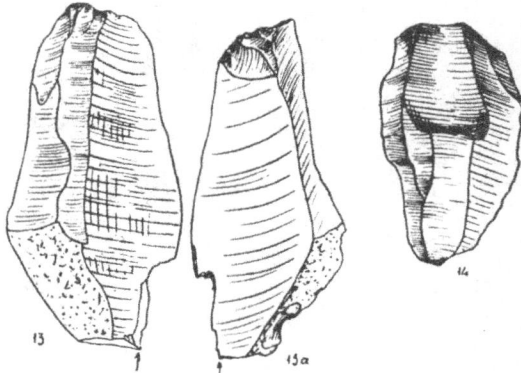

Fɪɢ. 80. Stone Industry of Borshevo II, Lower Layer. After L. Sawicki.

Dating.—According to the excavator, the lower and middle horizons of Borshevo II are quite similar to Gontzi, and are of late Magdalenian affinities.

The upper horizon, however, while having the same similarity to Gontzi, surpasses the latter in the size and regularity of the forms, almost approaching Epipalaeolithic. Ephimenko dates it as final Magdalenian, corresponding to the upper horizon of Cyrill Street (page 386) and Zhuravka (page 405).

16. Cyrill Street, Kiev

Location.—This site is located in the heart of Kiev on Cyrill Street, Podol section, near Yordan Church, which is at the foot of one of the branches of hills, that run along the Dnieper River. Two large gullies separate it from the river.

Discovery.—It was discovered, in 1893, by V. V. Khvoiko,[1] who made excavations in 1893, 1894, 1895, 1896, and 1899, in which V. B. Antonovitch[2] and P. Y. Armashevsky[3] participated. Unfortunately, this extremely interesting and enormous site was very badly investigated. Excavations of Khvoiko in 1893–1896 were done on a very small scale due to the expense involved in clearing the overlying layers, which were several hundred feet

[1] Khvoiko, V. V., "The Stone Age of the Middle Dnieper Region," Trudi of the XI Archaeological Congress, Vol. I; also, "The Kiev-Cyrill Street Site and the Magdalenian Culture," Archaeological History of the South of Russia, 1903, No. 1.

[2] Antonovitch, V. B., "The History of Habitations in the Territory of Kiev from Palaeolithic Times to the Beginning of Christianity," "Kievskoe Slovo," 1896, No. 2927.

[3] Armashevsky, P. Y., "The Palaeolithic Site of Cyrill Street in Kiev," Izvestia of the XI Archaeological Congress, Vol. II, pp. 142–144.

thick.   Later, in 1899–1900, the owner of the land made large excavations of spondyl clay for his brick factory.   During this time about nine thousand square meters of the site were opened in the presence at Khvoiko whose observations, of necessity, could not be very detailed.

Geology.—According to Armashevsky, as quoted by Gorodzov,[1] the geological structure of the site is as follows:

I. Humus.

II. A series of formations of undoubtedly Quaternary age, with a total thickness of 32 meters as follows:

1. Loess, which formed during the third interglacial period, but whose diluvial shifting did not take place until the fourth glaciation.   The loess, being light, is very susceptible to diluvial processes of shifting by precipitation waters.   During the fourth (Würm) glaciation, when atmospheric precipitation was increased, a considerable shifting of loess took place from the higher points to the slopes, as is evident in the present section.

2. Various marl formations of post-glacial origin.   Gorodzov supposes that the marl layers were deposited by the waters of the third glaciation.   The height of this horizon, which lies twelve hundred and eighty feet above the present level of the Dnieper, shows that during the period of its formation the river did not have the present deeply furrowed bed.

3. A layer of yellow-red-gray moraine clay, containing strata of coarse sands.   The clay abounds in large and small boulders of granite, gneiss, diorite, quartzite, and limestone, carried over from the northern mountains of Scandinavia, Finland, and Valday.   (Gorodzov states that this deposit was made by the third (Riss) glaciation.)

4. A layer of marl loam with a thin layer of dark reddish-gray clays similar to those of group III (see below).   In this horizon are found remains of fresh-water molluscs of the genus *Limnaeus*, *Planorbia*, etc., indicating that this layer was deposited by river or lake waters.

III. A layer, forty five feet thick, containing a series of red-gray spotted kaoline pottery clays.   The age of this stratum is undetermined but it is thought to be post-Tertiary.

IV. A layer of gray-white and then pure white quartz sands, sixty-three feet thick. No fossils were found in it, but because of the proximity to the greenish sands, this layer is attributed to the same Tertiary age, having been deposited by the drying and receding sea.

V. Greenish sands 45 feet thick, are called by Armashevsky "ameroid" because of the presence of pieces of amber in the upper layers.   This amber is the pitch of ancient pine trees that grew on the shores of the sea which deposited the green sands.   Besides amber, stems and roots of sea plants are found in this layer.

VI. At the base are spondyl clays with a total thickness of one hundred feet, of which some seventy-five or eighty feet are above the level of the Dnieper.   They are bluish in color; green when dry.   The clay is very plastic and contains separate bones and even whole skeletons of fish and many marine shells, one of which, *Spondylus Buchii*, is re-

---

[1] Gorodzov, V. A., "Archaeology, The Stone Age," Moscow, 1925, Vol. I, pp. 232–39.

sponsible for the name. Sponges, starfish, corals, plants of the lentil family, pines and tropical palms are also found. The presence of the latter indicated that the formation took place in sea waters near a very fertile continent.

During the retreat of the Würm glacier, the raised waters of the Dnieper washed off all these deposits down to the level (VI) of spondyl clay. The most ancient cultural deposits of the Cyrill Street site were found under the steep slope of the shore on the surface of the exposed layer of spondyl clay.[1] This layer was in the form of a terrace some 450–475 feet wide. At its base were the waters of the Dnieper, on the other side was the steep shore of Kiev. This spot was an ideal habitation, and man used it. As time passed, the surface of this terrace-like platform was covered with material falling from the steep hilly shore. There were boulders of various crystalline formations from the moraine clay covering the high bank, and loose sands, among which were multi-colored sands of Tertiary and Quaternary deposits, that gave to this new layer a greenish-gray color.

Man remained under these new conditions, as the greater part of the kitchen refuse was found at the base of this debris. The process continued until the sand layers reached a thickness of 28–30 feet. Some of the layers became ferrous sandstone, others acquired a schistous structure. Only after this, the diluvial loess was deposited, reaching a thickness of forty feet. All these enormous changes took place during the IV (Würm) glaciation. Thus, the cross section of the site at the place of finds shows:

1. Humus.
2. Loess.
3. Clayish sands.
4. Sand.
5. Sand.
6. Sand.
7. Fine sands with pebbles.
8. Gray sands with three horizons of culture finds.
9. Blue spondyl clay.

The geologist, N. I. Krishtafovitch, agrees with Armashevsky that the age of the whole geological complex covering the cultural remains, including the gray sands containing them, is of post-glacial times. He thinks, however, that the gray sands with the cultural remains and mammoth bones in the section of Kiev, as described by Armashevsky, were deposited during a period synchronous with the last glaciation [2]

Fauna (according to Gromov [3]).—

1. *Elephas primigenius* (many).
2. *Rhinoceros tichorhinus.*
3. *Bison* (*priscus*).
4. *Ovibus moschatus.*
5. *Alces alces.*
6. *Gulo gulo.*
7. *Felis leo.*
8. *Marmota* sp.
9. *Spalax microphtalmus.*
10. *Ellobius talpinus.*

[1] Armashevsky attributes all formation of the terrace deposits to washing out, disregarding the deposits laid by the river during the abatement of its course.

[2] Krishtaphovitch, N. I., "The Post-Tertiary Formations in the Neighborhood of Novo-Alexandria," Warsaw, pp. 65–66.

[3] Gromov, V. I., "Geology and Fauna of the Palaeolithic Period in U.S.S.R.," Problems of the History of Material Culture, No. 1–2, Leningrad 1933.

An important feature of the Cyrill Street site is the fact that its cultural remains are found not in one, but in several horizons, showing that we are dealing here with several succeeding habitations.

The Lower Horizon.—The oldest cultural horizon lies on the spondyl clays. In it were found traces of two fireplaces,[1] containing ashes, charcoal, charred wood, bones, a considerable number of flint tools, several worked bones of mammoth, pieces and even whole trunks of petrified cedar and pine, a quantity of mammoth bones, most of which were tusks, and one skull of *Rhinoceros tichorhinus*.

The best preserved and investigated hearth was long and irregular in shape, forming a large thick layer, fifty meters long, fifteen meters wide and forty to fifty cm. thick.[2] It consisted of ashes, charred bones and wood, and large quantities of mammoth bones (jaws with tusks, leg bones, shoulder blades, and vertebrae) on its periphery. On one side there were more tusks than bones which sometimes lay in piles. Only few bones of other animals were found. Altogether, there were fifty-three mammoth jaws and over one hundred tusks, constituting, according to Volkov, the remains of sixty-seven individual animals.[3] Toward the hill the quantity of ashes and bones decreased. At that end were found one large bone of some animal, and three trunks of coniferous trees, two meters long and 30 cm. thick, one with part of the roots preserved, all of them with burned ends.[4]

In other hearths charcoal was more often found, sometimes in nests of one and a half meters in diameter, always containing an agglomeration of large charred bones and ashes.

The flint implements are in the form of small sharp flakes, such as are often found in neolithic workshops (Plate XXXIII).[5] The larger type of flint tools is absent. Among the bone work are notable:

1. A young mammoth's tusk about six feet long with a groove chiseled along its length, covered by a quantity of notches, made by heavy blows of a flint tool [6] (Fig. 82).

2. Part of another tusk, seven inches long, the surface of which was engraved with a design partly obliterated by time. Some investigators distinguish among ornamental figures the outline of a bird with the eye and the figure of a turtle [7] (Fig. 81, Plate XXXI).

The find of mammoth bones associated with the bones of *Rhinoceros tichorhinus* in the lower layers of this site testifies, in the opinion of Gorodzov, to the considerable antiquity of the site.

[1] Three according to Spitzin, A. A., "The Russian Palaeolithic Period," Zapiski of the Imperial Russian Archaeological Society, Slavic Division, Vol. XI, Petrograd, 1915, pp. 132–172.

[2] Spitzin, A. A., "The Russian Palaeolithic Period," Zapiski of the Imperial Russian Archaeological Society, Slavic Division, Petrograd, 1915, Vol. XI, pp. 132–172, Fig. 12.

[3] Volkov, Th. (Th. Vovk), "The Prehistoric Finds in Cyrill Street, Kiev," Zapiski of Scientific Society of Shevchenko in Lvov, No. 1, 1898.

[4] According to Gorodzov; V. ("The Results of Archeological Activities in U.S.S.R., During 1917–1930." Manuscript), one of them, with part of the roots, had traces of notches, evidently made by a sharp tool.

[5] Illustrations available are far from being satisfactory, and the separations of horizons is not always clear. (E. G.)

[6] Volkov, Th., "Magdalenian Art in Ukraine," Zapiski of Scientific Society of Shevchenko in Lvov, Lvov, 1902, Vol. XLVI, pp. 1–13, Table I.

[7] According to Spitzin, A. A. ("The Russian Palaeolithic Period," Zapiski of the Imperial Russian Archaeological Society, Slavic Division, Vol. XI, Petrograd, 1915), it was later found accidently some distance from Khvoiko's excavations.

According to Gorodzov,[1] some of the bones were worked into implements (?) and with them were found flint implements in the form of scrapers and knife-like blades.

The bone tools were made of mammoth tusks and leg bones. Heavy clubs were made of tusks and gigantic adze-like implements of leg bones. In the latter case one end was cut obliquely and sharpened. The surface of some of them was covered with notches and other depressions.[2]

The Upper Horizon.—The upper horizon lies in the layer of gray sand. It may be further subdivided into two sections:

1. Some three and a half feet over the lower horizon were found remains of a charcoal layer, with a very small quantity of bones, mostly broken and charred.

2. Seven feet higher were found groups of hearths of a different character.[3] They were found in nests, round or oval in plan, and marked by ashes, charcoal, charred bones and very many small flint implements, hammerstones and flint pebbles stored for the manufacture of tools. About twenty such cultural nests were found, situated from six to sixteen feet above the lowest horizon and twelve to twenty-two feet beneath the foundation of loess. Mammoth bones were found rarely and only in small fragments, but with them were remains of:

> *Leo spaeleus.*[4]
>
> *Hyena spaelea.*
>
> *Ursus spaeleus.*

None of the above animals was found in the lower horizon.

The Flint Industry.

The material found in the Cyrill Street site is not very plentiful, considering the area excavated. There is quite a difference of opinion in evaluating it. The older writers, like Spitzin, felt that one should treat the different horizons separately as belonging to different cultures. Gorodzov (1925) states that all flint industry of the Cyrill Street is of one cultural horizon.

The illustrations available [5] (Plates XXXII, XXXIII) do not always distinguish the horizons, but it seems advisable to consider the latest opinion of P. P. Ephimenko,[6] who, undoubtedly, is one of the best Russian specialists. He definitely distinguishes two periods:

[1] Gorodzov, V. A., "Archaeology," The Stone Age, Vol. I, Moscow, 1925, pp. 232–239.

[2] Unfortunately there are no illustrations of these unparalleled implements. It is possible that Gorodzov refers to objects illustrated on Figs. 81, 82. (E. G.)

[3] According to Vishnevsky, B. N. ("Prehistoric Men in Russia," Appendix to translation of Osborn's "Men of the Old Stone Age," Leningrad, 1924, p. 462), "One gets an impression that on definitely outlined areas of hearths were shelters under which fires were built."

[4] According to Vera Malycheff, "Le Loess," Revue de Géographie Physique, 1929–30, Vol. II, III, also *Gulo* sp.

[5] Volkov, Th. ("The Prehistoric Finds in Cyrill Street, Kiev," Zapiski of Scientific Society of Shevchenko in Lvov, No. 1, 1898), gives a photograph of the lower, and some drawings of the upper horizon.

Zelizko, J. V. ("The Settlement of Diluvial Man in Kiev," Casopisi Vlasten Muzej Spolku, v. Olomouci, 1907, cislo 95–96, Olomouci, 1907), gives two plates of photographs without indicating the horizon, probably the upper.

According to Spitzin, the best illustrations are given by Grushevsky, "History of Ukraine," but not available to me. (E. G.)

[6] Ephimenko, P. P., "Pre-Clan Society," Leningrad, 1936, pp. 433–437.

Older Horizon.—The older horizon is characterized by a small number of tools (200 chips, flakes and unfinished objects). The majority of finds consists of small flint blades of the Upper Palaeolithic character, among which are some *burins*. All tools are made out of dark gray flint of chalk origin. The general aspect of the industry according to him is very near that of Kostenki I and Karacharovo.

From the point of view of dating, the objects made of mammoth tusk are of more importance. Thus, in this horizon were found the following:

1. A tip of a mammoth tusk 170 cm. long, and 13 cm. in diameter at the butt. Along the inner side of the curvature, starting from the butt and running three quarters of the length, there is a deep incision, triangular in cross section.

2. A fragment of the mammoth tusk, 21 cm. long and 4.5 cm. in diameter at the middle, has an engraved line with twenty-three smaller lines perpendicular to the first one, running at about equal intervals (Fig. 82, no. 2).

3. A larger fragment 21.5 cm. long and 6 cm. in diameter, had two parallel lines running lengthwise with nine perpendicular lines connecting them. Volkov suggests the possibility of a hunting record (Fig. 82, no. 1).

4. The most interesting object is the end of a mammoth tusk, 30 cm. long, covered with the engraved design (Plate XXXI) of rather complex and undeterminate character. Mr. Khvoiko distinguishes there (Fig. 81) a bird (*a*) and a turtle (*b*). J. V. Zelizko sees a lake landscape with a figure of the bird just alighting on the shore of the lake. Volkov pointed out the similarity of the ornamental motives to the examples of Magdalenian art in Western Europe. We have, according to Volkov, the following main elements:

A long line with perpendicular notches, similar to objects illustrated in Fig. 82, 1 and 2, which sometimes form a ladder design, (*c*). A zig-zag line, (*d*). Rows of semi-curves, (*e*). Curved lines with perpendicular notches, (*f*).

The technique of Cyrill Street engravings, according to Ephimenko,[1] is typically Magdalenian, but the high schematization and stylization of the subject matter resemble very much the female engraving of Predmost, and geometrical tendencies of Mezine. All this makes Ephimenko date this horizon as early Magdalenian.

The Upper Horizon.—The flint industry of this horizon is much richer than that of the lower one. Besides a very large quantity of the flint chips and flakes, there were found fine regular blades, nuclei, and large pieces of unworked flint as a supply for manufacturing, which was done on the spot.

Scrapers.—Scrapers are small in size and resemble very much the similar forms of Borshevo II and Gontzi.

Burins.—Burins are often made on small blades and are of medial and *bec de flûte* type; others are made on small pointed blades with finely retouched edges, having at times triangular and other geometrical forms, quite similar to microlithic types.

Ephimenko [2] considers the upper horizon of the Cyrill Street of very late Magdalenian, almost a transition to the Azilian. Gorodzov, however, places both horizons of Cyrill Street in the time corresponding to the Solutrean in Western Europe. He definitely states that, "The east European cultures of the middle period of the Palaeolithic (our Solutrean) are absolutely alien to the corresponding cultures of Western Europe." According to

[1] Ephimenko, P. P., "Pre-Clan Society," Leningrad, 1936, p. 437.
[2] Ephimenko, P. P., "Pre-Clan Society," Leningrad, 1936, pp. 448–450.

Vishnevsky,[1] the German investigators called it Aurignacian on the basis of the similarity of the design on the tusk to the bone engravings of Predmost.[2]   Volkov comparing it with the Madgalenian of France derives this culture from the West.   Khvoiko derives it, together with the mammoth and the reindeer, from the northeast.   Vishnevsky seems to take the

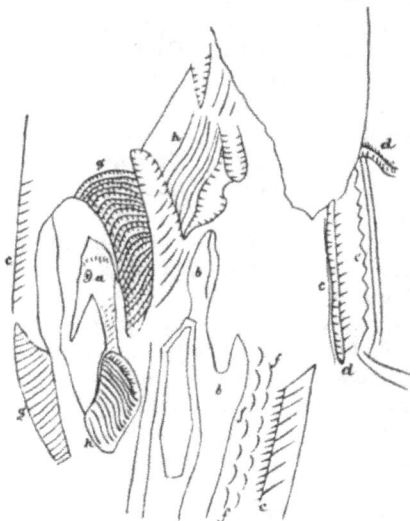

FIG. 81.   Enveloped Design of the Engraved Mammoth Tusk from Cyrill Street.   After Volkov.

FIG. 82.   Mammoth Tusk with Engraved Notches from Cyrill Street.   After Volkov.

[1] Vishnevsky, B. N., "Prehistoric Man in Russia," Appendix to translation of Osborne's "Men of the old Stone Age," Leningrad, 1924.

[2] Ebert, Max ("Reallexicon der Vorgeschichte," p. 32), says that the stone industry "suggests early Aurignacian."

PLATE XXXI.

ENGRAVED MAMMOTH TUSK FROM CYRILL STREET. After Khvoiko.

PLATE XXXII.

STONE INDUSTRY OF CYRILL STREET. Probably Upper Horizon. After V. Želízko.

PLATE XXXIII.

STONE INDUSTRY OF CYRILL STREET. Probably Lower Horizon. After V. Želízko.

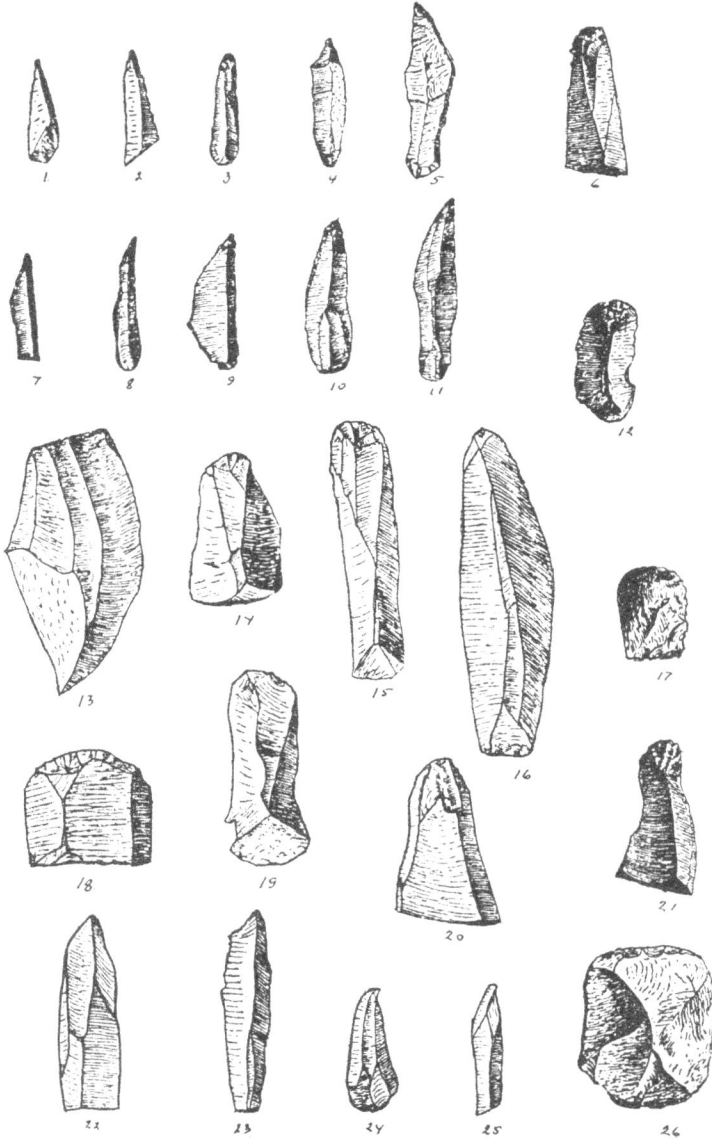

FIG. 83. Stone Industry of Cyrill Street. Excavations of 1898. Probably Upper Horizon. After Volkov.

point of view of Spitzin that the remains of Cyrill Street belong to at least three different periods of Palaeolithic time.

17. Suren II

Location and History.—Suren II was discovered by K. S. Merezhkovsky at the same time as Suren I in 1897. It is situated next to the latter (see p. 289).

Merezhkovsky collected [1] a rich flint industry of small implements, a great part of which were small end scrapers, often made of very short blades and flakes; gravers (middle and side), perforators (?), and peculiar flint points, either accidentally pointed, slightly retouched blades, or much better made ones, with pressure flaking along the edge.

Merezhkovsky adds very little about this site, beyond indicating that he found cultural layers similar to those in Suren I. The industry seems to be very like that of the upper horizon of Suren I.

The excavations of G. A. Bonch-Osmolovsky in 1924, added the following facts to our knowledge of this site:

He describes [2] Suren II as a rock shelter, very similar to Suren I. The stratigraphy corresponds to Merezhkovsky's:

1. A shallow contemporary layer with large slabs fallen from the ceiling, 25 cm.
2. A thick deposit formed by similar slabs of limestone with incrustations of grey clay in the upper part and yellow clay in the lower. Here, at a depth of 50 cm. from the surface, was found a cultural layer with a "hearth" black in color, 10–15 cm. thick.
3. Beneath this at a depth of 3.70 meters is a sterile layer of larger slabs of limestone with streaks of bright grey, wet clay.

Fauna.—The bones of the following mammals were found:

1. *Equus caballus.*	13. *Lepus europeus.*
2. *Equus asinus.*	14. *Ochotona pusilla.*
3. *Cervus elaphus.*	15. *Castor fiber.*
4. *Cervus megaceros* sp.	16. *Cricetus cricetus.*‡
5. *Saiga tataricus.*	17. *Cricetus (cricetulus) migratorius.*‡
6. *Canis lupus.*	18. *Mus sylvaticus.*‡
7. *Canis familiaris.*[3]‡	19. *Microtus arvalis.*‡
8. *Vulpes vulpes.*	20. *Lagurus luteus.*‡
9. *Vulpes corsac.*‡	21. *Ellobius talpinus.*‡
10. *Meles meles.*‡	22. *Alactaga jaculus.*‡
11. *Felis leo.*	23. *Alactaga elater.*‡
12. *Felis lynx.*	24. *Scirtopoda telum.*‡

[1] Ephimenko, P. P., Some Results of the Study of Palaeolithic Period in U.S.S.R., "Chelovek," No. 1, 1928, Leningrad, p. 57.

[2] Bonch-Osmolovsky, G. A., "Le Paleolithique de Crimée," Bulletin of the Quaternary Committee, No. 1, 1929.

[3] The fauna marked ‡ were added from Gromov, V. I., "Geology and Fauna of the Palaeolithic Period in U.S.S.R.," Problems of the History of Material Culture, No. 1–2, Leningrad, 1933.

Fish:

1. *Rutilus frisii.*   2. *Lucioperca luciperca* (L.)

Flora.

    Populus (tremula)?

Flint Industry.

Flint tools and chips of the same character were found over and under the hearth layer. This fact can be explained by the cavities in between the slabs, where the tools may have fallen. Many skeletons of rodents indicate the possibility of another explanation.

The flint industry is represented by upward of one hundred tools and many chips. Most of them are made of long and backed blades, which gives a fine aspect to the industry. We have various gravers, drills, scrapers, points (arrowheads?). The polyhedric massive gravers and nuclei-like scrapers are totally absent; the ordinary gravers and scrapers made on ends of blades, are small and more blunt. The most typical forms are definitely geometrical; less numerous are those of the segment type and arrow points (?) with an almost Solutrean retouch on the ends. One is impressed by the fact that each tool is quite skillfully made and highly specialized. These points which are called by Polish archaeologists "points of Font-Robert type of special kind" are very common in Mesolithic cultures of Poland and Western Russia.

Dating.—In his earlier work, Bonch-Osmolovsky (1925), was inclined to date Suren II as belonging either to the end of the Magdalenian or to Azilian. Finally, in his later summary (1929) he definitely assigns it to Azilian times in accordance with the local conditions of the Crimean Palaeolithic. He adds that, "In its general aspect the industry of Suren II approaches that of the lower horizon of the cave Urkusta, which is a little more perfected, but has arrow points completely analogous to those of Suren II." [1]

No illustrations of flint material are available.

18. TIMONOVKA

Location.—Timonovka is situated on the right bank of the Desna River, in the village of Timonovka, 4 km. below the town of Briansk and 4 km. above the village of Suponevo (see page 401).

The site is located on a narrow plateau, formed by a deep gulley "Pereulok" on one side, and the steep slope of the Desna River on the other. This plateau is roughly triangular in shape, and on the narrow end of it are orchards and buildings of the section "Strelitza" of the village.

History.—It was discovered independently, in 1927, by M. V. Voevodsky and the expedition of V. A. Gorodzov. Later it became the object of an intensive study by V. A. Gorodzov, who conducted large excavations there in 1929–1931 and 1933.

In 1928 Voevodsky [2] made five soundings and a trial excavation 4 by 0.5 meters and collected some surface material. His preliminary work showed that the cultural layer lay immediately under the humus and was composed of gray compact sandy clay 7.17 cm. thick, also rich in humus. This layer was literally stuffed with flint implements.

[1] Bonch-Osmolovsky, G. A., "Le Paleolithique de Crimée," Bulletin of the Quaternary Committee, No. 1, 1929.

[2] Voevodsky, M. V., "The Palaeolithic Site of Timonovka," Russian Anthropological Journal, Vol. XVIII, pt. 1–2, 1929.

Geology.—A more detailed study of the geological conditions of Timonovka was made by G. F. Mirchink. According to him [1] the main bank shows the following cross section:

I. Humus.

II. Loess, which Mirchink dates as Würmian.

III. Ancient humus (badly preserved).

IV. Fluvio-glacial formations consisting of a layer of sands and loess-like sandy clay of Riss age.

V. Chalk-like marls, chalks, and glauconite sands.

According to Mirchink, previous to the formation of the fluvio-glacial Riss deposits (IV), or partly during the process of this formation, the terrace-like slope of the Desna River was cut by a series of deep gullies, one of which borders the village from the north, the other separating Timonovka from the Svensk monastery. These gullies were partly filled by fluvio-glacial formations and then were covered by the deposit of loess in Würmian times, which was later washed away.

The cross section at the head of the smaller gully "Pereulok," where it lies 30 meters above the level of the river, shows:

I. Humus.

II. Fluvio-glacial formations of loess-like sandy clay, red-gray in color—4 meters thick (Riss).

III. Yellow, fine-grained quartzite sand, 2 meters in thickness.

IV. Marls and glauconite sands.

On the surface of layer II were found the cultural remains. Mirchink states that at the place of the site, as in many other spots examined, the layers of loess and ancient humus were washed away (II, and III of the first cut). He concludes, therefore, that the Timonovka site is either younger than the time of the deposition of the Würmian loess, or is synchronous with the last stages of its formation. The gully "Pereulok" did not exist at that time and was formed later.

Fauna.—Most of the bones found were in such a state of decomposition that only greasy spots of dark soil revealed former places of bone accumulation. Only one tooth of mammoth was preserved.

Flint Industry of Voevodsky's Excavations (Plates XXXIV, XXXV, XXXVI).

The stone artifacts found were made of fine, dark gray flint, still found in the vicinity some 10 km. from the site. The local flint, which crops out in many places on the shore of the Desna, is of very poor quality and was not used.

All tools found are patinated, with light blue film when it is thin, and bright-white, when patination is deep. Some artifacts show remnants of a yellow chalky crust.

Nuclei (Plate XXXV, no. 1).

Blades (Plate XXXV, nos. 2, 3, 4) are usually short 4–5 cm., quite wide and irregular. Only a few long blades with parallel edges were found.

Flakes are found in large quantity of various sizes. Some are large and massive.

[1] Mirchink, G. F., "The Geological Conditions of Timonovka Near Briansk," Russian Anthropological Journal, Vol. XVIII, Part 1–2, Moscow, 1929, pp. 57–58.

PLATE XXXIV.

THE HAMMERSTONES FROM TIMONOVKA. After Voevodsky. 1, 2, hammerstones (*percuteurs*); 3, the intermediare tool (*utile intermédiaire*).

End-scrapers of Two Types:

1. Scrapers made on the end of a blade (Plate XXXV, nos. 6, 7, 8) slightly convex, with a steep working edge, formed by a fine regular retouch.   In one case both edges are retouched (no. 8).

2. Scrapers made on a flake (Plate XXXV, nos. 5, 9–15) have many different forms such as round scrapers with the peripheral retouch (no. 9), and square ones (nos. 11, 14).

Burins.—

Burin d'angle-grattoir (Plate XXXV, no. 16; Plate XXXVI, no. 5).

Burins d'angle (Plate XXXVI, no. 7) have a well retouched end, some of them approaching microlithic proportions (Plate XXXVI, nos. 6–7).   An example of *burin d'angle* with transversal revivification blow is shown on Plate XXXVI, no. 9.

Burin bec de flûte (Plate XXXVI, nos. 8, 20).

Multifacetted *burins*, and double *burins* (nos. 10, 19).

Lames with retouched ends in which Voevodsky sees unfinished *burins d'angle* minus the *burin* stroke (nos. 13, 14).

Lamelles á dos rabbatu (no. 17).

Lame retouched on both sides (no. 18).

Nucleiform *burin* (no. 21).

The tool represented on Plate XXXVI, no. 15, is, according to Voevodsky, a new type of *burin*, first established by P. P. Ephimenko in the neighboring site of Suponevo (see page 401).   The burin facet is formed by lateral retouch of the flake and transversal blow.[1]

Graver flakes (Plate XXXVI, nos. 11, 12).[2]

Hammer stones (Plate XXXIV, no. 3).

Intermediate tool (Plate XXXIV, no. 3), according to L. Sawicki, quoted by Voevodsky, was used between the nucleus and hammerstone to obtain flakes (?).

Voevodsky gives the following table of the quantitative relations of the types found.

Nuclei, 11.

Knife-like blades (*lames*), 170.

Scrapers on the end of knife-like blade (*grattoirs sur le bout de lame*), 14.

Scrapers on flake, 22.

Side gravers (*burins lateraux*), 71.

Side-middle gravers (*burins medans lateraux*), 3.

Gravers on the corner of broken blade (*burins à angle de lame brisée*), 14.

Other gravers, 7.

Prepared flakes for the side gravers (*ebauche pour les burins lateraux*), 21.

Flakes from the gravers (*eclats lateraux*), 14.

Dulled back blades (*lames à dos rabbatu*), 2.

In general, the Timonovka flint industry seems to be decadent and similar to such stations as Gontzi, or the upper horizon of Borshevo II, both of which are late Magdalenian according to P. P. Ephimenko.

[1] Possibly no. 16 is the same type as "*burin transversal.*"   (E. G.)

[2] The first according to P. Kelly is *burin de noaille.*

This decadent tendency is especially seen in the knife-like blades which are rather short and wide; there is an absence of regular long blades, so characteristic for the earlier stations of Kostenki I and Mezine. It is especially accentuated by the fact that many tools are made not on blades but on rather irregular flakes. The technique seems to be very negligent. The paucity of forms, the absence of drills, scarcity of blades with retouched backs, the small size of the average tool, the absence of the "finger retouch" which is so characteristic of the Russian Solutrean cultures, the absence of the multifaceted gravers, a certain geometrical character of scrapers, all speak for the Magdalenian period of the Upper Paleolithic.

Further investigations were made by V. A. Gorodzov, on a much larger scale in 1928, 1929, 1930, 1931 and 1933, resulting in much additional material and the discovery of four Palaeolithic habitations in the shape of large underground huts. Unfortunately, the results of Gorodzov's excavations are as yet unpublished, and only a very summary account based on a short article,[1] personal conversations with the excavator, and the notes on his lecture in 1933 is given below.[2]

Gorodzov's Excavations.—The total area of the site is estimated by Gorodzov as equal to 10,000 square meters. Up to now only 1,118 square meters or about one-eighth of the total area is excavated.

Fauna.—Though Gorodzov also reported on the poor preservation of bones in some sections of the site, the other produced sufficient quantity to enable Mme M. B. Pavlova to identify the following animals:

Mammoth.	Antelope.	Wolf.
Reindeer.	*Bos* (sp.).	Arctic fox.
Elk.	Bear.	Rodents.

The majority of bone remains are found in the ashes of the fireplaces, these in loess are completely disintegrated.

Flint Industry.

Fine flint of chalk origin, found in the vicinity of the site, furnished the principal material for the flint industry. Judging by the quantity of the chips and stone artifacts found, the manufacturing of tools took place on the spot. During the five years of investigation approximately 105,000 flint flakes and finished tools were found. Out of the 42,000 found during the season of 1933, 8,297 are finished tools. Gorodzov does not give any description of the tools beyond indicating the quantitative distribution of the general types. Thus, out of 24,842 finish-implements found there were:

Nuclei	1,114
Rabot	38
Knife-like blades	10,478
Gravers	7,984
Scrapers	1,386
Hammerstones	9[3]
Other types	3,733

[1] Gorodzov, V. A., "Preliminary Account of the Investigations of the Palaeolithic Site of Timonovka, Near Briansk," Soobschenia of the State Academy for the History of Material Culture, No. 11–12, pp. 55–57, Leningrad, 1932.

[2] See also Golomshtok, E. A., "Trois gisements du paléolithique supérier russe et sibérien," L'Anthropologie, t. XLIII, 1933, pp. 311–327.

[3] The unusually small number of hammerstones is unexplicable. (E. G.)

PLATE XXXV.

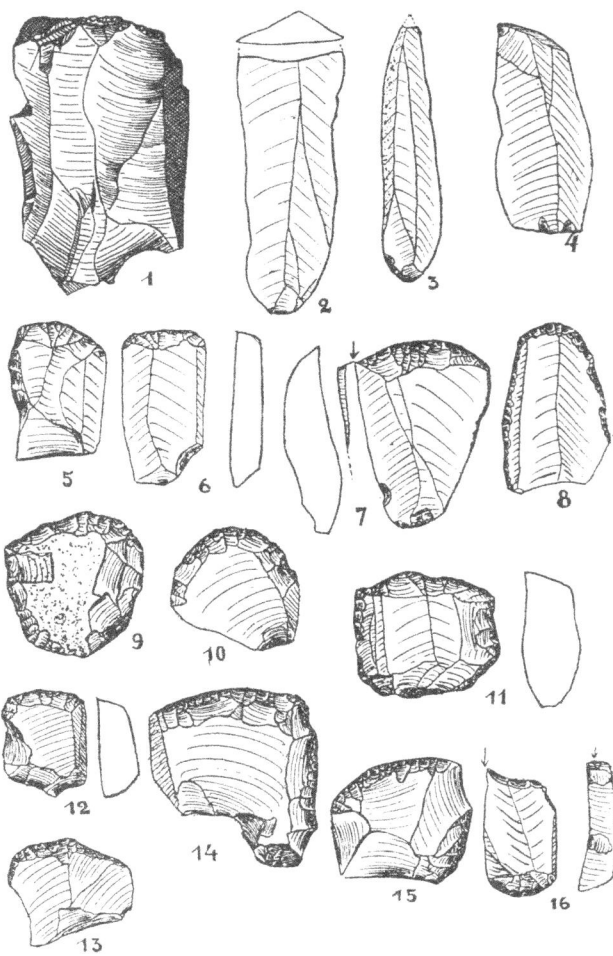

STONE INDUSTRY OF TIMONOVKA. After M. Voevodsky.

It is interesting to point out that outside of the area occupied by the dwellings the finds of flint and bone implements were very negligible.   Thus, during the excavations of 1931, there were 36,000 objects found, 25,000 of which came from one hut.   Altogether some forty thousand flint tools and chips were found in these dwellings during four seasons of work.

Hut no. 1   5,000
Hut no. 2   5,590
Hut no. 3 . . . .   (not completely excavated)
Hut no. 4   30,864

A very large number of gravers was found, hut no. 4 yielding 3,253; the record quantity for all Russian Palaeolithic stations.

The stone industry is represented by nuclei, knife-like blades, scrapers on the end of a flake or blade, side gravers, side-middle gravers, gravers made on the end of a broken blade.   In general the industry is characterized by the decadence of technique, a smallness of size, and a similarity of some of the forms to the Azilian.   Many tools are made negligently on irregular flakes.

Bone Industry.

A number of bone objects were found, the most interesting of which are shaftings for flint implements made of the tusk of a young mammoth; one of them is of a large size and is covered with a finely engraved geometrical design.   On the fragment of the tusk of a mammoth was found an engraving representing a small fish, with well-outlined scales, fins and tail.   Several pieces of large mammoth tusks found had deep notches made by the gravers, for the purpose of separating the splinters for manufacturing of small objects such as needles, awls, etc.   A number of bone fragments covered with engraved geometrical designs, bone awls made of the tubular bones of the arctic fox, perforated bone needles, the fragment of a needle container with ownership marks, and a number of carved objects were found.

The presence of a large number of gravers, almost 20 per cent. of the tools found, and such auxiliary tools as stone polishers and bone sharpeners made of flat slabs of quartzite and sandstone, likewise numerous, speak of a well developed bone industry.

Of quite unusual character are the small stone lamps made of soft gray stone.   Two smaller ones are whole; the third and largest is in fragments.[1]

The Dwellings.—The most interesting part of the Timonovka site is the discovery of well-pronounced remains of semi-subterranean huts.   During the five years of excavation, the remains of six semi-subterranean huts, rectangular in their ground plan, were uncovered. Gorodzov distinguishes two general types:

(*a*) With the entrance at the long end (Fig. 84, no. 1).
(*b*) With the entrance at the middle of the long side (Fig. 84, no. 2).

[1] Unfortunately no illustrations of these important finds are available to me.   (E. G.)

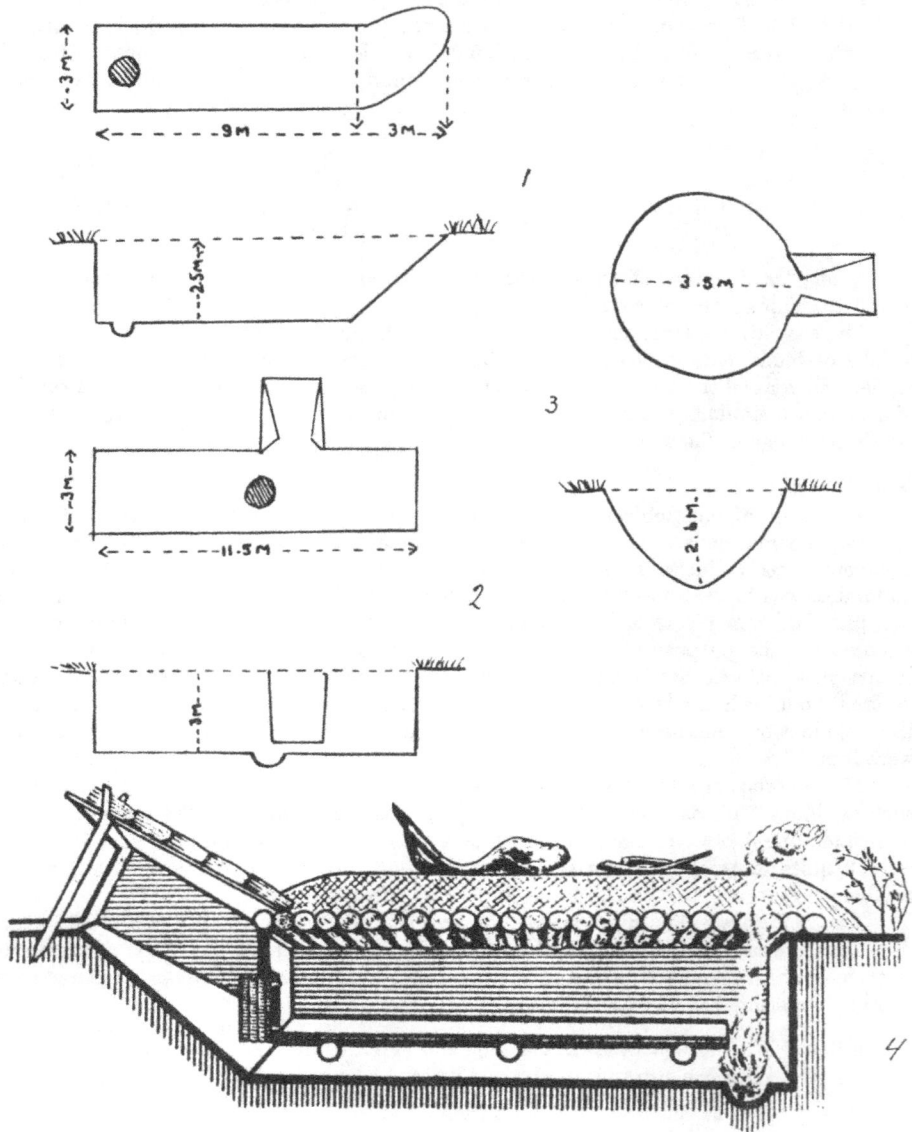

FIG. 84. Plans and Reconstruction of the Timonovka Dwellings. After V. Gorodzov.

PLATE XXXVI.

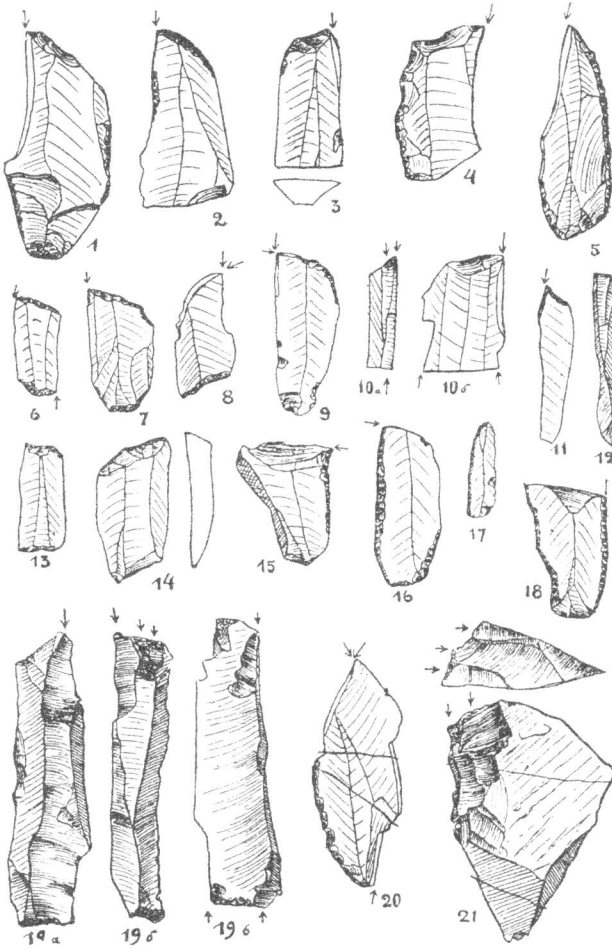

STONE INDUSTRY OF TIMONOVKA. After M. Voevodsky.

The following dimensions are given:

Hut no. 1, 12 x 3 x 2.5 meters.
Hut no. 2, 12 x 3.5 x 3 meters.
Hut no. 4, 11.5 x 3.5 x 3 meters.
Hut no. 5, ? x 3.5 x 2.8 meters.   Excavations not completed.
Hut no. 6, 11.5 x ? x ? meters.   Excavations not completed.
Hut no. 7, 11.5 x 3 x 3 meters.   Excavations not completed.

The huts no. 1 and 6 have hearths in the shape of round pits 0.70 and 1.10 meters in diameter and 0.25 and 0.50 meters in depth respectively.   Both fire pits were dug in the main floor of the dwellings and were filled with ashes and small bits of charcoal.   Over the hearth of dwelling no. 6, there was a chimney in the roof made of bark covered with clay, conical in shape, 0.70 meters high and 0.60 meters at the upper diameter and 1.10 meters at the base.   The roofs, in the case of hut no. 1 and 6, were flat, made of horizontally laid logs, (see reconstruction Fig. 84, no. 4) and covered by earth over which was accumulating the kitchen refuse.

Besides the remains of the huts, Gorodzov indicated the presence of storage pits (Fig. 84, no. 3), round and conical in shape, and bearing the following dimensions:

Pit no. 1, 1.60 meters in diameter; 2 meters deep.
Pit no. 2, 3.50 meters in diameter; 2.15 meters deep.
Pit no. 3, 4.50 meters in diameter; 2.00 meters deep.
Pit no. 4, (discovered in 1928–1929; dimensions undetermined).

In association with the dwellings Gorodzov found two work shops containing a very large number of chips, raw material, caches of nuclei and many "intermediate tools," such as hammerstones, anvils, retouchers, etc.

Dating.—Gorodzov is more or less indefinite in setting the time of Timonovka.   He feels that the microlithic tendency, decadence and geometrization of some of the forms, coupled with the palaeontological evidence discovered by him, places this culture in the Magdalenian period, in spite of the fact that some of the implements show Aurignacian tendencies in such form as *Châtelperron* blades, *burins busqué*, etc.[1]

19. SUPONEVO

This site is located on the right bank of the Desna River in the village of Suponevo, thirteen kilometers south of Briansk.   It was discovered in 1925 when the first excavations were made by Dr. S. S. Deev, curator of the Briansk Museum.   Under the river deposits, 25 meters thick, a number of chipped flint implements and bones of animals of the glacial period were unearthed.   In 1926 the excavations were continued by P. P. Ephimenko, G. F. Mirchink, B. S. Zhukov, S. S. Deev, Z. G. Grinberg and V. I. Gromov.

The rich cultural contents of this site may be judged by the fact that, according to

[1] Sawicki, L. ("Materials for the Study of Russian Archaeology," Przeglad Archeologicznii, Vol. III, No. 2, 3, Poznan, 1926–28), overlooking the fact that strong traces of Aurignacian tradition are encountered in undisputedly later stages of the Upper Palaeolithic of Russia, is inclined to date the majority of the sites including Timonovka, as Aurignacian.   The comparative study of the material from the Kostenki-Borshevo region with that of the Timonovka-Suponevo area seems to support the opinions of Voevodsky and Gorodzov.   (E. G.)

Ephimenko,[1] "from June 21st to July 12th many thousands of flint chips, blades, over one thousand finished implements, a quantity of bone objects, (mammoth tusks and reindeer horn) as well as remains of a very interesting late glacial fauna were obtained."

According to Gromov [2] the bones of the following animals were found:

Elephas primigenius	numerous
Rhinocerous thichorhinus	few
Equus equus	X
Bos sp.	few
Rangifer tarandus	X
Canis lupus	X
Alopes lagopus	numerous
Lepus sp.	X

Ephimenko [3] mentions also Arctic fox, and usual carnivorae.

The investigators are inclined to assign this site to the Magdalenian, but the quality of the flint technique, in the opinion of V. A. Gorodzov,[4] indicates perhaps an earlier period.[5]

In 1927 the excavations were continued by B. S. Zhukov, and in 1928–29 by V. A. Gorodzov. The material obtained is very large. Besides a rich flint industry, objects of bone covered with geometrical designs are reported by Gorodzov.

The whole site awaits detailed publication [6] and previous to that, the precise dating of it would be difficult, although Ephimenko believes that in Suponevo one finds strong traces of the influence of the Magdalenian culture of the Gontzi type, which in his opinion existed quite early and continued for a long time, spreading over a large territory.[7] He mentions [8] the presence of house pits similar to those found in Kostenki I and dates it as early Magdalenian.

20. ELISEVICHI

Location and History.—This site is located near the village of Elisevichi on the upper right bank of the Sudai River, a tributary of the Desna.

The finds of fossil animal bones were reported from this locality as nearly as 1870, but no investigation of the site was made. The report of the finds of animal bones with flint tools by peasants, who were digging a pit for domestic purposes, attracted the attention of K. M. Polikarpovich [9] who, during the season of 1929–30, was making an archaeological

[1] Ephimenko, P. P., "The Expeditions for the Study of Palaeolithic Cultures," Soobchenia of the State Academy for the History of Material Culture (GAIMK), Leningrad, 1926, Vol. I, pp. 319–20.

[2] Gromov, V. I., "The Geology and Fauna of the Palaeolithic Period in U.S.S.R." Problems of Material Culture, 1933, No. 1–2, p. 28.

[3] Ephimenko, P. P., "Pre-Clan Society," Leningrad, 1936, p. 439.

[4] Gorodzov, V. A., "The Results of Archaeological Activities in U.S.S.R. during 1917–30." Manuscript.

[5] Ephimenko, P. P. ("Pre-Clan Society," Leningrad, 1936, p. 438), states that many types characteristic for Kostenki II, III, and IV, are found in Suponevo.

[6] See also Gromov, V., "New Palaeolithic Site," "Priroda," No. 9–10, Leningrad, 1926, p. 65.

[7] Ephimenko, P. P., "Some Results of the Study of Palaeolithic Period in U.S.S.R.," Chelovek, No. 1, 1928, Leningrad, p. 56.

[8] Ephimenko, P. P., "Pre-Clan Society," Leningrad, 1936, p. 291.

[9] Polikarpovich, K. M., "Palaeolithic Sites on the Sudai River," Pratzi of Archaeological Section of Bielorussian Academy of Sciences, Vol. III, Minsk, 1932, 153–166.

survey of the Desna River region.   The investigation revealed that the finds were made on the steep slope, eleven meters above the Sudai River, between two deep gullies, where the spring floods washed out the steep banks.   Around a sand pit nearby was found a quantity of patinated flints and fossil bones which prompted further investigation.

Geology.—The trial pit (0.75 by 0.8 meters) made by Polikarpovich showed the following stratigraphy:

1. Humus.. 1–1.06 cm.
2. Loess... 1.06–1.56 cm.
3. Upper interlayer............................... 1.56–1.58 cm.
4. Cultural horizon............................... 1.58–1.66 cm.
5. Lower interlayer............................... 1.66–1.69 cm.
6. Loess... 1.69–1.91 cm.

This loess, according to geologists, is of eolian origin and was deposited after the formation of the upper Riss-Würm terrace.   The end of the formation of loess preceded the end of the process of accumulation of aluvial deposits on the lower terrace which corresponds to the end of the Bühl stage.   This, according to Polikarpovich, definitely dates this loess as belonging to Würm age.

The cultural layer which is some 10 to 12 cm. thick was found in the layer of loess, 1.58 cm. from the surface.   It consists of ashes, broken bits of bones, and a number of flint chips and finished tools.

Fauna.
1. *Elephas primigenius.*
2. *Vulpes* (sp.) (*alopex?*).
3. *Canis* (*lupus?*).

Flint Industry.

The majority of implements are made of dark, almost black, or gray flint of chalk origin, partly patinated.

There are five nuclei ranging from 3.4 cm. to 7.4 cm. in length (Fig. 85, nos. 16, 17, 19; four-fifth natural size.)

A large quantity of small chips and flakes was collected.   Among these were 74 blades (22 whole and 51 broken).   They are all characterized by regularity of form and smallness of size.   Of the set of 22 whole ones, 6 are 2.96 cm. by 3.65 cm.; 7 are 4.04 cm. by 4.52 cm.; 9 are 5 cm. by 7.4 cm. (Fig. 85, nos. 9–10).   Only three out of the 74 have patination.   Some blades (four of the whole ones and seven of the broken ones) have small notches, mostly semi-circular in form (Fig. 85, no. 7).   Polikarpovich sees in them special implements for bone and wood work.   Some blades (Fig. 85, nos. 12, 14) have a small secondary retouch along the edges and may have served as scrapers.   One blade is reminiscent of the type "*à dos rabattu.*"   One implement is especially interesting.   It is 5.1 by 2.6 cm. (Fig. 85, no. 18) and has no trace of striking platform or bulb of percussion, but shows on one side a semi-circular notch made with side straight blow from top to bottom.   Polikarpovich sees a parallel to *pièces esquillees* of Kostenki I and Predmost.

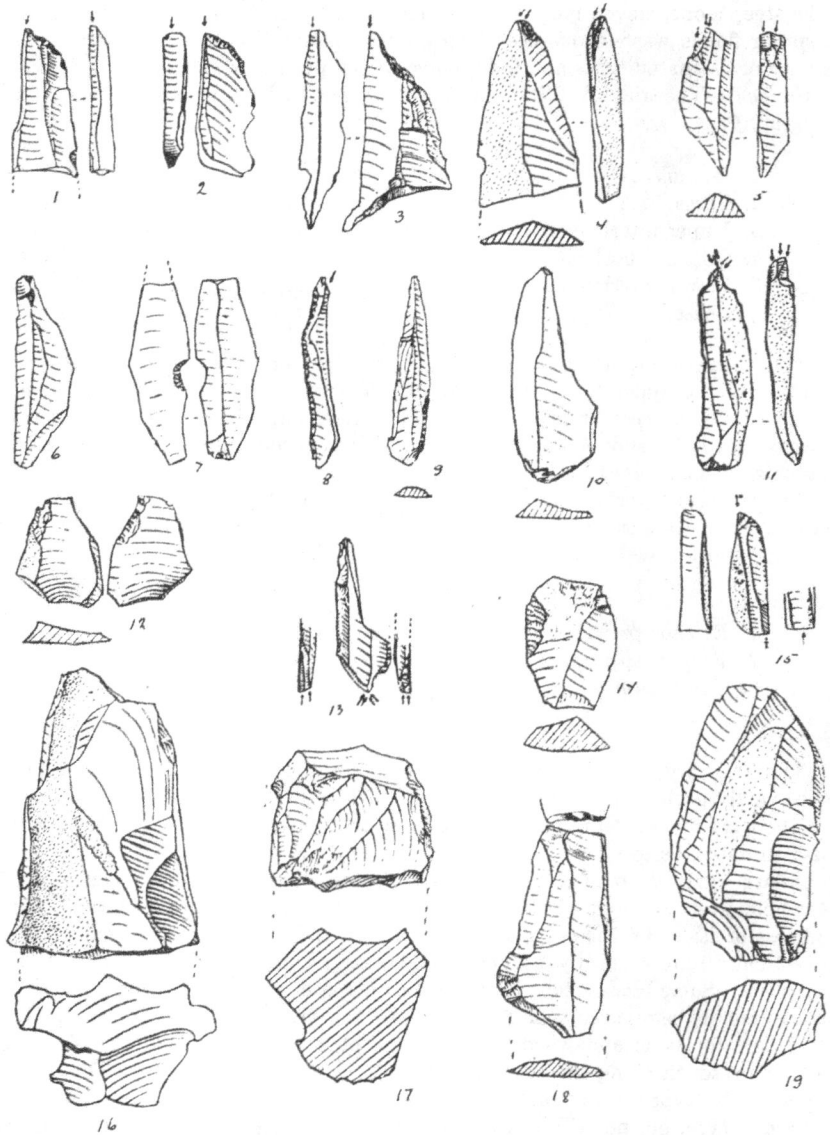

Fig. 85. Stone Industry of Elisevichi. After L. Polikarpovich.

Gravers constitute the next largest group.   There are thirteen of them.   Fig. 85 shows:

 1 *Burin d'angle and bec de flûte.*
 4 *Burin d'angle*, multifacetted.
 2 *Burin d'angle*, single.
 3 *Burin bec de flûte.*
15 *Burin d'angle* and *bec de flûte.*
 5 *Burin*, multifacetted.
11 *Burin*, multifacetted, nucleiform.
13 *Burin*, multifacetted, nucleiform.
 8 *Éclat de coup de burin.*

Dating.—On the basis of geological and palaentological evidences [1] as well as the "microlithic" character of the stone industry, with the total absence of large tools, and the general similarity to Timonovka, Suponevo, Mezine and Gontzi, Polikarpovich places Elisevichi in the first part of Magdalenian epoque.

21. Zhuravka

Location and History.—The site of Zhuravka is located in the vicinity of the village of Zhuravka, 25 km. from the town of Priluk. It was discovered in 1927 by A. Voronii, who was studying the formation of the terrace of the left bank of the Udai River. The local peasants working at the clay pit northwest of the village in the so called "Zarechiye," 120–125 meters from the entrance of the village from the Priluki-Ladin direction, reported the find of flint chips.

M. Rudinsky and Mme. Muskat arrived at the site at the request of Voronii and established the existence of a cultural layer with bones of fossil animals.

After the preliminary investigation in 1927, this site was thoroughly excavated during the years of 1928 and 1929 by the expedition of the Ukrainian Academy of Sciences under the general direction of M. Rudinsky, in which A. Voronii, V. Krokos, and others participated. Results of their work were published year after year in the Annual reports of the Laboratory of Theodor Vovk.[2] The following resumé is based on these various publications.

Geology.—The village Zhuravka is situated on the left bank of the Udai River, on the typical "one-loess" terrace of the old river valley. This terrace is well pronounced here

---

[1] Mammoth are absent in the late Magdalenian sites of European Russia. (E. G.)

[2] Rudinsky, M., and Voronii, A., "The Finds in the Village Zhuravka, Priluk County," Anthropologia, Annual Report of the Anthropological Laboratory of T. Vovk for 1927, Kiev, 1928, pp. 65–73.

Rudinsky, M., "The Finds in Zhurava," *Ibid.* for 1928, Kiev, 1929, pp. 140–151.

Krokos, V., "The Conditions of the Palaeolithic Deposit in Zhuravka," *Ibid.*, 1928, Kiev, 1929, p. 135.

Rudinsky, M., "Zhuravka," *Ibid.*, for 1929, Kiev, 1930, pp. 97–122.

Voronii, A., "Materials for the Study of Geology of the Palaeolithic Site of Zhuravka," *Ibid.*, for 1929, Kiev, 1930, pp.123–132.

Riznichenko, V., "The Geological and Geomorphological Conditions of the Palaeolithic Site of Zhuravka," *Ibid.* for 1930, Kiev, 1931, pp. 183–190.

Pidoplichka, J., "Rodents and Carnivorae from Zhuravka," *Ibid.*, for 1929, Kiev, 1930, pp. 133–147.

Field, H., and Prostov, E., "Recent Archaeological Investigations in the Soviet Union," A. A., Vol. 38, No. 2, 1936, p. 277.

and offers fine sections for the study which was done by Voronii (*op. cit.*), Krokos (*op. cit.*) and Riznichenko (*op. cit.*).

The stratigraphy of the site appears to be as follows:

I. Rough humus.
    (*a*) Upper layer 0.60 m.   Riznichenko distinguishes four layers here.
    (*b*) Lower layer 0.60 m.
II. Loess—2.60 m.
    (*a*) Light yellow loess, not stratified, porous, with a number of traces of rodents, and rich in CO_2 (eolian according to Riznichenko), 1.10 m.
    (*b*) Fine stratified sandy loess quite compact, of dark gray-yellow color with admixture of clay in the lower part, 0.85 m. (river deposit, according to Riznichenko).
    (*c*) Loess-like loam, very sandy with the thick interlayers of sand. This layer contains the cultural horizon.[1]

III. A layer of very compact greenish, sandy, coarse loam, 0.30 m.
IV. Fine stratified sands, 0.60 m.
V. Greenish loam with a thin layer of sand, 1.00 m.
VI. Yellow, medium grained sand, 1.00 m.
VII. Greenish sandy loam.

The cultural remains were found at the approximate depth of 3.80 m. from the surface, right on the top of the layer III, covered by the lower strata of layer II.

Fauna.

Fauna of this site is represented by the following list according to Pidoplichka:

1. *Marmota bobac.*
2. *Citellus rufescens.*
3. *Spalax microphtalmus.*
4. *Citellus* sp.
5. *Cricetus cricetus.*
6. *Microtus arvalis.*
7. *Vermela perugusna.*[2]
8. *Antelope* sp.

The Stone Industry.

The material for the stone industry of Zhuravka was almost exclusively dark gray flint of very good quality and bluish patination, which was evidently imported.

Summarizing the results of the three years of excavation one can say that the most characteristic forms for Zhuravka are backed points and blades with the straight or slightly curved edge and rounded long end. Rudinsky sees here variations from the type of *Châtelperron* (Fig. 86, no. 3) to *La Gravette* (Fig. 88, no. 2) and to the *point à cran atypique* (Fig. 91, no. 1). There is a large number of *lames* and *lamelles à dos rabbatu* or with partial side retouch (Fig. 91). Quite a few examples have either straight or oblique end retouch (Figs. 86, 91). In the "microlithic" series (by which he simply means implements of small size) the most characteristic are the *lamelles à retouche terminale* and *lamelles* and *éclats retouchés* (Figs. 91, 92). Among the other forms a series of *burins* (Figs. 89–91)

---

[1] The description of this layer by Voronii differs from the above given on the basis of Rudinsky in as much as, according to the former, this layer is "loess, very clayish and a bit sandy."

[2] *Vermela perugusna* and *Citellus* sp. may be accidental according to Pidoplichka. In general the character of the fauna is that of the steppe, the forest forms being notably absent. Pidoplichka is inclined to believe that the fauna may be of somewhat later character than Aurignacian times, as assigned by Rudinsky.

(*busqué*(?), *à coche*, *d'angle*(?), *bec de flûte*, single and multifacetted of various sizes) and several rather atypical *grattoir carenée* (Fig. 90) and *rabots* (Fig. 90, no. 4) should be mentioned.

The stone implements were not evenly distributed over the excavated area. They had a tendency to be localized in a few spots, one of which seemed to be the remains of a hearth, filled with fragments of charred bone. A lump of reddish matter was found which, according to Rudinsky, proved to be ochre. No bone implements were found.

The illustrations with the explanatory notes of the excavations will give the best idea of the industry.

Fig. 86.   4/5.   *Excavation of 1927.*

> No. 1. Fragment of blade with vertical retouch and beak-shaped point.
> No. 2. Fragment of blade with fine retouch on right side.
> No. 3. *Lame de canif.*
> Nos. 4–5. Backed blades.
> No. 6. Flake with retouch on oblique upper end.
> Nos. 7–16. Flakes of various sizes.
> Nos. 17–21. *Nuclei*

Fig. 87.   1/1.   *Excavation of 1928.*

> Nos. 1, 2, 3. Long blades.
> No. 4. Triangular blade of Aurignacean type with the round end retouched.
> Nos. 5, 7, 9. Long blades.
> No. 6. Thick blade with transversal flaking.
> No. 8. *Lamelle* from *coup de burin.*
> No. 10. A point with the *burin* stroke, on conchoidal end.
> No. 11. *Rabot*, nucleiform.
> No. 12. *Burin* with trace of vivification.

Fig. 88.   1/1.

> No. 1. Backed point.
> No. 2. Backed point of *La Gravette* type.
> No. 3. Broken backed point.
> No. 4. Point in transition to type "*à cran.*"
> No. 5. Backed point with end retouched at right angle.
> No. 6. Broken backed point.
> No. 7. Small flake.
> No. 8. Lame with base of graver—*coup de burin* (?) (*Burin*).
> No. 9. Polyhedric *burin bec de flûte.*
> No. 10. A thin blade.
> No. 11. Flake.
> No. 12. *Burin en bec de flûte* with simple facet.
> No. 13.

Fig. 89.   4/5.

> No. 1. *Burin busqué à coche* made on thick end (?).
> No. 2. *Burin d'angle à troncature* without retouch.
> No. 3. *Burin d'angle à troncature* without retouch.[1]
> No. 4. *Burin* on nucleus of *bec de flûte* type.
> No. 5. *Burin* triple on rough flake.

[1] Simple multifacetted burin, as there is no retouch on the end.   (E. G.)

No. 6. *Burin polyedrique*—broken.
No. 7. *Burin en bec de flûte atypique* made on flake.
No. 8. *Burin en bec de flûte* atypique made on flake.

Fig. 90. 4/5.

No. 1. *Grattoir carénée.*
No. 2. *Grattoir carénée* fragment.
No. 3. *Éclat transversal de nucleus en d'un rabot revivé.*
No. 4. Nucleus-like scraper (*rabot*).
No. 5. Nucleus-like scraper.

Fig. 91. 1–13, 17 are 4/5; all others 1/1.

No. 1. Large flake with secondary backing retouch.
No. 2. Flake with vertical side retouch, type *La Gravette* (broken).
No. 3. Small *La Gravette* point.
No. 4. *Lame à dos rabattu.*
No. 5. *Lamelle à dos rabattu.*
No. 6. *Lame à dos rabattu.*
No. 7. *Lame à dos rabattu.*
No. 8. *Lamelle à dos rabattu.*
No. 9. *Lamelle* with partial side retouch.
No. 10. Flake with partial side retouch.
No. 11. Double *burin* on end of broken blade.
No. 12. *Lamelle* with the straight end retouched.
No. 13. *Lamelle* with partial side retouch.
No. 14. *Lamelle* with partial side retouch.
No. 15. Fragment of *La Gravette* point.
No. 16. Backed blade (broken).
Nos. 17–21. *Lamelles* and *éclats allongues.*
No. 22. *Burin d'angle à troncature retouche oblique* (?) (Lateral, ? The troncation does not show the
retouch.   Perhaps a simple *burin* with the side retouched. E. G.)
No. 23. *Éclat au bord droit retouché.*
No. 24. Backed blade, fragment.
No. 25. Backed blade.
Nos. 26–30. *Lamelle* and *éclat allongue.*
No. 31. *Lamelle* with the oblique end retouched.
Nos. 32–34. *Lamelles*, fragments.
No. 35. Triangular *lame à arete ecrasée* (*lame arete dorsal du nucleus*).
No. 36. Flake.
No. 37. Flake.
No. 38. *Burin d'angle sur lame appointée* (*burin bec de flûte* ?).
No. 39. *Lamelle* with partial side retouch.
No. 40. *Burin en bec de flûte.*
No. 41. *Burin.*

Fig. 92. 1/1.

No. 1. *Lamelle* with the straight end retouched and some retouch on both sides.
No. 2. *Lamelle* with the oblique end retouched.
No. 3. *Lamelle* with oblique retouch on both sides forming a point.
No. 4. Fragment of *La Gravette* point.
No. 5. *Lamelle* with partial side retouch.
No. 6. *Lamelle* with the round end retouched.
No. 7. Flake with retouch on both sides.[1]

[1] Perhaps in this case the retouch on the lower side is due to the reduction of the bulb to make the base thinner. (E. G.)

No. 8. *Lamelle* with the round end retouched.
No. 9. *Lamelle* with partial side retouch.
No. 10. *Lamelle de coup de burin.*
No. 11. *Burin de fortune.*
No. 12. *Lame à dos rabattu.*

Dating.—Discussing the various technical peculiarities of the Zhuravka stone industry, Rudinsky stresses the presence of so many small forms. In fact he quite often uses the term "microlithic," indicating, however, that no true microlithic forms were found. This impression is somewhat weakened if one considers that quite a few of his small forms are but fragments of originally larger pieces, a fact which he does not always recognize. While it is true that the largest specimen is but 7 cm. long, and the largest flake only 9 cm., could it not be simply due to the size of the available (imported) material?

It may be also suggested that *La Gravette* points may be relatively rough in appearance due again to the quality of the material. Rudinsky's assertion of the definite transition from *Châtelperron* (rough forms) to *La Gravette* should not influence the dating. He is evidently unduly influenced by the presence of small tools and goes to great length to prove Aurignacian tendencies as against possible (?) microlithic ones. Still he states that there are two methods of retouch: first, coarser—Aurignacian, and the second—finer, which he calls "microlithic." Here again the limitation of material may be considered.

In conclusion, Rudinsky points out the significance of Zhuravka as another site in the general region of Gontzi (page 367) the industry of which, in his opinion, is quite similar to Zhuravka. He states that it is also quite similar to Borshevo I, and considers it as representing the Upper Aurignacian culture of South Russia.[1]

Quite fortunately somewhat detailed geological researches enable us to consider the temporal dating of the Zhuravka deposit. According to Riznichenko,[2] there are five well-pronounced terraces in the Dnieper region:

1. The contemporary.
2. Sandy terrace of Würm II without loess, and covered by ancient dunes.
3. The terrace of Würm I with one layer of loess, which covers the old alluvial deposits.
4. The terrace of Riss with two horizons of loess over the fluvio-glacial deposits of Riss glaciation.
5. The oldest Mindel terrace with two horizons of loess (the first is divided into two sections) over the moraine of Riss glaciation; the layer of loess under the moraine, and the layer of sands of Mindel-Riss age and fluvio-glacial sands of Mindel glaciation. (The age of Mindel-Riss sands is well documented by the fauna (*Elephas trogontherii*).) The base of this last terrace was formed during Mindel epoch.

According to Riznichenko the terrace in which the Zhuravka finds were discovered is of Würm I, the geological history being reconstructed as follows:

[1] Ephimenko, P. P., ("Pre-Clan Society," Leningrad, 1936) considers Zhuravka contemporaneous with Cyrill Street (p. 386) and Borshevo II. All of these he considers final Magdalenian.

[2] Riznichenko, V., "The Geological Conditions and Geomorphological Conditions of the Palaeolithic site of Zhuravka," Anthropologia, Annual Report of the Laboratory of T. Vovka for 1929, Vol. IV, 1930, pp. 183–190.

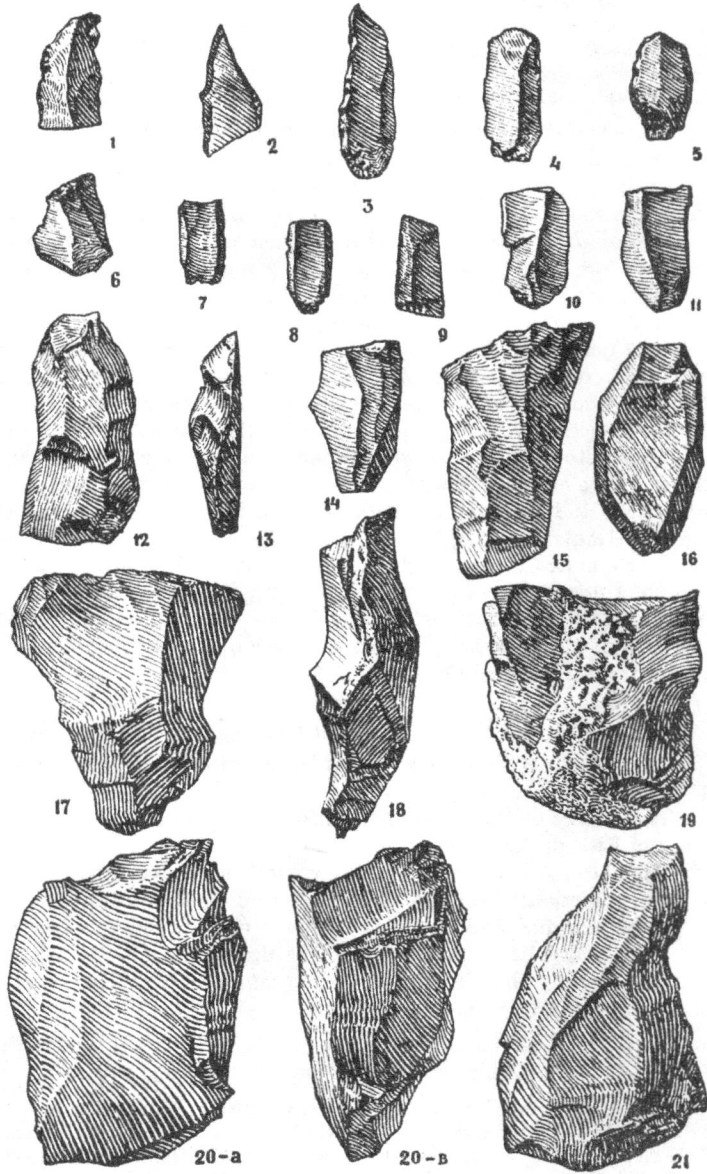

FIG. 86. Stone Industry of Zhuravka. Excavations of 1927. After M. Rudinsky.

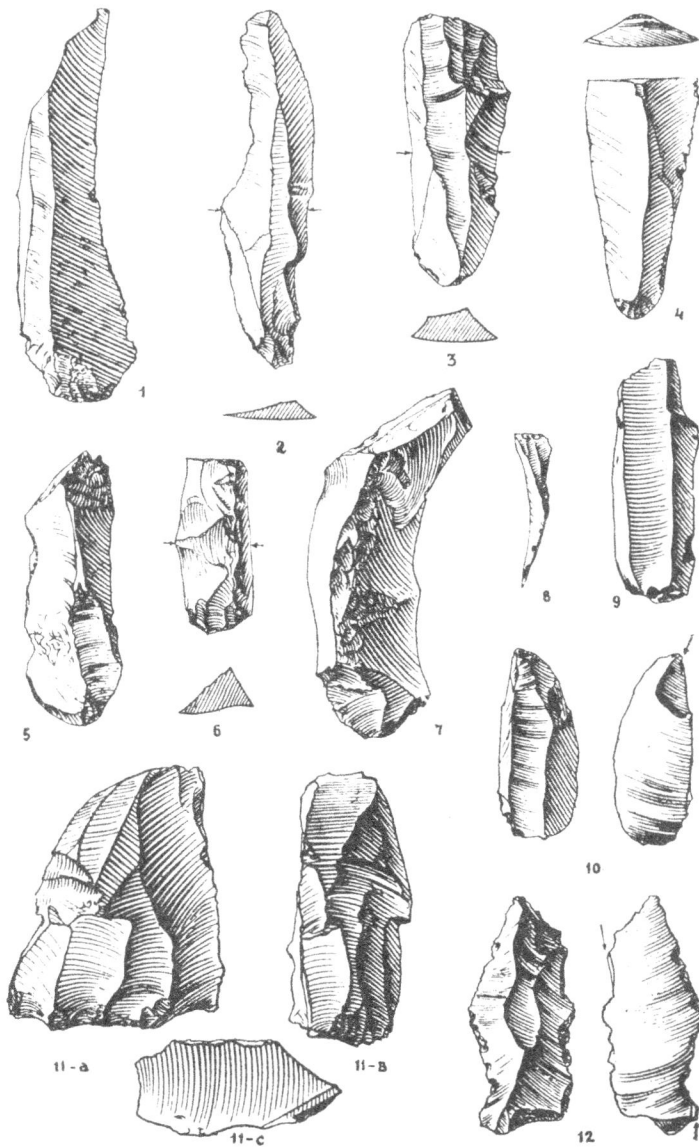

FIG. 87.   Stone Industry of Zhuravka.   Excavations of 1928.   After M. Rudinsky.

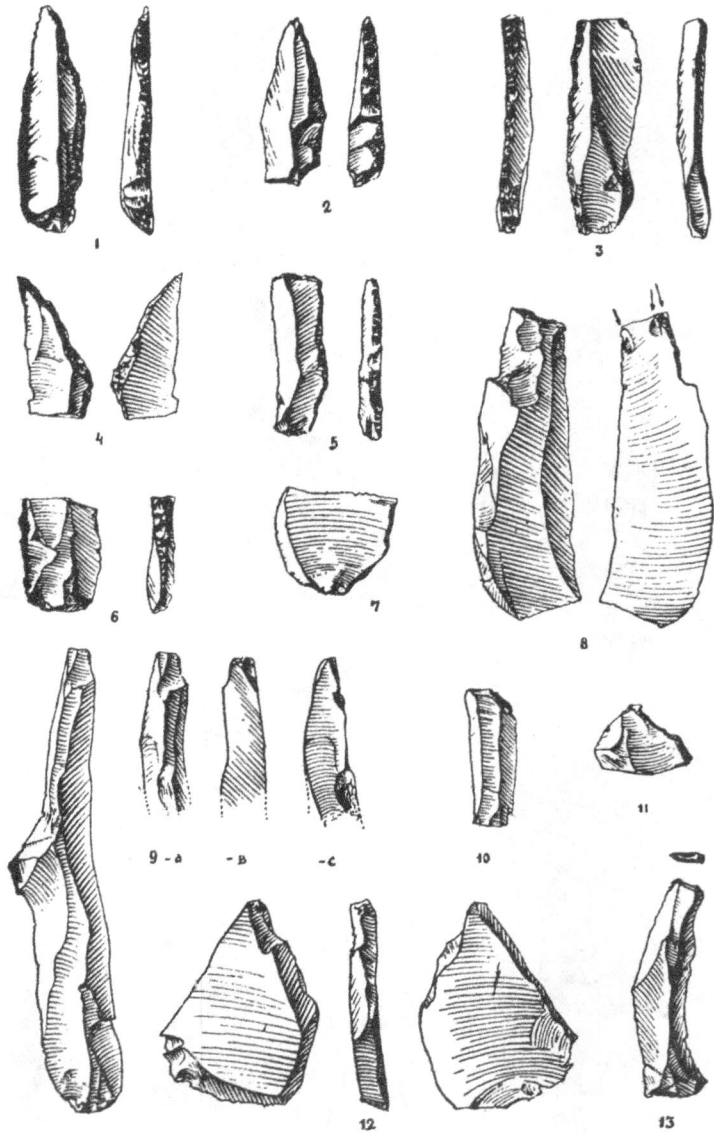

FIG. 88.   Stone Industry of Zhuravka.   Excavations of 1928.   After M. Rudinsky.

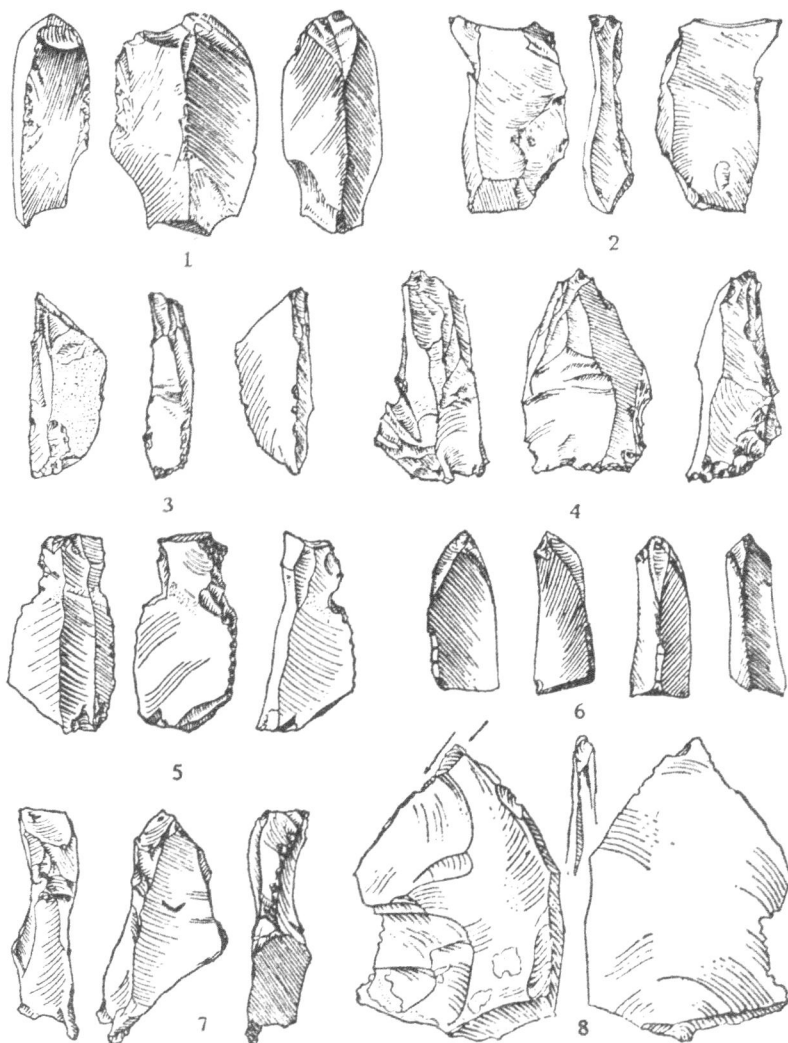

FIG. 89.   Stone Industry of Zhuravka.   Excavations of 1929.   After M. Rudinsky.

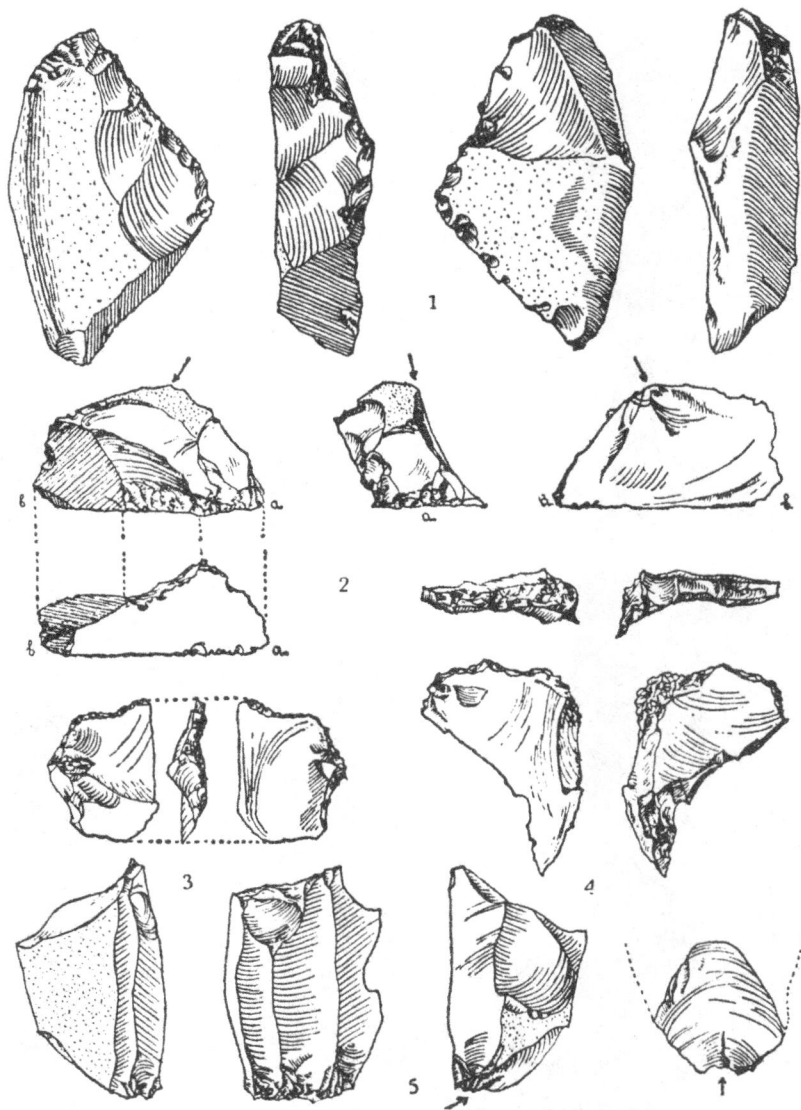

FIG. 90.   Stone Industry of Zhuravka.   Excavations of 1929.   After M. Rudinsky.

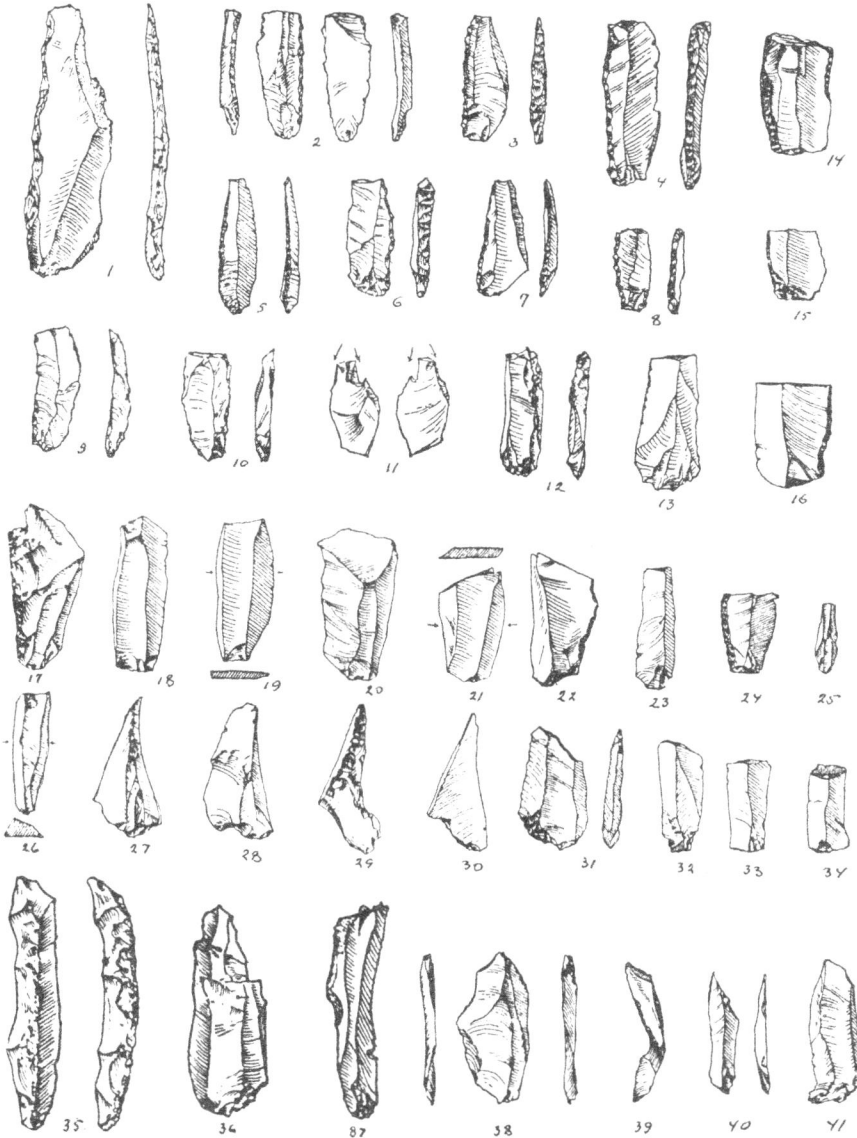

FIG. 91.   Stone Industry of Zhuravka.   Excavations of 1929.   After M. Rudinsky.

During Würm I, due to great humidity, the abatement of the course of the Udai River was very rapid and resulted in the deposit of a sand layer (IV of the stratigraphy given above). The dry period which followed during the end of Würm I permitted these deposits to emerge, to dry and be covered by the layer of eolian loess (III of the stratigraphy). The process, which continued during the interval between Würm I and Würm II, was accompanied by very dry climate. In the cross-section this layer of eolian loess cannot be observed, being destroyed and partly transformed to the diluvial loess by the water action during the next period of Würm II, which possessed a humid climate. The results of the water action during this period was the deposit of the clay-like layer II and later, more sandy layers in which the cultural horizon is situated.

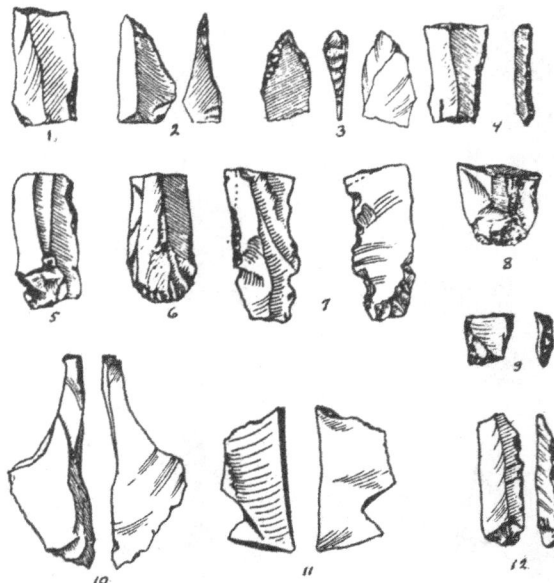

FIG. 92. Stone Industry of Zhuravka. Excavations of 1929. After M. Rudinsky.

During the dry climate of the next part of Würm II the fluvial process ceased and was replaced by the eolian processes which continued for a certain period through Würm II and were the causes of the formation of the eolian loess. The upper part of this formation, the most recent layer of loess of the postglacial deposits indicates more moist conditions, and the process of formation of the soil of the steppes.

In conclusion, Riznichenko states that the Palaeolithic man of Zhuravka inhabited the terrace at the end of the Würmian interval and at the beginning of Würm II. The piece of charcoal identified as *Picea excelsa* proves that during this time the climate was not warm but moist and cold, corresponding to the beginning of the advance of the ice during Würm II.

22. Dubova Gully (Dubova Balka)

Location.—This site is situated on the left shore of the Dnieper River, at the village of Svistunovo-Petrovskoye, near the town of Dniepropetrovsk, on the low terrace of the high slope of the Dubova Gully, 236 meters from the river shore.

History.—It was discovered by T. T. Teslja during the investigations of the left bank of the Dnieper inaugurated by the archaeological expedition for the study of the region of Dnieper's rapids in conjunction with the activities of Dnieprostroi, in July, 1931.

Geology.—The excavations carried by T. T. Teslja[1] and J. Arrachenko disclosed the presence of cultural remains in two distinct geological deposits:

1. In the lower part of the diluvial loess-like sandy "limon" which, according to Riznichenko belongs to the end of Würm Interglacial and to the first part of Würm II.

2. In the upper part of the schistous sands of alluvial origin, the more definite age of which is hard to establish.

Fauna.

According to preliminary investigations made by V. I. Gromov,[2] the remains of the following animals were found:

1. *Bison priscus.*   4. *Lepus* sp.
2. *Equus* sp.   5. Others to be identified.
3. *Canis lupus.*

Cultural Finds.—The excavations uncovered eight distinct cultural horizons, varying in thickness from 10–130 cm., and separated by the interlayers of loess-like sandy "limon" with schistous sands. The uppermost layer No. 1 is 4.51 meters from the surface; layer No. 8 is 7.20 meters from the surface.

Not all of the cultural layers have distinct and characteristic material. Two layers, No. 1 and No. 8, are represented only by remains of fireplaces containing a large quantity of charcoal, and pieces of broken and charred bone, but no flint material. Layer No. 5 produced the richest finds.

The raw material for the stone industry is of various types. The most commonly used is flint of chalk origin. A small number of tools are made of pebbles. The majority of finished tools is made of light brown transparent flint. A number of nuclei, flakes and blades were found.

The leading types of the flint industry are represented by:

1. Points of *La Gravette* type (ten specimens found).
2. *Lames à dos rabbatu.*
3. *Percoirs medians.*
4. Several double end scrapers with steep lameloid retouch.
5. Nucleiform *burin* found together with several transversal flakes belonging to it.
6. Blades with notch, one of them with a notche on both edges, resembling the type of *"lame étrangée."*

[1] Teslja, T. T., "Das Palaolith in Bereich der Dnieprostrom-schnellen," Die Quartarperiode, Vol. IV, 1932, Ukranische Academie der Wissenshaften, Kiev, 1932, pp. 83–85.

Field, H., Prostov, Eugene, "Recent Archeological Investigations in Soviet Union," A. A., Vol. 38, No. 2, 1936, p. 277.

[2] Gromov, V. I., "Geology and Fauna of the Palaeolithic Period in U. S. S. R.," Problems of GAIMK, No. 1–2, Leningrad, 1933.

7. *Rabot*-scraper with nine flakes belonging to it, permitting one to determine the manner in which it was made.

A large number of small blades and tiny flakes, ranging from 6 x 5 to 1 x 2 cm., characterizing the technique of flint work, was collected.   Often they were found in nests.

Material from other horizons is very similar to layer No. 5, the *La Gravette* point being the characteristic tool.

Bone Industry.
    1. Spear points.
    2. Awls.
    3. Cut tubular bones with fine engraving.
    4. Animal teeth with the roots cut off; perhaps objects of adornment.

A number of perforated shells of *Cardium, Buccintum* and *Nerita fluviatilis* were found. Some of them were colored red, possibly by ochre.   Red, and, to a lesser degree, yellow coloring matter, was found in all layers.   Fireplaces were found in all horizons, the small ones usually oval in form.   These fireplaces were filled with layers of ashes, pieces of charred wood and bone, fragments of bone, etc.   Around the hearth the ground was covered with fragments of bone and flint chips, the quantity of which decreased toward the periphery.

No high bone piles of the type found in Mezine and Gontzi were found here.   Teslja considers this as indicative of the fact that fireplaces were built in the open.   Their small size denotes the presence of a rather small population.   In one layer there were two fireplaces.

Dating.—According to Teslja, the character of the finds indicates that we are dealing in Dubova Gully with nomadic hunters who returned to the site repeatedly, and who were masters of stone technique.   On the basis of the development of the stone technique and the presence of certain leading forms, he assigns this culture, in all its horizons, to the Upper Aurignacian.

23. DOVGINICHI (DOLGINICHI)

This site is located ten klm. south of the village of Ovruch, Volyn district.   It lies on the Goryn River, on the left slope of the gully which cuts the south slope of Slavechansk-Ovruch plateau.   I. F. Levitsky [1] found here, in 1921, the remains of an upper Palaeolithic culture.   According to Krokos,[2] his excavations in 1924 and 1928 showed the following stratigraphy:[3]

    I. Loam—0 to 270 cm. from the surface.
    II. Loess—270 to 1395 cm. from the surface.
    III. Cultural layer with animal bones and flint industry.
    IV. Loess—1395 to 1650 cm. from the surface.
    V. Loam—1650 to 1945 cm. from the surface.

[1] Levitsky, I. F., "The Palaeolithic Site of Dovginichi," Anthropologia, Annual Report of Laboratory of Th. Vovk (Volkov) for 1929, Kiev, 1930, p. 155.
[2] Krokos, V., "The Stratigraphy of the Upper Palaeolithic Site of Dovginichi in Volyn," Die Quaterperiode.   The Ukranian Academy of Science, Vols. 1–2, 1930, Kiev, 1931, pp. 27–35.
[3] The cross-section of Levitsky (Fig. 93) differs slightly from the data of Krokos.

Krokos considers the upper layer of loess as belonging to Würm II, thus placing the layer containing the cultural remains in the Würmian interglacial stage.

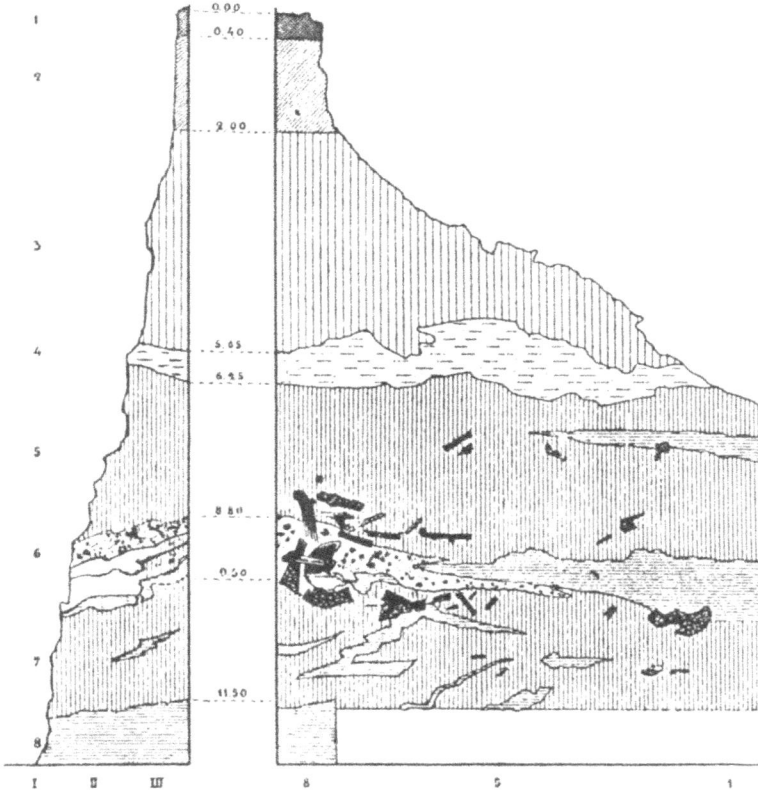

FIG. 93.   Cross-section of the Site of Dovginichi.   After Levitsky.   *1*, humus; *2*, loam; *3*, gray-yellow loess; *4*, gray loess-like strata; *5*, gray-yellow loess with occasional animal remains; *6*, loess-like sandy loam with cultural horizon; *7*, same as layer 5; *8*, blue clayish loam with streaks of loess and washed sands.

Polikarpovitch [1] lists the following fauna: [2]

1. *Elephas primigenius.*
2. *Rhinoceros tichorhinus.*
3. *Rangifer tarandus.*

4. *Equus* sp.
5. *Canis lupus.*

[1] Polikarpovich, K. M., "The Palaeolithic and Mezolithic Sites of Bielorussia," Pratzi of the Archaeological Section of the Bielorussian Academy of Sciences, Vol. III, Minsk, 1932, pp. 218–221.

[2] Field, H., and Prostov, E. ("Recent Archaeological Investigations in the Soviet Union," A. A., Vol. 38, No. 2, 1936, p. 276), add to the above list *Sus scrofa ferus*, *Citellus suslicus* and identify *Equus* sp. as *Equus–caballus*.

Stone Industry.

Stone artifacts are not numerous.   Flint of moraine origin in the vicinity served as material for the tools.   It is mostly black, occasionally dark gray, and of very good quality.
Fig. 94, after Levitsky, reproduces the following implements:

    1. Nucleiform high scraper, made of dark flint.
    2. High scraper.
    3. Notched scraper.
4–10. End scrapers in various degrees of finish.
  11. Flake with some chipping of the lower face.
  12. Fragment of a point with partial retouch on the lower face.
  13. Small flake with steep retouch forming a point.
  14. Part of the type *Châtelperron.*

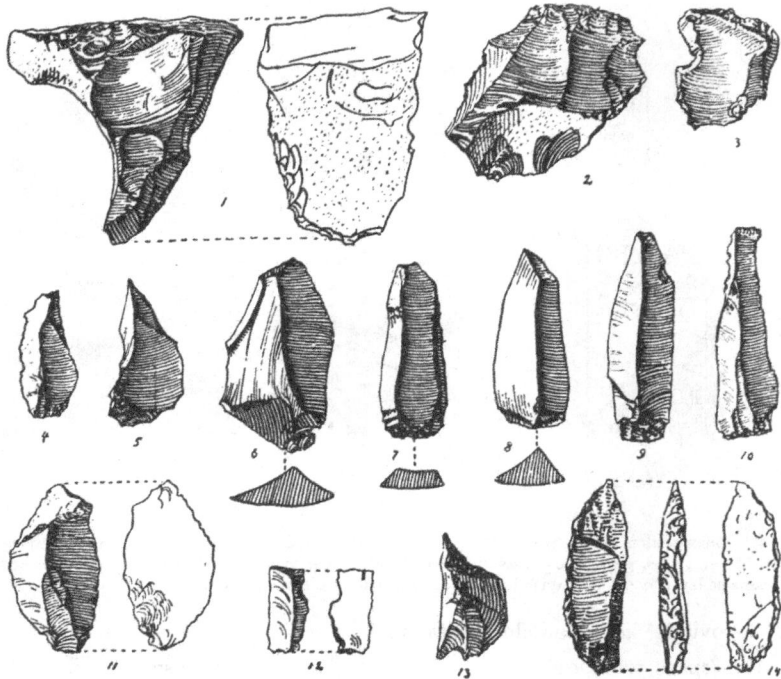

Fig. 94.   Stone Industry at Dovginichi.   ½ nat. size.   After Levitsky.

Dating.—Polikarpovich, stating that the site is rather poor industrially, considers it as belonging to the Upper Palaeolithic, while Levitsky, perhaps on the basis of presence of *Châtelperron,* and point No. 13, which he calls of the type of *point á cran,* calls it Aurignacian.   According to H. Field and E. Prostov,[1] "Stratigraphical and paleontological

    [1] *Op. cit.,* p. 276

data, combined with a technological study of the industry, ascribe this station to the Upper Aurignacian period."

24. Kostenki IV (Alexandrovka)

This site was discovered by S. N. Zamiatnin, during the reconnaissance work of 1927. It is situated in the village of Kostenki, Voronezh region, at the south of the main Alexandrovka gully, almost at its junction with the valley of the Don.

Stratigraphically it is very similar to Borshevo II, which is situated several kilometers lower down the Don.

In its cultural aspect it is also quite similar to Borshevo II. The cultural layer, though rather thin, has a considerable spread. It is strongly and evenly colored, and almost totally lacking in large bones or even fragments. Its industry, however, differs considerably from other sites of this region. "The blades with dulled backs" in many variants, form the largest groups of Kostenki IV tools. Very interesting is the almost total absence of gravers, which are replaced here by a tool, the cutting edge of which is formed by bilateral chipping of the upper end. This form, found in Kostenki II and III, as well as consideration of the stratigraphy, permits the dating of the site [1] as the Aurignacian-Solutrean of Ephimenko.[2]

25. Kostenki V

Kostenki V is located in the village of Kostenki, Voronezh region, on the bank of the Don River. In the same locality as Kostenki II (see page 357), but near the mouth of the gully and lower on its slope, trial excavations revealed traces of a hearth older than Kostenki II. The flint implements are represented by: _

(a) Massive multi-facetted gravers.

(b) Laurel-leaf points with retouch on the lower sides, several large blades.

All tools are made of uniform dark flint of calceous origin, very different from the spotted pebble material of which the tools of Kostenki II were made. The investigations were not finished.[3]

26. Kastrova Gully

Kastrova Gully is situated on the left shore of the Dnieper River, near the village of Svistunovo-Petrovskoye, on the lower terrace on the steep slope of the deep gully Kastrova. It was discovered in July, 1931, by T. T. Teslja [4] during the investigations of the left bank of the Dnieper initiated by the archeological expedition for the study of the Rapids region of the Dnieper in connection with "Dnieprostroi," [5] and studied by A. Dobrovolsky who found three different cultural deposits in different horizons.[6] The first cultural

[1] Zamiatnin, S. N., "Expeditions for the Study of Palaeolithic Cultures in 1927," Soobschenia of the State Academy for the History of Material Culture, Vol. II, p. 211, Leningrad, 1929.

[2] Later ("Pre-Clan Society," Leningrad, 1936, p. 346) Ephimenko dates it as Early Magdalenian.

[3] Zamiatnin, S. N., "Expeditions for the Study of Palaeolithic Cultures in 1927," Soobschenia of the State Academy for the History of Material Culture, Vol. II, Leningrad, 1929.

[4] Teslja, T. T., "Das Palaolith in Bereich der Dnieprostrom-schnellen," Die Quartarperiode, Vol. IV, 1932. Ukrainische Akademie der Wissenschaften, Kiev, 1932.

[5] Dnieper River construction project.

[6] Field, H., and Prostov, E., ("Recent Archaeological Investigations in the Soviet Union," A. A., Vol. 38, No. 2, 1936, p. 275), consider these deposits as four different sites.

deposit (upper ?) was found in the middle part of the loess layer, and, according to the investigator, contains rather untypical industrial types of Magdalenian character.

The second deposit was represented by a series of horizons in the loess-like "limon" and contained Aurignacian industry.

The third deposit is also in loess-like sandy "limon," but the industry is devoid of leading or identifying forms.

Bone and stone implements were found, and are in the process of being studied.[1]

These deposits are assigned by V. Riznichenko to the Würm interglacial period, or the first half of Würm II.

27. LUGANSK

This site is situated on the sandy surface of the flood plain terrace of the Donetz River, near the town of Lugansk.[2]

In 1921, S. A. Loktushev found here bones of the following animals:

1. *Elephas primigenius.*
2. *Rhinoceros tichorhinus.*
3. *Equus caballus.*

4. *Bos priscus.*
5. *Rangifer tarandus.*

According to Gorodzov,[3] flint implements and a very well preserved spearhead were found with the bone remains. The spearhead, made of mammoth bone, 25 cm. long, is characteristically Magdalenian of the Sagai type.[4]

The investigations were continued in 1926, and published by Loktushev. This publication was not available to me.

Dating.—Gorodzov considers the site Magdalenian; Sawicki,[5] Aurignacian.

28. GVARDZHILAS-KLDE

This cave is situated near the village of Rgan, Shoporan District, Kutais region of Caucasus, in the deep gully of the stream Chirula, which is the right tributary of the Kvirilla River.

The cave faces southeast and lies fifteen meters above the level of the Chirula. It is semi-oval in ground plan, being 30 meters long, 12.5 meters wide at the entrance, and 3–3.5 meters wide (Fig. 95).

Gvardzhilas Klde was investigated by S. Krukovsky[6] in 1910. He found it very

[1] No illustrations of cultural finds are available. (E. G.)

[2] Rudinsky, M., "The Scientific Work on Paleolithic and Mezolithic Culture in Ukraine during the Revolution," Anthropologia, Annual report of the Anthropological Laboratory of T. Volkov, 1930, Kiev, 1931.

[3] Gorodzov, V., "Archeological Excavations and Investigations in Soviet Russia from 1919–1923," "Drevnii Mir," Part I, Moscow, 1924, p. 2.

[4] According to Field, H., and Prostov, E., ("Recent Archeological Investigations in the Soviet Union," A. A., Vol. 38, No. 2, 1936, p. 36), the fauna list does not include the mammoth. This report further states that S. Loktushev, V. Gorodzov and L. Sawicki attribute this site to the Aurignacian period.

[5] Sawicki, L., "Materials for the Study of Russian Archaeology," Przeglad Archeologicznii, Vol. III, No. 2, 3, Poznan, 1926–28.

[6] Krukovsky, S., "The Cave of Gvardzhilas Klde in Rgan," Bulletin of the Caucasian Museum, Vol. X, Nov., 1916, Tiflis, pp. 1–7.

much disturbed by the treasure hunting activities of the local population.   In one section which, apparently, was not disturbed, he found the following stratigraphy:

1. Humus covering the limestone debris, with modern potsherds, iron objects, etc.
2. Layer of calcite breccia, with a few fragments of bones.
3. A dry layer of marl containing a few fragments of bones, and, at the lower part, flint chips.
4. First cultural layer, located in a dark dry stratum, with limestone debris. Many broken bones and flint implements were found here.
5. A layer of limestone material which fell from the ceiling.
6. Second cultural layer, of dark color due to the presence of charcoal, which was found in larger pieces.   It contained broken animal bones and numerous flint implements.
7. Same as layer 5, but much thicker.
8. Bottom of the cave.

Gromov cites the following fauna: [1]

Cervus elaphus.	*Ursus arctos.*
Gulo gulo.	*Ursus spelaeus* (?).

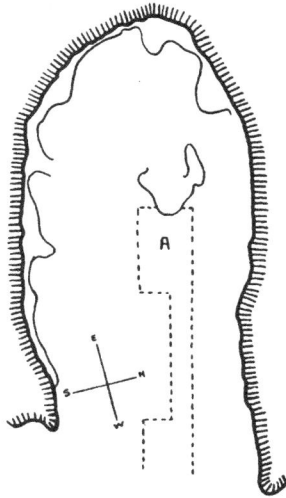

FIG. 95.   Cave of Gvardzhilas-Klde.   Ground plan, 1 : 200.   *A*, excavation of Krukovsky.   After S. Krukovsky.

Dating.—Ephimenko [2] states that the bulk of the stone industry consists of small blades, retouched on the back, and the microlithic triangular points, typical for the late Capsian, in which he sees the first geometrical microliths.   Of interest also are small, roughly chipped handaxes, which are found here and there in western Europe, in Azilian

[1] Gromov, V. I., "Geology and Fauna of the Palaeolithic Period in U. S. S. R.," Problems of the History of Material Culture, No. 1-2, Leningrad, 1933, pp. 23-33.
[2] Ephimenko, P. P., "Pre-Clan Society," Leningrad, 1936, p. 493.

and Tardenoisian sites. A few simple tools of bone, and a harpoon of Azilian type,[1] concludes the list. On the basis of the industry, Ephimenko dates Gvardzhilas-Klde as Azilian,[2] while Gorodzov [3] considers it Magdalenian.

29. KOUKREK

The rock shelter of Koukrek is situated 3 kl. south of the cave Kiik-Koba (see pages 240–257) some 20 km. from Siempheropol, Crimea. There is a cultural layer 1–1.5 meters from the surface under the structure of clay and coarse gravel of diluvial origin.

Above, at the depth of 0.6 meters is a second layer with several typical Tardenoisian trapezoids. Due unquestionably to unusually bad conditions for preservation, no charred wood or bone fragments were found.

Gromov [4] lists the following fauna, evidently for the lower horizon:

1. *Sus scrofa.*
2. *Cervus elaphus.*
3. *Canis lupus.*
4. *Canis familiaris.*
5. *Vulpes vulpes.*
6. *Vulpes corsac.*
7. *Meles meles.*
8. *Martes foina.*
9. *Felis lynx.*
10. *Felis silvestris.*
11. *Lepus europeus.*
12. *Citellus rufescens.*
13. *Cricetus cricetus.*

According to Bonch-Osmolovsky,[5] Koukrek industry occupies an intermediate place between Suren II (see page 394) and Tardenoisian, approaching the first by the presence of *burins* and by the Solutrean-like retouch on the lower side of blades, and to the second by the perfection of the blades and the presence of *grattoirs arrondis* and of *lames à encoches.* The geometrical forms are absent. Instead, only obliquely pointed, small blades with lateral retouch exist.

Koukrek is very much like Kizil-Koba (see page 433) but represents, apparently, an earlier stage than Tardenoisian.

30. KOLODIAZHNOE

This site is located near the village of Kolodiazhnoe, Miropol district, Volyn region, on the right bank of the Sluch River, where it is joined by its tributary Tuihterevka.

In 1924, the attention of I. Levitsky [6] and Gamchenko, who were excavating neolithic deposits in the vicinity, was attracted by the find of several objects from the different horizons in the near-by stone quarry.

[1] Unfortunately no illustrations of stone and bone industry are available to the author. (E. G.)

[2] Ephimenko, P. P., "Some Results of the Study of the Palaeolithic Period in U. S. S. R.," "Chelovek," No. 1, 1928, Leningrad, p. 57.

[3] Gorodzov, V. A., "New Archaeological Discoveries in Soviet Russia," Manuscript, 1931.

[4] Gromov, V. I., "Geology and Fauna of the Palaeolithic Period in U. S. S. R.," Problems of the History of Material Culture, No. 1–2, Leningrad, 1933.

[5] Bonch-Osmolovsky, G. A., "Le Paléolithique de Crimée," Bulletin of the Quaternary Committee, No. 1, 1929.

[6] Levitsky, I., "The Finds of Kolodiazhnoe," Bulletin of the Anthropological Laboratory of Th. Volkov, Vol. I, Kiev, 1925, pp. 35–39.

PLATE XXXVII

ENGRAVED REINDEER BONE FROM KOLODIAZHNOE. After I. Levitsky.

The finds consisted of:

    (a) Animal bone with both ends "sawed off."

    (b) Teeth of fossil horse.

    (c) Fragment of long bone with traces of engraving (Plate XXXVII).

In 1926, S. Gamchenko found the remains of mammoth and horse bones under the thick layer (9 inches?)[1] of loess, in the strata of sand with nests of greenish clay (27 mm.). No further finds were made.

Levitsky distinguishes on the engraved bone

    (a) The head of a deer (41 x 28 mm.) on the left side of the illustration.

    (b) A human head (23 x 17.5 mm.) on the right side of the bone.

The present author is inclined to question the Palaeolithic age of this find especially on the basis of the second design, the execution of which is unparalleled in that age, and the technique of which is too modern.

31. POSTNIKOV GULLY

This site is situated on the northern outskirts of the town of Samara, three and one-half km. from the Samara-Voskresenskii Spusk, (see page 437), in the corner formed by the bank of the Volga and the Postnikov Gully. According to V. A. Gorodzov,[2] it was discovered in 1924 by M. G. Matkin, and in 1925 investigated by Mme. V. V. Golmsten and P. P. Ephimenko.

V. V. Golmsten[3] reports that "The cultural layer containing the remains lay in loess-like clay. A small trial excavation revealed here a 'hearth' and a number of objects: a fragment of bone needle, a piece of shell disk, flint points retouched on both sides, scrapers of high form, perforators, gravers with oblique edge, etc. Among the bones, remains of the mammoth were found."

The following objects are reproduced here from illustrations given by Golmsten (Fig. 96):

    1. Bone needle.

    2. Shell disk.

    3. Object of stone (?).

    4. Fragment of stone point.

    5. Fragment of stone point.

    6. Fragment of stone point.

    7. *Grattoir carenée.*

    8. End scraper.

    9. End scraper.

    10. Small nuclei-form tool.

    11. Small nuclei-form tool.

[1] Field, H., and Prostov, E., ("Recent Archeological Investigations in the Soviet Union," A. A., Vol. 38, No. 2, 1936, p. 276), indicate "nine *meters* of deposit which yielded worked bone objects attributed to the Upper Palaeolithic period."

[2] Gorodzov, V. A., "The Results of Archaeological Activities in U.S.S.R. during 1917–30." MSS.

[3] Golmsten, V. V., "The Archaeological Remains at Samara Province," Russian Association of the Institutes of Social Science Research, Vol. IV, 1928, pp. 125–129.

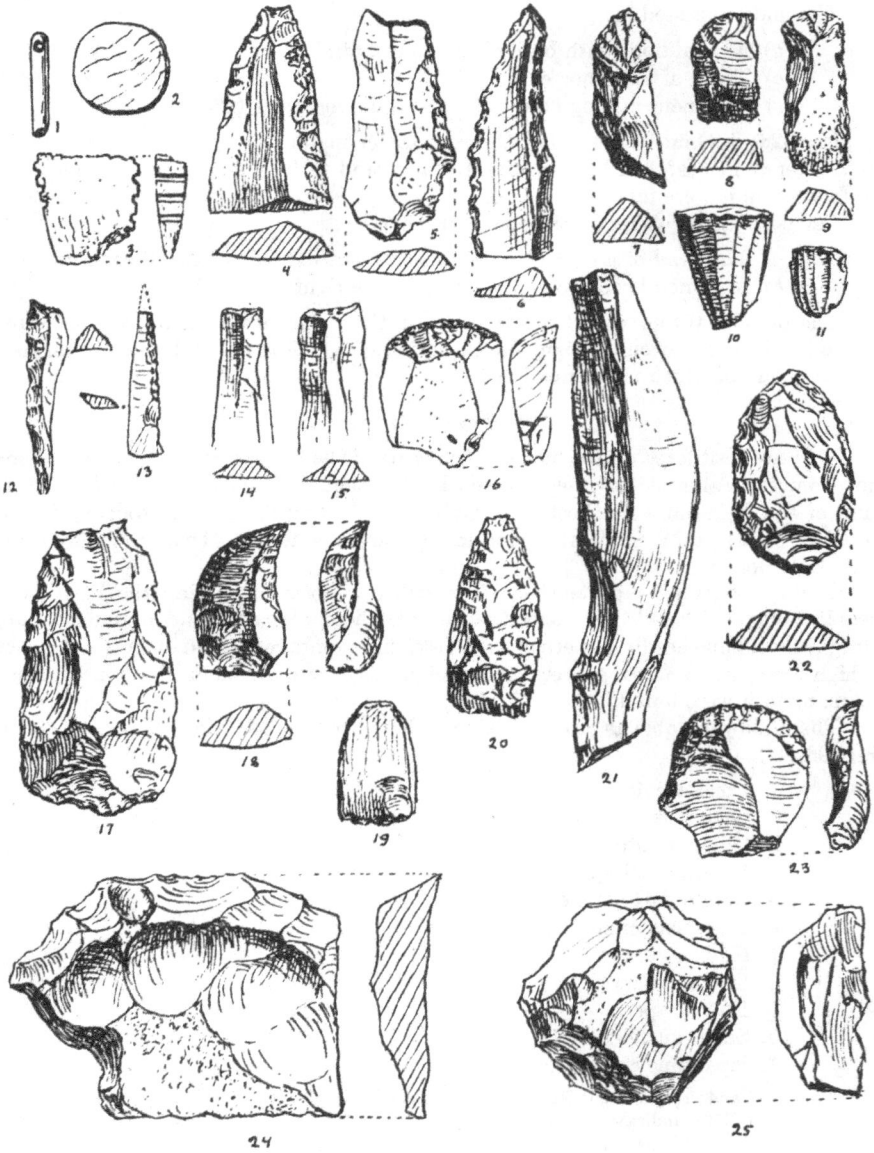

FIG. 96. Stone and Bone Industry from Postnikov Gully. After Golmsten.

12. Flint chip, retouched on one side.
13. Narrow point of *La Gravette* type.
14. *Burin d'angle* (?).
15. *Burin d'angle* (?).
16. Small scraper with round retouch.
17. Stone tool of hand-axe type.
18. Fragment of point with curved edge, two-third natural size.
19. Flake (?).
20. Finely retouched point, one-half natural size.
21. Long, rough flake with notch, two-third natural size.
22. Point retouched on one side (?), one-half-natural size.
23. Scraper on wide flake with fine circular retouch, two-third natural size.
24. Round scraper on thin flake, one-half natural size.
25. Round scraper on thick flake, one-half natural size.

V. A. Gorodzov, who made further excavations in 1929, states [1] that the cultural layer was at a depth of 2.48 meters in the first (flood-plain) terrace of Volga. The site was largely destroyed by the river.

According to him, the fauna of the site is represented by the mammoth, smaller animals, and fish. The flint industry is not described by him in any detail. He points out that tools are of small size, that gravers are rare, and that there are examples of the end scraper (*encoche*) made of a narrow knife-like blade, which in his opinion is identical with the "encoche" of Bateni (Siberia). He mentions also objects of bone and shell without further describing them, and assigns the date of the site to the Late Magdelenian.

P. P. Ephimenko, who upon the invitation of V. V. Golmsten and the Samara Archaeological Society joined the former in 1926 in check-up investigations, states [2] that his excavations showed, "that in spite of the archaic character of the flints, both (?) horizons of this site belong to the early Neolithic." It seems that he discovered two horizons:

(a) The lower with a very primitive microlithic industry
(b) The upper (younger) with a characteristic macrolithic industry.

Without explaining this, he states that the Postnikov Gully marks the eastern border of the distribution of the Early Neolithic in South Russia.

Thus we have two different opinions and at present it is impossible to decide in favor of one or the other. It seems strange that Gorodzov, who worked later, in 1929, agrees in general with Golmsten, and that neither of them mention the "macrolithic" and the two horizons of this site.[3]

32. KHERGULIS-KLDE

The cave of Khergulis-Klde is situated near the village of Vachevi, Chiaturi region, Caucasus (Fig. 20, no. 7).

It was first investigated by Krukovsky in 1918; later Zamiatnin [4] made a detailed

[1] Gorodzov, V. A., "The Results of Archaeological Activities in U.S.S.R. during 1917–30," MSS.

[2] Ephimenko, P. P., "Expeditions for the Study of Palaeolithic Cultures," Soobschenia of the State Academy for the History of Material Culture, Vol. I, p. 319, Leningrad, 1929.

[3] In my opinion the industry, judging by illustrations, does not warrant the Palaeolithic dating of the site. (E. G.)

[4] Field, H., and Prostov, E., "Recent Archaeological Investigations in the Soviet Union," A. A., Vol. 38, No. 2, 1936, pp. 262–263.

study of Caucasian Palaeolithic cultures and examined the collection made by Krukovsky in the Museum of Georgia, Tiflis.

The industry includes "characteristic Upper Palaeolithic implements with a large percentage of Late Mousterian forms, prepared from broad triangular flakes, sharp-pointed instruments, and scrapers distinguished for their perfection of retouch."

The illustrations available (Fig. 97) seem to represent:

1. End scraper made on biface.
2. Side scraper make on broad triangular flake, one end forming a point.
3. Crude side scraper made of long flat pebble (?).
4. Oval scraper on thick blade, with retouch on three-fourths of periphery.
5. Graver of *bec de flûte* type.
6, 7. Notched end scrapers.
8. Multi-facetted *burin*.
9, 10, 11. Small backed blades triangular in cross-section, with parallel sides.

Dating.—Zamiatnin considers this site early Upper Palaeolithic.

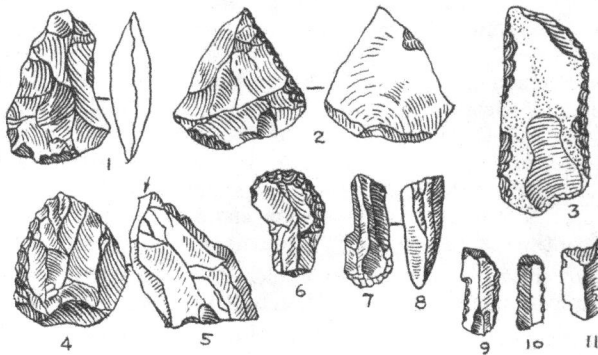

FIG. 97. Flint Industry of Khergulis-Klde. After Field and Prostov.

33. TARO-KLDE

The cave of Taro-Klde is situated near the village of Shukruti, Chiaturi region, Caucasus (Fig. 20, no. 7).

It was investigated by Krukovsky in 1918, and material obtained was studied later by S. N. Zamiatnin.

Like Khergulis-Klde, it yielded "characteristic Upper Palaeolithic implements, as well as a large percentage of Late Mousterian forms."

Field and Prostov[1] reproduce, after Zamiatnin, the stone and bone industry given below (Fig. 98).

[1] Field, H., and Prostov, E., "Recent Archaeological Investigations in the Soviet Union," A. A., Vol. 38, No. 2, 1936, pp. 262–263.

Judging from illustrations we have the following forms represented:

1. Small *biface*, with additional chipping along the edges.
2. Point of Mousterian type made on thick triangular flake. Both edges retouched.
3. Multi-facetted *burin bec de flûte*.
4. Double end scraper on triangular blade.
5. Double end scraper (*Grattoir mouseau ?*).
6. End scraper made on thick flake; the upper part preserves the original surface of the pebble.
7-8. Nuclei-form *burin*.
9-10. Backed blades.
11. Point made on thick long blade, triangular in cross-section, edges bear fine retouch.
12-13. Bone points.

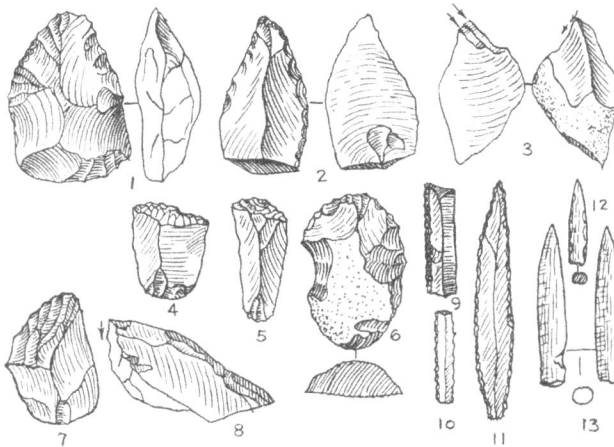

Fig. 98.   Flint and Bone Industry of Taro-Klde.   After Field and Prostov.

34. Studenitza

Location.—This site is located near the village of Studenitza, Ushitzk district, Podol region. Two large gullies join the bank of the river Dniester, to form a valley in which the village is situated.[1] The river flows one hundred and forty feet below the top of the hills. The shore is divided by these gullies into three separate mountains, Gorshkovataya, Ganusisko, and Bielaya, which surround the valley on three sides.

On the top of Bielaya mountain, as well as on the floor of the caves situated on the river slope, were found, according to Antonovitch, "several dozens chipped flint implements, knives, points, scrapers, nuclei and axes (?). The tools were roughly worked by

[1] Spitzin, A. A., "The Russian Palaeolithic Period," Zapiski of the Imperial Archaeological Society, Slavic Division, Vol. XI, Petrograd, 1915, pp. 140–143.

large facets and improved by secondary retouch along the edges."[1]   Antonovitch surveyed five caves.   Cave No. 4 is fifteen meters deep, the chambers of cave No. 5 are seven, ten and twelve meters deep.   According to Spitzin, the floor of the caves is covered by debris falling from the ceiling and has a cultural layer, perhaps of palaeolithic times.

In 1904, Studenitza was investigated by Professor Krishtaphovich.[2]   He gives the following cross-section of the shore here, beginning from the bottom.

(a) Whitish-gray clay with fresh water molluscs—12 feet.

(b) Loess with terrestrial molluscs with streaks of gravel and conglomerate, the lower part of which contains fresh water molluscs. The thickness of the conglomerate is about seven feet, but towards the head of the gully the loess increases to 28–30 feet, and contains as many as five and six horizons of conglomerate.

(c) Diluvial loam and humus.

In the loess of the region of the Bielaya and the Ganusisko Mts. were found many animal bones, mostly those of mammoth, often purposely broken, together with flint implements.   Similar tools were found on the flat top of Bielaya mountain.   Krishtaphovich states that Studenitza culture was contemporaneous with the deposition of loesses, but his short description of the latter does not give any material for dating the separate horizons. No description is given of the small quantity of flint tools, part of which were found in the caves.

Henry Field and Eugene Prostov[3] list Studenitza among several sites excavated in 1927–28, revealing Aurignacian industry and fragments of bones of small animals, belonging to the final phase of the Quaternary period in Eastern Europe.

Ephimenko,[4] however, dates it as Early Magdalenian on the basis of the industry, which, according to him, is identical with Kostenki III and Karacharavo.   Thus, 80 per cent. of all finds are gravers, with the predominance of rough gravers made on thick flakes, reminiscent of the primitive Aurignacian types.   Considerably fewer in number are the usual end scrapers as well as perforators and other tools.

35. CHERKESS-KERMEN

In the same general area as Suren I and II, Merezhkovsky found traces of human habitation in the two caves of Cherkess-Kermen.

The industry, according to Bonch-Osmolovsky,[5] had a microlithic character.   In the small collection preserved in the Anthropological Department of the Museum of Anthropology and Ethnography in Leningrad, there is one specimen of trapezoid form, but the large size and rudeness of chipping makes it a prototype of Microlithic industry.

[1] Antonovitch, V., "About Rock Caves on the Bank of the Dniester River in Podol Province," Trudi of the Odessa Archaeological Congress, Vol. 1, pp. 93–95.

[2] Krishtaphovich, N. I., "Geological Investigation of the Palaeolithic Sites during the Summer of 1904," Drevnosti, Trudi of the Moscow Archaeological Society, Vol. XXI, Part II, pp. 178–181.

[3] Field, H., and Prostov, E., "Recent Archaeological Investigations in the Soviet Union," A. A., Vol. 38, No. 2, p. 276.

[4] Ephimenko, P. P., "Pre-Clan Society," Leningrad, 1936, p. 346.

[5] Bonch-Osmolovsky, G. A., "Prehistoric Cultures of Crimea," Krim, No. 2.

According to Ephimenko,[1] we have here the following forms: small scrapers, small oblique points, typical geometric microlithics in the form of a rough trapezoid, or double oblique points.

The blades have very regular sides, unlike those of earlier cultures. On the basis of the general character of the industry and the fact that gravers are very rare here, Ephimenko dates this site as late Azilian.

36. Novo-Alexandria

The site of Novo-Alexandria is located near the town of the same name, in the vicinity of the village of Gora-Pulavskaya on the left bank of the river Visla. The discovery was made by N. I. Krishtaphovich in 1894.[2]

The river Visla runs northward, cutting a deep valley in the chalk plateau south of Novo-Alexandria. In its southern part, the bank is high, steep, and rocky. The river deposits are absent, having been washed away by the river, but where they are preserved, they appear in the form of well-pronounced terraces.

The investigator distinguishes three kinds of terraces:

(a) Basic (ancient).

(b) Ancient, diluvial.

(c) New, alluvial.

The remains of a palaeolithic site were discovered in the deposits of the second diluvial terrace which adjoins the basic terrace and is forty-two feet above the river surface. Its surface is even; only in spots is it furrowed by the gullies.

A very good natural cross-section can be observed on the left bank of the valley above the village of Gora-Pulavskaya, where a number of bones of fossil animals were collected which are now preserved in the Museum of Novo-Alexandria.

We have here the following picture:

1. The surface layer of humus—10–12 inches.

2. Quartzite or sometimes clayish sands, alternatly fine and coarse—10 feet.

3. A layer of glacial boulders which lay at the base of the sand layer; here and there the boulders are displaced by gravel.

4. Loess-like loam. Its upper part is sandy, the lower part more clayish, gradually changing into the dark gray, almost black clay, of the next layer. The whole layer, especially the upper part, contains an admixture of partly decayed roots, stems of plants, bits of shell, and parts of mammal bones. It is twelve to fourteen feet thick.

5. A dark gray, almost black clay containing in spots many shells of terrestrial molluscs (*Pupa muscorum, Lucena oblonga, Limnophysa truncutula, Helix*, etc.), the remains of plants and trees, and bones of birds and mammals. Thin steaks of debris of local chalks, sandstone, and pebbles of northern crystalline formations, are observed in this layer. The total thickness of it is 2 or 2.5 m.

[1] Ephimenko, P. P., "Some Results of the Study of Palaeolithic Period in U.S.S.R.," Leningrad, 1928, "Chelovek," No. 1, p. 58.

[2] Krishtaphovich, N. I., "Post-Tertiary Formations in the Vicinity of Novo-Alexandria," Warsaw, 1896, pp. 47–69. Also, "The Old Palaeolithic Sites on the Territory of European Russia and Their Geologic Age," Trudi of the 10th Congress of Russian Naturalists, 1898, 1902.

In one of the lowest horizons of the last layer were found remains of a Palaeolithic site:

(a) Flint implements of various types.
(b) Broken bones of—

Elephas primigenius.	*Bos priscus.*
Rhinoceros tichorhinus.	*Sus scrofa, fossilis.*
Equus caballus, fossilis.	

All this was mixed with charcoal and was imbedded in the clay in irregular layers. The bones were numerous, but very much broken and split. The fact that light pieces of charcoal were found together with heavy stones, indicates, according to A. A. Spitzin,[1] that the material was found *in situ.*

Krishtaphovich, according to Spitzin, believes that the gray clay was formed during the interglacial period, as it lies over the stratum of pebbles. He thinks that at the end of this period the layers of lake and other water type of schistous loesses were formed.

He further maintains that the character of this formation, together with the absence of any traces of water action on any of the objects, so different in weight, indicates definitely that the objects became imbedded in the clay at a time when it was not covered by water. Considering the topography of the clay layer and its relation to the location of the finds, it is highly probable that during the period of habitation this clay was temporarily above water, forming the shore of the basin. This supposition is confirmed by the numerous remains of terrestrial molluscs and plants found in this layer.

In the horizon of the ancient diluvial terrace described above, special attention should be paid to the presence of the remains of two glacial moraines, the oldest of which is characterized by the debris of local formations and northern crystalline boulders. It forms the foundation of all post-tertiary formations of the Visla valley and, therefore, is older than the terrace itself and the remains of the palaeolithic station found there. The remains of the second and later moraine (layer No. 3) were deposited above the cultural clay level and represent a later phenomenon.

Ascribing the remains of the moraines to the last glaciations (Third Riss and Fourth Würm) formerly covering Poland, Gorodzov agrees with the geologist-investigator in considering the age of this site as belonging to the Riss-Würm Interglacial period.

Unfortunately, the writer has no available material for judging the nature of the flint industry outside of a very sketchy description given by Gorodzov[2] and Spitzin. The original work by Krishtaphovich could not be obtained.

"The flint industry," writes Spitzin, "is characterized by a predominance of chips and flakes indicating that the tools were manufactured on the spot. There are knives of rather careless technique, sometimes combined with scrapers; and many needle-like thin splinters which have hardly been used. Only several knives bear a secondary retouch."

According to Krishtaphovich this industry is analogous to the lower layer of the Oitzov cave.

Gorodzov, who is inclined to stress geological conditions for purposes of dating, feels that in this case there is sufficient basis for dating this site as Middle-Aurignacian. He

[1] Spitzin, A. A., "The Russian Palaeolithic Period," Zapiski of the Imperial Russian Archaeological Society, Slavic Division, Petrograd, 1915, Vol. XI, p. 139.

[2] Gorodzov, V. I., "Archaeology, The Stone Age," Moscow, 1925, Vol. I, pp. 191–192.

believes that the collected flint material confirms his conclusions.   The pressure flaking technique and the presence of both large and microlithic forms, point to the existence of a culture corresponding to that of the Middle-Aurignacian.   He finds that the tools are similar to those from the Polish cave finds of earlier date, but that their general character shows that this culture is different from contemporary cultures of Western Europe.

This difference is especially manifest in the absence or perhaps scarcity of retouching by the counter-stroke technique, which is widely distributed and characteristic of the West-European cultures of the "early period." [1]

37. KIZIL-KOBA

This site is located on one of the Yayla mountains, Crimea, near the village of Kizil-Koba, on the Alma River.   It was discovered by K. S. Merezhkovsky [1] in 1879–80.   Later, it was visited by S. A. Zabnin, N. L. Ernest and G. A. Bonch-Osmolovsky who made further excavations.

Geology.—According to Merezhkovsky,[2] the cultural remains were found under the humus on the top of the hills, and along the slope of the gully in the layer of dark gray clay. *Flint Industry.*

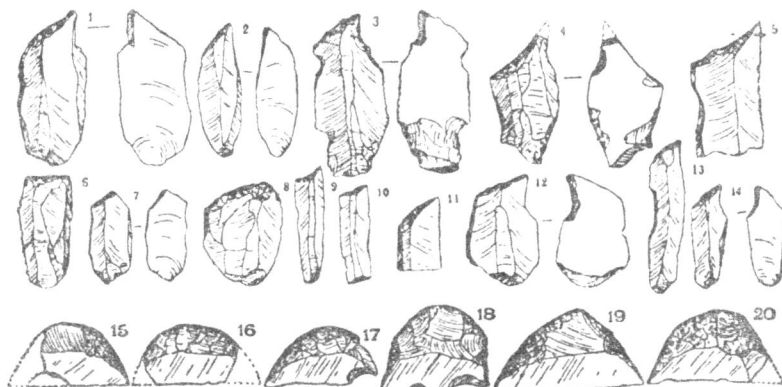

FIG. 99.   Stone Industry of Kizil-Koba.   After Ephimenko.

According to Ephimenko,[3] we are dealing with a workshop.   There is a mass of nuclei, chips, flakes, unfinished and finished tools.   Over 10,000 flakes and chips, as well as approximately 1,000 tools, were found.   The tools are made of fine brown flint of different shades.

Nuclei.—The nuclei are conical and prismoidal in form, and number several thousand. They may be divided into two groups:

[1] Gorodzov, V. I., "Archaeology, The Stone Age," Moscow, 1925, Vol. I, p. 192.

[2] Merezhkovsky, K. S., "The Report of the Preliminary Investigations of the Stone Age in Crimea," Izvestia of Imperial Russian Geographical Society, 1880, Vol. 2, pp. 106–176.

[3] Ephimenko, P. P., "Small Flint Implements of Geometrical and Other Peculiar Forms in Russian Sites of Early Neolithic Age," Russian Anthropological Journal, Vol. 13, Part 3–4, pp. 211–228.

1. The larger, 5–6 cm. long, served to prepare the usual microlithic implements.
2. The miniature nuclei, out of which only the smaller points were made.

Scrapers.—The scrapers are of various forms but mostly round, discoidal, semi-cylindrical, and made on the end of a blade (Fig. 99, nos. 8, 15–20).

Burins.—The burins are of *d'angle* type (Fig. 99, nos. 1, 2, 3).

Blades.—Blades are retouched obliquely, forming more or less sharp points (Fig. 99, nos. 5, 9, 11).

As a whole the industry is characterized by the presence of peculiar microliths of geometrical forms, together with Upper Palaeolithic tools such as gravers and scrapers. The small size of the tools, which average from 1.5–4 cm., together with the almost mathematical regularity of the forms of trapezoids, triangles, and segments, as well as the fineness of the finish, is extraordinary. P. P. Ephimenko considers it as belonging to an earlier part of the Microlithic epoch than the other Crimean finds.

38. ISKOROST

Location and History.—In 1911 V. V. Khvoiko discovered a Palaeolithic station near the village of Iskorost, Ovruchev County, Volyn Province Palaeolithic implements were found in the earth which formed the portion of two mounds above the ground, which led to the continuation of the excavations below the surface of the ground.

According to Grushevsky, as quoted by Spitzin:[1] "At a depth of 0.5 m. in a layer of undisturbed ground flint chips and tools varying in shape and size were found. Further, in the same layer, at a depth of 0.8 m. were found several hearths, 2.5–5 m. in diameter, somewhat apart from each other. They consisted of a layer of ashes and charcoal, under which was an area of burned red soil. All of them were surrounded over a considerable area by horizontal layers of cultural deposits, formed by decayed organic matter, mostly food remains, a large quantity of different flint tools, and flint pebbles of various sizes. The material was *in situ*, above a horizontal layer of diluvial origin." The largest part of the site was covered by a mound. The excavation revealed a series of separate hearths of different sizes, distinguished by spots of ashes, charcoal, burned earth, and layers of cultural deposits with decayed animal bones. The hearths were surrounded by an unusually large quantity of flint tools, gravers of special form and size, spear points, knives, scrapers, and also some tools of unusual form.

Spitzin writes further: "As this site was situated in one horizon and consisted of complex hearths, it is evident that it belongs to one time and one family or clan. The fact that it was located far from the water is inexplicable. As the hearths are small and the bone remains insignificant, it is doubtful whether it served as a place of long habitation.[2]

On the basis of the large number of chips and the small quantity of finished tools, Spitzin concludes that this was a workshop. We cannot explain the even distribution of large chips of flint around the hearths, and he suggests that a clear space would have been preserved here and there for lying and sitting.

The collection brought by Khvoiko to the Imperial Archaeological Committee consists of a considerable number of nuclei, knives, chips, and a few scrapers.

[1] Spitzin, A. A., "The Russian Palaeolithic Period," Zapiski of the Imperial Russian Archaeological Society, Slavic Division, Vol. XI, Petrograd, 1915, pp. 142–144.

[2] *Ibid.*

According to Vishnevsky,[1] "The high degree of workmanship of stone tools indicates a high cultural development."

This site requires further investigation.[2]

39. OSOKOROVKA

This site is located on the left shore of the Dnieper, near the village of Svistunovo-Petrovskoye, on the lower terrace, near the Osokorovka River. The site was discovered in July, 1931, by T. T. Teslja,[3] and studied by I. Levitsky.

Several cultural horizons were found. The lowest one is in the loess and, according to investigators, contains Aurignacian industry. The horizon above it is in the "limon," which is in turn covered by the layers of loess. This upper horizon is Magdalenian. Over it are the layers deposited in interglacial intervals and containing the Epipalaeolithic industry.[4]

The remains of hearths and habitations are of especial interest. No further details are available as the material is being studied and prepared for publication.[5]

40. MELTINOVA

This site is situated on the bank of the Donetz river, near the village of Meltinova, Belevsk district, Tula region.[6] It was partly investigated by V. A. Gorodzov.[7] A cross-section of the ground at the site presents the following picture:

1. Gray-yellow loess-like clay of recent origin—3 feet.
2. Gray clay with remains of vegetation—0.75 feet.
3. Gray clay with short streaks of coarse washed sand, especially abundant in the lower part, with many remains of vegetable matter and shells—3.5 feet.
4. Thin layer of yellow clay with spots of ochre—0.25 feet.
5. Bright yellow clay—3 feet.
6. Limestone.

The yellow clay is of a Quaternary age. The flint chips and bones of fossil animals were found in the third layer and are attributed tentatively to the post-glacial period. Further details are not available.

[1] Vishnevsky, "Prehistoric Man in Russia," Appendix to the translation of Osborn's "Men of the Old Stone Age," Leningrad, 1934, p. 465.

[2] Skutil, J., ("Ossarynce und Iskorost, zwei palaeolitische stationen in Osteeuropa," Eiszeit und Urgeschichte, Vol. IV, part 5, 1928, pp. 46–48) gives additional information.

[3] Teslja, T. T., "Das Paläolith im Bereich der Dnieprostrom-schnellen," Die Quartarperiode, part 4, Ukrainische Akademie der Wissenschaffen, Kiev, 1932, pp. 83–85.

[4] According to V. V. Riznichenko, quoted by Teslja, the lower horizons are in the loess layer of Würm I, the middle horizon in the deposits of Würm Interglacial, and the upper horizons in the post-glacial formations.

[5] It should be noted that H. Field and E. Prostov (op. cit. A. A. Vol. 38, p. 276) refer, presumably, to the same site as Osokorovka II in the body of the article, and as Osokorovka in the list of sites. The description, however, so tallies with the above in all details, as to permit the author to assume the Roman II after the name of the site to be a typographical error.

[6] Spitzin, A. A., "The Russian Palaeolithic Period," Zapiski of the Imperial Russian Archaeological Society, Slavic Division, Vol. XI, Petrograd, 1915.

[7] Gorodzov, V. A., "Materials for the Archaeological Map of the Valley and Shores of the Oka River," Trudi of the XII Archaeological Congress, Vol. I, p. 540.

41. URKUSTA

During the excavation of the Tardenoisian burial in two grottos of Urkusta, in the valley of Baidar, Crimea, G. A. Bonch-Osmolovsky [1] discovered the remains of a Palaeolithic site with several cultural layers, containing a flint industry, objects of bone, and fossil remains.

According to the investigator, the cultural contents of these layers differ considerably. Thus, in the lower layer, was found a number of gravers and scrapers made on short blades, as well as a quantity of geometrical forms rather thick in cross-section. The blades here appear to be quite massive.

In the upper layers the blades are thinner and more regular, becoming almost typically Tardenoisian in style. The gravers disappear; the scrapers are made only on round flakes. Here backed blades and blades with lateral notch appear in large numbers, while the geometrical forms are like the fine trapezoides of the Tardenoisian.

No further information is available to the author.

42. STENINA

According to Spitzin,[2] a palaeolithic site was discovered in 1900 near the village of Stenina, Kozelsk district, Kaluga Region. On the bank of the Torstenka Creek several split mammoth bones were found in association with charcoal. The trial excavations produced a considerable quantity of bones, flint tools, and small bone points. The small excavations made in 1902 produced bones of elk and bison and about twenty implements (scrapers and knives) made of bone, horn, flint, and limestone. In the same year Krishtaphovich visited this site and found that the layer containing bones and bone tools was of lake-river origin, and that part of the objects were *in situ*, while part had been moved about within the borders of the same lake depression. The bones of the mammoth, reindeer, and *Bos primigenius*, were identified.[3] This site is especially interesting because of its bone work, which may belong to another period. Spitzin does not venture to give any dating.

43. KACHINSK

Kachinsk rock shelter, situated in the southern part of the Crimean peninsula, was discovered by K. S. Merezhkovsky. According to Bonch-Osmolovsky,[4] the cultural layer contained flint industry of earlier character than that of Suren II (see page 394). However, the presence of scrapers made on the end of a blade suggests a relation to Magdalenian culture.

[1] Bonch-Osmolovsky, G. A., "Le Palèolitique de Crimèe," Bulletin of the Quaternary Committee, No. 1, 1929, p. 32.

[2] Spitzin, A. A., "The Russian Palaeolithic Period," Zapiski of the Imperial Russian Archaeological Society, Slavic Division, Vol. XI, Petrograd, 1915.

[3] Chetirkin, I., "The Discovery in Kozelsk County near the Village of Stenina of Remains of the Palaeolithic Period," Izvestia of Kaluga Archaeological Committee, 1902, pp. 2–5.

Krishtaphovich, N. I., "Geological Investigation of the Palaeolithic Site during the Summer of 1904," Drevnosti, Trudi of the Moscow Archaeological Society, Vol. XXI, part 2, p. 176.

Krishtaphovich, N. I., "The Location of Mammoth Bones in Kozelsk County," Year Book of Russian Geology, Vol. IV, part 1, pp. 173–176.

[4] Bonch-Osmolovsky, G. A., "The Prehistoric Cultures of Crimea," Krim, No. 2.

The bones of a ten-year-old child were found in the same layer in association with the flint implements. The materials await further study.

44. CHATIRZHDAG

A culture similar to that of Suren I was discovered by K. S. Merezhkovsky[1] in the cave of Chatirzhdag (Binbash-Koba and Sula Koba) in Crimea. No materials are preserved and therefore dating is impossible.

45. SERGEYEVKA

The site of Sergeyevka, is located in the Gadatsky district of the Poltava region. In 1921 a local teacher discovered bones of mammoth and flint flakes in a deep gully of the Gadatsky district. This site was visited by V. Scherbakovsky,[2] but not investigated.

46. KHOTIANOVKA

This site is situated on the bank of the Desna River in the village of Khotianovka, 15 klm. southwest of Trubchevsk, Briansk region. It was discovered by the representatives of the Academy of Sciences of U.S.S.R. in 1929,[3] who made the preliminary investigations.

The materials obtained are now preserved in the Geological Museum of the Academy of Sciences in Leningrad. They consist of mammoth bones and a small number of flint chips. Of special interest is the rib of a mammoth covered with a design in the form of deep notches rhythmically arranged in two rows. One end of the rib is broken. The function of this tool is difficult to determine. The results of investigations are as yet unpublished.

47. VOSKRESENSKII SPUSK

This site is almost opposite the town of Samara at the base of the first (flood-plain) terrace on the left bank of the Volga River.[4] The discovery was made in 1926, when sewers were laid at a depth of 3.15 meters. According to Golmsten,[5] attention was attracted by the mammoth tusk found at a depth of 3 m. Next to it were found flint chips and artificially broken bones.

The first investigator was the Samara archaeologist, M. G. Matkin, who placed in the Samara Archaeological Museum the material collected in the sewer trench. The more detailed investigation of V. A. Gorodzov in 1929 resulted in finds of the bones of the mammoth, Siberian rhinoceros, and a large species of *Bos*, a fauna which points to a considerable antiquity for this site. No further details pertaining to the industry are available to the author.

[1] Bonch-Osmolovsky, G. A., "The Prehistoric Cultures in Crimea," Krim, No. 1.

[2] Scherbakovsky, V., An article in Zbornik of the Local Ukrainian Scientific Society in Kiev, Kiev, 1921, p. 153.

Scherbakovsky, V., "Bemerkungen über neue wenig-bekannte Palaolithische Stationen in der Ukraine," Eiszeit IV, 1927.

Rudinsky, M., "The Scientific Work on Paleolithic and Mesolithic Cultures in Ukraine during the Revolution," Anthropologia, Annual Report of the Anthropological Laboratory of T. Vovk for 1930, Kiev, 1931, p. 148.

[3] Gorodzov, V. A., "The Results of Archaeological Activities in U.S.S.R. during 1917–1930," MSS.

[4] Gorodzov, V. A., "The Results of Archaeological Activities in U.S.S.R. during 1917–1930,".MSS.; also "Diary of Archaeological Excavations in Samara in 1929," MSS.

[5] Golmsten, V. V., "Archaeological Remains of Samara Province," RANION, Vol. IX, 1928, p. 129.

48. Kamenetzk-Podolsk

Very little is known about this site, which is near Kamenetzk-Podolsks, Podol region, beyond the fact that according to Uvarov and others [1] during dam construction, 2.5 km. from the Dniester River bones of a mammoth were found, which were later brought to Kiev University.   A student of V. B. Antonovitch who reported this find to Uvarov, unfortunately arrived on the spot of the find after most of the loess layer had been removed. In the remaining loess, besides the mammoth tusk and several long bones, were found two flint scrapers.   No further investigations were made.

49. Protasov Yar

According to Krishtaphovich,[2] mammoth bones and small flint tools with traces of a hearth were discovered in 1903 near the railway station in Kiev, in the so-called Protasov Yar (Protasov Gully).

The cultural remains were in the layer of sand, underlain by the glauconite clays at the depth of 56 feet below the surface.   The geological conditions are said to be identical with those at the site of Cyrill Street (see pages 386–393).   The series of layers of gray sands and loesses cover the cultural horizon.   No illustrations of industry or further details are available to the author.

50. Viazovka

Spitzin,[3] quoting Kaminsky,[4] indicates that in the gully in the village of Viazovka, near the town of Luben, Poltava Province, a flint tool of Mousterian type was found, in association with a fragment of mammoth jaw.

51. Shapovalovka

This site is located near the village of Shapovalovka, Koroton District, Chernigov region, in the basin of the river Seim.   On the slope of the lake shore bones of a mammoth and flint knives [5] were found in 1879 at a depth of three and a half feet.

Further investigations were made there in 1929 and, according to Rudinsky,[6] so far it is impossible to ascribe a Palaeolithic age to this site.

[1] Uvarov, A. S., "The Archaeology of Russia," Moscow, 1881, Vol. 1, pp. 111.

Spitzin, "The Russian Palaeolithic Period," Zapiski of the Imperial Russian Archaeological Society, Slavic Division, Vol. XI, Petrograd, 1915, p. 140.

Vishnevsky, B. N., "Prehistoric Man in Russia," appendix to translation of Osborn's "Men of the Old Stone Age," Leningrad, 1924.

[2] Krishtaphovich, N. I., "New Traces of Prehistoric Man in Kiev," Appendix to the Izvestia of the Imperial Archaeological Committee, St. Petersburg, Vol. I, p. 40.

[3] Spitzin, A. A., "The Russian Palaeolithic Period," Zapiski of the Imperial Russian Archaeological Society, Slavic Division, Vol. XI, Petrograd, 1915.

[4] Kaminsky, F., "The Traces of the Most Ancient Period of the Stone Age Along the River Sula and its Tributaries," Trudi of III (Kiev) Archaeolog. Congress, Kiev, 1878, Vol. I, pp. 147–152.

[5] Spitzin, A. A. ("The Russian Palaeolithic Period," Zapiski of the Imperial Russian Archaeological Society, Slavic Division, Vol. XI, Petrograd, 1915), quotes from the article of Samokvasov in Anthropological Viestnik, Vol. III, pp. 3–38.

[6] Rudinsky, M., "The Finds in Shapovalovka," Die Quartarperiode, Vols. 1–2, No. 10, 1930, pp. 37–42.

52. Selis'che

Selis'che [1] is situated near the village of Selis'che, Kanev district, Kiev region. The only mention of it is in a short notice of Krishtaphovich,[2] who states that this station, as well as the Kiev site, is situated on the main bank of the Dnieper valley under the layer of loess-sand deposits above the moraine clay. Animal bones and flint tools were found there. This site was further investigated from a geological point of view by Krishtaphovich in 1904.

53. Schurii Rog

This site is situated in the village of Schurii Rog in the Izum district, Kharkov region. The discovery of mammoth bones and flint industry was made in 1920, and confirmed in 1923 by O. Pheodorovsky.

According to Rudinsky,[3] there is a small unpublished collection in the Kharkov Museum, consisting of implements made of rose-colored flint, among which are several burins of poor technique and without retouch, similar to those of Zhuravka.

54. Kolachkovichi

M. Rudinsky [4] states that in the village Kolachkovichi, of the Kamenetzk district, Podol region, he discovered a site with late Aurignacian industry. The material is to be published. [5]

55. Kosheleva

M. Rudinsky [6] states that in the village Kosheleva, of the Kamenetzk district, Podal region, he discovered a site with late Aurignacian industry. The material is to be published.[7]

56. Questionable Sites

M. Rudinsky [8] lists the following locations and suggests the possibility of discovering there, with further investigation, palaeolithic remains.

(a) *Skakalka*,[9] near the village of Gorodische, Kremenchug district, Poltava region, where in 1922 were found remains of mammoth bones and a wide flint blade, heavily patinated.

[1] Spitzin, A. A., "The Russian Palaeolithic Period," Zapiski of the Imperial Russian Archaeological Society, Slavic Division, Vol. XI, Petrograd, 1915.

[2] Krishtaphovich, N. I., "Geological Investigation of the Palaeolithic Site during the Summer of 1904," Drevnosti, Trudi of Moscow Archaeological Society, Vol. XXI, part 2, p. 178.

[3] Rudinsky, M., "Scientific Work in Palaeolithic and Mesolithic Culture in Ukraine during the Revolution" (Annual Report of the Anthropological Laboratory of Th. Volkov for 1930, Vol. IV, Kiev, 1931, p. 148).

[4] Rudinsky, M., op. cit., p. 149.

[5] Field, H., and Prostov, E., ("Recent Archaeological Investigations in the Soviet Union," A. A., Vol. 38, No. 2, 1936, pp. 275–276), list it as Kolach Kovtsy I–II. According to them both sites are Aurignacian "with implements and fragments of bones of small animals belonging to the final phase of the Quaternary period in eastern Europe." (E. G.)

[6] Rudinsky, M., op. cit., p. 149.

[7] Field, H., and Prostov, E., op. cit., spell it Kuzheleva. (E. G.)

[8] Rudinsky, M., "Scientific Work in Palaeolithic and Mesolithic Culture in Ukraine during the Revolution," Annual Report of the Anthropological Laboratory of Th. Volkov for 1930, Vol. IV, Kiev, 1931, p. 150.

[9] A notice in the Annual Report of the Anthropological Laboratory of Th. Volkov, Vol. I, Kiev, 1925, p. 27.

(b) *Babanka*,[1] Uman district, Kiev region, where ribs of large fossil animals were found in loess.

(c) *Gai*,[2] Romni district, Poltava region, where bones of *Rhinoceros tichorhinus* and rough flint flakes were found.

(d) *Gorbov and Yuhnovo*,[2] Novgorod-Sieversky district, Chernigov region, where mammoth bones and flint flakes were found.

Vishnevsky,[3] probably after Spitzin,[4] lists also:

(e) *Degtiarevo*, Chernigov region, where a pile of split mammoth bones were found at a depth of six feet.

(f) *Umrichino*, Kursk region, where mammoth bones were found in association with flint tools. Spitzin doubts the Palaeolithic age of the site.

57. Additional Sites

The following are additional sites discovered during the last few years according to the information compiled for H. Field and E. Prostov,[5] through the cooperation of the Director of the All-Ukrainian Academy of Sciences (VUAN), F. A. Kozubovsky:

Sokol, on the river Dnieper, Podol region.
Vrublintsy, on the river Dnieper, Podol region.
Kitaigorod I and II, on the river Ternava, Podol region.
Demshin, on the river Dniester, Podol region.
Bakota, on the river Dniester, Podol region.
Staraya Ushitza, on the river Ushitza, Podol region.
Kalus, on the river Dniester, Podol region.
Semenki, on the river Boog, Podol region.
Gorodok, on the river Goryn, Volyn region.
Korosten near Iskorost, on the river Uzh, Volyn region.
Pushkari, on the river Desna, Chernigov region.
Novgorod Sieversky, on the river Desna, Chernigov region.
Krivoi Rog, on the river Ingulets, Dniepropetrovsk region.
Kodak village (location not indicated), Dniepropetrovsk region.
Miorka (discovered in 1932), Dneproges region.
Yamburg (discovered in 1934), Dneproges region.

Unfortunately the above is all the information given by Field and Prostov, so that any idea of the dating and cultures must be deferred for the future.

According to Field and Prostov,[6] a series of new sites was discovered by the Abkhazian Expedition of the Institute of Anthropology and Ethnography of Leningrad under the

[1] *Ibid.*, p. 22.

[2] Semenchik, M., "Some Prehistoric Finds in Romni District," Annaul Report of the Anthropological Laboratory of Th. Volkov, for 1929, Vol. III, Kiev, 1930, p. 209.

[3] Vishnevsky, B. N., "Prehistoric Man in Russia" Appendix to translation of Osborne's "Men of the Old Stone Age," Leningrad, 1924, pp. 441–508).

[4] Spitzin, A. A., "The Russian Palaeolithic Period," Zapiski of the Imperial Russian Archaeological Society, Slavic Division, Vol. XI, Petrograd, 1915, pp. 132–172.

[5] Field, H., and Prostov, E., "Recent Archeological Investigations in the Soviet Union," A. A., Vol. 38, No. 2, 1936, pp. 274–276.

[6] Field, H., and Prostov, E., "Archaeology in the Soviet Union," American Anthropologist, Vol. 39, No. 3, July–September, 1937, pp. 459–460.

leadership of S. N. Zamiatnin, who began the survey in 1934, in cooperation with a geological expedition of the Russian Academy of Sciences.

The following list of sites is given, most of which are situated on the eastern shore of the Black Sea. (See map, Fig. 20.)

1. Byrts, southeast of Yashtukh Mountain. Acheulean.
2. Gvard, south of Byrts. Acheulean.
3. Apiancha Mountain, near Tsebelda, east of Sukhum. Acheulean.
4. Yagish Mountain, on the left bank of Madzharkl River. Acheulean.
5. Akhabiuk, near Mikhailovskoe, north of Sukhum. Mousterian.
6. Kelasuri, on Kelasuri River, five km. southeast of Sukhum. Mousterian.
7. Esheri, on the right bank of Gumista River, seven km. west of Sukhum. Mousterian.
8. Okum, near railroad, seven km. southeast of Ochemchiri. Mousterian.
9. Chuburiskhindzhi, twelve km. southeast of Gali, near Sanzhio Mountain. Mousterian.
10. Lichkop, two km., north of Sukhum. Mousterian and Upper Palaeolithic.
11. Gali, Asheulean, Mousterian, and Upper Palaeolithic.
12. Achikhvari, north of Ochemchiri.
13. Kolkhida, five km. southeast of Gagry.
14. Barmish, five km. east of Kolkhida.

It is indicated that during the summer's work some twenty Palaeolithic sites were discovered, though no further names of sites are given. We have no idea of their industry beyond the statement that these sites represent, " mainly Aheulian, Clactonian, Levalloisian and Mousterian types, while Upper Palaeolithic implements were also excavated." [1]

It is important to remember that, as a whole, the terms "Acheulean" and "Mousterian" are used by Russian Archaeologists in a temporal, rather then cultural sense, though Field and Prostov specifically mention that, "at several sites, for the first time in the U.S.S.R., Lower Palaeolithic *coups de poing* were found."   The basis for the dating is not indicated beyond stating that, "in many cases it was possible to establish the connection between the Palaeolithic implements of various periods and the corresponding sea terraces."

Further details pertaining to faunal remains, types of industry, and conditions of deposition, are needed in order to form a clearer conception of these important discoveries.

---

[1] *Ibid.*, p. 460.

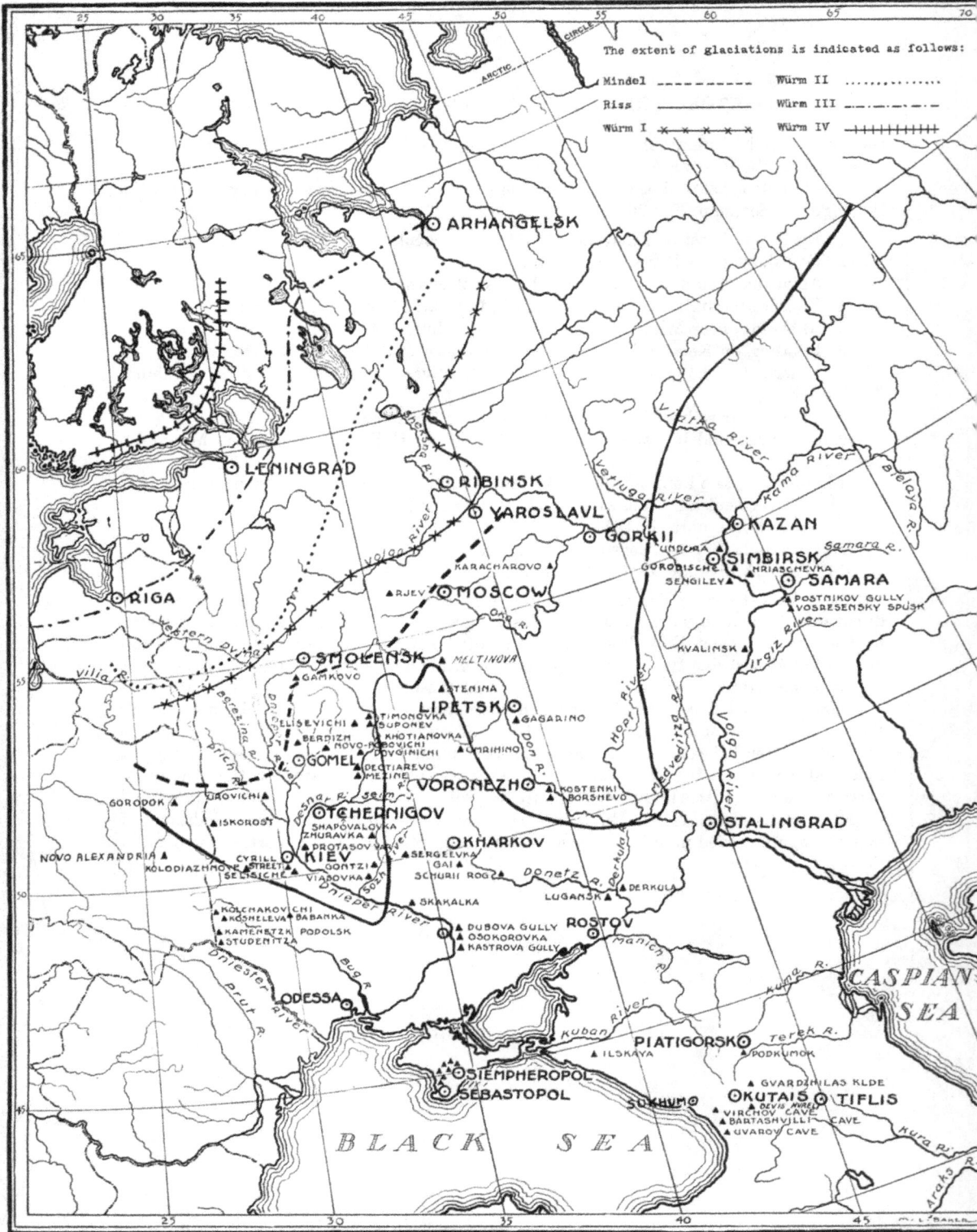

The extent of glaciations is indicated as follows:

Mindel	– – – – –	Würm II	··········
Riss	————	Würm III	–·–·–·–
Würm I	× × × ×	Würm IV	+++++++

ARHANGELSK

LENINGRAD

RIBINSK

YAROSLAVL

GORKII

KAZAN

SIMBIRSK

GORODISCHE ▲ MRIABCHEVKA ▲
SENGILEY ▲

SAMARA
POSTNIKOV GULLY ▲
VOSRESENSKY SPUSK ▲

RIGA

KARACHAROVO ▲

RJEV ▲ MOSCOW

KVALINSK ▲

SMOLENSK MELTINOVA

GAMKOVO ▲

STENINA ▲

LIPETSK ○ GAGARINO ▲

TIMONOVKA ▲
ELISEVICHI ▲ SUPONEV ▲
BERDIZM ▲ KHOTIANOVKA ▲
NOVO-ROBOVICHI ▲
DOVGINICHI ▲

UMRIHINO ▲

GOMEL ○ DEGTIAREVO ▲
MEZINE ▲

VORONEZH ○ KOSTENKI ▲
BORSHEVO ▲

GORODOK ▲ UROVICHI ▲

ISKOROST ▲

TCHERNIGOV ○
SHAPOVALOVKA ▲
ZHURAVKA ▲

STALINGRAD

NOVO ALEXANDRIA ▲
KOLODIAZHNOYE STREET ▲
SELISICHE ▲

CYRILL ○ PROTASOV VAR ▲
GONTZI ▲
VLADOVKA ▲ KIEV

SERGEEVKA ▲
GAI ▲
SCHURII ROG ▲

KHARKOV

DONETZ R. ▲ DERKUL

LUGANSK ▲

SKAKALKA ▲

KOLCHAKOVICHI ▲ BABANKA ▲
KOSHELEVA ▲
KAMENETZK PODOLSK ▲
STUDENITZA ▲

DUBOVA GULLY ▲
OSOKOROVKA ▲
KASTROVA GULLY ▲

ROSTOV

ODESSA

PIATIGORSKO ○
ILSKAYA ▲ PODKUMOK ▲

SIEMPHEROPOL ○

SEBASTOPOL ○

SUKHUM ○

GVARDJILAS KLDE ▲
KUTAIS ○ DEVIS HVREL ▲
VIRCHOV CAVE ▲
BARTASHVILLI CAVE ▲
UVAROV CAVE ▲

TIFLIS ○

CASPIAN SEA

BLACK SEA

CONCLUSIONS

The question of the evaluation and dating of Palaeolithic cultures of European Russia presents considerable difficulties.

The size of the territory involved, the divers geological and topographical features, and the inadequacy of information about various determining factors, make the application of criteria, evolved, tested, and valid for Western Europe, at least a dangerous procedure here.

Another difficulty is met, it must be reluctantly admitted, in the fact that Russian archeologists have not as yet established any definite conception of the historical processes involved, even among themselves.[1]

In a preface to Dr. D. A. Garrod's work on the Upper Palaeolithic Age in Britain, M. L'Abbe Breuil, in speaking about English pre-historians, says: "Leur vocabulaire pour designer les formes industrielles est trop personnel, et leur connaissance de ces dernieres et de leur signification bien souvent tout a fait rudimentaire."[2]

This severe, but just, accusation may be applied to a great many authors of works on the Palaeolithic in other regions, not excluding Russia. Not only is "individualism," in purely technical terminology, very abundant (see Gorodzov's one hundred and seven types of burins) but even the general understanding of such terms as "Aurignacian," "Solutrean," and "Magdalenian," vary more than should be permissible, even for such a relatively young science as Prehistory.

This last situation is augmented by the fact that most Western European specialists employ the subdivisions of Mortillet, not only with temporal significance, but also in a cultural sense, associating them with specific ethnic groups. Thus, we may find the same author speaking of Aurignacian or Solutrean "race"; as well as Aurignacian, Solutrean, or Magdalenian "times."

While one is somewhat justified in using these terms in both senses in application to Western Europe, or more specifically to France, difficulties arise the moment one leaves the territory of the type sites. Well realizing that not all peoples of the world had precisely the

[1] Attempts to apply the Marxian historical method to the reconstruction of the history of early man, occupy a place of prime importance with Soviet Prehistorians. The following are some examples of the articles devoted to discussions on these problems.

Zhakov, M. P., "The Problem of the Genesis of Human Society." Problems of the History of Pre-Capitalistic Societies. Leningrad, May, 1934, No. 5, pp. 29–46.

"Labor, Technique, and Industrial Relation in Early Society," *Ibid.*, June, 1934, No. 6, pp. 9–45.

Ravdonikas, V. I., "The Periodisation in the History of Pre-Clan Society," *Ibid.*, July–August, 1934, No.7–8, pp. 72–87.

"To the Question of Dialectics of the Development of Pre-Clan Society," *Ibid.*, February, 1934, No. 2, pp. 25–48.

"The Marxian History of Material Culture," Izvestia of the State Academy for the History of Material Culture, Vol. VII, Part 3–4, Leningrad, 1930, pp. 1–94.

Meschaninov, I. I., "Paleonthology and Homo Sapiens," *Ibid.*, Vol. VI, Part 7, pp. 1–36.

Boriskovsky, P. I., "The Historical Premises of the Origin of the So-Called 'Homo Sapiens,'" Problems of the History of Pre-Clan Societies, 1935, No. 1–2, pp. 5–46.

Sagatsky, A. P., "Labor and the Origin of Society," *Ibid.*, 1935, No. 1–2, pp. 177–192.

[2] Garrod, D. A., "The Upper Palaeolithic Age in Britain," Oxford, 1926, p. 5.

same culture at a given period, European students stress the ethnical meaning of these terms and are busy tracing the influences and explaining the differences between the "mother" cultures and their derivatives. And yet, in many cases, they take as a basis for comparison the "classical" culture, which often is not the original "mother" complex underlying the subsequently developed local variations, but the specific manifestation, locally developed, sometimes tinted by other influences, and often of a very short duration.

The "classical" Solutrean, which in its pure form has such limited distribution both in space and time, may serve as an example. Outside of the very restricted area of central France, Cantabria, and Catalonia, the investigator, in describing the culture studied, has to point out either the absence of the basic elements, or the presence of new ones. Consequently, in the majority of cases, so little is left from the contents of the "basic" culture, that the exceptions constitute the bulk of the traits. The net result is often so complicated that new terms, such as Proto-Solutrean, Pseudo-Solutrean, Upper and Lower Solutrean, are invented, only to make matters further involved.

Perhaps sensing this difficulty, the Russian archaeologists tried to avoid it by assigning only temporal significance to the above terms. Their effort to tie up the archaeological divisions with specific types of social structure and all its resulting complexes, led them, however, into another set of complications.

Before discussing the Palaeolithic industries of European Russia, it is thought advisable to outline briefly some of the basic assertions of Soviet Prehistorians, which often serve as the foundation for their dating and interpretation of cultural finds.

One of the most important points of divergence is the acceptance of the universal evolutionary scheme of cultural development with the material basis as the main determining factor. This results in a fairly well outlined succession of the phases of social structure, corresponding to the different stages of industrial development. It may be illustrated, though perhaps subject to modifications and some corrections, with the scheme given by V. I. Ravdonikas,[1] to which the present author adds the concepts of N. Y. Marr, on the evolution of modes of thinking, writing, and language.

THE DEVELOPMENT OF SOCIETY

As is seen from the chart (page 445) the center of gravity lies in the method of production, (the basis of industry) determining the appearance of "additional values" which, in turn, determine the type of social structure and all its super-structures (religion, art, language, etc.).

The Soviet archaeologists postulate this universal scheme of development as an inevitable course of evolution, recognizing, however, that "the stages long passed by a given historical society, which is today on a higher plane of development, may still be experienced by other societies in different geographical regions."[2] Likewise, they recognize the possibility and occurrence of "survivals" of archaic forms in later periods.

At the same time, they very strenuously object to other methods of explanation of similarities or differences in cultural phenomena, such as "migrations," "borrowing," and

---

[1] "The Question of the Sociological Periodisation of the Palaeolithic Period in Connection with the Opinion of Marx and Engels on Primitive Society." Izvestia of the State Academy for the History of Material Culture, Leningrad, 1931, Vol. IX, p. 2.

[2] Ravdonikas, V. I., op. cit., p. 3.

THE DEVELOPMENT OF SOCIETY.

Geological epoch.	Archaeological period.	Physical type.	Technique of tool making.	Tools—indicators of production.	Main productions	Gain	Production relationship.	Form of society.	Ideology.	Modes of communication.	Modes of thinking.	Marriage.	Writing.	Type of speech.
RECENT	IRON	Modern Man	Casting, welding, soldering, etc.	Weapons. Armor. etc.	Complex agriculture and industry.	Basis of society.	Kings. Chiefs. Organized trade.	Capitalism.	Organized religion.	Sound language.	Modern. (Realistic — Micro-cosmic.)	Patriarchal family.	Alphabet.	Flective.
	BRONZE		Welding. Casting.	Weapons. Sickles.				Feudalism.				Father-right ↔ Classical mother-right.	Ideogram and syllabical sign.	
	COPPER		Cold and hot method.	Axe. Hoe. Bow and arrow.	Agriculture. Herding.									Agglutinative.
	NEOLITHIC		Drilling and polishing of stone. Pottery.		Hoe-culture. Beginning of domestication of animals.	Permanent supplies.	Separation of organizers. Social labor divisions. First barter.	Father's clan.					Pictogram. ↔ Ideogram.	Amorphic-synthetic.
WÜRM	PROTO-NEOLITHIC	Cro-Magnon Man / Grimaldi Man	Work on Bone. Pressure-flaking.	Dart and harpoon points. Needles. Knives. Engraving tools.	Organized hunting and fishing.		Sex division of labor. Appearance of leaders in process of collective hunting.	Mother's clan. Proto-clan society. Age-sex groupings.	Animism and cult of ancestors.	Kinetic language.	Ethnographical totemism. (Mythological — Micro-cosmic.)	Pair family.	Picture-idea. ↔ Pictographs.	Transition to Amorphic-synthetic.
	MAGDALENIAN		Blade detaching.						Production magic.			Origin of clan. Early exogamy. Two class system. Matrilocality. Matrilineal descent. Group marriage. Pair family.		
	SOLUTREAN		Engraving.											
	AURIGNACIAN		Bifacial chipping.											Amorphic
RISS-WÜRM	MOUSTERIAN	Homo-Neanderthalensis	Flakes detached. One side retouched.	Point. Scraper.	Hunting.	Beginning of permanent supplies.	Primitive division of labor according to sex and age.	Stable group of hunters with permanent center.			Ethnological totem.	Beginning of consanguin family.	Diffused sign.	
RISS	ACHEULEAN	Homo-Heidelbergensis		Hand axe as universal tool.	Gathering and small game hunting.	Absence of permanent supplies.	Simple coöperation.	Primitive communism. Collective with artificial tools.			Cosmic thinking.	Sexual promiscuity.		
MINDEL-RISS	CHELLEAN													
	PRE-CHELLEAN													
MINDEL	EOLITHIC	Pekin-Man. Java-Man.	Natural tools.	Eoliths.	Gathering.			Collective of gatherers with natural tools. Horde			Horde consciousness.			

"superior races," very much used, and, one should confess, often misused, by the Euro-American scientists.

The biological development of man is also pictured as one uninterrupted chain of evolution. Meschaninov, one of the ablest followers of N. Y. Marr, father of the science of Japhetology, expresses this point of view in his work "Palaeontology and Homo Sapiens" (*op. cit.*). He considers Cro-Magnon man not a different race from Neanderthal, but a direct descendant of the latter. He even adds that, "we have no facts for the assertion that this transformation did not take place in Europe." [1] The usual conception that the new race came and displaced the old one is vigorously denied as one contradicting the "unity of the process of development." [2]

The fact that we are dealing not only with two cultural divisions (Mousterian and Upper Palaeolithic) but also with two different physical types, is very ingeniously dealt with by P. P. Ephimenko. He refutes the hypothesis of the wholesale destruction of Neanderthal Man, with Mousterian culture, by the new "superior race" of Cro-Magnon, with Aurignacian culture, and considers this explanation comparable to the idea of "cataclysms," abandoned by science over a hundred years ago. [3]

Following the accepted Soviet conception of the unilineal development of culture and their anti-migration tendency, he considers that such sites as Abri-Audi and Shaitan-Koba in Crimea, conclusively prove the uninterrupted chain of development from Mousterian to the Upper Palaeolithic. Thus he derives the Châtelperron point made on a blade, from the Mousterian point made on a flake.

Likewise, the appearance of bone industry which takes place, in its embryo form, in Mousterian culture (La Quina, Kiik-Koba, Chokurcha, etc.), and especially the new method of detaching of blades from the prismatic nucleus, instead of triangular flakes from the discoidal one, are, in his opinion, the most important technical inventions during Upper Palaeolithic times. Again, faithful to the Marxian interpretation of history, he attributes these changes to a material basis. Such factors as the change of climate, followed by the change of fauna, necessitated a transition from the primitive horde, consisting of small groups of game hunters very much scattered over an enormous territory with very little intercommunication, to the stable, larger, and more or less permanent hunting society.

The small size of the Mousterian groups led, in the opinion of Ephimenko, to incestuous mating and the degeneration, which manifested itself in such physical characteristics as low forehead, very much developed supraorbital ridges, etc. This retarded the growth of population and, in general, had an unfavorable influence on the physical development of Neanderthal Man. [4]

At this point Ephimenko finds himself in a dilemma. On one hand he tries to refute the usual conception of Neanderthal Man as a physically less endowed individual than his successor. In fact, he sees in the comparatively poor inventory of Mousterian industry, a proof of the tenacity and ingenuity of Neanderthal Man, who with such simple tools was able to maintain an existance. To make this more convincing, he even attempts to reconstruct the conditions of the Glacial epoch as being far less severe than it is usually depicted. Al this is done with one idea in mind: to refute the assertion that Neanderthal Man is of a

[1] Meschaninov, I. I., "Palaeontology and Homo Sapiens," *op. cit.*, p. 11.
[2] *Op. cit.*, p. 12.
[3] Ephimenko, P. P., "Pre-Clan Society," Leningrad, 1936, p. 261.
[4] *Ibid.*, p. 213.

different, inferior race, non-ancestoral to modern man.   However, the facts show that the skeletal remains of Mousterian Man have very primitive characteristics; hence, the "degeneration," based on close intermating, is brought up to explain away this glaring inconsistency.

What remains, then, is still another set of facts: the superior physical characteristics of Cro-Magnon Man, and his superior industry.   "Only during the next stage of the development of primitive industry," he writes,[1] "did the new conditions of existence change the character of the social structure, creating, instead of closed primitive groups, more complex combinations of intercommunicating hordes, connected with each other.   Only in these conditions could have taken place the rapid cultural growth and formation of the new human type, which entered the history of Europe, during the Upper Palaeolithic epoch, as the Cro-Magnon race."

This eugenic explanation of the physical characteristics of Cro-Magnon and Neanderthal Man is not accepted by all Soviet archaeologists.   Thus, Bonch-Osmolovsky writes: "Neanderthal Man, in the light of present-day knowledge differed physically from modern man, and, in comparison with the latter, was very much limited in his functional possibilities."[2]   On the basis of the study of arm bones from Kiik-Koba, he asserts that the structure of the hand prevented Neanderthal Man from making long knife-like blades, the detaching of which required a very accurate blow.   For the same reason, he could not have used the spear for throwing, and hence knew only its piercing use.   He sees in Neanderthal Man a certain stage of "perfection process" of a humanized animal, and not a degenerate group resulting from close intermarriage.   Bonch-Osmolovsky warns some of his colleagues about underestimating the importance of physical limitations in the development of culture.   "We have every reason to assert that whole series of the working processes were unattainable to Neanderthal Man due to his anatomical imperfection," he writes, explaining the appearance of Cro-Magnon Man as the continuation of the process of perfection on the basis of his social development.   "It is necessary to say that there could be no purely biological evolution of man, as all his physical changes were dictated by the gradual development of social life, and that the present physical conditions and physiological processes are the result of social influences on the organisms of his animal ancestor."[3]

The often invoked "migrations" and vainly sought and much fought for "centers of origins of cultures," form another target of attack for Soviet archaeologists as the unnecessary evasions of much simpler and more certain explanations.   To quote Ravdonikas again, the Kultur-Krise group "presupposes the basis distinction of the cultural ability of people, the division into the chosen, or leading groups and the delinquent ones, which are predestined to be the 'manure of history,' the idea on which the whole colonial policy of the imperialistic bourgeoisie is based."[4]

"The tracing of migrations," writes Meschaninov,[5] "inevitably leads to the so-called 'cultural centers,' which are often established by the deliberate desire of the investigator. . . . Besides, these migrations are found to be especially active in the prehistoric times

[1] *Ibid.*, p. 214.

[2] "The Problems of the Complex Study of the Quaternary Period," Soobschenia of GAIMK, No. 3–4, Leningrad, 1932, p. 47.

[3] *Ibid.*, p. 48.

[4] "The Question of the Sociological Periodisation . . .," p. 4.

[5] "Palaeontology and Homo Sapiens," p. 7.

which are not documented by written sources. This makes us see the 'activity' not on the part of the tribes, but in the minds of investigators themselves, who cannot comprehend the complex forms of the studied objects, and, therefore, resort to much simpler explanations, by tabulating on a map the supposed routes of movements from the original cradle places, designated as cultural centers."

While not all of the present Soviet archaeologists subscribe *in toto* to these views, and many of them are constantly being criticised and corrected for deviations, or indulgence in the erroneous bourgeois theories, the above opinion could be taken as a sufficiently accurate presentation of the accepted ideas on the subject and may serve to explain the differences in dating and interpretations of the Palaeolithic sites.

The present writer, being a product of the American School of Anthropology, does not believe that the uninterrupted scheme of cultural evolution has been proven beyond a doubt. He also makes use of the arrivals of new races, migrations, and "cultural centers," as explanations when warranted by the material at hand. Yet, there is a great deal of truth in this perhaps too vigorous criticism of our methods. Too often some studied group, dear to the heart of the investigator, is elevated to the position of a "superior race" or conclusions are allowed to rest upon arbitrarily invoked migrations which, unwarranted by either fact or logic, serve only to cover up the ignorance of the investigator, much as the proverbial term, "ceremonial," is used to explain the use of some puzzling object.

There are some hundred and thirteen localities listed in the present survey. (See map, Fig. 100.) From ten sites which produced skeletal material, only in two (Kiik-Koba and Devis-Hvreli) do we have an accompaning stone industry. Thirty-eight sites are so insufficiently described as to preclude any possibility of forming an idea of their age. The remainder are more or less adequately documented and permit certain generalizations.

LOWER PALAEOLITHIC INDUSTRIES

European Russia seems to be very poor in remains of Lower Palaeolithic cultures. No true Chellean and Acheulean sites are found. Two accidental finds of *coup de poing* type of implement are questionable. The Rzhev specimen is very reminiscent of Neolithic tools, and the Volga find, while "classical" Acheulean in appearance and technique, is too much like a real French implement both in form and patination. One wonders if it was not "imported" by some amateur archaeologist from France and subsequently lost. At any rate, it would be very rash to assert the existence of a culture on the basis of two isolated finds, both very imperfectly documented.

The reported finds of a series of Acheulean sites in the Caucasus by Zamiatnin in 1934, may change the above conclusions. It is stated that Lower Palaeolithic hand-axes were found in several sites, some of which, apparently, could be geologically dated by correlation with sea terraces. Pending further information as to the industrial types and fauna, evaluation at present is impossible.

MIDDLE PALEOLITHIC INDUSTRIES

The so-called "pre-Mousterian" cultures, the "atypical" of Bonch-Osmolovsky and the "Tyasian" of M. Breuil, such as are found in the lower layer of La Micoque and La Ferrassie, are represented in European Russia by the lower layer of Kiik-Koba. There were tendencies, on the part of some Russian investigators (Gorodzov and others), to doubt the existence here of two separate cultural layers. The examination of the collections and reproductions of M. Bonch-Osmolovsky and especially the drawings made by M. Breuil seem to substantiate the excavator's assertion.

The dating of Kiik-Koba, however, seems to be far from settled. Besides the known controversy which exists among European scholars as to the exact placing of La Micoque, the situation becomes still more complicated due to Bonch-Osmolovsky's assertion that his atypical industry is co-existent with the Chellean;[1] and that his second horizon corresponds to La Micoque, which he considers synchronal with Acheulean (Acheulean V–VI?).

An examination of the flora and fauna of Kiik-Koba layers points to the increasing cold during the second period of occupation of the cave. Thus, such animals as *Elephas primigenius*, *Rhinoceros tichorhinus*, *Ursus spelaeus* and *Hyena spelaea*, do not figure in the lower layer and appear in the upper. The quantitative analysis of the flora also shows four times more remains of *Juniper* sp. in the upper layer. In the same time there is nothing to indicate the possibility of an interglacial period for the lower layer, as Bonch-Osmolovsky supposes,[2] in spite of the fact that this would correspond to the opinion of M. Breuil, who places La Micoque with the warm fauna of Riss-Würm.

It would indeed be stretching the point, to consider the fauna of the lower layer of Kiik-Koba as that of a warm climate. Not only is the warm fauna of the Riss-Würm in Western Europe, such as *Elephas antiquus*, *Rhinoceros merkii*, *Hippopotamus*, etc., absent, which could be accounted for by local conditions, but out of eleven species listed for the lower layer of Kiik-Koba, only one, *Equus hemionus*, is not found among the animals of Suren I, which is also in the Crimea, and which, according to Bonch-Osmolovsky, belongs to Aurignacian times, or the Würm glaciation.

The only remaining explanation is that the first settlement of Kiik-Koba took place during the beginning of the Würm, and the second layer somewhat later, when the developed Würnian glaciers drew nearer and further lowered the temperature.

It seems to be also significant that in Kiik-Koba, (horizon not indicated) in Chokurcha, (Mousterian) and in Shaitan Koba, (Late Mousterian), Bonch-Osmolovsky found traces of the use of bone, similar to those described by H. Martin for the Late Mousterian site of La Quina.

In view of the above, and considering the fact that the fauna of Chokurcha differs from that of Kiik-Koba only in the absence of *Rhinoceros tichorhinus*, the dating of Kiik-Koba's upper layer as synchronous with the Acheulean in the old sense[3] cannot be accepted. Ephimenko[4] calls it Pre-Mousterian, or Archaic Mousterian, and Gromov,[5] Mousterian with Acheulian tradition.

The Micoquian type of industry, where small bi-facial tools are found in association with fine points and scrapers made on flakes, is also represented by Chokurcha and Wolf

[1] Bonch-Osmolovsky, G. A., "The Evolution of the Old Paleolithic Industries," "Chelovek," No. 2–4, Leningrad, 1928, p. 184.

[2] Bonch-Osmolovsky, G. A., "The Evolution of the Lower Paleolithic Industries," Chelovek, No. 2–4, Leningrad, 1928, footnote to page 178.

[3] Bonch-Osmolovsky, G. A., "The Cuts and Notches on the Animal Bones of Palaeolithic Period," Soobschenia of the State Academy for the History of Material Culture, No. 8, August, 1931, pp. 25–27.

[4] Ephimenko, P. P., "Pre-Clan Society," Leningrad, 1936, pp. 157–159, 169.

[5] Gromov, V. I., "Geology and Fauna of the Palaeolithic Period in U.S.S.R.," Problems of the History of Material Culture, No. 1–2, Leningrad, 1933, p. 29.

Grotto in Crimea, and Ilskaya in the Caucasus. While in the case of the upper layer of Kiik-Koba and the lower layer of Chokurcha the similarity to La Micoque is pronounced, this tendency decreases in Wolf Grotto and appears still less in Ilskaya. The latter site seems to be a hybrid of "classical" Mousterian of Derkula type with the Mousterian of Acheulean tradition, as represented by the above-mentioned caves of Crimea.

A transitional position between the Middle Palaeolithic and the Upper Palaeolithic corresponding to that of Abri Audi in Western Europe, is occupied by the site of Shaitan-Koba, with its large *bifaces* and the appearance of a true blade industry.

In general, the lower and middle Paleolithic sites are not numerous in European Russia. The total, even counting Podkumok,[1] is less then twelve.

Gromov [2] dates these sites geologically as corresponding to the end of the formation of the ancient alluvials of the second terrace; that is to say, from the second half of Riss-Würm and the first part of Würm, to the time of the maximum development of this glaciation. In the open sites they are characterized by the steppe fauna, *Bovidae*, and mammoth; in the caves of the Crimea by the large quantity of *Megaceros*.

UPPER PALAEOLITHIC INDUSTRIES

According to Gromov,[3] the Upper Palaeolithic sites of European Russia correspond in time to the formation of the second terrace (counting as the first the present flood-plain terrace in the valleys of the Dniester and Don), which took place in Würmian times in the dating of Mirchink.

Gromov states further, that early Aurignacian sites belong to the time of the early stages of the formation of this terrace; Middle and Late Aurignacian, Solutrean and Magdalenian to the middle stages of this formation; and Late Magdalenian to the last phases of it. Geologically, it would correspond to the period after the maximum development of Würm and the beginning of its retreat. As a rule, cultural remains are found in the middle horizons of the old alluvial deposits of these terraces, or in contemporaneous formations, which fill the gullies cutting the earlier terraces.

Zamiatnin, on the basis of comparative study, asserts that the older sites are usually situated at the mouths of the gullies, as in the case of Kostenki I, Gargarino, and Borshevo I; while the younger sites, such as Kostenki II, III, Borshevo II, Gontzi, etc. are in the actual valley of the river. Thus, according to him, the mere location of the site sometimes gives a clue for the determination of its date.

Much additional help is derived from the faunal complexes, which, according to Gromov, seem to have such definite character that, as in the case of Gagarino, the cultural stage can be determined by an analysis of the fauna. It is especially helpful for the cave sites of the Caucasus and Crimea, where the usual geological evidences are lacking.

Unfortunately, in European Russia there is practically a total absence of properly documented stratigraphy of distinctly different cultural layers. The case of Suren I,

[1] The newly discovered sites in the Caucasus (pp. 440–441) are not being considered here, due to inadequacy of available information. If confirmed, they would increase the above figure by some twenty additional sites of Acheulean and Mousterian type of culture. (E. G.)

[2] Gromov, V. I., "The Geology and Fauna of the Paleolithic Period in U.S.S.R.," Problems of the History of Material Culture, No. 1–2, Leningrad, 1933, p. 29.

[3] Gromov, V. I., "Geology and Fauna of the Palaeolithic Period in U.S.S.R.," Problems of the History of Material Culture, No. 1–2, Leningrad, 1933.

where Bonch-Osmolovsky distinguishes three horizons, does not warrant major cultural subdivisions, beyond the phases of the Aurignacian. The two horizons of Cyrill Street, are very poorly documented; the three layers of Borshevo II present a practically uniform cultural horizon. This scarcity of superimposed cultural layers, so valuable for comparative dating, may be due to the fact that in European Russia cave sites with Upper Paleolithic industries are very rare; the bulk of the finds are made in open sites.

Consequently, in the majority of cases, we are dealing with the open camp sites of hunting peoples, where two different cultural complexes can be distinguished. First, there are the large accumulations of animal bones, definitely segregated, often referred to as "hearths," and very much like the well-known bone piles from Predmost, Moravia. These accumulations of bones are variously explained as representing kitchen refuse, supplies for bone work, or fuel. As a rule, very few stone implements are found among these piles. Typical Russian examples of these bone accumulations are found in Borshevo I, Berdizh, Kostenki I, Gagarino, Timonovka, Gontzi and others.

The second complex, according to most Russian investigators, represents the place of a more or less permanent habitation of these hunting groups. It is characterized by the red pigmentation of the ground, due to the presence of the ochre; the remains of open fires, filled with ashes and charred bones; and, usually, by a large quantity of stone and bone tools.

In several cases (Kostenki I, Timonovka, Gagarino) definite traces of semi-permanent dwellings were found, sometimes with stone slab foundations (Gagarino), well-preserved fire-places and specially made hearths with "shoulders" (Kostenki I), and various types of entrances, post holes, chimneys, etc. (Timonovka).

Inside the dwellings, investigators often found copings, niches, in which objects were kept, and caches of flint and bone tools. Near-by were found storage pits and primitive "ovens," in the shape of coneformed pits.

Timonovka, judging by the enormous quantity of flints in every degree of completion, from the rough chunks of stone to the thousands of finished implements, evidently represents a large work shop.

The flint industry of the Upper Paleolithic of European Russia presents almost every variety of form known in Western Europe, except the very specialized types of "classical" Solutrean.

The "leading" bone forms, such as the Aurignacian cleft base point and harpoons, or any examples of cave art, which form the foundation of industrial dating for Western Europe, are totally absent in European Russia. Consequently, this necessitates an entirely new set of criteria for comparative dating, applicable, perhaps, only to this region.

The usual terms for the subdivision of Upper Palaeolithic industries are employed by Russian scientists, but, due to the general idea of the unilineal development of culture, with somewhat different implications.

Accepting, perhaps, too literally the Western European scheme of the succession of industries as a definitely evolutionary series, they use the terms Aurignacian, Proto-Solutrean, Solutrean, and Magdalenian, in the temporal sense, to designate certain stages of the development of culture, rather then the specific cultures.

The usual criteria of Western Europe being absent, they attach the utmost importance to anything suggesting bifacial flaking (even if it is only a reduction of the bulb) and consider the presence of *point à cran*, even in its most atypical form, as indicating Solutrean temporal affinities.

Establishing this as a middle point for the dating, they class the industries where these characteristics are less prominent, as older, or Aurignacian; other industries which show the general decadence of technique or, for any other reason, are thought to be younger, as Magdalenian.

It may be significant that Ephimenko finds the true Aurignacian represented in Russia only by one site: Suren I in Crimea.[1] The rest of the Upper Palaeolithic sites, excluding the transitional group with microlithic industries, he divides into two groups: Aurignacian-Solutrean and Solutrean-Magdalenian, stating that there are no true Solutrean sites known in European Russia.

Ephimenko, Zamiatnin, and others, more or less agree to consider as Proto-Solutrean (Aurignacian-Solutrean or Russian Solutrean) the type of culture encountered in the loess in sites such as Predmost, Willendorf, Kostenki I and others, where we have more or less permanent camps of mammoth hunters characterized by large accumulations of mammoth bones, the presence of ochre pigmentation, and by the stone industry of the developed Aurignacian type coupled with the *point à cran* and some Solutrean-like bifacial retouch. Here, contrary to the usual assertions, art is fairly prominent; numerous carved stone and bone objects appear, and female statuettes are frequently found. Traces of semi-subterranean huts, in the shape of pits, sometimes reinforced with stone slabs, complete the list.

This scarcity of Aurignacian sites is due to the fact, as it was pointed out above, that every time *point à cran*, even in its atypical form, makes its appearence in a given industry, the site is dated as Aurignacian-Solutrean by Russian scientists.

The term Solutrean-Magdalenian is applied to all those sites which, either geologically or by some characteristic of the industry, appear to be younger than the first group. As "classical" Madgalenian, with its cave art, various harpoons, etc., is not found in European Russia, it is extremely difficult to distinguish between the developed Aurignacian and Magdalenian deposits. Since the usual criteria for dating are lacking, Russian archeologists have resorted to such general characterizations as microlithic tendency, atypical character, variety of form, decadence, and negligence of technique.

Thus, Ephimenko dates Kostenki II and III as Magdalenian on the basis of "general archaic or decadent character of the industry," while Menghin describes them as representing the "degeneration of Aurignacian technique." If we accept the absence of the Solutrean as established, then there is, in fact, no more contradiction between Ephimenko and Menghin, than in the case of Karacharovo, which Zamiatnin dates "later than final Aurignacian" and Ephimenko, as "early Magdalenian."

Basically, we seem to have in European Russia, the Early, Middle, and Late Upper Palaeolithic industries, all characterized by the blade technique, which starts very much like the Aurignacian in Western Europe, develops, aquires some peculiar traits suggesting Solutrean influence, and then slowly degenerates, finally reaching the stage of a microlithic industry.

The origin of this Aurignacian culture still remains unsettled. Burkitt's theory of a southern source may find added confirmation in the fact that in its purest form it is found only in the south of Russia—Crimea.

[1] It should be borne in mind, that in his most recent work (Pre-Clan Society) Ehimenko analyzes only about one fourth of the total number of sites, discussing, naturally, only the richest and the best known. (E. G.)

The Solutrean influences, as seen in some bifacial chipping and *point à cran*, seem to have their origin in Hungary.   There was, perhaps, sufficient connection between the Aurignacian cultures of Central Europe and the Solutrean in Hungary, to facilitate the spread and interchange of cultural traits.   Much in the same way as the peripheral site of Yankovichi [1] receives the cultural elements of the Aurignacian in the form of the typical cleft-base bone point, in contrast with the boneless Solutrean cultures of other sites in Hungary, the series of North German, Moravian, Polish, and Russian sites may have received some of the Solutrean influences from Hungary.

The gap between Hungary and central Russia is particularly well filled by the Polish sites.   Thus, in the cave of Nietoperzowa, near Ojcow [2] (excavated by Krukovsky and further analysed by Sawicki) we seem to deal with two horizons.   The lower horizon has crude laurel leaf points, with partial lower face retouch, as well as some typical Aurignacian forms employing fine Aurignacian retouch.   In the upper layer the "Solutrean" influences increase, resulting in a series of regular laurel leaves with a fine flat retouch, usually covering the upper face completely, and the lower face only partially.

Further west, in European Russia, in such sites as Kostenki I, Borshevo I, Berdizh, etc., this distant echo of the Hungarian Solutrean is found only in the appearance of the retouch on the lower face, and then only partially, and in *point à cran* of a very atypical form.   If the supposition of the Hungarian origin of these influences is accepted, the "cultural lag" will necessitate dating these Russian sites as younger than corresponding cultures of Poland and Moravia.

It is hard to say whether this more widely spread culture, which often is referred to as Proto-Solutrean, is the basic stratum and the Hungarian and French Solutrean only its local specializations, or whether, as was indicated above, the Moravian, Polish, and Russian sites represent the peripheral regions, which have received only a partial impetus from the original Hungarian source and have never developed further.

The presence of art works and bone industry, so alien to "classical" Solutrean sites, seems to indicate the plausability of the latter theory of the basic Aurignacian stratum, which received cultural influences of various strength, diminishing with the increase of the distance from the center.

Following the example of Polikarpovich, the present writer feels that in the absence of other terminology, one may be justified in speaking of Early, Middle, and Late Upper Palaeolithic in European Russia, supplementing the industrial characteristics with the geological data, as the only true criterion for dating.

The Russian material seems further to strengthen the necessity of using such terms as Solutrean and Magdalenian, not as temporal subdivisions, but more in the ethno-cultural sense; thus, the Solutrean must be considered as a specific cultural episode of comparatively short duration, and the "classical" Magdalenian only a local ethnic variation of that basic Upper Paleolithic facia of human development which is much more universal, and which we usually call Aurignacian.

[1] Breuil, H., "Notes de Voyage Paléolitique en Europe Central," L'Anthropologie, t. XXXIII, 1923, p. 338.

[2] Sawicki, I., "La Grotto Nietoperzowa à Jerzmanowice près Ojców, district Olkusz," Revue Archologique Polonaise, Poznan, 1926, Vol. III, Pt. 1, pp. 1–8.

BIBLIOGRAPHY

ANTEVS, E. Russia and Siberia. (Bulletin of the Geological Society of America, Vol. 40, 1929.) E.

The Last Glaciation. (American Geographical Society, Research Series, No. 17.) E.

Maps of the Pleistocene Glaciation. (Bulletin of the Geological Society of America, Vol. 40, 1929.) E.

ANTONOVITCH, V. B. About Rock Caves on the Bank of the Dniester River in Podol Province. (Trudi of the Odessa AC, Vol. I, pp. 93–95.) R.

Remains of the Stone Age Found in Kiev during the Last Three Years. (Trudi of the Xth AC in Riga, pp. 83.) R.

The History of Habitations in the Territory of Kiev from Palaeolithic Times to the Beginning of Christianity. ("Kievskoe Slovo," 1896, No. 2927.) R.

ARMASHEVSKY, P. Y. The Geological Survey of Chernigov Province. (Zapiski of the Society of Naturalists, Kiev University, Vol. VII, Pt. 1.) R.

The Palaeolithic Site of Cyrill Street in Kiev. (Izvestia of the XIth AC, Vol. II, pp. 142–144.) R.

ARMASHEVSKY, P. Y., AND ANTONOVITCH, V. B. The Geological Structure of Kiev Region. (Public Lectures on Geology and History of Kiev, 1897, pp. 21–25.) R.

BOGOLUBOV, N. N. Materials for the Geological History of the Kaluga Region in the Glacial Period. (The Yearbook of Geology and Mineralogy of Russia, Edited by N. N. Krishtaphovich. Vol. VII, pp. 111–119.) R.

BONCH-OSMOLOVSKY, G. A. A Palaeolithic Site in Crimea. (RAJ, Vol. XIV, Pt. 3–4, Moscow, 1926, pp. 81–87.) R.

Crimean Prehistory. (An article in the Guide for Crimea. Siempheropol, 1925.) R.

Archaeology of Crimea. (An article in "Priroda," No. 2, 1926.) R.

Le Paléolithique de Crimée. (Bulletin of the Quaternary Committee, 1929. No. 1, pp. 27–34.) F.

The Prehistoric Cultures of Crimea. ("Krim," No. 2, Moscow, 1926.) R.

The Question of the Evolution of Lower Palaeolithic Industries. ("Chelovek," 1928, No. 2–4.) R.

Shaitan-Koba, Crimean Site of the Abri-Audi Type. (Bulletin of the Quaternary Committee, Leningrad, 1930, pp. 61–82.) R.

The Prehistoric Cultures in Crimea. ("Krim," No. 1, Moscow, 1926.) R.

The Exposition of the Quaternary Period in the Academy of Sciences of U.S.S.R. (Soobschenia of GAIMK, Leningrad, 1932, No. 11–12, pp. 32–54.) R.

The Problems of the Complex Study of the Quaternary Period. (Soobschenia of GAIMK, Leningrad, 1932, No. 3–4, pp. 44–49.) R.

The Notches and Cuts on Animal Bones from Palaeolithic Sites. (Soobschenia of GAIMK, 1931, No. 8, pp. 25–27.) R.

BONDARTSCHUK, W. J. Die Fauna der Quartaren Ablagerungen der Ukraine S.S.R. (Die Quatarperiode, Ukrainische Akademie der Wissenschaften, 1932, No. 4, pp. 49–60.) G.

BORISKOVSKY, P. I. The Historical Premises of the Origin of the So-called "Homo Sapiens." (PHPS, Leningrad, 1935, No. 1–2, pp. 5–46.) R.

The Question of the Stages in the Development of the Upper Palaeolithic. (Izvestia of GAIMK, Vol. XIV, Pt. 4, Leningrad, 1932, pp. 1–40.) R.

BREUIL, H. Etudes de Stratigraphie Paléolithique dans le Nord de la France, la Belgique et l'Angleterre. (L'Anthropologie, Vol. XLI, 1931.) F.

Les Stations Paléolithiques en Transylvanie. (Bulletin de la Société des Sciences de Cluj, t. II, 1926, pp. 192–217.) F.

Notes de Voyage Paléolithique en Europe Central. (L'Anthropologie, 1923, Vol. XXIII, pp. 323–46; Vol. XXXIV, pp. 515–52; Vol. XXV, pp. 271–291.) F.

Le Paléolithique Ancient en Europe Occidentale et sa Chronologie. (Bulletin de la Société Préhistorique Française, No. 12, 1932.) F.

Les Subdivision du Paléolithique Supérieur et leur Signification. (Congrès International d'Anthropologie et d'Archéologie Préhistorique, Compte Rendu de la XIVme Session. Genève, 1912.) F.

La Préhistoire. (Revue des Cours et Conférences, 30, 1929, XII, Paris, 1930, pp. 1–20.) F.

Review of "Les Fouilles de la Station Paléolitique (Magdalenien) dans le Village de Hontzi, District de Loubne, Gouv. Poltava (Ukraine) en 1914 et 1915." (L'Anthropologie, t. XXXIV, p. 427.) F.

BOULE, M. Fossil Man. (London, 1925.) E.

BURKITT, M. C. Archaeological Work in Ukraine by Prof. Scherbakovsky. (The Antiquaries Journal, Vol. V, No. 3, July, 1925, pp. 273–277.) E.

Prehistory. (Cambridge University Press, 1925.) E.

Some Reflections on the Aurignacian Culture and its Female Statuettes. (ESA, Vol. 9, pp. 113–22.) E.

CAPITAN, L. "Etudes sur les Collections Rapportées de Russie par M. le Baron de Baye." (Bulletin de la Société d'Anthropologie de Paris, t. X (IVᵉ serie), fasc. 4, 1899, pp. 322–327.) F.

CHERNISCHEV, TH. Aperçu sur les Dépôts Posteriérs en Connection avec Trouvailles des Restes de la Culture Préhistorique au Nord et á L'est de la Russie d'Europe. (Congrès International d'Archéologie Préhistorique et d'Anthropologie, Moscow, 1892, Vol. VII.) F.

CHETIRKIN, I. The Discovery in Kozelek County near the Village of Stenina of Remains of the Palaeolithic Period. (Izvestia of Kaluga AC, 1902, pp. 2–5.)

COLEMAN, A. P. Ice Ages, Recent and Ancient. (New York, 1926.) E.

DE BAYE, G. Au Nord de la Chaine du Caucase. (Revue de Géographie, July–August, 1899.) F.

DE BRUIN. Travel in Moscowie. (1701.) R.

DOKUCHAEV, V. V. The Archaeology of Russia by A. E. Uvarov. (Trudi of the St. Petersburg Society of Naturalists, Vol. XIII, Pt. 1, St. Petersburg, 1882, pp. 1–54.) R.

DUBINSKY, S. A. Bibliography of the Archaeology of Bielorussia and Adjacent Regions. (Minsk, 1933.)

EBERT, MAX. Südrussland im Altertum. (Leipzig, 1921.) G.

Reallexikon der Vorgeschichte. (Leipzig, 1925.) G.

EPHIMENKO, P. P. Female Statuettes of Aurignacian and Solutrean Epochs. (Soobschenia of GAIMK, July, 1931, pp. 4–5.) R.

Kostenki I. (Soobschenia of GAIMK, No. 11–12, 1931, pp. 58–60.) R.

Pre-Clan Society. (Leningrad, 1936.)

Small Flint Implements of Geometrical and Other Special Forms in Russia Sites of Early Neolithic Age. (RAJ, Vol. 13, Pt. 3–4, pp. 211–228.) R.

Some Results of the Study of the Palaeolithic Period in U.S.S.R. ("Chelovek," No. 1, Leningrad, 1928, pp. 45–59.)

The Expeditions for the Study of Palaeolithic Cultures. (Soobschenia of GAIMK, Vol. I, Leningrad, 1926.) R.

The Palaeolithic Site of Kostenki. (Annual Report of RAS, Vol. V, St. Petersburg, 1915.) R.

The Statuette of Solutrean Epoque. (Materials for the Ethnography of Russia, The State Russian Museum, Vol. 3, pt. 1, 1926, pp. 139–142.) R.

The Stone Implements of the Palaeolithic Site in the Village of Mezine. (Annual Report of the RAS, Vol. IV, St. Petersburg, 1913, pp. 62–102.) R.

Traces of Mousterian Culture in South Russia. (Baghalei Memorial Volume, pp. 286–301.) U.

The Results and Perspectives of the Study of the Palaeolithic Period in U.S.S.R. (Soobschenia of GAIMK, No. 3, 1931, pp. 7–9.) R.

The Investigations of the Palaeolithic Period in U.S.S.R. During Recent Years. (Soobschenia of GAIMK, No. 9–10, Leningrad, 1932, pp. 24–27.) R.

ERNST, N. L. Excavation of the Palaeolithic Site Chokurcha Groto near Siempheropol. (Izvestia of Tavrida Society of History, Archaeology and Ethnology, Vol. III, 1929.) R.

The Archaeological Investigations in Crimea in 1921–30. (Izvestia of the Tavrida Society of History, Archaeology and Ethnography, Vol. 4, Siempheropol, 1931, pp. 70–92.) R.

FIELD, H., AND PROSTOV, E. Recent Archeological Investigations in the Soviet Union. (AA, Vol. 38, No. 2, pp. 260–290.) E.

Archeology in the Soviet Union. (AA, Vol. 39, No. 3, pp. 457–490.) E.

GARROD, D. A. A. The Upper Palaeolithic Age in Britain. (Oxford, 1926.) E.

GMELIN, S. G. Travel through Russia for Investigations of the Three Kingdoms of Nature. (1769.) R.

GOLMSTEN, V. V. Archaeological Remains in the Samara Region. (RANION, Vol. 4, Moscow, 1928, pp. 125–137.) R.

GOLOMSHTOK, E. Trois Gizements du Paléolithique Supérieur, Russe et Sibérien. (L'Anthropologie, Vol. XLIII, 1933, pp. 333–346.) F.

Anthropological Activities in Soviet Russia. (American Anthropologist, Vol. 35, No. 2, 1933, pp. 301–327.) E.

Le Trompe du Mammouth Sibérien. (L'Anthropologie, Vol. XLIII, No. 5–6, 1932, pp. 548–550.) F.

GORODZOV, V. A. Archaeological Excavations and Investigations in Soviet Russia from 1919–23. (Drevnii Mir, Pt. 1, Moscow, 1924.) R.

Archaeology, The Stone Age. (Vol. I, Moscow, 1925.) R.

Contributions to the Question of Dating the Cave Kiik-Koba. (Izvestia of Tavrida Society of History, Archaeology and Ethnology, Vol. 2, 1928, pp. 33–38.) R.

Diary of Archaeological Excavations in Samara, 1929. (Manuscript.) R.

Investigations of Quaternary Conditions in the Region of the Lower Volga in 1929. (Manuscript.) R.

Materials for the Archaeological Map of the Valley and Shores of the Oha River. (Trudi, XII AC, Vol. I.) R.

Preliminary Account of the Investigations of the Palaeolithic Site of Timonovka, Near Briansk. (Soobschenia of GAIMK, No. 11–12, Leningrad, 1932, pp. 55–57.) R.

Technique and Typological Classification of Gravers from Suponev Timonovka, Palaeolithic Site, Excavations 1928–29. (Trudi of the Archaeological Section of RANION, Vol. 5, Moscow, 1930, pp. 15–43.) R.

The Investigations of the Palaeolithic Site of Gontzi in 1915. (RANION, Vol. I, Moscow, 1922, pp. 92–110.) R.

The Results of Archaeological Activities in U.S.S.R. during 1918–1930. (Manuscript.) R.

The Determination of Age and Several Peculiarities of Yenisey Palaeolithic. (Northern Asia, Vol. I, Moscow, 1929.) R.

GREMIATSKY, M. A. The Podkumok Cranium and Its Morphological Peculiarities. (RAJ, Vol. XII, Pt. 1–2, Moscow, 1922, pp. 92–110.) R.

Remains of the Lower Jaw and Teeth of Podkumok Man. (Appendix to RAJ, Vol. 19, pp. 1–2, Moscow, 1925, pp. 91–95.) R.

GROMOV, V. I. Fauna of the Palaeolithic Site of Berdizh. (Pratzi of the ACBA, Vol. II, Minsk, 1930, pp. 7–30.) B.

Geology and Fauna of the Palaeolithic Period in U.S.S.R. (Problems of the History of Material Culture, No. 1–2, Leningrad, 1933.) R.

Geology and Fauna of the Palaeolithic Period in U.S.S.R. (Bulletin of Quaternary Committee, No. 3–4, 1932, pp. 37–53.) R.

New Palaeolithic Site. ("Perioda," No. 9–10, Leningrad, 1926, p. 65.)R.

HAMMERMANN, A. Kohlenreste aus dem Paläolithikum der Krim, Hohlen Ssjuren I and II. (BQC, No. 1, 1929, pp. 39–42.) G.

KAMINSKY, F. Traces of the Most Ancient Period of the Stone Age along the Sula River and Its Tributaries. (Trudi of the III AC in Kiev, Vol. I, Kiev, 1878, pp. 147–152.) R.

KEITH, A. New Discoveries Relating to the Antiquity of Man. (London, 1931.) E.

KIELSIEV, A. F. The Palaeolithic Kitchen Refuse of The Stone Age along the Sula River and Its Tributaries. (Trudi of the III AC in Kiev, Vol. IX, pp. 154–179.) R.

The Palaeolithic Kitchen Refuses in Village Kostenki, Voronezh Province. (Trudi of the Moscow Archaeological Society, Drevnosti, Vol. IX.) R.

KHVOIKO, V. V. Decouvertes Paléolitique Recemment Faites en Russie. (L'Anthropologie, Vol. XII, pp. 158–159.) F.

The Kiev-Cyrill Street Site and the Magdalenian Culture. (Archeological History of the South of Russia, No. 1, 1903.) R.

The Stone Age of the Middle Dnieper Region. (Trudi of the XI AC, Vol. I.) R.

KLEOPOV, J. D.   Uber des Alter der Relikte der Ukraine in Konnex mit dem Sukzession in ihrer Vegetation im Laufe der Quartarzeit.   (Die Quartarperiode, Ukrainische Akademie der Wissenschaften, No. 4, 1932, pp. 17–25.)   G.

KOSLOWSKY, L.   The Old Stone Age in Poland.   (Poznan, 1922.)   P.

KRISHTAPHOVICH, N. I.   Geological Investigations of a Palaeolithic Site During the Summer of 1904. (Drevnosti, Trudi of the Moscow Archaeological Society, Vol. XXI, Pt. 2, pp. 178–181.)   R.

New Traces of Prehistoric Man in Kiev.   (Appendix to the IX Volume of the Izvestia of the Imperial Archaeological Committee, St. Petersburg, pp. 40.)   R.

The Location of Mammoth Bones in Koselsk County.   (Year Book of Russian Geology, Vol. IV, Pt. 1.)   R.

The Old Palaeolithic Sites of the Territory of European Russia.   (Trudi of the X Congress of Russian Naturalists, 1898, 1902.)   R.

The Post-Tertiary Formations in the Vicinity of Novo-Alexandria.   (Warsaw, pp. 47–69.)   R.

KROKOS, V.   The Conditions of the Palaeolithic Deposit in Zhuravka.   (ALV for 1928, Kiev, 1929, pp. 135.)   U.

The Stratigraphy of the Upper Palaeolithic Site of Dovginichi in Volyn.   (Die Quartarperiode of the Ukranian Academy of Science, Vol. 1–2, 1930, Kiev, 1931, pp. 27–31.)   G.

Stratigraphie der Quarteren Ablagerungen der Ukraine.   (Die Quartarperiode, V. 4, 1922, Kiev, 1932, pp. 1–4.)   G.

Materials for the Study of Quarternary Deposits on the Ukraine.   (Materials for the Study of Ukrainian Soils, Vol. 5, pp. 1–326, Kharkov, 1927.)   R.

KRUKOVSKY, S.   The Cave Gvardzhilas-Klde in Rgan (Caucasus).   (Bulletin of the Caucasian Museum, Vol. X, Nov., 1916, Tiflis, pp. 1–7.)   R.

LAVDANSKY, A. N.   Archaeological Investigations of the Palaeolithic Site near Gamkovo, District of Smolensk.   (Pratzi of Archaeological Committee of Bielorussian Academy of Sciences, Vol. II, Minsk, 1930, pp. 495–498.)   B.

LEVITZKY, I.   The Find in Kolodiazhnoe.   (Bulletin of the Anthropological Laboratory of Th. Vovk, Vol. I, Kiev, 1925, pp. 35–39.)   U.

The Palaeolithic Site of Dovginichi.   (Anthropologia, ALV for 1929, Kiev, 1930, pp. 135–160.)   U.

MACCURDY, G. G.   Human Origins.   Vol. 1 and 2, 1924.   E.

MALYCHEFF, VERA.   Le Loess.   (Review de Géographie Physique et de Geologie dynamique, Vol. II, III, Paris, 1929–30.)   F.

(Map of the Quaternary Deposits of the European Part of the U.S.S.R. and the Adjacent Regions, Explanatory Note, 1932.)   R.

MENGHIN, O.   Die Weltgeschichte des Steinzeit.   G.

MEREZKOVSKY, K. S.   The Report of the Preliminary Investigations of the Stone Age in Crimea.   (Izvestia of Imperial Russian Geographical Society, Vol. XVI, pp. 106–142, 1880; Vol. XVII, pp. 104–115, 1881.)   R.

Station Mousterienne en Crimeé.   (L'Homme, 1884.)   F.

MESCHANINOV, I. I.   The Theory of Migrations in Archaeology.   (Soobschenia of GAIMK, No. 9–10, Leningrad, 1931, pp. 33–39.)   R.

Palaeontology and Homo Sapiens.   (Izvestia of GAIMK, Vol. VI, Pt. VII, Leningrad, 1930, pp. 1–36.) R.

MIRCHINK, G. F.   Epirogenetic Movements of European U.S.S.R. During the Quaternary Period.   (The Second International Conference of the Association for the Study of the Quarternary Period of Europe, Vol. II, Moscow, 1933, pp. 153–165.)   R.

The Geological Condition of the Palaeolithic Site of Berdizh on the River Sozh, District of Gomel. (Pratzi of the ACBA, Vol. II, Minsk, 1930, pp. 1–6.)   B.

Geological Conditions of Timonovka, near Briansk.   (RAJ, Vol. 18, Pt. 1–2, Moscow, 1929, pp. 57–58.) R.

Geological Correlation of River Terraces and Palaeolithic Sites in the Basin of the Desna and Sozh Rivers.   (Bulletin of the Moscow Society of Naturalists, Section of Geology, Vol. VII, Moscow, 1920, pp. 1–2.)   R.

Number of Glaciations of Russian Plains.   ("Priroda," No. 7–8, 1928.)   R.

MORTILLET, A. La Pré-Histoire. (Paris, 1883.) F.
   L'Epoque de la Madeleine en Russie. (L'Homme, Vol. 12, Paris, 1884, pp. 367–369.) F.
NIKITINE, S. Sur la Constitution des dépôts quaternaire en Russie et leurs Relations aux Trouvailles
   Resultant de L'Activité de l'Homme Préhistorique. (VII Congrès International d'Archéologie
   Préhistorique et d'Anthropologie, Moscow, 1892.) F.
NIORADZE, G. The Palaeolithic Industry of the Cave Devis-Hvreli. (Travaux de Musée de Géorgie, Vol.
   VI, Tiflis, 1933, pp. 1–109.) Gg. Résumé in F.
   Das Paläolithicum Georgiens. (Transactions, Second International Conference on the Study of the
   Quaternary Period in Europe. Leningrad, 1935. Facsimile 5, pp. 225–236.) G.
OBERMAIER, H. Fossil Man in Spain. (Yale University Press, 1925, p. 222.) E.
OSOKOV, P. A. An article in the Addition to the Minutes of a Meeting of the Moscow Society of Naturalists
   in 1913. R.
PALIBIN, J., AND HAMMERMANN, A. Kohlenreste aus dem Paläolithikum der Krim, Höhle Kiik-Koba.
   (Bulletin of the Quaternary Committee, No. 1, 1929, pp. 35–37.) G.
PAVLOV, A. P. Fossil Man Contemporaneous with Mammoth in Eastern Europe and Fossil Man of Western
   Europe. (Supplement to the RAJ, Vol. XIV, Pt. 1–2, Moscow, 1925, pp. 5–36.) R.
   Epoques Glaciaires et Interglaciaires de l'Europe et Leur Rapport à L'histoire de L'homme Fossil.
   (Bulletin de la Société des Amis des Sciences Naturelles, d'Anthropologie et d'Ethnographie, 1925.)
   F.
   Dépôts Néogenes et Quaternaires de l'Europe Mériodioale et Orientale. (Memmoirs de la Section
   Géologique de la Société des Amis des Sciences Naturelles, d'Anthropologie et d'Ethnography, Vol.
   V, Moscow, 1925, pp. 1–215.) F.
PHEOPHILAKTOV, K. M. The Location of the Finds of Stone Flint Tools with Bones of Mammoth in the
   Village Gontzi, on the Udai River. (Trudi of the III Archaeological Congress in Kiev, Kiev, 1878,
   Vol. I, pp. 153–159.) R.
PIDOPLITCHKA, J. G. Die Fauna der Quartaren Saugetiere der Ukraine. (Die Quartarperiode, Ukrainische
   Akademie der Wissenchaften, No. 4, 1932.) G.
   Rodents and Carnivorae from Zhuravka. (AVL for 1929, Kiev, 1930, pp. 133–147.) U.
POLIAKOV, I. S. Anthropological Trip to Central and Eastern Russia. (Zapiski of the Imperial Academy
   of Sciences, Vol. XXXVII, Pt. 1, St. Petersburg, 1880.) R.
   Investigations of the Stone Age in Olonetzh Province, in the Valley of the Oka River and the Upper
   Part of the River Volga. (Zapiski of the Russian Geographical Society, Division of Ethnography,
   St. Petersburg, 1882, Vol. IX, pp. 1–164.) R.
POLIKARPOVICH, K. M. The Palaeolithic Site of Novo-Bobovichi. (Pratzi of ACBA, Vol. II, Minsk, 1930,
   pp. 491–494.) B.
   Excavation of the Palaeolithic Site of Urovichi in 1929. (Pratzi of ACBA, Vol. II, Minsk, 1930, pp.
   499–501.) B.
   Palaeolithic Sites on the Sudai River. (Pratzi of ACBA, Vol. III, Minsk, 1932, pp. 153–166.) B.
   Prehistoric Sites of the Middle and Lower Part of the Sozh River. (Pratzi of ACBA, Minsk, 1928,
   Vol. I.) B.
   The Palaeolithic and Mesolithic Sites of Bielorussia. (Pratzi of ACBA, Minsk, 1932, Vol. III, pp. 218–
   221.) B.
RAVDONIKAS, V. I. The Marxian History of Material Culture. (Izvestia of GAIMK, Leningrad, 1930,
   Vol. VII, Pt. 3–4, pp. 5–94.) R.
   To the Question of the Sociological Periodization of the Old Stone Age in Connection with the Views
   of Marx and Engels on Primitive Society. (Izvestia of GAIMK, Leningrad, 1931, Vol. IX, Pt. 1–2,
   pp. 1–33.) R.
   The Periodisation in the History of Pre-Clan Society. (PHPS, Leningrad, July–August, 1934, No.
   7–8, pp. 72–87.) R.
   To the Question of Dialectics of the Development of Pre-Clan Society. (PHPS, Leningrad, February,
   1934, No. 2, pp. 25–48.) R.
RENGARTEN, V. P. The Age of Deposits Containing the Remains of Podkumok Man. (RAJ, Moscow,
   1922, Vol. XII, Pt. 1–2, pp. 193–195.) R.

REINGARD, A. L. The Number of Glaciations in the Caucasus. (Second International Conference for the Study of the Quaternary Period in Europe, Moscow, 1933, pp. 3–14.) R.

RICHTHOFEN, BOLKO, F. VON. Altsteinzeitliche Funde in Weissrussland und dem Grossrussichen Nachbargebiet. (ESA, Helsingfors, Vol. VIII, pp. 161–174.) G.

RIZNICHENKO, V. The Geological Conditions and Geomorphological Conditions of the Palaeolithic Site of Zhuravka. (Anthropologia, ALV for 1930, Kiev, 1931, Vol. IV, pp. 185–190.) U.

Geological and Geomorphological Conditions of the Palaeolithic Site of Mezine. (Die Quartarperiode, Kiev, 1931, Pt. 1–2, pp. 13–24.) U.

RUDINSKY, M. Archaeological Investigation in 1926. (Zvidomlenia of All-Ukranian Archaeological Committee for 1926, Kiev, 1927, pp. 9.) U.

Bone Industry of the Palaeolithic Site of Mezine in the Interpretation of T. Volkov. (Ukrainian Academy of Sciences, ALV, Kiev, 1931, pp. 1–65.) U.

The Archaeological Collection of the State Poltava Museum. (Poltava, 1928, pp. 7–8.) U.

The Finds in Zhuravka. (ALV for 1928, Kiev, 1929, pp. 140–151.) U.

The Scientific Work on Palaeolithic and Mesolithic Cultures in Ukraine during the Revolution. (ALV for 1930, Kiev, 1931, pp. 145–184.) U.

Zhuravka Excavation of 1929. (Anthropologia, ALV for 1929, Kiev, 1930, pp. 97–122.) U.

Die Derzeitige Sachlage Betreffend Palaeolithologishe Studien in der Ukranian SSR. (Die Quartarperiode, Vol. IV, 1932, Kiev, 1932, pp. 79–82.) G.

The Finds in Shapovalovka. (Die Quartarperiode, Vol. 1–2, No. 10, 1930.) R.

The Palaeolithic Finds in Shapovalovka, Konotop District. (The Quaternary Periode, Vol. 1–2, Kiev, 1930, pp. 37–42.) U.

RUDINSKY, M., AND VORONII, A. The Finds in the Village Zhuravka, Priluk County. (Anthropologia, ALV for 1927, Kiev, 1928, pp. 65–73.) U.

SAGATSKY, A. P. The Labor and the Origine of Society. (PHPS, Leningrad, 1937, No. 1–2, pp. 177–192.) R.

SAMOKVASOV, D. Y. (Article on Shapovolovka, Anthropological Viestnik, Vol. III, pp. 3–38.) R.

SAWICKI, L. La Grotto Nietoperzowa à Jerzmanowice près Ojcow, District Olkusz. (Revue Archeologique Polonaise, Poznan, 1926, Vol. III, Pt. 1, pp. 1–8.) F.

Materials for the Study of Russian Archaeology. (Przeglad Archeologicznii, Poznan, Vol. III, Pt. 2–3, 1926–1928.) P.

The Lower Palaeolithic Site of Gorodok in the Volyn Region. (Ziemia, Vol. XII, No. 3, Warsaw, 1927, pp. 36–42.) P.

SCHERBAKOVSKY, V. (Article in Zbornik of the Local Ukranian Scientific Society in Kiev, Kiev, 1921.) U.

Bemerkungen über Neue Wenig Becannte Paläolithische Stationen in der Ukraine. (Eiszeit, IV, 1927.) G.

The Excavations of the Palaeolithic Settlement in the Village Gontzi, Luben County, Poltava Region in 1914–1915. (Zapiski of the Ukranian Society for the Study and Preservation of Antiquities in Poltava Region, Poltava, 1919, pp. 62–75.) U.

SCHMIDT, R. R. Russland in Diluvialle Vorzeit. (Wurtenberg, Anthro. Verein.) G.

SEMENCHIK, M. Some Prehistoric Finds in Romni District. (ALV for 1929, Kiev, 1930, Vol. III, p. 209.) U.

SKUTIL, J. Ossarynce und Iskorost, Zwei Paläolithishe Stationen in Osteuropa. (Eiszeit und Urgeschichte, Vol. IV, Pt. 5, 1928, pp. 46–48.) G.

Les Trouvailles Mousteriennes en Crimée. (L'Homme, 1926, pp. 251–258.) F.

SPITZIN, A. A. Article in Otchet of Imperial Archaeological Committee, 1905. R.

Russian Palaeolithic Period. (Zapiski of the Imperial Russian Archaeological Society, Slavic Division, Vol. XI, Petrograd, 1915, pp. 132–172.) R.

SOBOLEV, D. N. Quaternary Morphogenesis in the Ukraine. (Second International Conference of the Association for the Study of the Quaternary Period of Europe, Vol. 11, Moscow, 1933, pp. 71–101.) R.

SOLLAS, W. J. Ancient Hunters. E.

TESLJA, T. T. Das Paläolith im Bereich dem Dnieprostrom-schnellen. (Die Quaterperiode, Ukrainische Akademie der Wissenschaften. PE, 4, Kiev, 1932, pp. 63–65.) G.

TICHII, M. Fische aus dem Paläolithikum der Krim. (Bulletin of the Quaternary Committee, Vol. I, 1929, pp. 43–46.) G.

UVAROV, A. S. Archaeology of Russia. (Two Volumes, Part I, The Stone Age, Moscow, 1881.) R.

VARDANIANTZ, L. A. The Number of Glaciations in the Caucasus. (Second International Conference for the Study of the Quaternary Period of Europe, Vol. II, pp. 15–20.) R.

VAUFREY, R. Le Palaeolithique Italien. (Archives de L'Institute de Paleontologie Humaine, Mem. 3, Paris, 1928.) F.

VISHNEVSKY, S. N. Prehistoric Man in Russia. (Appendix to the Translation of Osborn's "Men of the Old Stone Age," Leningrad, 1924, pp. 441–508.) R.

VOEVODSKY, M. V. The Palaeolithic Site of Timonovka. (RAJ, 1929, Vol. XVIII, Pt. 1–2, pp. 59–69.) R.

VOLKOV, TH. Annales Archeologique de la Russie du Sud, 1889. R.

    Decouvertes Préhistoriques de Kiev. (L'Anthropologie, No. 3, 1903.) F.

    Defense du Mammoth Gravée du Gisement Paleolithique de Kiev. (Bulletin et Memoires de la Société d'Anthropologie, Vol. V, series VI, Paris.) F.

    Magdalenian Art in Ukraine. (Zapiski of Scientific Society of Shevchenko in Lvov, Vol. XLVI, Lvov, 1902, pp. 1–13.) U.

    Materials for Ukrainian Ethnology. (Vol. I, 1899.) U.

    Nouvelles Découvertes Dans la Station Paleolithique de Mezine. (Congrés International d'Anthropologie et d'Archeologie Préhistoriques, Compte Rendu de la XIV Session, Vol. I, Geneve, 1912, pp. 415–428.) F.

    The Palaeolithic Site in the Village of Mezine, Chernigov Province. (Trudi of XIV AC, Moscow, 1911, Vol. 3, pp. 260–270.) R.

    The Prehistoric Finds in Cyrill Street, Kiev. (Zapiski of the Scientific Society of Shevchenko in Lvov, No. 1, 1898.) U.

VORONII, A. Materials for the Study of the Geology of the Palaeolithic Site of Zhuravka. (Anthropologia, ALV for 1929, Kiev, 1930, pp. 123–132.) U.

ZABNIN, S. A. A Newly Discovered Palaeolithic Site in Crimea. (Izvestia of the Tavrida Society of History, Archaeology, and Ethnography. Siempheropol, 1928, Vol. II, pp. 146–157.) R.

ZHAKOV, M. P. The Problems of the Genesis of Human Society. (PHPS, Leningrad, May, 1934, No. 5, pp. 29–46.) R.

    Labor, Technique, and Social Relations in Early Society. (PHPS, Leningrad, June, 1934, No. 6, pp. 9–45.) R.

ZAMIATNIN, S. N. The Excavation of the Palaeolithic Site of Berdizh in 1927. (Pratzi of the ACBA, Minsk, 1930, Vol. II, pp. 479–490.) R.

    Expedition for the Study of Palaeolithic Cultures in 1927. (Soobschenia of GAIMK, Leningrad, 1929, Vol. II, pp. 209–214.) R.

    Notes on Prehistory of the Voronezh Region. (Voronezh, 1922, pp. 1–16.) R.

    The Palaeolithic Site of Karacharovo. R.

    Station Moustérienne à Ilskaya, Province de Kouban, Caucase du Nord. (Revue Anthropologique, No. 7–9, 1929.) F.

    Gagarino. (GAIMK, 1935.) R.

ZELIZKO, J. V. The Settlement of Dilluvial Man in Kiev. (Casopisu Vlasten muzej spolku, Vol. Olomouci, 1927, cislo 95–96.) P.

ZHIRMUNSKY, A. M. The Question of the Limits of Glaciation on the Russian Plains. (BQC, 1929, Vol. I, pp. 21–26.) R.

INDEX

Numbers in **bold type** indicate pages of main reference; those in *italics* refer to pages on which the subject is illustrated.

461

www.ingramcontent.com/pod-product-compliance
Lightning Source LLC
Chambersburg PA
CBHW081332190326
41458CB00018B/5973